Comparative anatomy
of the vertebrates

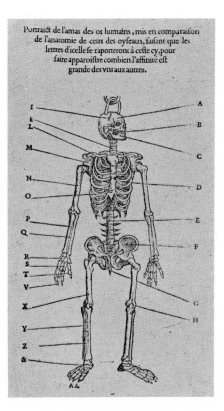

Portraict de l'amas des os humains, mis en comparaison de l'anatomie de ceux des oyseaux, faisant que les lettres d'icelle se raporteront à ceste cy, pour faire apparoistre combien l'affinité est grande des vns aux autres.

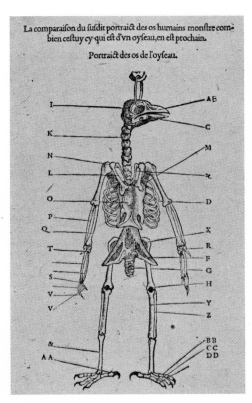

La comparaison du susdit portraict des os humains monstre combien cestuy cy qui est d'vn oyseau, en est prochain.

Portraict des os de l'oyseau.

Woodcut by Belon, 1555

Comparative anatomy of the vertebrates

GEORGE C. KENT

Department of Zoology and Physiology,
Louisiana State University,
Baton Rouge, Louisiana

FOURTH EDITION

with 833 illustrations including 204 in color

THE C. V. MOSBY COMPANY

Saint Louis   1978

COVER ILLUSTRATION DRAWN FROM A PHOTOGRAPH COURTESY OF
The American Museum of Natural History

FOURTH EDITION

The C. V. Mosby Company
11830 Westline Industrial Drive, St. Louis, Missouri 63141

Library of Congress Cataloging in Publication Data

Kent, George Cantine, 1914-
    Comparative anatomy of the vertebrates.

    Bibliography: p.
    Includes index.
    1. Vertebrates—Anatomy. 2. Anatomy, Comparative.
I. Title. [DNLM: 1. Vertebrates—Anatomy and his-
tology. 2. Anatomy, Comparative. QL805 K37c]
QL805.K43 1978   596'.04   77-13588
ISBN 0-8016-2650-1

CB/CB/B 9 8 7 6 5 4 3 2 1

Preface

Emphasis in this edition is on basic patterns of vertebrate structure and development and on the function of these structures in terms of adaptive significance. I have repeatedly expressed or implied the concept that recent patterns are modifications of older patterns, that the adult is a modification of the embryo, that individual differences exist, and that structure and development are broadly determined by inheritance and adaptively modified by natural selection. This approach evokes spontaneous examination of the concept of organic evolution, which, devoid of theories of *how* or dogma of *why*, may be reduced to the axiom that the organisms of the world have been changing and that the present is a reflection of the past. A simple statement of what organic evolution does and does not necessarily imply has been added in Chapter 18.

I have used some English plurals that have gained wide acceptance, but in response to the expressed preference of many teachers I have reinstated Latin plural endings for vertebrae, larvae, centra, foramina, and a number of other anatomical terms.

The unconventional glossary entitled Anatomical Terms has proved popular and has been expanded. Intentionally, it lacks formidability so that it may have maximum appeal. Selected reading lists continue to include works of historical value and some that may be consulted for comprehensive bibliographies.

Prefaces to earlier editions have documented my indebtedness to many persons who have offered suggestions and criticism in the past. To the list I am privileged to add J. A. A. Harák, University of Tasmania, and Randy Price, New York University, for information incorporated in the present edition; Robert Bowker, for providing a live iguana for Fig. 16-12; Jeanne Robertson, who redrew a number of the illustrations from earlier editions and prepared some new ones; Mike Turner, for photographs not otherwise acknowledged; and Karen Westphal, for her drawing of the shark's brain. The kindness of these and other individuals, including many students, has contributed immeasurably to this work. To each I extend my most sincere thanks. Frontispiece courtesy Macmillan Company, Ltd., from *A History of Comparative Anatomy* by F. J. Cole.

George C. Kent

v

Contents

Prologue

Cavemen had some knowledge of the internal organs of mammals, as did the Babylonians and ancient Egyptians, who practiced surgery and embalming. Embalming, which involved removal of most of the internal organs, was an advanced art. Aside from a small number of Egyptian medical papyri dating to about 3000 B.C., the oldest anatomical works were written during the last 400 years B.C. by Greek philosophers and physicians. These works were incomplete, mostly superficial, and often imaginative. Anatomy of the classical era culminated in the works of Galen, a Greek philosopher-physician who practiced in Rome between A.D. 165 and 200. He assembled all available Greek anatomical writings, supplemented them with his own dissections of apes from the Barbary Coast (human dissection was at that time prevented by public opinion and superstition), and, in addition, wrote more than 100 treatises on medicine and human anatomy. Shortly thereafter, scholasticism took over, and during the next 1300 years Galen's descriptions were considered infallible and dissent was punishable. Subservience to authority became so accepted that (as has been said, probably partly in jest) if a scholar wanted to know how many teeth horses have, he saddled up and rode 100 miles, if necessary, to the nearest library to see what Galen said. (No doubt the peasants looked into the horse's mouth—an application of the experimental method.)

It was not until the fifteenth century that Leonardo da Vinci (1452-1519) and other Italian artists began to make anatomical observations of their own (at the peril of excommunication from the church), and a renaissance in anatomy began. In 1533 a young medical student named Vesalius at the University of Paris attended public demonstrations where Galen's works were read (by a "Reader") while the professor tried, often with embarrassing lack of success, to harmonize Galen's descriptions with the dissection. (Recall that many of Galen's descriptions of man were written from dissections of apes.) After 3 years Vesalius quit Paris, earned a degree at Padua, stayed on to teach, and recorded his own anatomical observations. In 1543 he published *De humani corporis fabrica (On the Structure of the Human Body)*, and anatomy entered the modern investigative period.

Twelve years after publication of Vesalius's anatomy, Pierre Belon (1555) published what has become a classical illustration of a human and bird skeleton side by side, showing that the parts correspond, bone for bone (frontispiece). The bird was drawn in an upright position facing the viewer, with wings hanging like arms. The bones of the bird and of man were similarly labeled. Belon, a botanist and a physician, was therefore also a pioneer in comparative anatomy.

It was another century, however, before anatomists became interested in cataloging other examples of what we today call "homologous structures"—structures in two different species that develop in the same way from the same embryonic precursor and are really the same structure even though they may not look alike or even perform the same function. Each newly discovered instance of an identical structure in two species was dutifully explained as being a manifestation of a basic architectural plan or archetype in the Creator's mind. The idea that these similarities might be the result of inheritance of a similar genetic code from a common ancestor, with modifications, had not yet been openly expressed. When it finally was, the idea met with the same almost universal condemnation in the Western world that Copernicus (1473-1543) encountered earlier when he announced that the earth revolves around the sun. Both theories were considered atheistic, which they are not, since neither concept in any way denies (or confirms) the validity of the concept of a Creator.

Comparative anatomy today is the study of structure, of the functional significance of structure, and of the range of variation in structure and function in different species. Its methods are descriptive and experimental. The data are employed partly to attempt to deduce the history of the different species on our planet and the environmental conditions under which they rose, flourished, and became extinct. The data also help to satisfy the curiosity of the human mind. Like other scientific disciplines, comparative anatomy has its roots in philosophy, and its aim is enlightenment.

1

The vertebrate body

In this chapter we will preview the basic architectural features of vertebrate animals. We will learn what happens to their embryonic notochords, examine the embryonic pharynx, and find out why brains and spinal cords are hollow from fish to man. Finally, we will note some additional features that, although not unique, are found in animals with backbones.

General body plan
  Head
  Trunk
  Tail
  Appendages
  Bilateral symmetry
Vertebrate characteristics: the big four
  Notochord and vertebral column
  Pharynx
  Dorsal, hollow central nervous system
Satellite characteristics
  Skin
  Metamerism
  Respiratory mechanisms
  Coelom
  Digestive organs
  Urinogenital organs
  Circulatory system
  Sense organs

A study of comparative vertebrate anatomy is, in a sense, a study of history. It is the history of the struggle of vertebrate animals for compatibility with an ever-changing environment. It is the history of the extermination of the unfit and the invasion of a new territory by those best equipped for survival. It is a study of history, just as is the study of man's conquests, political fortunes, and social evolution.

The study of vertebrates is, by definition, a study of man, although not of man alone. It leads to better understanding of man's past and to an assay of his present state. As for predicting the future, a most important, though often neglected, objective of history, the biologist can predict that neither the earth nor that which grows on the earth will remain unchanged. The prediction is based in part on the fact that there has been a succession of animals and plants on the earth and that the species of today are not the same species that would have been seen 300 million years ago. On the basis of probability, it can be predicted that they will be still more different tomorrow. Since there has been a succession of species, and since all life seems to come from preexisting life, logic tells us that the species have been changing. This phenomenon is the premise of the discipline.

The discipline has an important function. It is not that of promoting the premise. It is, instead, one of continually seeking additional facts, of finding new interrelationships, of periodically reevaluating our tentative conclusions, and of drawing new ones. When comparative anatomy ceases to be a search for the truth, it will have surrendered its status as a science and will have become a body of meaningless facts. The facts are important, but their meaning is immeasurably more so. To the study of no discipline is the dictum of *Proverbs* more applicable, ". . . in all thy getting, get understanding."

The premise that the vertebrate species have been changing is strengthened by the observation that all vertebrates, past and present, are constituted in accordance with a basic architectural pattern. This phrase has two implications. The first is that vertebrates conform quite closely to a **uniform pattern of anatomical structure.** This is revealed by dissection and study of adult vertebrates. The second implication is that there is a **uniformity of developmental processes,** which is revealed by studies of embryos. It is the purpose of this book to assist the reader to discover this basic architectural pattern and to show in what directions the pattern has been modified. The modifications appear in all the body systems—integumentary, skeletal, muscular, digestive, respiratory, circulatory, urinogenital, nervous, and endocrine.

GENERAL BODY PLAN

The vertebrate body is divided into head, trunk, and tail. Paired appendages are associated with the trunk in all but a few vertebrates. A neck develops in reptiles, birds, and mammals.

Head

Far down in the animal kingdom bilaterally symmetrical animals begin to concentrate sense organs on that portion of the body that first enters a new milieu. Many invertebrates "creep up" on their environment. An earthworm goes a short distance forward, tests the new environment carefully with sense organs at the cephalic end, and if the environment is friendly, proceeds deeper and deeper into the area ahead. A friendly environment minimally contains food and a mate. It may also provide shelter. If the environment appears inimical, the earthworm withdraws. An inimical environment is one that might destroy the individual and endanger the species. Thus an advantage is gained by increasing the number of monitoring devices at the anterior end of the body. Concentration of sense organs in the head is accompanied by increase in brain size. Jaws and respiratory mechanisms add to the complexity of the anterior end.

Trunk

The trunk is the part containing the body cavity, or **coelom** (Fig. 1-1). Surrounding the coelom is the **body wall** (Fig. 1-2, *H*), covered by skin on its outer surface and by **parietal peritoneum** on its inner surface. In addition to skin and parietal peritoneum, the body wall consists chiefly of muscle, vertebral column, and ribs and sternum when present. The body wall must be cut open to expose the viscera. The latter are covered by **visceral peritoneum,** which is continuous with the parietal peritoneum via dorsal and ventral mesenteries. Those few visceral organs that do not develop dorsal mesenteries lie against the dorsal body wall just external to the parietal peritoneum. In this position they are said to be retroperitoneal.

The neck is a narrow anterior extension of the trunk lacking a coelom. It consists primarily of vertebrae, muscle, nerves, and elongated tubes—esophagus, long arteries and veins, lymphatics, trachea—connecting head and trunk.

Tail

The tail begins at the anus or vent. In fishes and tailed amphibians it is used chiefly for locomotion (Fig. 1-3). It consists almost exclusively of a caudal continuation of the body wall muscles for power, of the nervous system for innervation, of the notochord or vertebral column for support, and of a caudal artery and vein. Frogs and toads exhibit a locomotor tail prior to metamorphosis when, as swimming larvae, they need it most. Modern birds have reduced the tail region to a nubbin, but the first birds had long tails (Fig. 3-20). Mammals have

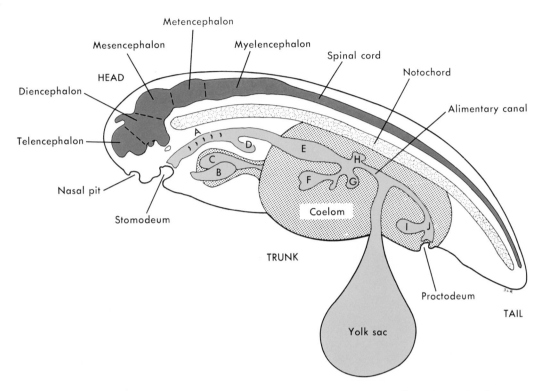

Fig. 1-1. Sagittal section of vertebrate embryo, illustrating basic pattern of vertebrate structure. **A,** Third pharyngeal arch of pharynx, lying between the second and third pharyngeal slits; **B** and **C,** ventricle and atrium of heart; **D,** diverticulum that gives rise to the lung in tetrapods and swim bladder in fishes; **E,** stomach; **F,** liver bud and associated gallbladder; **G,** ventral pancreatic bud; **H,** dorsal pancreatic bud; **I,** urinary bladder of tetrapods; **J,** cloaca. The sto- modeum is separated from the pharynx by a thin oral plate. The proctodeum is separated from the cloaca, **J,** by a cloacal membrane. The brain has five major subdivisions: telencepha- lon and diencephalon (forebrain), mesencephalon (midbrain), and metencephalon and myel- encephalon (hindbrain).

prehensile tails (monkeys), foreshortened tails (hamsters), flyswatters (cattle), balanc- ers (squirrels), and organs of defense (porcu- pines), all of which keep these animals from coming to an unnecessary end. Even man exhibits a tail in early embryonic life (Fig. 1-11). Vestiges may be observed on any hu- man skeleton as the lowest three, four, or sometimes five caudal (coccygeal) verte- brae.

□ Appendages

Vertebrates typically have paired anterior and posterior appendages—fins in fishes, limbs in tetrapods. In addition, fishes usu- ally have median (unpaired) dorsal and ven- tral fins. The skeleton of paired fins is variable, but such outwardly different limbs as wings, flippers, and hands are con- structed in accordance with a basic archi- tectural pattern.

□ Bilateral symmetry

Vertebrates exhibit three principal body axes: an anteroposterior (longitudinal) axis, a dorsoventral axis, and a left-right axis. With reference to the first two, structures found at one end of each axis are different

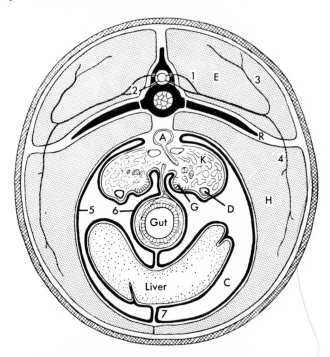

Fig. 1-2. Typical vertebrate body in cross section. **A,** Dorsal aorta, giving off renal artery to kidney; **C,** coelom; **D,** kidney duct; **E,** epaxial muscle; **G,** future gonad (gonadal ridge); **H,** hypaxial muscle in body wall; **K,** kidney; **R,** rib. **1,** Dorsal root of spinal nerve; **2,** ventral root; **3,** dorsal ramus of spinal nerve; **4,** ventral ramus; **5,** parietal peritoneum; **6,** visceral peritoneum; **7,** ventral mesentery. A remnant of the notochord lies within the centrum of a vertebra (immediately dorsal to **A**). The spinal cord lies above the centrum surrounded by a neural arch.

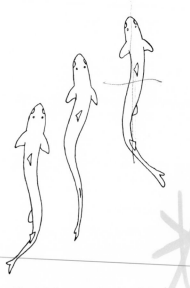

Fig. 1-3. Locomotion in a fish.

from those at the other end. The left-right axis terminates in identical structures on each side. Thus the head differs from the tail and the dorsum differs from the venter, but right and left sides are mirror images of each other. An animal with this arrangement of body parts exhibits bilateral symmetry.

It is sometimes convenient to discuss parts of the vertebrate body with reference to three principal anatomical planes. Two axes define a plane. The transverse plane is established by the left-right and the dorsoventral axes. A cut in this plane is a cross section (Fig. 1-4). The frontal plane is established by the left-right and longitudinal axes. A cut in this plane is a frontal section. The sagittal plane is established by the longitudinal and dorsoventral axes. A cut in this

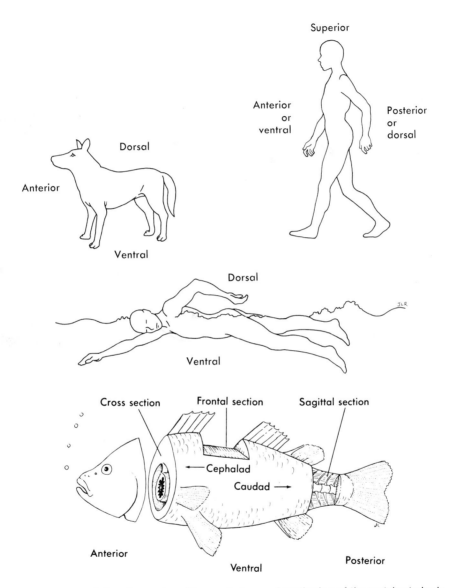

Fig. 1-4. Terms of direction and position and planes of sectioning of the vertebrate body.

plane is a sagittal section. Sections parallel to the sagittal plane are parasagittal. Acquainting oneself with these concepts constitutes a simple exercise in anatomy and logic.

■ VERTEBRATE CHARACTERISTICS: THE BIG FOUR

Vertebrates constitute a subphylum (**Vertebrata,** or **Craniata**) in the phylum Chordata. They exhibit four definitive structural characteristics: (1) the presence of a notochord, at least in the embryo; (2) the presence of a pharynx with pouches or slits in its wall, at least in the embryo; (3) the occurrence of a dorsal, tubular nervous system; and (4) a vertebral column. These are the "big four" vertebrate characteristics. The first three are chordate characteristics and are found also in protochordates. Other features associated with vertebrates but not necessarily unique among them will be discussed as satellite characteristics.

□ Notochord and vertebral column

The notochord is the first skeletal structure to appear in vertebrate embryos. At its peak of development it is a rod of living cells located immediately ventral to the central nervous system and dorsal to the alimentary canal extending from the midbrain to the tip of the tail (Fig. 1-1). The location is explained partly by its origin from the roof of the archenteron, or embryonic gut (Fig. 4-7). During later development the part of the notochord in the head becomes incorporated in the floor of the skull, and, except in agnathans, the part in the trunk and tail becomes surrounded by cartilaginous or bony rings called **vertebrae.** These provide more rigid support for the body than does a notochord alone. A typical vertebra consists of a centrum that is deposited around and within the notochord, a neural arch that forms over the spinal cord, and various processes. In the tail a hemal arch may surround the caudal artery and vein (Figs. 7-1, *C* and *D*, and 7-3).

The fate of the notochord in adult vertebrates is variable. In almost all fishes it persists the length of the trunk and tail, although usually constricted within each centrum (Fig. 7-4). The same is true in many urodeles and some primitive lizards. However, in modern reptiles, birds, and mammals the notochord is almost obliterated during development. A vestige remains in mammals within the intervertebral discs separating successive centra (Fig. 7-2, *D*). The vestige consists of a soft spherical mass of connective tissue called the **pulpy nucleus.** Modern reptiles and birds lack even this vestige.

In protochordates and agnathans the notochord has a different fate. In an amphioxus it continues to grow as the animal grows and never becomes surrounded by vertebrae. Therefore it remains throughout life as the chief axial skeleton. In urochordates the notochord is confined to the tail, and disappears at metamorphosis when the tail is resorbed (Fig. 2-7). In agnathans the notochord grows along with the animal, but paired **lateral neural cartilages** are perched on the notochord lateral to the spinal cord (Fig. 1-5). These cartilages are reminiscent of neural arches, but whether they are primitive vertebrae, vestigial vertebrae from an ancestor that had a typical vertebral column, or entirely different structures is not known. When a notochord persists as an important part of the adult axial skeleton, it develops a strong outer elastic and inner fibrous sheath (Fig. 1-5).

It is apparent that the notochord has been disappearing as an adult structure in recent vertebrates. But the development of a notochord in every vertebrate embryo—even the embryo of man—is a reminder that all vertebrates are built in accordance with a basic architectural pattern.

□ Pharynx

The pharynx is the region of the alimentary canal exhibiting pharyngeal pouches in the embryo (Fig. 1-6). The pouches may

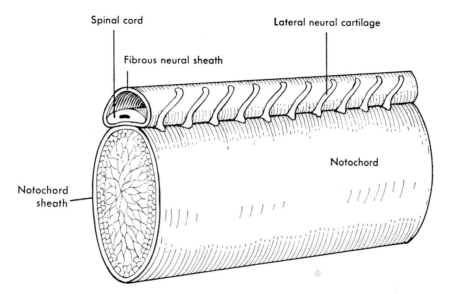

Fig. 1-5. Lateral neural cartilages of a lamprey.

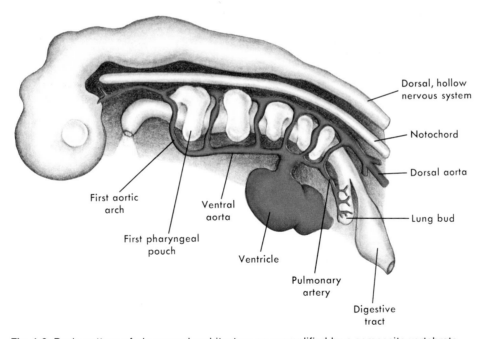

Fig. 1-6. Basic pattern of pharyngeal architecture as exemplified by a composite vertebrate embryo. The notochord lies ventral to the nervous system and extends from the midbrain caudad. A series of pharyngeal pouches have evaginated from the lateral walls of the digestive tract. Six aortic arches (red) connect the heart and ventral aorta with the dorsal aorta. (Typically, the first aortic arch disappears before the sixth one has formed.) The anterior end of the dorsal, hollow nervous system is enlarging to form the brain. Although a lung does not form in all vertebrates, it is an ancient structure and is represented by a swim bladder in most fishes.

rupture to the exterior to form pharyngeal slits. These slits may remain throughout life, or they may be temporary. If they remain throughout life, the adult pharynx is the part of the alimentary canal having slits. If the slits are temporary, the adult pharynx is the part of the alimentary canal connecting the oral cavity and esophagus.

PHARYNGEAL POUCHES AND SLITS

The basic pattern of the vertebrate pharynx is expressed in all vertebrate embryos. A series of paired **pharyngeal (visceral) pouches** arise as diverticula of the pharyngeal endoderm (Figs. 1-6 to 1-8). The pouches invade the pharyngeal wall and grow toward the surface of the animal. Simultaneously, an **ectodermal groove**

grows toward each pharyngeal pouch (Figs. 1-7 and 1-8). Soon only a thin **branchial plate** separates the ectodermal groove from the pharyngeal pouch. When the branchial plate ruptures, as it usually does, a passageway is formed between the pharyngeal lumen and the exterior. This embryonic passageway is a **pharyngeal (visceral) slit.** The slits may be permanent or temporary.

Pharyngeal slits are permanent in adults that live in water and breathe by gills (Fig. 1-9). For example, in embryos of the spiny dogfish shark *(Squalus)* six pharyngeal pouches form and all six rupture to the exterior. Highly vascularized gill surfaces develop in the walls of the last five pouches of *Squalus,* and these are therefore **gill pouches.** In the anterior wall of the first em-

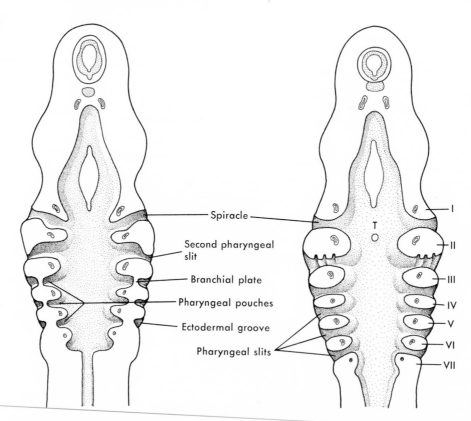

Spiracle
Second pharyngeal slit
Branchial plate
Pharyngeal pouches
Ectodermal groove
Pharyngeal slits
T

Fig. 1-7. Pharyngeal arches (**I** to **VII**) and slits in embryonic shark, frontal section, looking down onto floor of pharynx. Early stage, *left;* later stage, *right. T,* Thyroid evagination.

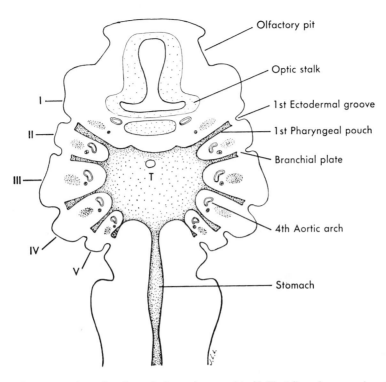

Fig. 1-8. Frontal section of embryonic frog pharynx. **I** to **V,** First five pharyngeal arches; **T,** thyroid evagination.

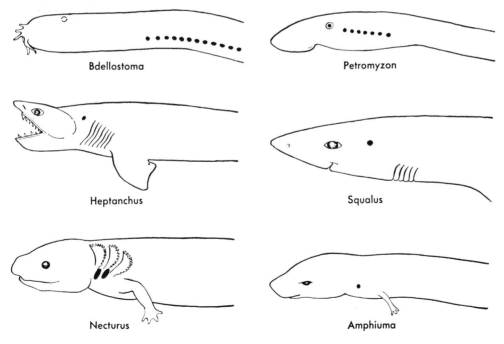

Fig. 1-9. Pharyngeal slits in selected adult aquatic vertebrates.

Fig. 1-10. Cervical fistula resulting from persistent pharyngeal slit.

bryonic pouch an abortive gill surface (**pseudobranch**) develops, and the pouch becomes the **spiracle.** Eight is the largest number of pouches that forms in any jawed vertebrate, and that many are found only in primitive sharks. Agnathans have as many as fifteen pouches and slits. Some urodeles retain one to three slits throughout life (Fig. 1-9, *Necturus, Amphiuma;* Table 3-1).

Pharyngeal slits are temporary if the animal is going to live on land. Of the six pharyngeal pouches that form in frog embryos, four give rise to gill slits in tadpoles. These slits close permanently when the tadpole metamorphoses into a frog. In reptiles, birds, and mammals no gill surfaces develop in the pharyngeal pouches, and the pharyngeal slits are transitory. Of the five pouches that develop in chicks, three rupture to the exterior, and these close again. Only one or two of the more anterior pouches of mammals may rupture. Embryologists tell us that the cervical fistulas occasionally seen in human beings (Fig. 1-10) are usually the result of the failure of the second or third pharyngeal slit to close.

Although the pharyngeal pouches of embryonic tetrapods rarely give rise to permanent slits, the first one becomes the auditory tube and middle ear cavity, and the second in mammals becomes the pouch of the palatine tonsil. The endodermal walls of several pouches give rise to endocrine tissue in all vertebrates (Fig. 17-16).

PHARYNGEAL ARCHES

Each embryonic pharyngeal pouch or adult pharyngeal slit is separated from the next by a column of tissue called a pharyngeal (visceral) arch (Figs. 1-7, 1-8, and 1-11). Each pharyngeal arch, whether in a fish or human being, typically contains four basic structures or blastemas from which these structures develop. They are (1) a pharyngeal (visceral) skeletal element (illustrated in an adult shark in Fig. 8-1), (2) branchiomeric muscle (Fig. 10-12, *A*), (3) branches of certain cranial nerves, and (4) an aortic arch (Fig. 1-6), which, at least in the embryo, directly connects the ventral and dorsal aortas. These basic components are also found in front of the first pouch and, with perhaps some omissions, directly behind the last. Therefore a pharyngeal arch is a column of tissue located between two successive pharyngeal pouches or slits, as well as in front of the first pouch or slit and immediately behind the last. It is covered externally by ectoderm and internally by endoderm.

The upper and lower jaws and associated muscles, nerves, and vessels constitute the first, or **mandibular, arch.** The second, or **hyoid, arch** is behind the first pouch or slit. The remaining pharyngeal arches are referred to by number.

The boundaries of a pharyngeal arch can be determined from the exterior when there are ectodermal grooves or pharyngeal slits to serve as landmarks, and only then. After the grooves disappear or the slits close, the boundaries of the arches are lost and the components become reoriented. In most tetrapods, therefore, pharyngeal arches are anatomical entities in embryos only.

The primitive vertebrate pharynx was evidently a device for filtering food out of a respiratory water stream as does the pharynx of an amphioxus. The modifications that oc-

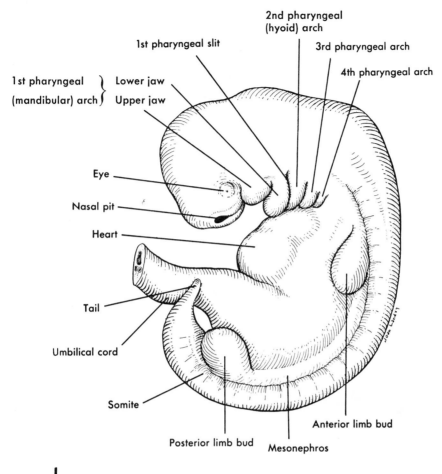

2nd pharyngeal
(hyoid) arch

3rd pharyngeal arch

1st pharyngeal slit

4th pharyngeal arch

1st pharyngeal } Lower jaw
(mandibular) arch } Upper jaw

Eye

Nasal pit

Heart

Tail

Umbilical cord

Somite

Posterior limb bud

Mesonephros

Anterior limb bud

| Actual crown-rump length

Fig. 1-11. Human embryo approximately 4½ weeks after fertilization (5 mm stage).

curred in the pharynx of those vertebrates that shifted from water to a land habitat is one of the fascinating chapters of vertebrate history. These changes will be described in later chapters.

☐ Dorsal, hollow central nervous system

The central nervous system in vertebrates consists of a brain and spinal cord and contains a central cavity, or **neurocoel.** Dorsal, hollow central nervous systems are found only in chordate animals. Their dorsal location and the presence of a cavity result from the fact that the central nervous system typically arises as a longitudinal **neural groove** in the dorsal ectoderm (Fig. 1-12), which subsequently sinks into the dorsal body wall to form a hollow **neural tube.** The tube is wider anteriorly, and this part becomes the brain with its ventricles (Fig. 15-5).

A few vertebrates (teleosts, ganoid fishes, cyclostomes) do not form a neural groove. Instead, a solid, ventrally directed **neural keel** separates from, and sinks beneath, the surface ectoderm (Fig. 4-7, teleost). It eventually develops a lumen and becomes tu-

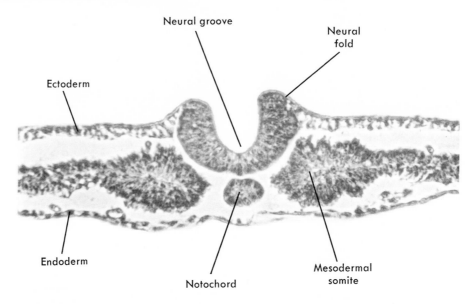

Fig. 1-12. Cross section of 24-hour frog embryo showing neural groove, notochord, and mesodermal somites.

bular, as though it had arisen from a groove.

Cranial and spinal nerves connect the central nervous system with the various organs of the body. The nerves, along with associated ganglia and plexuses, constitute the peripheral nervous system. The spinal nerves of most vertebrates are metameric (Fig. 15-7), arising at the level of each body segment and passing to the skin and muscles of that segment and to the viscera. Ten cranial nerves arise from the brain in fishes and amphibians and twelve in reptiles, birds, and mammals. The extra two nerves in higher vertebrates are spinal nerves that have become "trapped" within the skull.

■ SATELLITE CHARACTERISTICS
□ Skin

The integument, or skin, of vertebrates consists of an epidermis of ectodermal origin and an underlying dermis of mesodermal origin (Figs. 5-5 and 5-13). The epidermis of animals that live in water is different from that of animals exposed to air,

and the dermis of ancient vertebrates was bony. Many types of glands develop from the skin and open on the surface, and the skin is modified locally to form such membranes as the transparent conjunctiva of the eye, mucous membranes of the lips, and respiratory surfaces. Light organs in deep-sea fishes are modified skin glands.

□ Metamerism

Serial repetition of body structures in the longitudinal axis is known as metamerism. Among invertebrates metamerism is sometimes apparent at a glance. The successive body segments of earthworms are readily seen externally as well as internally. In crayfish the metamerism is obvious in the caudal region, but in the trunk it is obscured dorsally by the carapace. Internally, the metamerism of earthworms and crayfish is expressed in many systems.

Vertebrates, too, exhibit a basic metamerism that is clearly expressed in embryos and is retained in many adult systems. Yet no external evidence is seen because the

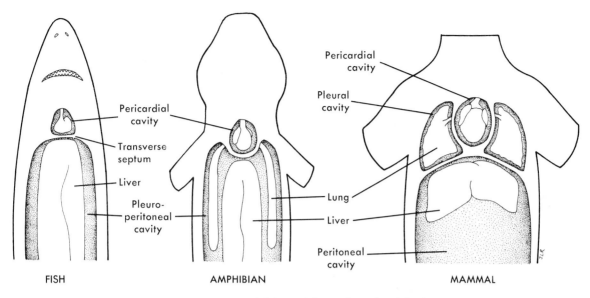

Fig. 1-13. Chief subdivisions of the coelom of vertebrates.

skin is not metameric. If, however, the integument is stripped from the body of fishes, amphibians, and even some reptiles (Fig. 10-3), a series of identical muscle segments is seen. A mere glance at the embryo of a bird (Fig. 15-5) or mammal will reveal the fundamental metamerism of these vertebrates. The serial arrangement of vertebrae, ribs, spinal nerves, embryonic kidney tubules, segmental arteries and veins, and body wall muscles is an expression of the fundamental metamerism of vertebrates.

□ Respiratory mechanisms

Most vertebrates carry on external respiration (exchange of respiratory gases between animal and environment) by means of highly vascularized membranes derived chiefly from the pharyngeal wall or floor. Internal gills are situated in gill pouches opening to the exterior via gill slits. External gills develop as outgrowths from a pharyngeal arch (Fig. 12-6). Lungs arise from a midventral evagination of the pharyngeal floor. The evagination, called a lung bud (Fig. 1-6), pushes into the coelomic cavity but remains

connected with the pharynx by an air duct.

Vertebrates sometimes carry on respiration by other devices such as skin, the buccopharyngeal lining, and (during embryonic life) special extraembryonic membranes that lie just under an eggshell or in contact with the mother's uterus. These will be discussed in later chapters.

□ Coelom

Like many invertebrates, vertebrates are built like a "tube within a tube," having a coelom between body wall and digestive tube. The coelom is subdivided in fishes, amphibians, and many reptiles into a **pericardial cavity** housing the heart and a **pleuroperitoneal cavity** housing most of the other viscera, including the lungs (Fig. 1-13). The pericardial and pleuroperitoneal cavities are separated by a fibrous **transverse septum.** In some reptiles and in birds and mammals the lungs become isolated in separate **pleural cavities.** The transverse septum is then supplemented by other septa, which may be muscular. In many male mammals, caudal outpocketings of the coelom house the testes, and these **scrotal cavities** are a fourth subdivision of the coelom.

Digestive organs

The digestive (alimentary) tract exhibits specialized regions for the acquisition, processing, temporary storage, digestion, and absorption of food and for the elimination of the unabsorbed residue. Typical are the oral cavity, pharynx, esophagus (which is as long as the neck), stomach, and intestine. The last is often coiled, which increases the absorptive area without increasing the body length. The tract exhibits a number of ceca, or diverticula, including a liver and pancreas.

The terminal segment of the digestive tract in all but a few vertebrates is the cloaca, which opens to the exterior via the **vent.** All vertebrate embryos develop a cloaca, but in modern fishes it becomes so shallow as to be nonexistent, and in mammals it becomes subdivided by partitions so that adult mammals other than monotremes lack one. The intestine in such cases opens directly to the exterior via an **anus.**

Urinogenital organs

The urinary and reproductive organs of vertebrates are closely related. Kidneys and gonads arise close together in the roof of the coelom (Fig. 1-2, *G* and *K*), and the two systems share certain passageways.

Kidneys (nephroi) are the chief organs for elimination of excess water in those species in which this is necessary. (It is not necessary in marine or desert animals.) They also assist in maintaining an appropriate electrolyte balance. In the most primitive vertebrates, fluid and metabolic wastes were removed from the coelom by microscopic kidney tubules resembling somewhat the nephridia of earthworms. In most vertebrates, however, the substances to be excreted are collected by the tubules directly from the blood. The tubules transmit the substances to a pair of longitudinal ducts that empty into the cloaca, urinary bladder, or to the exterior.

Reproductive organs include gonads, ducts, glands, storage chambers, and copulatory mechanisms. Early in development all vertebrate embryos are bisexual, having gonadal and duct primordia for both sexes. If the animal is genetically constituted to become a female, the gonad primordia develop into ovaries, and the embryonic female ducts differentiate. If the animal is constituted to become a male, the gonad primordia become testes, and the embryonic male ducts differentiate. The duct system associated with the opposite sex largely disappears. Cyclostomes lack reproductive ducts, and sperm and eggs pass into the coelom and then to the exterior via two pores in the caudal body wall.

Circulatory system

Whole blood is confined to arteries, veins, capillaries, and sinusoids. A single, ventral, multichambered, muscular pumping organ, the heart, is located ventral or caudal to the pharynx. Blood flows from the heart into a ventral aorta and then through aortic arches to the dorsal aorta. The latter conducts the blood caudad. It is to the blood vascular system laden with vital oxygen that vertebrates owe their moment-to-moment existence. Vertebrates also have a lymphatic vascular system.

Although almost all adult vertebrates have red blood cells, at least three species of fishes (*Champsocephalus gunnari, Pseudochaenichthys georgianus*, and *Chaenocephalus acertus*) have none. They are native to the waters of South Georgia Island in the South Atlantic.

Sense organs

Vertebrates have a wide variety of general and special sense organs (**receptors**) that monitor the constantly changing external and internal environment. Receptors will be discussed in Chapter 16.

☐ Chapter summary

1. The vertebrate body is divided into head, trunk (elongated anteriorly as a neck in amniotes), and tail. The tail is characteristic of all vertebrate embryos and appears to have functioned primitively for locomotion in water. Two pairs of lateral appendages (fins in fishes, limbs in tetrapods) are typical.

2. Vertebrates have a notochord, dorsal, hollow nervous system, and pharynx with pharyngeal pouches in the embryo at least. In addition, except for agnathans, they have cartilaginous or bony vertebrae consisting of centra, arches, and processes.

3. The notochord is the most primitive skeletal structure of vertebrates and the first skeletal structure to appear in embryos. In agnathans it is surmounted by lateral neural cartilages of unknown homology. In other vertebrates the notochord becomes surrounded by cartilaginous or bony vertebrae and is either constricted within each vertebra, or it disappears completely during embryonic life except for occasional intervertebral vestiges such as the pulpy nucleus of mammals.

4. The pharynx exhibits a series of pharyngeal pouches that rupture to the exterior to form temporary or permanent pharyngeal slits in most species. The slits serve as gill slits in fishes and most aquatic amphibians. They are temporary in other tetrapods.

5. In front of and behind each embryonic pharyngeal pouch or adult slit lies a pharyngeal arch containing a skeletal element for support, branchiomeric muscle for moving the arch, nerves for the innervation of the muscles and sense organs, and an aortic arch connecting the ventral and dorsal aortas. Pharyngeal arches are readily distinguishable as long as ectodermal grooves or pharyngeal slits remain. They are not identifiable externally after the slits close.

6. The nervous system consists of central nervous system (brain and spinal cord) and peripheral nervous system composed of nerves and associated ganglia and plexuses. The central nervous system is dorsal and hollow because it typically arises as a dorsal ectodermal groove that sinks into the body wall to form a tube. In a few fishes it arises as a solid keel. There is a variety of general and special sense organs.

7. Vertebrates are metameric with respect to numerous systems. They have a two-layered skin. They usually respire by gills or lungs, sometimes assisted by skin, buccopharyngeal lining, and (in embryos) extraembryonic membranes.

8. There are at least two coelomic chambers (pericardial and pleuroperitoneal cavities) and sometimes four (pleural, pericardial, peritoneal, and scrotal cavities).

9. A typical digestive tract consists of oral cavity, pharynx, esophagus, stomach, intestine, cloaca or its derivatives, and numerous glandular or nonglandular diverticula.

10. The *urinogenital* system includes kidneys (nephroi), gonads, ducts, and accessory reproductive organs.

11. The blood circulatory system includes a ventral, multichambered, muscular heart, arteries, veins, capillaries, and, in restricted locations, sinusoids. A lymphatic circulatory system is also present.

LITERATURE CITED AND SELECTED READING

Wessells, N. K.: An essay on vertebrates. In Vertebrate structures and functions, San Francisco, 1974, W. H. Freeman and Co., Publishers.

2

Vertebrate beginnings and some simple chordates

In this chapter we will meet the oldest known vertebrates. We will speculate concerning their invertebrate predecessors and possible kinship with protochordates. Then the best known protochordates—sea squirts and the amphioxus—will be described. Finally, we will examine the structure of the ammocoete, a vertebrate larva that bears a remarkable resemblance to protochordates.

■ OSTRACODERMS AND THE ORIGIN OF CHORDATES

The oldest vertebrates that we have knowledge of are the ostracoderms (Fig. 2-1). These fishes lived chiefly in fresh water from the early Ordovician to the late Devonian periods. They had no jaws, no paired fins, and are thought to have been mostly filter feeders. This means that they filtered food particles out of the stream of respiratory water that was continually flowing into their pharynx and over their gills. Broad plates of bone were embedded in the dermis of their head and anterior trunk, and the more caudal parts of the body had smaller bony scales. These bony plates and scales provided a protective armor that inspired their nickname, "armored fishes." The oldest ostracoderms belong to the order Heterostraci, class Agnatha. We know about these fishes because their bony skin made it possible for them to become fossilized and for mankind to examine them nearly 500 million years later.

The broad outlines of vertebrate history after the ostracoderms have been remarkably well determined. The jawless ostracoderms were followed by placoderms—jawed fishes (Fig. 3-4)—which were followed by the rest of the known jawed fishes and, eventually, by tetrapods (Fig. 3-10). The perplexing problem is, "Who preceded the ostracoderms?"

Although we can follow vertebrate history forward from ostracoderms with reasonable confidence, we can do little but speculate as to the most probable invertebrate ancestors of ostracoderms. This is because there are no fossil links in pre-Ordovician rocks that might suggest the nature of transitional forms. The absence of fossilization suggests the absence of mineralized tissues in ostracoderm ancestors (fossilization takes place

17

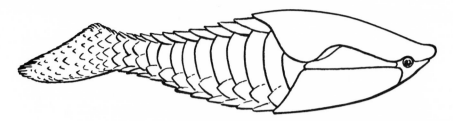

Fig. 2-1. Ostracoderm, a very ancient armored, jawless fish.

most readily in mineralized tissue). There-fore, dermal bone may have appeared with comparative suddenness in soft-bodied in-vertebrates that already possessed a noto-chord, dorsal nervous system, and pharyn-geal slits. It is not necessary to design such a soft-bodied animal out of pure imagina-tion; similar animals already exist. We call them protochordates, and they include uro-chordates and the amphioxus (Fig. 2-2, *A* and *B*). Perhaps they provide some clue, however meagre, to the ancestors of ostra-coderms. It may be that some protochor-date-like organism gave rise to the known protochordates and to a divergent line that led to ostracoderms.

We are not at a dead end in our thinking when we derive protochordates and verte-brates from a common ancestor. We find genetic codes very similar to those of larval chordates in acorn tongue worms and larval echinoderms. Acorn tongue worms (Fig. 2-15) have pharyngeal slits and a dorsal nerve cord, like chordates, and a ventral nerve cord, like invertebrates. However, they have no recognizable notochord. Some acorn tongue worms have ciliated larvae, and all are filter feeders. These hemichor-dates evidently have affinities with chor-dates. They also have affinities with echi-noderms. Larval echinoderms are so similar to larval acorn tongue worms that the latter were at one time mistaken for starfish larvae. Echinoderms, like ostracoderms, have min-eralized tissue in their mesoderm (not in their ectoderm, as in molluscs, for instance), and, like amphioxus, they form their meso-derm and coelom as outpocketings of their archenteron (Fig. 4-6). Finally, echino-derms, acorn tongue worms, amphioxuses, urochordates, and vertebrates are all deu-terostomes, meaning that they convert their blastopore, the original opening into their archenteron, into an anus and develop a new mouth. This trait, shared with only one other invertebrate group (Chaetognatha), is fur-ther evidence of a genetic affinity among these organisms.

Perhaps the most convincing evidence for genetic ties between vertebrates on the one hand and protochordates, hemichor-dates, and echinoderms is the ammocoete, a free-swimming, filter-feeding larval stage of lampreys. The ammocoete larva resembles protochordate larvae very closely (Fig. 2-2).

A phylogenetic flow chart summarizing possible phylogenetic relationships among echinoderms, hemichordates, protochor-dates, and vertebrates is given in Fig. 2-3. It derives vertebrates from a sessile or semi-sessile, bilaterally symmetrical, filter-feed-ing, deuterostomous stem chordate having a dorsal hollow nerve cord, pharyngeal gill slits, and a notochord confined to the larva where it stiffened the muscular tail for locomotion. The presence of a notochord throughout life in any chordate would then be a neotenic condition. (Neoteny means the prolonged retention of a larval charac-ter.) While speculative, this hypothesis seems less so than earlier ones that derived chordates from upside-down annelids (if an earthworm is turned upside down, its ven-tral aorta then flows cephalad and its dorsal

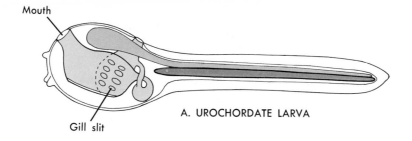

Mouth

Gill slit

A. UROCHORDATE LARVA

Mouth

Gill slit

B. AMPHIOXUS LARVA

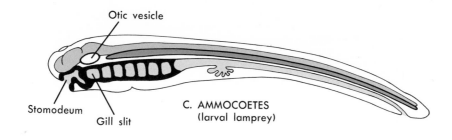

Otic vesicle

Stomodeum

Gill slit

C. AMMOCOETES
(larval lamprey)

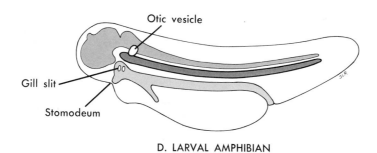

Otic vesicle

Gill slit

Stomodeum

D. LARVAL AMPHIBIAN

Fig. 2-2. Chordate larvae showing basic architectural pattern. *Dark red*, notochord; *medium red*, dorsal nervous system; *light red*, alimentary canal.

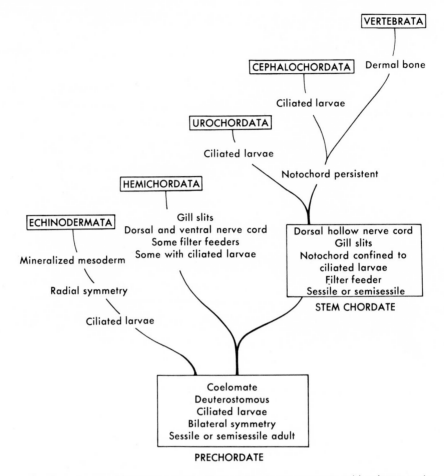

Fig. 2-3. One of several possible hypothetical relationships between echinoderms, enteropneusts, protochordates, and vertebrates.

aorta flows caudad) or from an arthropod whose solid ventral nerve cord became entwined around its stomach and intestine and who therefore ended up with a hollow nerve cord (and, providentially, a new digestive tract).

One of the unsettled problems about the origin of vertebrates is whether they originated in fresh or salt water. Paleontologists who think that they originated in fresh water point to the absence of vertebrate fossils in *marine* Ordovician rocks. This is a formidable argument. Others believe that such fossils may well exist but have not yet been found because at that time vertebrates were few in number and therefore fossils would be rare. Renal physiologists point out that kidney glomeruli are largest in freshwater fishes and smallest and sometimes absent in saltwater fishes; from this they deduce that the original function of glomeruli was to eliminate fresh water. This, too, is an opinion. Arguments supporting both sides have been reviewed by Stahl.[10]

■ AMPHIOXUS AND LESSER PROTOCHORDATES

In 1874 Ernst Haeckel established the phylum Chordata, incorporating the subphyla Urochordata, Cephalochordata, and

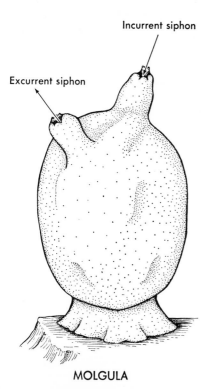

MOLGULA

Fig. 2-4. Sea squirt.

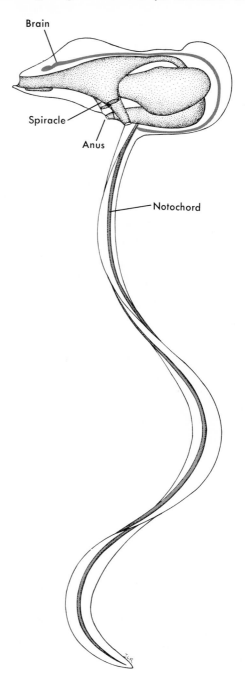

Fig. 2-5. A larvacean. The neural tube continues into the tail along with notochord.

Vertebrata. The phylum was erected to accommodate all organisms having a notochord, pharyngeal slits, and a dorsal, hollow nervous system. Members of the two lower subphyla have come to be known as protochordates. They are all marine organisms. Our interest in them stems from the hypothesis that they share a common ancestor with animals with backbones.

□ Urochordates

The notochord in urochordates is confined to the tail of larvae. Urochordates that metamorphose, that is, the **sea squirts** (Fig. 2-4), lose the notochord during metamorphosis and usually become sessile adults. A second group of urochordates, the tiny **larvaceans** (Fig. 2-5), remain free-swimming larvae throughout life.* The third group of

---

*This is an instance of paedogenesis, the attainment of sexual maturity while in the larval state.

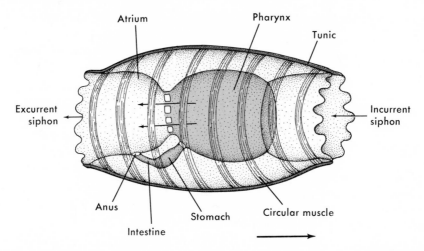

Fig. 2-6. Thaliacean showing direction of respiratory current through gill slits (small arrows) and direction of locomotion (large arrow). Digestive tract (red) is seen through semitransparent tunic and body wall.

urochordates, the **thaliaceans** (Fig. 2-6), have no tail at all and, therefore, no notochord. They are propelled forward by a stream of water that is forcefully expelled from their excurrent siphon. All urochordates are surrounded by a tough cellulose-like tunic that is often beautifully colored and usually transparent. This tunic earned them their alternate name, tunicates.

SEA SQUIRTS

Larval sea squirts (Fig. 2-7, A) are free-swimming organisms about 6 mm long. They have no separate head, so the nerve cord commences in the trunk in a brain-like swelling containing a ventricle. Respiratory water enters the pharynx and passes through pharyngeal gill slits into the atrium, a chamber surrounding the pharynx that also receives the end of the digestive tract. From the atrium, water and digestive tract residues are flushed to the exterior via an atriopore.

At metamorphosis the larva attaches to a substrate via its three adhesive papillae, the tail is resorbed, the notochord disappears, and a rearrangement of internal organs takes place. The larval mouth becomes an incurrent siphon, and the atriopore becomes an excurrent siphon. A water stream laden with food particles and oxygen passes via the incurrent siphon into the pharynx (now the largest organ in the body). Here, food is filtered out of the water stream and ensnared in mucus secreted by the endostyle, a glandular groove in the pharyngeal floor. The particles are then moved by papillae and ciliary action into the stomach while water passes over the gills and into the atrium. The animal is, therefore, a filter feeder. The gametes also are shed into the atrium during reproductive seasons. The forceful discharge of water from the atrium via the excurrent siphon whenever the animal is irritated inspired the descriptive name, sea squirt.

In adults the nervous system is reduced to a solid elongated neural ganglion, a remnant of the "brain" of the larva, with nerve strands radiating to all parts of the body. There are no known special sense organs. Arising from each end of the heart, located

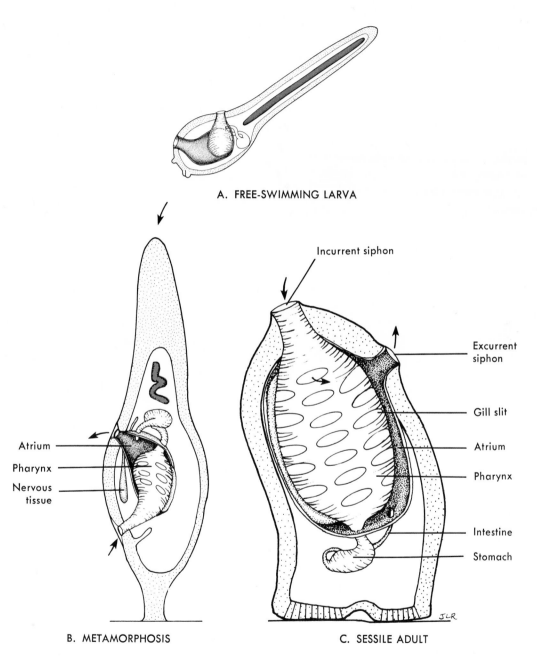

A. FREE-SWIMMING LARVA

Incurrent siphon

Excurrent siphon

Gill slit

Atrium

Pharynx

Intestine

Stomach

Atrium

Pharynx

Nervous tissue

B. METAMORPHOSIS

C. SESSILE ADULT

Fig. 2-7. Metamorphosis in a sea squirt. In **B** and **C,** the atrium has been opened to show pharyngeal slits opening into atrium, and arrows indicate direction of water flow. The notochord (red) disappears during metamorphosis.

near the pharynx, is a vessel. Blood is propelled first into one vessel for several pulsations and then into the other.

There are several genera of sea squirts. *Ciona*, a long, slender, yellow-green polyp, is found off the coast of southern California. *Styela* is a brown polyp that attaches to the substrate by a stalk. *Ascidia ceratodes* has green blood cells, vanadium being substituted for iron in the respiratory pigment.

□ Amphioxus

Amphioxus means "sharp at both ends." Any member of the subphylum Cephalochordata may be called an amphioxus, or lancelet (little spear), but the correct generic name for the lancelet commonly studied in the laboratory is *Branchiostoma* (Fig. 2-8). *Asymmetron* is the only other genus in the subphylum.

Lancelets are found a short distance out from sandy beaches throughout most of the globe. They quickly burrow into sand with eel-like movements, make a U-turn, and then emerge until only the oral hood area

is protruding. Adults vary from less than 2 cm to more than 8 cm in length, the largest being *Branchiostoma californiense*. Off the coast of China, amphioxus is collected in quantity and sold as a table delicacy.

An amphioxus is semitransparent but becomes opaque when immersed in preserving fluids. The body is practically all trunk and tail, for there is almost no cephalization. A pair of longitudinal ridges of unknown function, the **metapleural folds,** hang along each side of the midventral line beneath the pharynx.

Notochord. The notochord extends from the tip of the rostrum to the tip of the tail (Fig. 2-8, *B*). It consists of muscular discs arranged like a long column of coins separated by fluid-filled spaces. The muscle fibers in each disc run transversely and have dorsal extensions that end near nerve terminals from which they evidently receive their motor innervation.[7] The muscle resembles that of invertebrates in that the protein is paramyosin. Contraction of the muscles increases the stiffness of the notochord,

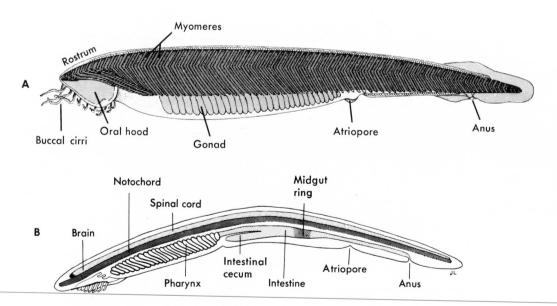

Fig. 2-8. Branchiostoma. **A,** Mature adult. **B,** internal structure of young specimen.

which may aid in swimming. Its continuation to the very tip of the rostrum—unlike in any other chordate—may be an adaptation for burrowing in sand (Fig. 2-9). Surrounding the notochord is a thick collagenous sheath. The only other skeleton in an amphioxus consists of fibrous rods that support the gill bars, buccal cirri, and fins.

Skin (Fig. 5-1). The skin of the amphioxus consists of a single layer of epidermal cells and a thin dermis. Interspersed among the epidermal cells are unicellular glands. Larval skin is ciliated but the cilia later disappear, and the epidermis secretes a cuticle resembling that of annelids. Immediately internal to the dermis is the body wall muscle.

Body wall musculature. The body wall musculature is metameric. It consists of an uninterrupted series of <-shaped muscle segments called **myomeres,** extending from the anterior tip of the body to the tip of the tail. Each myomere is separated from the next by a connective tissue partition, the **myoseptum.** Since the myomeres are <-shaped, cross sections of the body wall include several successive myomeres (Fig. 2-

10). The myomeres are the muscles of locomotion.

Nervous system. The amphioxus exhibits a hollow central nervous system resembling that of vertebrates in basic structure. However, only two brain subdivisions can be detected: an anterior **prosencephalon,** containing a single ventricle, and a more posterior **deuteroencephalon.** The prosencephalon is lined with cilia and long filamentous projections of the ependymal cells, demonstrable only with electron microscopy.

Attempts to homologize the parts of the brain of the amphioxus with those of a vertebrate have not been entirely successful. In an amphioxus the notochord extends anterior to the brain. Does this indicate the absence of a forebrain? The answer must await further research. Whether or not the nerves that supply the gills should be considered cranial nerves complicates the problem. If the branchial nerves are omitted, there are seven cranial nerves (including the apical, or terminal, nerve). If the branchial and oral nerves are included, there are thirty-nine. The absence of semicircular

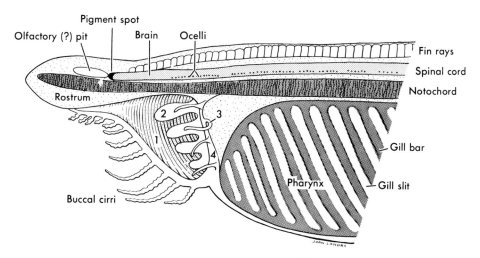

Fig. 2-9. Cephalic end of an amphioxus shown in sagittal section. **1,** Vestibule bounded by oral hood; **2,** part of the wheel organ projecting into vestibule; **3,** velar tentacle; **4,** velum.

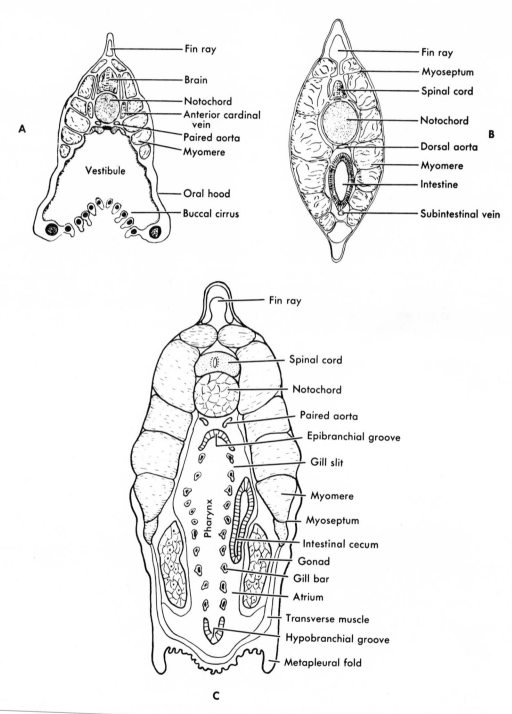

Fig. 2-10. Cross sections of an amphioxus. **A,** Anterior to mouth. **B,** Posterior to atriopore. **C,** Level of pharynx.

canals, eyes, lateral-line system, and foramen magnum deprives us of landmarks that would be helpful. Because of these difficulties, it is not possible to decide at this time just where the brain ends. We must be content to say that it merges imperceptibly with the spinal cord.

The canal within the spinal cord is lined by nonnervous supporting elements called **ependymal cells.** Near the caudal end of the cord the nervous elements disappear, and the cord is composed of ependymal cells alone. A similar situation occurs in vertebrates. A single membrane (**meninx**) surrounds the brain and cord.

Spinal nerves emerge from the cord metamerically. They have dorsal roots that contain sensory and motor fibers, the latter supplying only visceral organs. The "ventral roots" are not nerve roots at all but tubes that conduct extensions of the body wall muscle cells into the spinal cord where they receive their innervation.[5] Since the somites of the two sides are not exactly opposite one another, the roots of the left and right sides do not arise directly opposite one another.

Special sense organs. The relatively small size of the brain is correlated with the paucity of organs of special sense. There are no retinas, semicircular canals, or lateral-line organs. It is doubtful whether an olfactory

epithelium is present. Chemoreceptors are particularly abundant on the buccal cirri and velar tentacles, where they monitor the incurrent water stream. They are also scattered on other surfaces of the body, the tail being more sensitive than the trunk. Touch receptors, which elicit withdrawal, occur over the entire body surface.

The most characteristic sense organs are the light-sensitive, pigmented **ocelli** embedded within the ventrolateral walls of the spinal cord (Fig. 2-11). Each ocellus consists of a receptor cell and a caplike melanocyte. The melanocyte lies between the receptor cell and the incoming light rays and is packed with large melanin pigment granules. A conducting process extends away from the base of the receptor cell. Ocelli probably assist in orienting the animal as it burrows in the sand.

Coelom and atrium. The coelom is almost crowded out by the large atrial chamber surrounding the pharynx. A compressed remnant occurs between the atrial wall and body wall, and other remnants are found adjacent to the gonads, around the ventral aorta, and in the metapleural folds. Development of the coelom is described on p. 81.

Filter feeding and respiration. The entire digestive tract is ciliated. The mouth is an opening in the velum and leads to the pharynx. Cilia on the pharyngeal surface of the gill bars create a steady flow of water through the mouth and into the pharynx. A set of stubby projections in the vestibule, the wheel organ (Fig. 2-9), is covered with sticky mucus that retrieves some of the heavier food particles that miss the mouth, and it directs these through the mouth along with the water stream. Buccal cirri partially strain the water as it enters the vestibule and monitor it chemically.

Food is processed as follows: In the pharyngeal floor there is a **hypobranchial groove,** or endostyle (Fig. 2-10, C). In the roof is an **epibranchial groove.** On the gill

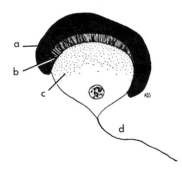

Fig. 2-11. Ocellus (light receptor) from spinal cord of an amphioxus. **a,** Melanocyte; **b,** apical border of receptor cell; **c,** receptor cell; **d,** process for conduction of impulse.

bars ciliated peripharyngeal bands connect the two grooves. The cells of these bands and grooves secrete mucus. Particles of organic matter are trapped in the mucus and incorporated into a stringy food cord that is propelled by cilia dorsally into the epibranchial groove and then caudad into the midgut behind the pharynx. Here it is temporarily arrested by the **midgut ring** (Fig. 2-8, *B*) and mixed with digestive juices. Some of the digesting foodstuffs then pass beyond the ring into the hindgut, and some are driven forward into the **intestinal cecum.** This is an evagination of the midgut that arises in the same way as the liver of vertebrates, although the two do not function alike. The cecum secretes enzymes, and its lining cells phagocytose the smallest food particles and digest them by intracellular digestion. Extracellular digestion takes place in other parts of the digestive tract. The intestine opens to the exterior via an anus.

Relieved of food particles, water passes between the gill bars into the atrium and then to the outside via an atriopore. The number of gill slits varies, but it exceeds sixty in adults. During metamorphosis each larval slit is divided into two by the down-

growth of a tongue bar (Fig. 2-12). Much respiration takes place through the skin.

Circulatory system. The amphioxus has the basic circulatory pattern of vertebrates. However, the heart consists only of a venous sinus (sinus venosus), and the colorless blood is pumped by two muscular pulsating vessels. These are the cecal vein leading to the sinus venosus and the ventral aorta emerging from it (Fig. 2-13). The remaining blood vessels have thin walls, and histologically the arteries, veins, and capillaries are alike. Arteries are vessels that carry blood from the sinus venosus to the gills and then to the body wall and viscera. Veins collect blood from these locations and return it to the sinus venosus.

Arterial blood courses forward beneath the pharynx in the muscular, contractile ventral aorta that commences in the sinus venosus. From the ventral aorta, afferent branchial arteries pass up the gill bars. Before joining the dorsal aorta, these arteries contribute to vascular channels supplying protonephridia (excretory organs). The blood then enters the paired dorsal aortas. These pass caudad above the pharynx and unite just behind it to form an unpaired aorta. This distributes blood by paired ves-

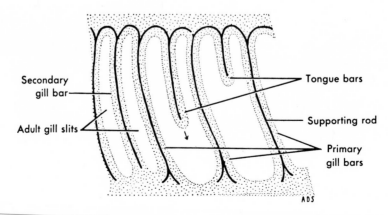

Fig. 2-12. Tongue bars in the pharyngeal wall of an amphioxus growing ventrad (arrow), subdividing the larval slits in two.

sels to the body wall and by median vessels to the visceral organs. The dorsal aorta continues into the tail as the caudal artery.

The venous channels (Fig. 2-13) are similar to the embryonic venous channels of vertebrates. From the capillaries of the tail a single caudal vein courses forward and divides into right and left posterior cardinal veins. These pass forward in the lateral body wall to a point just behind the pharynx. Here the posterior cardinals meet anterior cardinals from the rostrum and pharyngeal wall. The blood then enters a common cardinal vein leading to the sinus venosus. Two parietal veins drain the dorsolateral body wall

caudal to the pharynx. These also terminate in the sinus venosus.

Drainage from the visceral organs is via a median subintestinal vein arising from the caudal vein. The subintestinal passes cephalad along the ventral surface of the intestine. There it breaks up into smaller channels, receives tributaries, and reconvenes to continue forward as a portal vein ending in the capillaries of the cecum. From the cecum the contractile cecal vein leads to the sinus venosus.

Urinogenital system. The amphioxus is **dioecious;** that is, the sexes are separate. Mature gonads are visible through the mus-

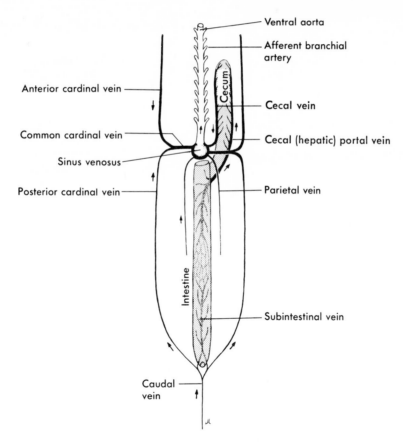

Fig. 2-13. Basic venous channels and ventral aorta of an amphioxus, dorsal view. The cecum has been rotated 90° around the long axis. The cecal portal vein in life is ventral to the cecum and the cecal vein is dorsal. The conventional term hepatic portal for the cecal portal vein is not appropriate.

cle and skin of the trunk. Sperm and eggs are shed directly into the water within the atrium.

Removal of coelomic wastes is accomplished by protonephridia lying beside the secondary gill bars. Each protonephridium consists of clusters of solenocytes that project into the coelom and a chamber that opens into the atrium via a small pore (Fig. 2-14). The flagellum causes a current of coelomic fluid to enter a solenocyte and pass down the stalk. The excretory mechanism

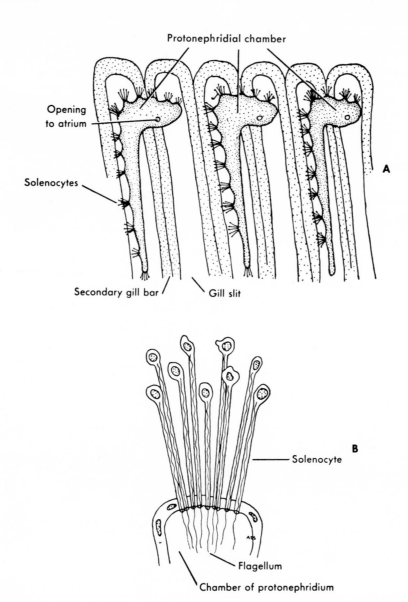

Fig. 2-14. Excretory organs of an amphioxus. **A,** Three protonephridia. **B,** A cluster of solenocytes. They project into the coelom at their free end and empty into the protonephridial chamber at their base.

resembles the nephridia of marine annelids and the flame cells of some other invertebrates.

☐ Amphioxus and the vertebrates

Although an amphioxus resembles a vertebrate in many respects, differences are evident. An amphioxus has almost no cephalization and no paired sense organs; it has a notochord, but no vertebral column; it has gill slits, but in large numbers, emptying into an atrium; it has a dorsal, hollow nervous system, but the brain lacks the major vertebrate subdivisions; it has a segmented musculature, but the segments extend to the anterior tip of the head; it has median fins, but no paired ones; it has a two-layered skin, but the outer layer is only one cell thick; it has arterial and venous channels similar to the basic channels of vertebrates, but no heart; it is coelomate, but the coelom is greatly restricted; liquid wastes are removed from coelomic fluid as in lower vertebrates, but the excretory protonephridia resemble those of nonchordates. A hypothetical relationship of amphioxus to other protochordates and to vertebrates is diagrammed in Fig. 2-3.

■ HEMICHORDATA: INCERTAE SEDIS

The term "incertae sedis" means of uncertain status. Despite years of deliberation, not all organisms have been assigned to a taxonomic group without dissent. Among animals of uncertain status are the acorn tongue (balanoglossid) worms, or enteropneusts, *Dolichoglossus* and *Balanoglossus* (Fig. 2-15).

In 1884, when the chordate phylum was celebrating its tenth birthday as a taxonomic

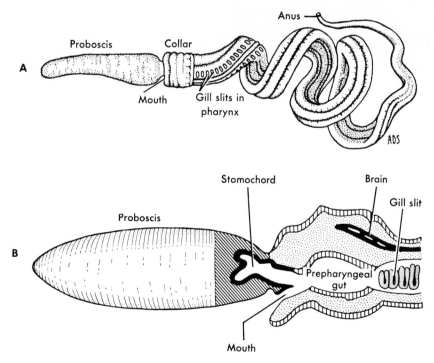

Fig. 2-15. An acorn tongue worm, *Dolichoglossus.* **A,** Entire worm. **B,** Head, sagittal section (except proboscis).

group, William Bateson added acorn tongue worms to the phylum Chordata as the subphylum Hemichordata. Bateson's reasons were:

1. Acorn tongue worms have a dorsal nerve cord consisting of a groove of epidermal cells, in addition to a ventral nerve cord. In the collar region the groove sinks under the surface, forming a tube that is the hollow brain (Fig. 2-15, *B*). The presence of the dorsal nerve cord as an ectodermal groove and tube suggests that hemichordates share a common ancestry with vertebrates.

2. Acorn tongue worms have gill slits in the pharyngeal wall, leading directly to the outside.

3. Acorn tongue worms have a short diverticulum of the gut, called a **stomochord,** extending forward into the proboscis.

Bateson considered the stomochord homologous with the notochord of chordates. Many investigators think otherwise and prefer to classify hemichordates as a phylum close to the echinoderms, since most hemichordates have a free-living larval stage (tornaria larva) resembling larvae of some echinoderms.

■ THE AMMOCOETE LARVA

Ammocoete is the name given to the larval lamprey. *Ammocoetes* was a genus at one time when these larvae were erroneously believed to be adult protochordates. The ammocoete, which lives from 2 to 6 years as a larva, is a filter feeder and greatly resembles the amphioxus, although it has more cephalization and lives in fresh water. The existence of such a larva in the lowest vertebrates strengthens the hypothesis that vertebrates share a common ancestry with protochordates.

Nerve cord, notochord, pharynx (Fig. 2-2, *C*). The dorsal hollow nervous system consists of a brain with three subdivisions—forebrain, midbrain, hindbrain—and spinal cord. The notochord underlies the hindbrain and extends almost to the tip of the

tail. The pharyngeal wall has seven gill pouches, each opening to the outside via a gill slit. A complicated subpharyngeal gland lies ventral to the pharynx and opens into it by a short duct (Fig. 17-12). The cells secrete mucus and thyroid hormone, although there are no thyroid follicles in the larva. Much of the gland degenerates when the larva metamorphoses, but cells that persist organize thyroid follicles like those of higher vertebrates.

Skin and body wall muscle. The epidermis contains many unicellular mucous glands but, unlike that of amphioxus, it is stratified (Fig. 5-4). The dermis is thin and fibrous. The body wall muscle consists of myomeres extending the length of the body. These constitute the chief mass of the lateral body wall (Fig. 2-16).

Special sense organs. From the median naris on the surface of the head a nasal canal slants downward and backward to open into a median olfactory sac, the receptor for smell. A pair of olfactory nerves connects the sac with the olfactory bulb, which is the anteriormost structure of the brain.

Two median eyes are located just beneath the skin, attached to the roof of the forebrain by a stalk. These are the pineal and parapineal organs (Fig. 16-13, *A*). Each consists of a simple lens and receptor cells. Nerve fibers connect these organs with the forebrain. The lateral eyes are deep in the head and functionless.

An otic vesicle is located at the cephalic end of the hindbrain above the first gill chamber on each side (Fig. 2-2, *C*). The vesicles arise as fluid-filled invaginations of the ectoderm and develop into membranous labyrinths (inner ears). Other fluid-filled receptors, the neuromast organs (Fig. 16-1), lie in series under the skin of the head and open onto the surface via pores.

Digestive tract. The most cephalic part of the early larval digestive tract is the stomodeum, a midventral invagination of the ectoderm of the head that grows inward until

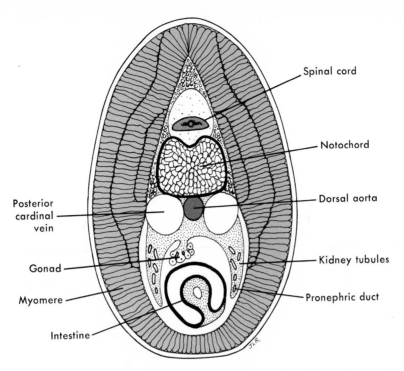

Fig. 2-16. Relatively old ammocoete in cross section just behind liver. Several myomeres are cut in a single section because they curve caudad in passing inward from the skin.

it meets, and opens into, the embryonic gut. In older larvae the stomodeum becomes a buccal funnel leading to the pharynx via the mouth opening. A velum at the entrance to the pharynx creates a respiratory current. Food particles in the incurrent stream are ensnared in mucus, as in amphioxus, and these pass into the esophagus and intestine (there being no stomach). Just behind the esophagus is a liver diverticulum. Embedded within the liver is a gallbladder, but it disappears at metamorphosis. The intestine opens directly to the exterior via an anus.

Circulatory system. The heart lies in a pericardial cavity just behind the pharynx. It has a sinus venosus, atrium, and ventricle. The pericardial cavity is broadly open to the main coelom in ammocoetes, although it becomes a separate cavity in adults. Blood is

pumped forward by the ventricle into the ventral aorta, which is unpaired near the heart but paired farther cephalad. From the ventral aorta, blood passes via afferent branchial arteries to the capillaries of the gills. Oxygenated blood then passes via efferent branchial arteries into the dorsal aorta, which courses caudad.

Venous blood returns from the head via a pair of anterior cardinal veins and from the trunk and tail by two posterior cardinal veins (Fig. 2-16). Anterior and posterior cardinals flow into a common cardinal that empties into the sinus venosus. Blood returns from the intestine via a ventral intestinal vein that ends in the capillaries of the liver. The ventral intestinal and its tributaries therefore constitute a hepatic portal system. From the liver, blood proceeds forward in a hepatic vein to the sinus venosus. The sinus venosus enters the atrium, which leads to the ventricle.

Urinogenital organs. Primitive kidneys are located retroperitoneally just behind the head. Each kidney consists initially of three to six tubules that drain the coelom by peritoneal funnels (nephrostomes). Each funnel collects coelomic fluid secreted by arterial tufts (glomeruli, Fig. 14-2, A). From each kidney a duct passes caudad to terminate in a median papilla that opens to the exterior behind the anus. Medial to the kidneys is an elongated gonadal ridge, which is paired in early larvae but unpaired in older ones. The ridge is the rudiment of the future unpaired gonad.

☐ Chapter summary

1. Ostracoderms are the oldest known vertebrates. Fossils are found chiefly in freshwater rocks of the Ordovician, Silurian, and Devonian periods, and there is formidable evidence that they arose in fresh water. Their dermis contained large bony plates and smaller bony scales. They had no jaws, generally lacked paired appendages, and were filter feeders. They have been placed in the class Agnatha.

2. Vertebrates may share ancestors with the echinoderms, hemichordates, urochordates, and cephalochordates. The common precursor may have been a filter-feeding, coelomate, deuterostomous, bilaterally symmetrical invertebrate with ciliated larvae and sessile or semisessile adults. The earliest chordates added a notochord that was probably confined to the larva, a dorsal, hollow nervous system, and pharyngeal slits.

3. Protochordates are marine invertebrates in the phylum Chordata. They have a notochord, gill slits, and a dorsal, hollow nervous system. There are two groups, urochordates (tunicates) and cephalochordates.

4. Urochordates include sea squirts, permanently larval larvaceans, and tailless, notochordless thaliaceans. They have a tunic and a notochord confined to the larval tail.

5. Cephalochordates are represented by the amphioxus. The notochord is chiefly muscle tissue and is retained throughout life.

6. Hemichordates are of uncertain taxonomic status. They have a stomochord rather than a notochord and are usually excluded from the phylum Chordata.

7. The ammocoete is the free-swimming, filter-feeding larva of lampreys. It closely resembles amphioxus and some protochordate larvae and provides convincing evidence for genetic ties between protochordates and vertebrates.

LITERATURE CITED AND SELECTED READINGS

1. Barrington, E. J. W.: The biology of Hemichordata and Protochordata, San Francisco, 1965, W. H. Freeman and Co. Publishers.
2. Berrill, N. J.: The tunicata, London, 1950, The Ray Society.
3. Berrill, N. J.: The origin of vertebrates, London, 1955, Oxford University Press.
4. Conklin, E. G.: The embryology of amphioxus, Journal of Morphology **54**:69, 1932.
5. Flood, P. R.: A peculiar mode of muscular innervation in amphioxus: light and electron microscopic studies of the so-called ventral roots, Journal of Comparative Neurology **126**:181, 1966.
6. Flood, P. R.: Structure of the segmental trunk muscle in amphioxus, Zeitschrift für Zellforschung und mikroskopische Anatomie **84**:389, 1968.
7. Flood, P. R.: The connection between spinal cord and notochord in amphioxus (*Branchiostoma lanceolatum*), Zeitschrift für Zellforschung und mikroskopische Anatomie **103**:115, 1970.
8. Grassé, P.-P., editor: Traité de zoologie, vol. 11, Echinodermes, stomochordes, procordes, Paris, 1948, Masson & Cie Editeurs.
9. Miller, R. H.: *Ciona*, Liverpool, 1953, Liverpool University Press.
10. Stahl, B. J.: Vertebrate history: problems in evolution, New York, 1974, McGraw-Hill Book Co.

3

Parade of the vertebrates in time and taxons

In this chapter we will look at the major groups of vertebrates, meet
some representatives of each group, learn something about the natural history
of a few, and note some opinions about who came from whom. Finally,
we will examine briefly the probable effect of geographical isolation on species.

There are 49,000 different species of animals exhibiting vertebral columns. Fortunately for Noah 30,000 of these species are fishes. However, many amphibians and some reptiles and mammals share with fishes the freshwater ponds or streams as their permanent abode, and a much smaller number share the seas. Amphibians that can tolerate salt water (certain frogs) are rare and not permanent marine residents. A few snakes, iguanid lizards that live on marine algae, turtles, and crocodiles are predominantly or permanently marine. Among mammals, whales, porpoises, sea cows, and most dolphins are marine. Although birds are not permanent residents of water, many are entirely dependent on marine organisms for food. All the foregoing animals other than fishes, however, are evidently descendants of terrestrial ancestors and have returned to the water.

The purpose of this chapter is to present to the reader vertebrates that are referred to in other chapters, and to introduce their probable ancestors and their nearest living relatives. The printed program for this parade is the abridged classification of vertebrates at the end of the book.

■ VERTEBRATE TAXONS

Natural classification is a device for lumping into groups (taxons) animals that are genetically similar. The chief vertebrate taxons are classes, subclasses, superorders, orders, suborders, families, genera, and species. The following classes are generally recognized, although Agnatha are sometimes given the status of a superclass or even a subphylum.

Aves		Mammalia
	Reptilia	
	Amphibia	
Chondrichthyes		Osteichthyes
	Placodermi*	
	Agnatha	

Agnathans lack jaws and hence are **agnathostomes;** all other vertebrates are **gnathostomes** (Fig. 3-1). Commencing with amphibians, vertebrates typically have four legs (sometimes modified as wings or paddles); hence they are **tetrapods.** Commencing with reptiles, vertebrates exhibit a special membrane that surrounds the developing embryo. This membrane is the **amnion,** and animals that possess it are said to be **amniotes.** Fishes and amphibians do not exhibit this membrane and are **anamniotes.**

Since the time of Carl von Linné (Latin, Linnaeus), a Swedish naturalist, taxonomic nomenclature has been latinized by agreement among zoologists of the world. This enables zoologists of all languages to under-

---

*This class is known only as fossils.

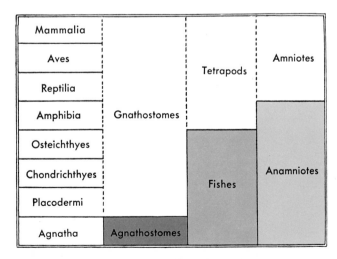

Fig. 3-1. Major categories of vertebrates.

stand one another without translation when the name of any animal is mentioned. For instance, not every zoologist knows what "un chat," "die Katze," or "el gato" is. All these words refer to the common domestic cat, but *Felis cattus* is the taxonomic name for this species. The generic name *Felis* separates certain cats—mountain lion, European wildcat, domestic cat, and others—from lions, tigers, leopards, and so forth, which have been placed in the genus *Panthera*. *Felis cattus* separates domestic cats from mountain lions *(F. concolor)* and from European wildcats *(F. sylvestrus).* The binomial designation for a species was introduced by Linnaeus in the tenth edition (1758) of his classic book *Systema Naturae.*

Below is one classification for the spiny dogfish, the spotted necturus, and the domestic cat. Many taxonomic names, when translated, characterize the taxon, making it easier to remember. Thus Chondrichthyes means "cartilaginous fishes," Carnivora means "flesh eaters," and *maculosus* means "spotted." Because of space limitations the anatomical terminology at the end of the book does not contain the stems from which all taxonomic names have been derived. Some, however, are included, and the rest can be found in an unabridged dictionary.

■ AGNATHA: THE FIRST VERTEBRATES

The class Agnatha includes two groups of jawless fishes—the ancient, bony **ostracoderms** and the boneless **cyclostomes**. Ostracoderms (Fig. 2-1) are the oldest known vertebrates. A bony armor was present in the skin of the entire body and, in addition, their head contained a deeper skeleton of bone and considerable cartilage. They attained a length of up to 30 cm. Ostracoderms are found in four extinct orders from the Ordovician, Silurian, and Devonian periods (p. 436).

Living cyclostomes include lampreys and hagfishes. They became separated from the mainstream of vertebrate evolution some 400 or more million years ago and, while retaining what are evidently primitive traits (no jaws, median nostril and olfactory sac, no paired fins, a pineal gland, only two semicircular ducts, large gill pouches with small round external openings) they lost all capacity to form bone in the skin or elsewhere, became eellike, developed a buccal funnel with a rasping tongue, and many became parasitic. Little is known of the axial skeleton of the trunk and tail of ostracoderms, but living cyclostomes retain a prominent notochord throughout life and have no recognizable vertebral column (Fig. 1-5). Although cyclostomes are the lowest living vertebrates taxonomically, it must not be assumed that, because of this, all systems are primitive. Many features are highly specialized.* Cyclostomes are sometimes placed in two orders (Petromyzontiformes and Myxiniformes), sometimes in one (Cyclostomata). Lampreys have been

---

*A discussion of *primitive, specialized,* and related terms is found on pp. 433 to 435.

	Spiny dogfish	Spotted necturus	Domestic cat
Class	Chondrichthyes	Amphibia	Mammalia
Subclass	Elasmobranchii	Lissamphibia	Theria
Superorder			Ferae
Order	Selachii	Caudata	Carnivora
Suborder	Squaloidea		Fissipedia
Family		Proteidae	Felidae
Genus and species	*Squalus acanthias*	*Necturus maculosus*	*Felis cattus*

around unchanged for a long time. A fossil lamprey from the Carboniferous period is seen in Fig. 3-2.

□ Lampreys

The ammocoete larvae of lampreys have been described on p. 32. Many larval traits are retained by adults. A large buccal funnel lined with horny denticles develops from the larval stomodeum (Figs. 2-2, *C*, and 5-8). It helps keep the adult lamprey, which is parasitic, attached to the host while the tonguelike rod covered with horny teeth rasps the flesh of the victim. A single nostril is located dorsally on the head, and a nasal canal leads from the nostril to the olfactory sac and then terminates blindly in a nasohypophyseal sac (Fig. 12-5). The seven pairs of gill pouches open separately via gill slits that conduct water in and out of the pouches, thus freeing the buccal funnel for feeding on the host.

Petromyzon marinus marinus is a species of anadromous lampreys. That is, members live in the sea but migrate upstream to lay their eggs. In 20 to 21 days the small, nonparasitic larvae emerge. After several years upstream the larvae metamorphose into immature adults and migrate to sea. There they attain sexual maturity and be-

Fig. 3-2. Fossil lamprey *Mayomyzon* from the Carboniferous period. (From Bardack and Zangerl.[2])

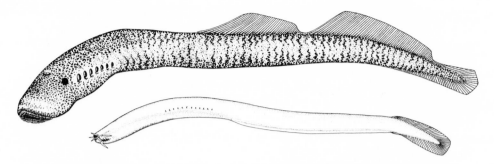

Fig. 3-3. Lamprey *Petromyzon* above; hagfish *Bdellostoma* below.

come physiologically adapted for the journey back to the spawning place. *Petromyzon marinus dorsatus* is a land-locked population inhabiting the freshwater Great Lakes between Canada and the United States. They, too, enter rivers to spawn. Most *Lampetra* are freshwater lampreys that do not migrate. They spawn on reaching sexual maturity, and promptly die without even eating!

☐ **Hagfishes**

Hagfishes are marine cyclostomes with a shallow buccal funnel lacking denticles. They feed on live and dead fish and a variety of small invertebrates. A fringe of stubby, fingerlike papillae surrounds the buccal funnel, and a single nostril is located just above the funnel. A canal leads from the nostril to the olfactory sac and then on to the pharyngeal cavity, carrying respiratory water. The eyes are vestigial.

Myxine glutinosa, the Atlantic hagfish, has six pairs of gill pouches (occasionally five or seven) that open into a common efferent duct (Fig. 12-4). *Bdellostoma stouti*, common off the coast of California, has ten to fifteen pairs of gill pouches opening directly to the exterior. Hagfishes are not anadromous, and the larvae stay within the egg membranes until metamorphosis. This may be an adaptation to the saltwater environment.

■ **PLACODERMI: THE FIRST JAWED FISHES**

Placoderms were armored fishes that were increasing in number in the Devonian fresh waters while the ostracoderms were disappearing. However, they were probably off the main line of vertebrate evolution. They had paired fins and bony jaws and were swift predators. The best-known placoderms are the arthrodires (Fig. 3-4, *Coccosteus*). They had a heavy bony dermal shield in the head and pharyngeal region and another in the anterior part of the trunk, the two shields meeting in a movable joint. The rest of the body was usually naked.

Another group of early armored gnathostomes, perhaps the oldest ones, are the little acanthodians (Fig. 9-9). Superficially they were sharklike, but they were bony and only a few centimeters long. Their fins, as many as five pairs, were hollow spines that supported webs of skin. Rhomboid scales formed a continuous dermal armor. Acanthodians, or "spiny sharks" as they are sometimes called, are sometimes included among placoderms and sometimes placed in a separate class Acanthodii.

■ **CHONDRICHTHYES: CARTILAGINOUS FISHES**

Chondrichthyes (Fig. 3-5) are cartilaginous fishes with no bone other than that in their scales and teeth. Their ancestors had a

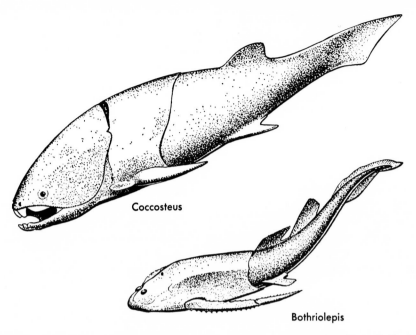

Fig. 3-4. Two Devonian placoderms, each about one-third natural size. (From Colbert.[4])

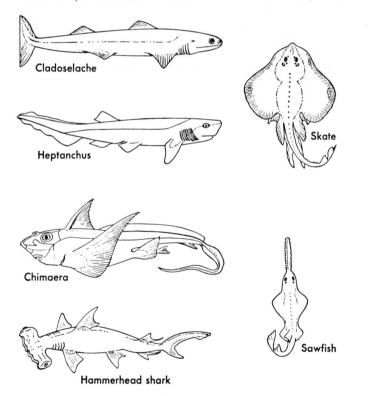

Fig. 3-5. Group of Chondrichthyes. *Cladoselache* is primitive and extinct.

bony skeleton, and the absence of bone is a specialization. The pelvic fins of males are modified as claspers for transfer of sperm to the female. They have placoid scales made of dentin and enamel.

Chondrichthyes are numerous at present but were far more so in ancient times. This is confirmed by the large number of fossil species known, some of which are identified only by a horny spine, by otoliths (calcareous concretions in the inner ear), or by their teeth. Let it not be assumed, however, that fragments alone remain of these ancient hordes. Many remarkably preserved cartilaginous fishes with intact viscera and muscle fibers have been uncovered within comparatively recent years, and it is likely that further discoveries will be forthcoming. Some of the ancient forms were heavily armored, as may be expected, and placoid scales are remnants of that armor. The most common cartilaginous fishes are elasmobranchs.

□ Elasmobranchs

Elasmobranch fishes are divided into **Cladoselachii,** all of which are extinct; **Selachii,** or sharks; and **Batoidea,** or rays, skates, and sawfishes. The first pharyngeal slit is small, contains only a miniature gill-like surface (**pseudobranch**), and is called a **spiracle.** The gill slits are "naked," that is, visible on the side of the pharynx rather than covered by an operculum. Except in primitive forms such as *Cladoselache* (Fig. 3-5), the mouth is on the ventral surface rather than terminal.

Sharks are of interest to students of vertebrate anatomy because of their generalized structure. If one were to seek a living blueprint of the vertebrate body—an architectural pattern that, by modifications here and deletions there, would serve for all vertebrates, the shark would be one. The anatomy of the eye is in all basic respects the anatomy of the human eyeball. The distribution of the nerves to the jaws, to the

olfactory epithelium, to the inner ear, and to the musculature is essentially the arrangement found in man. The arrangement of the visceral skeleton, the aortic arches, the chief venous channels, and the urinogenital system of sharks is, in essential features, comparable with the arrangement of these same structures in the embryos of higher vertebrates. Thus, knowledge of the anatomy of the shark constitutes a point of departure for the study of higher vertebrates, culminating, if one wishes, in the study of man. Often studied is the spiny dogfish of the Atlantic, *Squalus acanthias,* named for the prominent spine associated with each dorsal fin. *Squalus suckleyi* is the Pacific spiny dogfish. *Mustelus* is the "smooth dogfish" in the sense that it lacks dorsal spines. These species have a spiracle and five gill slits. Hexanchid sharks have six gill slits, plus a spiracle, and heptanchid sharks have seven. This seems to be a primitive number for sharks, and is the largest number found in any jawed fish.

Rays, skates, and **sawfish** are elasmobranch fishes whose bodies are more or less flattened. If one could grasp the lateral body wall of a dogfish just above the gill slits and stretch the body wall laterad to form a "wing" without interfering with the location or shape of the coelomic cavity, this would produce the body form of a ray or skate. The mouth and gill slits are ventral (Fig. 12-20), but the spiracle is dorsal. Movement is accomplished by undulation of the winglike lateral body wall. The ventral location of the mouth and dorsal location of the spiracle are adaptations to the fact that skates and rays, unlike sharks, are bottom feeders and their chief incurrent flow of respiratory water is via the spiracle. They scoop up food with their ventral mouth and sometimes stir up mud and sand while doing so. The spiracle is removed from this debris.

The tail in sting rays has become an organ of defense and offense. In some rays it becomes an electric organ capable of delivering a high-voltage electric discharge. The giant ray of tropical waters weighs nearly half a ton and has a pectoral finspread of more than 20 feet! Sawfish (Fig. 3-5) do not become as flattened as rays and skates, but the gill slits are ventrally located.

☐ Holocephalans

Holocephalans are the chimaeras (Fig. 3-5), an atypical group of Chondrichthyes that lack scales on most of the body. The gill slits are covered by a fleshy operculum, and the spiracle is closed. The upper jaw, unlike in elasmobranchs, is solidly fused with the cartilaginous braincase. Instead of teeth there are hard, flat plates on the jaws.

■ OSTEICHTHYES: BONY FISHES

Osteichthyes have a skeleton composed partly or chiefly of bone, the gill slits are covered by a bony operculum that grows from the second visceral arch, and the skin has scales with more or less bone. Most bony fishes also have a gas-filled swim bladder. The cloaca in all but lungfishes is so shallow that it is practically nonexistent and there are no more than five gill apertures. Osteichthyes are either ray-finned or lobe-finned. Some are very old and some are very modern.

☐ Ray-finned fishes

Ray-finned fishes (**Actinopterygii**) are bony fishes in which slender rays are essentially the sole support for the fins and internal nares are lacking. (A few species have internal nares, but these are not homologous with those of lobe-finned fishes or tetrapods.) There are three groups: chondrosteans, holosteans, and teleosts.* Of the three, the lower two, known as **ganoid fishes,** have seen better days. Time was when these ganoids, well supplied with a bony internal skeleton, protected on the head by bony plates, and covered over the entire body by

---

*Other classifications are also employed.

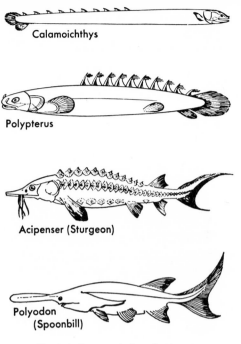

Fig. 3-6. Group of chondrosteans.

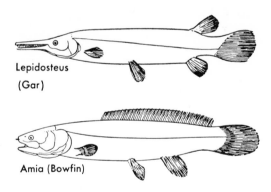

Fig. 3-7. The sole living holosteans.

bony ganoid scales, were the dominant fishes. They are now represented by only a few survivors. Teleosts are more recent and are referred to as modern bony fishes.

Chondrosteans. Chondrosteans are represented today by only the sturgeons, spoonbills, and two aberrant African genera (Fig. 3-6). Sturgeons and spoonbills have lost the ganoin on the scales and have undergone a general loss of ossification. Although ancient ganoids had a bony endoskeleton, recent sturgeons and spoonbills have an endoskeleton composed primarily of cartilage.

Polypterus and *Calamoichthys* are African freshwater ganoids. They are sometimes placed in a superorder of their own.

Holosteans. Only two genera of holosteans have survived until today (Fig. 3-7)—*Lepidosteus* (gars) and *Amia* (bowfins*). Both are freshwater fishes that breathe air. Gars are completely covered with typical ganoid scales, but bowfin scales have lost their ganoin. In the head, bony dermal plates lacking ganoin overlie a cartilaginous braincase (Fig. 8-9).

Teleosts. Unless biology students are in-

---

*The bowfin is also known as a choupique, cypress trout, mudfish, grindle, blackfish, and beavertail, according to the locality.

terested in ichthyology they may gain little notion of the enormous variety of teleosts (Fig. 3-8). More than 25,000 species (about 95% of all living fishes) have displaced ostracoderms, placoderms, chondrosteans, and holosteans, in that order.

There are long, slim teleosts without paired appendages; short, fat ones with sails; transparent ones; fishes that stand on their tails; fishes with both eyes on the same side of the head; fishes that carry lanterns, that climb trees, that carry their eggs in their mouth, that appear to be smoking pipes, that possess periscopical eyes; and hundreds of other bizarre genera. They inhabit the abyssal depths far out from the continental shelf, they cavort in modest brooks, and some make nightly sorties onto land. They run the gamut of colors, although a relatively small number of pigments assisted by myriads of light-dispersing crystals are responsible for all hues.

The skeleton of teleosts is well ossified,

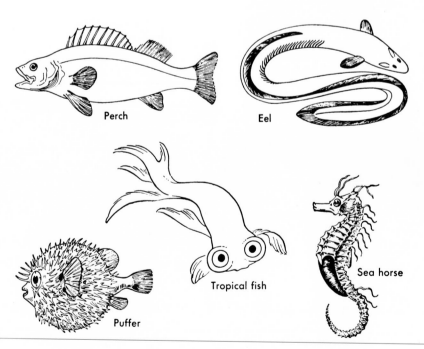

Fig. 3-8. Group of teleosts.

but their overlapping cycloid and ctenoid scales are flexible because they have only a thin layer of bone. The pelvic fins are often far forward and there is no spiracle. These are only a few of the many characteristics of teleosts. With the exception of the fishes already discussed and the lobe fins, any fish caught on a hook or seen in an aquarium or in the marketplace is a teleost.

☐ **Lobe-finned fishes** (Fig. 3-9)

Lobe-finned fishes (**Sarcopterygii**) have a fleshy lobe at the base of their paired fins.

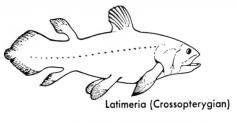
Latimeria (Crossopterygian)

Neoceratodus (Dipnoan)

Fig. 3-9. Lobe-finned fishes.

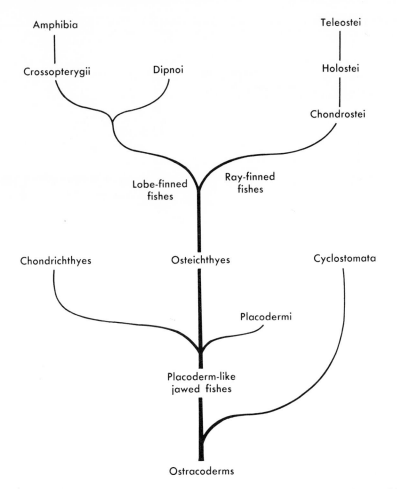

Fig. 3-10. Theoretical phylogenetic lines leading to the major groups of fishes and to amphibians. The distribution of fishes in geological time is given in Fig. 3-33.

The lobe contains part of the fin skeleton. They also have internal nares that open into the oral cavity. There are two orders, Crossopterygii and Dipnoi. These were differentiated by the start of the Devonian.

Crossopterygians. With the exception of *Latimeria*, crossopterygians are all yesterday's animals. They are of special interest because of their resemblance to early amphibians. The skeletal elements within the fin lobe corresponded closely to the proximal skeletal elements of early tetrapod limbs (Fig. 9-29). The skull was similar to that of the earliest amphibians (Fig. 8-10). They had swim bladders that some may have used as lungs, and most of them had internal nares, although these were probably not used for breathing. Because of these and other traits, crossopterygians (probably the freshwater rhipidistians) are thought to have been the stem from which amphibians evolved.

Dipnoans. There are three living genera of these true lungfishes. These are *Protopterus* from Africa, *Neoceratodus* from Australia, and *Lepidosiren* from Brazil. *Protopterus* and *Lepidosiren* have inefficient gills and suffocate if held under water, but *Neoceratodus* relies on gills for oxygen except when the oxygen content of the water is low. During the wet season these animals enjoy life in their freshwater habitats; but when the sun dries the streams, the African and Brazilian species dig deep burrows in the moist, muddy banks and spend the dry, hot season in a state of lowered metabolism (aestivation). This minimizes water loss and reduces the need for nutrients and oxygen.

Lungfishes and amphibians are derived from ancient crossopterygians and have undergone a number of similar mutations. In both the swim bladder became supplied by a branch from the sixth aortic arch instead of from the dorsal aorta as in their crossopterygian ancestors, the atrium of the heart became partially divided into two chambers, both usually have a larval stage with external gills, both have internal nares, and in both the swim bladders or lungs have ducts leading to the pharynx. The approximate phylogenetic lines leading to the major groups of fishes and to amphibians are given in Fig. 3-10.

■ AMPHIBIANS: THE LOWEST TETRAPODS
□ Who they are and where they came from

The oldest known amphibians are the **labyrinthodonts** (Fig. 3-11). They were discovered in rocks from the end of the Devonian or the beginning of the Carboniferous period in Greenland. Besides bearing resemblances to later tailed amphibians, they had many fishlike features not found in today's amphibians. Among these were minute bony scales in the skin (today found only in burrowing amphibians) and a tail structured for swimming and resembling that of a water-dwelling newt, but containing fin rays that newts do not have. Their skulls were so similar to certain crossopterygian skulls that the two are easily confused (Fig. 8-10). Grooves in the bones just

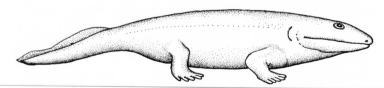

Fig. 3-11. A labyrinthodont, *Ichthyostega,* from the Devonian period.

under the skin of the head show that labyrinthodonts had a sensory canal system of neuromast organs, which means that they lived primarily in the water. Today's aquatic amphibians still have this system, but terrestrial species lose it at metamorphosis. Labyrinthodonts were as small as today's newts and as large as crocodiles. Most authorities consider them to be the ancestors of modern amphibians and of the first reptiles.

A second major group of Paleozoic amphibians are the **lepospondyls** (subclass Lepospondyli). Attempts to trace their ancestry have not been successful because of lack of fossil evidence. There are authorities who, although in the minority, think that lepospondyls arose from some crossopterygian completely independently of labyrinthodonts or even from an ancient dipnoan, and that they gave rise to urodeles, whereas labyrinthodonts gave rise to anurans, apodans, and cotylosaurs. This viewpoint is based on interpretation of anatomical evidence alone and is known as the diphyletic theory of the origin of tetrapods. The monophyletic theory is that tetrapods are derived from a common crossopterygian ancestor, probably a rhipidistian, and that lepospondyls are side branches from this common stem. Arguments for and against these viewpoints have been summarized by Stahl.[11]

The remaining amphibians are in the subclass **Lissamphibia,** which consists of three "modern" orders, Anura (frogs and toads), Caudata or Urodela (modern tailed amphibians) and Apoda (wormlike burrowing amphibians). Modern amphibians usually have an aquatic larval stage with external gills and sensory canals that detect certain waterborne stimuli. Without special adaptations, therefore, amphibians that live on land must return to the ponds and streams to lay their fishlike eggs. At metamorphosis the external gills are normally lost even in aquatic species, although in the latter the sensory

canals are retained. Amphibians are the first vertebrates to exhibit a middle ear cavity with an ear ossicle (columella) for transmitting airborne sound waves. The middle ear complex is a contribution from the branchial apparatus of crossopterygian forebears. The skin has lost all traces of ancestral bony scales except in apodans, and the surface layer tends to develop a stratum corneum of keratinized (cornified) cells that help prevent desiccation in air. In aquatic forms this layer is thin and the skin contains many mucous glands, but in terrestrial forms the cornified layer becomes thick and mucous glands become sparse. The limbs exhibit the usual tetrapod skeleton, which has the potential of supporting the weight of the body for locomotion on land. Nevertheless, aquatic urodeles still swim like fishes by sinusoidal movements of the trunk and tail. The pelvic girdle has become modified to brace the hind limbs against the vertebral column by articulating with a single trunk vertebra, called sacral. The first vertebra (now called cervical) has been modified to articulate against the two occipital condyles of the amphibian skull, which enables amphibians to move their heads up and down as in saying "yes." (Shaking their head "no" is anatomically impossible.) The skull exhibits many uniquely amphibian traits. These and other amphibian features are discussed in later chapters.

☐ Caudata (Urodela)

Tailed amphibians resemble, in body form, the ancient amphibians from which they are descended. Some retain external gills throughout life, even though lungs develop, and these are neotenic.* Plethodontidae lose their gills but do not develop lungs. The eight urodele families of the world are listed in Table 3-1 and represen-

---

*Neoteny is the prolonged retention of larval characters. When the larval trait retained is external gills, the animal may also be said to be "perennibranchiate."

tative genera are discussed below and illustrated in Fig. 3-12.

Necturus, the mud puppy, is the only genus of Proteidae in the United States and Canada, where there are six species and subspecies. Its European counterpart is *Proteus*, a blind cave dweller. As a 2-cm larva, *Necturus* has tiny imperfect tetrapod appendages, the belly is distended by the presence of yolk, and the tail is keeled and has dorsal and ventral fins. There are three pairs of external gills and three pairs of gill slits. As the larva grows, toes become more pronounced, the skin darkens, and the yolk is used up. At about 5 years of age *Necturus* attains 20 cm in length and is sexually mature. Lungs have developed, but the external gills and two pairs of slits remain. The tail is still keeled and has fins. Thus *Necturus* is neotenous, but unlike neotenous populations of *Ambystoma*, *Necturus* cannot be induced to discard its gills by administration of thyroxin. One or more enzymes necessary for thyroxin to bring about

Table 3-1. Distribution of gills, pharyngeal slits, and lungs among adult urodeles

| Family | Representative genera | Number of pairs | | Lungs |
		Gills	Slits	
Proteidae	*Necturus*	3	2	Yes
Amphiumidae	*Amphiuma*	0*	1	Yes
Hynobiidae	*Hynobius*	0*	0	Occasionally
Cryptobranchidae	*Cryptobranchus*	0*	1	Yes
Salamandridae	*Notophthalmus*	0*	0	Yes
Ambystomatidae	*Ambystoma*	0*	0	Yes
Plethodontidae	*Plethodon*	0*	0	No
Sirenidae	*Siren*	3	3 to 1	Yes

*Some species or individuals are perennibranchiate (see footnote at bottom of p. 47).

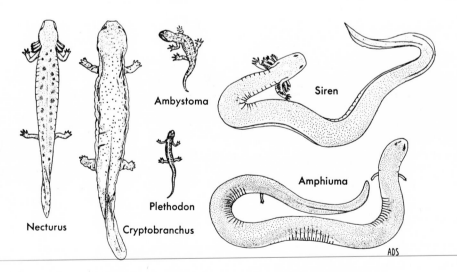

Fig. 3-12. Representatives of six families of urodeles.

metamorphosis are either lacking in *Necturus* or are blocked by other metabolic substances.

Amphiuma (family Amphiumidae) is a large, aquatic, North American, eel-like urodele that attains a length of 1 m. One pair of slits remains in adults (Fig. 1-9). The appendages are very small and entirely inadequate for bearing weight. There are two species, *Amphiuma pholeter* and *A. means*. *A. pholeter* has one digit, *A. means means* has two, and *A. means tridactylum* has three.

Hynobius and *Ranodon* are genera in the family Hynobiidae, which are Asiatic land salamanders. The family has primitive caudate characteristics.

Cryptobranchus is a genus in the family Cryptobranchidae that is found through Asia and North America. It looks ferocious because of its broad, flattened head; compressed tail with a thin, deep dorsal keel; longitudinal wrinkles on its chin; and a wrinkled, fleshy fold along the entire trunk on each side. *Cryptobranchus* attains a length of 70 cm. One gill slit usually remains on each side, sometimes concealed beneath folds of skin.

Salamandra and *Notophthalmus* are Salamandridae (Eurasian and American aquatic and terrestrial salamanders). *Notophthalmus viridescens* typically exhibits three phases of postembryonic life (Fig. 3-13). The larva lives in water. After several months gills and slits are lost, four legs appear, and the animal, now an **eft,** emerges from the water to commence a term of residence on land, often ascending to 4,000-foot altitudes. The skin develops a thick stratum corneum that blocks the sensory canals and skin glands. The body gradually assumes a bright orange-red color, and a series of black-bordered red spots develop along the dorsolateral aspect. The land phase lasts 1 to 3 years, depending on the locality. It ends as the eft approaches sexual maturity under the stimulus of gonadotropic hormones. As the time for mating draws near, the efts commence mass migrations down the hills, through the lowlands and meadows, toward the freshwater ponds. The migration is a manifestation of the water drive brought about by the pituitary hormone prolactin. The thick stratum corneum is shed, exposing the mucous glands and sensory canals. The skin

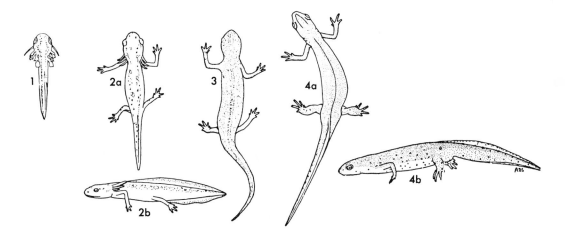

Fig. 3-13. Life history of the salamander *Notophthalmus*. **1,** Newly hatched larva (7 mm); **2a,** fully formed larva (30 mm); **2b,** fully formed larva, lateral view; **3,** red eft (70 mm); **4a,** male newt (95 mm); **4b,** male newt, lateral view.

loses its brilliant color and by the time the animals enter the ponds, many have assumed the adult coloration, olive green on the back and light yellow ventrally. The tail changes from round to laterally compressed and again develops dorsal and ventral fins. The animal is now a sexually mature waterphase individual known as a **newt.** In some localities the larvae remain in the water and retain vestigial gills throughout life. *Salamandra atra* is viviparous.

Ambystoma belongs to a family (Ambystomatidae) of terrestrial North American urodeles. In certain localities they retain external gills and remain aquatic throughout life. However, these reluctant individuals can be caused to discard their gills by being provided with thyroid hormone or iodine. Perhaps the most widely known species is the neotenic *Ambystoma mexicanum,* commonly known as the Mexican axolotl.* Even among neotenic populations some individuals metamorphose spontaneously.

Plethodon belongs to the American and European family of terrestrial or aquatic urodeles, Plethodontidae. At metamorphosis most of them lose their gills but none develop lungs. They breathe via their skin. Plethodons live at considerable distances from bodies of water and, except for some tropical forms, lay their eggs in moist places such as the undersides of logs or in damp caves. The larvae hatch with legs already formed and may never enter water. In such instances larval gills are of no advantage and are discarded a few days after hatching.

Siren and *Pseudobranchus* are two genera in the American family Sirenidae, which lives in muck, or swamplands. They are perennibranchiates that never develop hind limbs. *Siren* usually has three gill slits, but not all of these may be open. *Pseudobranchus* retains one gill slit.

---

*Xolotl was an Aztec god of twins and monsters.

□ **Anura**

Frogs and toads are tailless amphibians in which the caudal vertebrae are fused into one elongated urostyle (Fig. 7-8). The adult anuran breathes by lungs and skin and lives on land or in fresh water. A few anurans, such *Rana cancrivora,* the crab-eating frog of the marshes of Thailand, and *Bufo viridis,* can tolerate salt water by hormonally maintaining a higher than usual level of salt in their blood and other tissues.

A few anurans do not live near water and therefore cannot deposit their eggs in water. But what would happen if larvae with external gills, no legs, and fishlike bodies for swimming were to hatch from their jelly envelopes only to find they were not in water? The species would almost surely perish. But such species survive because of an adaptive mutation. The larval stage of robber frogs occurs within the jelly envelopes and miniature adults emerge to assume life on land. (Remarkable as this seems, reptiles and birds regularly do essentially this!) Some thirty species in the family Hylidae (tree frogs) do not lay their eggs near water either. Instead, they carry developing eggs in a brood pouch under the skin of the back. Later, fully metamorphosed young frogs escape through a small posteriorly directed opening in the skin. A similar condition is found in the Surinam frog, *Pipa pipa.* The East African toad *Nectophrynoides vivipara* is viviparous. As many as 100 young develop in the female reproductive tract and are born alive.

The earliest fossil frog, or prefrog, if you prefer to call it that, is *Triadobatrachus* (= *Protobatrachus*) from the lower Triassic. It had a skull quite similar to today's frogs and toads and the body was shortened as a result of reduction in the number of trunk vertebrae, approaching the condition in modern anurans. It had a definite tail with separate caudal vertebrae, the ribs were longer than in modern anurans and were not fused with the vertebrae as they are today,

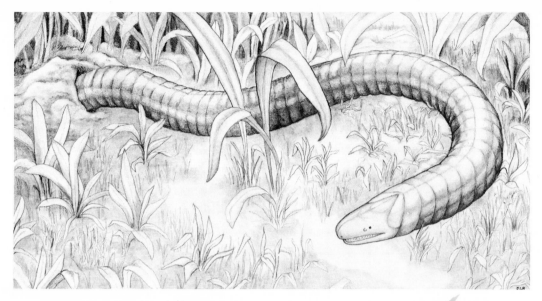

Fig. 3-14. An apodan. The annular structure is a specialization that evidently aids in burrowing.

the tibia and fibula of the leg were not fused as they are today, and the abdomen was covered with bony scales. Whom they came from is unknown. Schmalhausen[10] cites evidence that anurans may have originated in the mountains, their small ancestors having retreated upstream in the face of threats from lowland enemies, and that they returned to the lowlands only after extinction of the large reptiles. He points out that the more primitive living anurans—*Ascaphus*, for example—live in the mountains today.

□ **Apoda**

Apodans (Fig. 3-14), or caecilians, are circumtropical limbless amphibians, which, except for a few aquatic species, live in burrows on land. Their eyes are small and sometimes buried beneath the bones of the skull. They have minute scales in their skin, vestiges of the dermal scales of ancestral amphibians. Some are a half meter long and have as many as 250 vertebrae. They have a very short tail with the result that the vent is almost at the end of the body. Burrowing species lay large yolky eggs, and the larval

stage is passed in the egg envelopes. Several aquatic genera are viviparous. Only one fossil apodan has been found.

■ **REPTILES: THE LOWEST AMNIOTES**

From the ancient labyrinthodonts there arose, if we read correctly Nature's handwriting in the earth, a group of tetrapods destined to be named, 300 million years later, the **cotylosaurs** (Fig. 3-15, *A*). These were the first, or stem, reptiles. The earliest cotylosaurs had made few advances over the labyrinthodonts, but from the cotylosaurs developed a distinguished and hoary group of descendants, the members of the class Reptilia.

The cotylosaurs have vanished. Gone with them are their descendants the dinosaurs, the flying reptiles (pterosaurs), and the aquatic, viviparous ichthyosaurs that had paddles for limbs. Vanished are hundreds upon hundreds of the mighty or weak, great or small, fast or lumbering reptiles, which one by one bowed out of the vertebrate parade while mammals were first raising their voices in the forests and at the

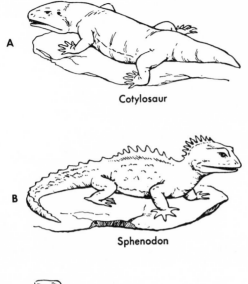

Fig. 3-15. A, Stem reptile. B, "Living fossil" reptile. C, Modern iguanid lizard.

edges of the bogs. Remaining with us are a few successful cotylosaur descendants: the turtles, an ancient group that thus far has doggedly persevered in the struggle for existence; *Sphenodon*, a "living fossil" lizard; modern lizards, more recent additions to reptilian society; snakes, which are lizards deprived of their appendages through some fortuitous circumstance; and crocodilians. These few survivors mirror the effect of mutations and the ravages of a changing environment on cotylosaurs and their descendants. Birds and mammals represent still other mutations of reptilian chromosomes, but they have become so different that they are no longer classified with reptiles.

Reptiles made significant advances over amphibians and fishes in the acquisition of three extraembryonic membranes (amnion, chorion, and allantois), which emancipated them and their descendants (birds and mammals) from the necessity of laying eggs in water. The amnion (Fig. 4-10) is a membranous sac filled with watery, salty, amniotic fluid. The embryo develops in this fluid just as the embryos of fishes and amphibians develop in the pond or sea. The fluid is secreted by the cells of the amnion. To state the situation fancifully, instead of the mother going to the water to deposit her eggs, the developing embryo has its own private pond. Thus reptiles became the first amniotes. The chorion and allantois usually constitute a vascular chorioallantoic membrane that lies against the porous eggshell (another reptilian innovation), taking the place of larval gills. In viviparous reptiles and in animals it performs the same function, absorbing oxygen from the uterine environment.

Because of these three extraembryonic membranes, oviparous reptiles found themselves able, and indeed forced, to lay their eggs on land. Their young hatch fully formed (Fig. 3-16), skipping the larval stage, ready to seek food on land. Not only were reptiles liberated from returning to water to lay their eggs, but aquatic oviparous reptiles must go onto the land to do so, since porous eggs would become water logged in an aqueous environment.

Reptiles have a thicker stratum corneum than amphibians, and the epidermis becomes especially scaly. These cornified epidermal scales are relatively impervious to water, which results in water conservation, an advantage to animals living in air and often remote from water. The digits are supplied with claws; a new kidney, the metanephros, has come into existence; the ventricle of the heart is partially or completely divided into right and left chambers; there is a single occipital condyle; and the

Fig. 3-16. *Anolis* in process of hatching. (Courtesy Carolina Biological Supply Co., Burlington, N.C.)

pelvic girdle articulates with two sacral vertebrae instead of one as in amphibians, thereby providing stouter bracing of the hind limbs against the vertebral column.

These then are reptiles: scaly, clawed, mostly terrestrial tetrapods lacking feathers and hair, which (except for a few viviparous forms) lay large, yolk-laden, shell-covered (**cleidoic**) eggs on land, the embryos of which are surrounded by an amnion, and the young of which are hatched fully formed. The temporal region of the skull (Fig. 8-29) is accorded special consideration in dividing the reptiles into groups. We will look briefly at selected representatives of the five subclasses.

☐ **Anapsida**

In addition to cotylosaurs this subclass includes turtles and tortoises. Turtles and tortoises are ancient reptiles that probably have remained relatively unchanged for 175 million years. They are identified by their shell of bony dermal plates, to which the ribs and trunk vertebrae are solidly fused. A loss of most trunk muscles occurred, but this was not detrimental, since the rigid shell would have rendered them useless anyway. Turtles and tortoises have also lost their teeth.

☐ **Lepidosauria**

Living lepidosaurians include *Sphenodon*, snakes, and modern lizards. *Sphenodon* (Fig. 3-15, *B*) is a 2- to 3-foot primitive lizard found only on islands off New Zealand. It is the sole survivor, a living fossil, of the primitive order Rhynchocephalia. It is called by natives the tuatara. Modern snakes and lizards are placed in the order Squamata because they are scaly (squamate). The largest lizard alive is the carnivorous Komodo Dragon from Indonesia. It reaches 2.75 m (about 9 feet) in length and 115 kg (about 250 lb) in weight. It has long claws, bright, beady eyes, and a long tongue that flicks in and out monitoring the "odors" in the environment. Amphisbaenians are small to medium sized (300 to 700 mm), wormlike, burrowing lizards that lack hind limbs and sometimes front ones. The horned "toad" of the North American desert is a lizard. Snakes are thought to have evolved from lizards. They lost both pairs of limbs and both girdles, although a few retain a trace of the pelvic girdle.

□ Archosauria

Archosaurs (Fig. 3-17) were the dominant land vertebrates during the Age of Reptiles. Teeth, when they were present, had roots and were in sockets in the jaw (thecodont teeth). Included are crocodilians, extinct flying pterosaurs, and dinosaurs. The earliest archosaurs are in the order Thecodontia.

Crocodilians, the sole surviving archo-

saurs, are large amphibious reptiles with bony plates under leathery, scaly skin of the back or back and belly (Fig. 5-35). They have a long secondary palate that separates the nasal passageway from the oral cavity all the way to the pharynx. The heart has a completely divided ventricle. Abdominal ribs (Fig. 7-12) are present. Crocodilians include crocodiles, alligators, caimans, and gavials. Crocodiles are common in the subtropical and tropical waters of Asia, Africa, the Americas, and Australia. Alligators occur in the warm southern regions of North America (*Alligator mississippiensis*) and China (*A. sinensis*). Caimans live in South America, and gavials in northern India. Crocodiles may be differentiated from alligators on the basis of the shape of the snout, which is more slender and triangular in crocodiles, broad and rounded in alligators; and by the fact that the enlarged fourth tooth of the lower jaw in crocodiles fits into a lateral notch on the upper jaw and can be seen in living

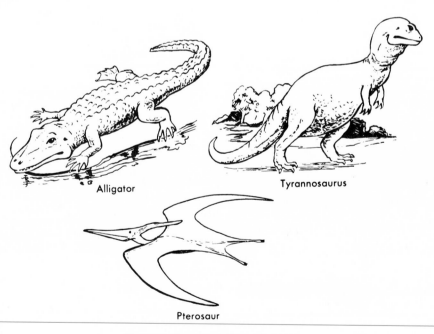

Fig. 3-17. Representative archosaurs.

crocodiles when the mouth is closed, whereas in alligators this tooth fits into a pit that is sufficiently deep and medial to the upper tooth line that it is hidden in living alligators when their mouth is shut. The snout of gavials is long and slender because the symphysis of the mandible extends from the first to the fifteenth tooth.

Pterosaurs (Fig. 3-17) were flying reptiles. Their bones were pneumatic, as in modern birds, but their wings were more like those of bats, since the wing membrane was supported by an elongated finger—the fourth, in pterosaurs. They had a tail, some very long, others quite short. Well-preserved fossils of one species, *Sordus pilosus*, show that they were insulated with a dense growth of either hair (why not?) or hairlike feathers and that they may have been endothermic.* The largest known pterosaur had a wingspread of about 17 m (55 feet).

Dinosaurs came in all sizes and many shapes, and the massive ones such as *Tyrannosaurus* (Fig. 3-17) were not at all characteristic of the group. Dinosaurs are found in two orders, Saurischia (with a reptilian pelvis) and Ornithischia (with a birdlike pelvis). Bakker[19] has summarized recent arguments that dinosaurs were endothermic

---

*Endothermy is the ability to produce internal heat sufficient to maintain a relatively stable temperature over a long period of time despite cyclic environmental temperature fluctuations. A less precise term is "warm blooded."

and that birds inherited endothermy from some small bipedal dinosaur.

☐ **Euryapsida**

This is a varied group that includes extinct large marine reptiles, the best known of which are the plesiosaurs and ichthyosaurs. They had an elongated snout armed with sharp teeth, and their limbs were modified as flippers. Plesiosaurs were up to 12 m long and had short tails and a very long neck that may have held the head above the sea while foraging for fish. Ichthyosaurs (Fig. 9-19) were smaller (up to 3 m or more) and were very fishlike in outward appearance with no visible neck.

☐ **Synapsida**

Synapsids are the reptiles, long extinct, from which mammals emerged. Like mammals, they had a single lateral temporal vacuity (Fig. 8-29, *B*). Early synapsids (pelycosaurs) had a parietal foramen, indicating the presence of a functional third eye. Later synapsids in the mammalian line (therapsids, Fig. 3-18) had two occipital condyles, like mammals, and a secondary palate and dentition consisting of incisors, canines, and grinding molars. The dentary was the largest bone in the lower jaw, presaging the mandible of mammals. Like other reptiles, synapsids had only one bone in the middle ear and a tiny braincase. Nothing is known of the skin. The distribution of selected reptiles in geologic time is shown in Fig. 3-19.

Cynognathus

Fig. 3-18. Mammallike (therapsid) reptile about the size of a large dog. (From Colbert.[4])

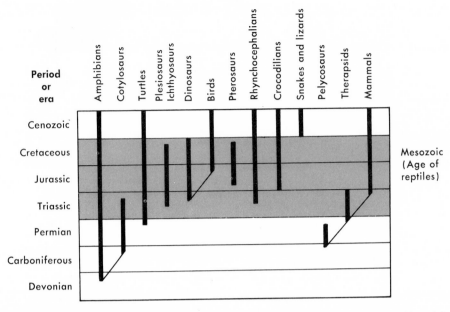

Fig. 3-19. Range of selected reptiles through time. Connecting lines indicate probable origins of cotylosaurs, birds, and mammals.

■ BIRDS: FEATHERED VERTEBRATES

Evidence indicates that birds, that is, endothermic vertebrates with feathers, arose from an archosaurian reptile, probably a small bipedal dinosaur. Dinosaurs could stand and run on their hind limbs, which freed the forelimbs for later flight. In time, birds lost such dinosaur traits as the long tail and teeth but retained claws, epidermal scales on legs and feet, a single occipital condyle, and a diapsid skull.

□ The earliest birds

There are two fossil genera we call birds because they have feathers, but most of their other characteristics were reptilian. These fossil birds, *Archaeopteryx* (Fig. 3-20) and *Archaeornis*, were recovered from slate deposits in Bavaria (West Germany) and have been placed in the subclass Archaeornithes. Whether they were in the direct line to modern birds is not known. They were about the size of a crow, had long reptilian tails with a row of feathers attached along the lateral borders, and had teeth. Their bones were solid instead of hollow as in modern birds. The wings were feebly developed, and the breastbone was small, indicating weak flight muscles. Probably they were incapable of sustained flight.

□ Later birds

The remaining birds are in the subclass Neornithes. It includes **odontognaths** (all extinct), **paleognaths** (ratites, incapable of flight), and **neognaths** (carinates).

Odontognaths, now all extinct, were discovered in Cretaceous marine deposits, chiefly in North America. They had many features of modern birds, including a foreshortened tail and hollow bones. More than a hundred specimens have been found in three families. *Hesperornis* (Fig. 3-21) and its immediate relatives had teeth and vestigial wings. *Ichthyornis*, a gull-like bird, had powerful wings but no teeth.

Paleognaths, or ratites, as they are called, are a comparatively small group of flightless

Fig. 3-20. Fossil *Archaeopteryx* embedded in limestone. Wings and their skeleton are near bottom of photograph, the long feathered tail is at top. (Photograph courtesy The Museum für Naturkunde der Humboldt-Universität, Berlin.)

Fig. 3-21. *Hesperornis,* a flightless toothed bird from the Cretaceous seas of Kansas in the United States.

weight and increased buoyancy. The skeleton of the wrist and hand has been greatly reduced by fusion and loss of parts (Fig. 9-17), the reptilian tail has been reduced to a stump called the **uropygium** (Fig. 3-22, *B*), the skull bones have become lightweight, the teeth have been lost, the large intestine has been shortened, the urinary bladder has been lost, and many bones have become hollow and contain diverticula of the lungs. The unique arrangement of the air ducts and lungs and some of the other adaptations associated with flight will be discussed in a later chapter. As a concession to bipedalism, the pelvic girdle became solidly fused with a long series of vertebrae to form a synsacrum (Fig. 7-15). A few carinates, such as the great auk and penguin, have lost the ability to fly.

Many carinates are annual migrants, passing part of a year in one geographical region and the remainder elsewhere, sometimes migrating great distances. The Arctic tern spends several months above the Arctic Circle and the remaining months in the Antarctic, traveling 22,000 miles round trip each year! During migration, birds move in mass flights, often at night and at an elevation of approximately 2,000 feet. Those destined for the same geographical location may pass over approximately the same routes (flyways) year after year. A great flyway is located directly over the Gulf of Mexico between the Yucatan peninsula and the Gulf Coast of the United States. Other flyways are located overland between Central America and the United States and from the West Indies via the Florida peninsula. Migration is associated in part with mating and is triggered by a specific hormonal balance.

birds with one exception, the tinamous. Many are known only as fossils, and all face extinction by man. They have small, incompetent wings but powerful leg muscles that permit them to run agilely. Among living ratites are the ostrich, kiwi, emu, rhea, and cassowary. One extinct wingless ratite from New Zealand, the moa, ranged up to nearly 4 m (13 feet) tall and laid eggs more than 30 cm (1 foot) long and 22 cm in diameter! Another, the elephant bird, was 3 m tall and weighed 360 kg. It is probable that the ancestors of ratites had wings that were capable of flight.

Neognaths are birds that are adapted for sustained flight. The sternum has developed a deep keel, or carina (Fig. 3-22, *B*), which provides a broad surface for attachment of the massive breast muscles for flight. Because of the carina, neognaths are called carinates. Feathers increase the surface area of the forelimbs, thereby providing an airfoil, and they conserve energy by insulating the body against heat loss. Numerous modifications have resulted in reduced body

■ MAMMALS: VERTEBRATES WITH HAIR

Mammals succeeded the pelycosaurs and therapsid reptiles (synapsids) in time (Fig. 3-19). They are vertebrates with hair and mammary glands. Distinguishing modern

Fig. 3-22. A, *Archaeopteryx,* the earliest known bird of the Jurassic period. **B,** A carinate bird (pigeon) for comparison. How many skeletal changes can you find? (From Colbert.⁴)

mammals from other vertebrates, too, are the single dentary bone on each side of the lower jaw articulating with the squamosal bone; three bones in the middle ear cavity; a muscular diaphragm separating thoracic and abdominal cavities; sweat glands (in most mammals); absence of an adult cloaca in all but the lowest order; heterodont dentition (except in toothed whales); two sets of teeth (milk teeth and a permanent set); marrow within the bones; biconcave, enucleate red blood cells that are circular (except in camels and llamas); loss of the right fourth aortic arch; a pinna, or sound-collecting lobe, accessory to the outer ear; a more specialized larynx; and extensive development of the cerebral cortex.

Because of the many variations in limb structure, mammals have been able to achieve a greater diversity of habitats than any other vertebrates except perhaps reptiles during their days of dominance. They burrow in the ground, hop, lumber, or gambol over the plains, placidly navigate mountain crags, swing through the trees, propel themselves through the air in true flight, and swim at great depths in the oceans—each activity made possible by modifications of body structure.

Mammals may be divided into two unequal groups, **Prototheria** and **Theria.** Prototheria are reptilelike mammals that lay heavily yolked eggs and have a cloaca. There is only one order, Monotremata, and it was the single cloacal opening that inspired their name. Theria, the remaining mammals, are all viviparous. If Prototheria did not originate independently from a different reptilian group than Theria, the two must have diverged very early in the evolution of mammals. Literature on the origin and history of mammals will be found at the end this chapter.

□ **Monotremata**

The platypus (Fig. 3-23) and two spiny anteaters, or echidnas, all from Australia and nearby New Guinea, are the sole surviving monotremes. These animals lay heavily yolked eggs and have a cloaca. The unspecialized mammary glands resemble sweat glands. There are no nipples, and the milk exudes onto tufts of hairs in shallow pits on the abdomen, from which it is licked up by the young.

The platypus lives in a burrow at the edge of a stream and has webbed feet, which adapt it to the water where it seeks its food. From one to three eggs are laid in a nest within a long burrow just above the water line. The eggs are nearly round, about 2 cm in diameter, and covered with a pliable white shell. They are incubated continuously by the female for about 2 weeks, after which they hatch.

Spiny anteaters are terrestrial and have a long sticky tongue that is used to capture insects. The body, except on the abdomen, is covered with strong sharp spines interspersed among coarse hairs. The single egg, about 4 mm in diameter, is incubated in a temporary pouch that develops as a thin flap of abdominal skin of females. The mammary

Fig. 3-23. The platypus *(Ornithorhynchus),* a monotreme.

glands are within the pouch and the young hatch and are carried in it until, after several weeks, they are able to seek their own food.

In addition to laying eggs and having a cloaca, monotremes are reptilelike in other respects. They have a ventral mesentery extending the length of the abdominal cavity and may have an adult abdominal vein. The testes are within the abdomen. The outer ear has no pinna. The malleus and incus of the middle ear are larger than in other mammals, resembling the articular and quadrate bones of reptiles. The brain lacks the great transverse fiber tract (corpus callosum) that connects the two cerebral hemispheres in other mammals. They are endothermic, but their body temperature is less stable than that of higher mammals, fluctuating as much as 13° C.

□ **Marsupialia**

Marsupials (Fig. 3-24) are primitive mammals in which the fetal yolk sac (in contact with the chorion) serves as a placenta. The young are born in almost a larval state and are transported, incubated, and nursed after birth in a maternal abdominal pouch (**marsupium**) of muscle and skin until they are old enough to become independent. The walls of the pouch are supported by two slender marsupial bones that project forward from the pelvic girdle. In several South American genera the pouch is incomplete or absent. Newborn young make their way to the pouch by squirming and wrig-

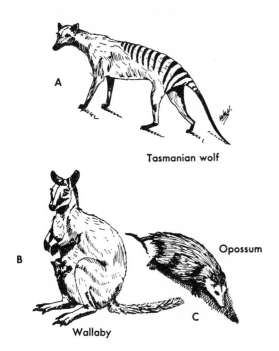

Tasmanian wolf

Opossum

B

Wallaby

C

Fig. 3-24. A and **B,** Marsupials from Australia. **C,** One of the few American marsupials.

gling and by using the claws on their fore-limbs, which are considerably larger than the hind limbs at birth. The lips are sealed at the angles of the mouth, which therefore consists of only a small circular opening. Once the young has taken a nipple into its mouth the tip of the nipple swells and the young cannot easily drop off.

All native Australian mammals except bats are marsupials or monotremes. Among Australian marsupials are kangaroos, the Tasmanian wolf, bandicoot rabbit, wallaby, wombat, Australian anteater, and phalangers. Australian marsupials resemble many true placental mammals such as wolves, foxes, bears, rabbits, mice, and cats in surprising details. Some phalangers resemble flying squirrels, and there are marsupial moles. Opossums are one of the few surviving marsupials in the Americas.

Why should there be no native mammal higher than a marsupial in Australia? It is now accepted that continental drift sepa-

rated Australia from Gondwanaland (South America, Africa, Australia, Antarctica), and it is theorized that higher (true placental) mammals had not yet reached Australia when the isolation of that continent occurred. As a result, only marsupials and monotremes would occur in Australia.

☐ Insectivora

Insectivores (Fig. 3-25) are the lowest mammals to use the allantois (in contact with the chorion) as a placenta. They are regarded as closely related to the primitive stock from which higher mammals arose. Although at one time abundant, they are represented today chiefly by small, retiring, often subterranean survivors: moles, shrews, shrew-like animals, tenrecs, and spiny hedgehogs. They subsist primarily on a diet of insects, worms, mollusks, and other small invertebrates.

Among their primitive characteristics are a flat-footed (plantigrade) gait; five toes; smooth cerebral hemispheres; small, sharp, pointed teeth with incisors, canines, and premolars poorly differentiated (Fig. 11-6, shrew); a large allantois and large yolk sac in the embryo; and sometimes a shallow cloaca. The testes are retained in the abdominal cavity in some genera (a primitive trait), and they never descend fully into scrotal sacs in any.

Moles have short, stout anterior limbs, with forefeet that are broad and more than twice as large as the hind feet, an adaptation for digging (Fig. 3-25). The neck is short and the shoulder muscles are so powerful that head and trunk seem to merge. Their tiny eyes are practically useless, but an acute sense of smell locates distant food, and the sensitive tip of the elongated snout tells them when they have encountered it.

Shrews superficially resemble mice. They are shy, busy little fighters with a keen sense of hearing. They have an elongated, be-whiskered, sensitive snout, and their incisor teeth are long and curved. The pigmy

Fig. 3-25. The mole, an insectivore.

Fig. 3-26. Tree shrew. (Courtesy Delta Regional Primate Research Center, Covington, La.)

shrew, weighing about 3 gm, is the smallest mammal on earth.

The tree shrew (Fig. 3-26) is classified by some authorities in the order Insectivora, by others as the most primitive primate. It has traits that bridge the gap between the two orders.

□ Dermoptera

Dermoptera are insectivorous mammals from Southeast Asia, East Indies, and the Phillipines. They are about the size of a domestic cat. A broad, muscular fold of skin (patagium) extends between the neck, forelimbs, hind limbs, and tail. There is only one genus, the "flying lemurs," and they do not fly, but glide, and are not lemurs. At one time or another they have been classified as insectivores and also as chiropterans, although the patagium is not well developed and the fingers are not elongated.

□ Chiroptera

Bats are the only known vertebrates in addition to pterosaurs and birds that have achieved true flight. This is possible because of the well-developed wing (patagium) extending between the neck, limbs, and tail and incorporating four greatly elongated, clawless fingers. The thumbs project from the anterior margins of the wings and bear claws, which aid in locomotion when not in flight. The five digits of the hind limbs all bear claws, which are used for hanging upside down, wings folded, from rafters or ledges in caves. Teats (usually two) are limited to the thoracic body wall. The pectoral muscles are strong, and the sternum is keeled, although not as much as in birds. All bones are slender, and those of the hand are greatly elongated (Fig. 9-18). Bats have large pinnas (external ear lobes), and the face and head glands are unusually numerous and enlarged. They constitute a large order and probably arose from an insectivore ancestor.

Bats are insectivorous, frugivorous (fruit-eaters), or sanguinivorous (subsisting on the blood of other mammals). Vampire bats have received attention because of sanguinivorous habits. Incisor teeth occur on the upper jaw only, and there is just one pair. They are razor sharp and point toward one another so they slit the skin of prey. As blood oozes from the wound, the bat licks it up without awakening the sleeping victim, which is often a domestic animal. Associated with the vampire habit of taking only fluid nourishment is the very small lumen of the esophagus, through which no solid food could possibly pass.

Vampire bats and Lamarckism. The adaptations of the teeth and the esophagus of vampire bats and the fluid diet inspire the speculative question whether a small esophageal lumen forced bats to adopt a fluid diet of blood (the only available fluid containing all the nourishment needed by a mammal) or whether the fluid diet during many generations was in any way a *cause* of the decrease in the size of the esophageal lumen. If the former is the case, it is fortunate that vampire bats hit on blood as a source of nourishment; otherwise they could not have survived the change. If, as thought by Lamarck, the change in the esophageal lumen resulted from disuse of this part for solid foods, modern genetics has thus far been unable to fathom the mechanism whereby the change in any generation became hereditary. The possibility also exists that sanguinivorous bats were already sanguinivorous before chance mutations altered the esophagus. If so, decrease in size of the esophageal lumen would have had no deleterious effect on these bats.

Convergent evolution. The evolution of "flying" lemurs (dermopterans), "flying" squirrels (rodents), "flying" phalangers (marsupials), and bats, all of which develop a wing membrane for aerial locomotion, is an illustration of convergent evolution. The term is applied when two species occupying the same kind of environment develop a similar adaptation when the two species do not have a common ancestor that could have contributed similar genetic information to both. The concept implies that unrelated species, long dissimilar, may, through adaptive mutations, approach each other with respect to some character. Another instance of convergence is development in whales and ichthyosaurs of flippers that superficially resemble fins.

□ Primates

Primates (prī-māʹ-tēz) are primarily arboreal mammals that arose as an offshoot of Cretaceous insectivore stock. One classification scheme divides them into **prosimians** and **anthropoids.**

Among specializations is the grasping hand so built that the thumb can be made to touch the ends of the other four fingers of the same hand. The big toe is also opposable in most primates. At least some of the digits are provided with nails instead of claws. Often there is a prehensile tail, which supplements the hands for grasping during arboreal locomotion. The cerebral hemispheres of the brain are larger than those of any other mammal. The snout has been shortened, with the result that both eyes can look forward. There is frequently only one pair of nipples, and these are on the thorax. Among primitive features are a flat-footed gait, five digits, a large clavicle, a central carpal of the wrist in many primates, and generalized dentition.

Prosimians. Prosimians include lemurs, lorises, and tarsiers. Lemurs receive their ghostly name from the nocturnal habit of swinging silently through the trees while most anthropoids, including man, are asleep. The long axis of the head is in line with the long axis of the body, as in most other mammals. The long tail is not prehensile. The second finger and toe have a claw instead of a nail. The uterus is duplex. The placenta is non-deciduate; that is, the fetal part of the placenta does not become rooted

Fig. 3-27. *Tarsius,* a prosimian.

into the uterine lining of the mother, and so there is no trauma of the uterus at birth. These are primitive traits.

Lorises include the nocturnal pottos and bush babies. They inhabit Africa, India, and the East Indies.

Tarsiers (Fig. 3-27) resemble anthropoids more than lemurs do. Their eyes are close together and directed forward so there is an overlap in the left and right fields of vision; the head is more nearly balanced at right angles to the vertebral column; all five fingers have nails and so do all toes except the second and third; and the placenta is deciduate.

Anthropoids. Anthropoids include three groups: **ceboids** (South American monkeys), **cercopithecoids** (Old World monkeys), and **hominoids** (apes and man). The head is at a right angle to the long axis of the vertebral column, the eyes are directed forward and are close together, the cerebral hemispheres are greatly developed, there is a bony ex-

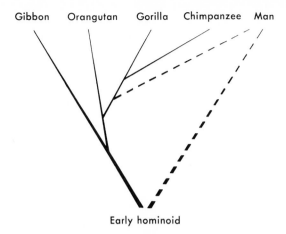

Gibbon   Orangutan   Gorilla   Chimpanzee   Man

Early hominoid

Fig. 3-28. Theoretical relationships of hominoids. Broken lines indicate alternative current hypotheses.

ternal auditory meatus in all but the ceboids, there are thirty-two teeth in the permanent set, the placenta is deciduate, and only one offspring is usually born at a time.

The best-known ceboids are *Cebus* (the capuchin), *Ateles* (the spider monkey), and *Alouatta* (the howler monkey). Howlers are named for their loud screeching cries, which are made with the greatly enlarged hyoid bone and larynx (Fig. 12-11). The best-known cercopithecoids are baboons, mandrills, and the macaque, or rhesus, monkey from which the symbol Rh (designating blood groups) was derived. The nostrils of cercopithecoids open downward and lie close together, as in man. The best-known hominoids are included in Fig. 3-28. Man probably diverged very early from other hominoids.

Since the discovery in 1856 of Neanderthal man, who lived in Europe 100,000 years ago, fossil parts of prehistoric man have been found in considerable numbers. The latest fossils indicate that man lived in Africa 4 million years ago. With each discovery the problem arises whether to include the new member in existing species, or to erect new species, or even genera. For example, Neanderthal man is sometimes

classified as a subspecies of modern man (*Homo sapiens neanderthalensis*) and sometimes placed in a separate species (*Homo neanderthalensis*). The rules for classifying other vertebrates are applied to the taxonomy of man. The oldest known manlike (hominid) remains are presently assigned to the genus *Australopithecus*. They used simple bone tools. *Homo erectus* is a more recent hominid. He fashioned tools and used fire. Modern man, *Homo sapiens sapiens*, who added atomic and nuclear energy to his array of tools, occupied Europe with Neanderthal man and replaced him either by competition or absorption.

In the emergence of man from an earlier anthropoid many changes occurred. An S-shaped curve in the vertebral column permitted an erect posture; the facial angle became less acute (Fig. 3-29); the teeth, especially the canines, became smaller; the frontal lobes of the cerebral hemispheres enlarged, resulting in an enlarged braincase and a more prominent forehead; the eyebrow ridges became reduced; the nose became more prominent; the tail became confined to embryonic stages; the arms became shorter; a metatarsal arch developed in the otherwise flat feet; the big toe moved in line with the other toes and ceased to be opposable; articulate speech appeared; and many other changes occurred.

It is to the massive development of the frontal lobes of the cerebral hemispheres, to the opposable thumb, and to articulate speech that man owes his present dominance in the animal kingdom. With his fingers he can construct instruments of offense and defense and machines to lighten his burden, and with them he can scrawl symbols that convey experiences and techniques to the corners of the earth and to generations unborn. With his voice he can communicate with contemporary fellow creatures and exchange ideas with delicate shades of meaning. With his brain he can associate sensory stimuli that are currently

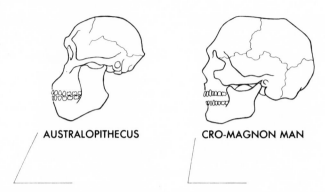

AUSTRALOPITHECUS   CRO-MAGNON MAN

Fig. 3-29. Facial angles of two hominids. The angles are shown beneath the skulls.

received with those recalled from earlier experiences, and after meditation elect a mode of action that he calls "intelligent." With his brain, too, he can enjoy the esthetic beauty of the imponderable universe, search for ultimate truth, and dream of a Utopia, which, but for his residual animal nature, might be his heritage.

□ **Carnivora**

Carnivores are a large and diverse group of flesh eaters that includes aquatic and terrestrial species. They have powerful jaws and, except in water forms, elongated, sharp canine teeth capable of spearing and tearing flesh (Fig. 11-7, *B*). The cerebral cortex is convoluted, and the animals are capable of considerable learning.

Terrestrial carnivores include "cats" of many kinds (domestic cats, panthers, lynx, etc.), dogs and related forms, bears and pandas, hyenas, and numerous species important economically for their furs such as racoons, mink, otters, skunks, and badgers. Terrestrial carnivores typically have five toes with sharp, sometimes retractable, claws.

Aquatic carnivores are sometimes placed in a separate order, **Pinnipedia,** which includes sea lions (Fig. 3-30), seals, and walruses. They have many adaptations for life in the water. They have webbed, paddle-like, and often nailless limbs that are more

or less included within the body wall. In "wriggling" seals the hind limbs are permanently bound to the tail, and movements on land consist of pulling themselves awkwardly along or flopping about. Despite their aquatic adaptations the young are born on land (in rookeries) and cannot even swim.

□ **Cetacea**

Whales, dolphins, and porpoises are moderately large or massive aquatic mammals, sometimes attaining 30 m (100 feet) in length and 150,000 kg (150 tons) in weight. The tail has a horizontal terminal fin (fluke) filled with fibrous tissue, which provides forward thrust by its up-and-down motion. A dorsal fin, similarly constructed, is sometimes present and serves as a rudder and stabilizer. The anterior limbs are paddlelike and serve chiefly as balancers. Although the fingers are not separated from the paddle, finger bones are present (Fig. 9-21). Posterior limbs and girdle are mere vestiges embedded in the trunk wall.

A few hairs, which can be easily counted, usually occur on the muzzle, otherwise the skin is smooth and naked. The nostrils are usually far back on the skull and are frequently united to form a single large "blowhole." The diaphragm is unusually muscular and the water spout (composed of vapor from the expired air) may last from 3 to 5

Fig. 3-30. Aquatic carnivores (sea lions). (Courtesy The American Museum of Natural History, New York, N.Y.)

minutes. Most cetaceans have teeth, but whalebone whales (blue whales and finbacks) have frayed horny sheets of whalebone (baleen) hanging from the roof of the mouth (Fig. 5-29). These immense carnivores strain several tons of small fish (up to 7 or 8 cm in length) and invertebrates out of the sea each day. A heavy layer of fat, called blubber, under the skin conserves body heat while the animal swims in the cold depths of the ocean. Because of their great bulk, they dare not approach too close to shore lest they become marooned and die, crushed by tons of their own flesh.

Cetaceans are the only completely adapted seagoing mammals. All other aquatic mammals must return to land to breed. Baby whales are born in the sea and when feeding, hang onto one of the two inguinal teats of the mother as she swims about. Whales have lost their olfactory nerves,* and the nostrils have valves that close during diving.

*A discussion of this mutation in terms of Lamarckism will be found on p. 432.

□ Edentata

Edentates are chiefly South American mammals that have diverged from insectivores. They include tree sloths, South American anteaters, and armored edentates, best known of which is the armadillo. Several giant armored fossil edentates are known. All teeth are absent in South American anteaters (hence the name "edentate"), but many other species have cheek teeth that lack enamel. Histological evidence of an enamel organ has been found, however.

□ Tubulidentata

Tubulidentates are another order with few teeth. There is only one living tubulidentate, the insectivorous aardvark from South America. It is a burrowing anteater about 1.5 m (5 feet) long with relatively few coarse hairs, an elongated piglike snout, and a long sticky tongue for capturing termites.

□ Pholidota

Pholidota is an insectivorous order lacking teeth. Here, again, there is only one pholidotan, the pangolin *Manis* (Fig. 5-24) from Africa and Southeast Asia. Pangolins are also called scaly anteaters because they have epidermal scales resembling those of lizards. Between the scales are scattered hairs.

□ Rodentia

Rodents are a large and diverse group of mammals that have a single pair of upper and lower incisor teeth. These teeth are long, curved, and covered with enamel on their outer surfaces only, which provides a chisellike edge for gnawing. The incisors grow throughout life. Since canine teeth are absent, there is a diastema, or stretch of jaw devoid of teeth, behind the incisors (Fig. 11-6). At the beginning of the large intestine there is a long, coiled cecum that houses commensal cellulose-digesting microorganisms. Rodents are the most numerous mammals on earth.

□ Lagomorpha

The order Lagomorpha was established for rabbits and hares, which were once classified as rodents. They differ from rodents in having two pairs of incisors on the upper jaw—a small pair lying immediately back of, rather than alongside of, a much larger pair. The front pair, like those of rodents, are long, sharp, deeply rooted, and grow throughout life. The smaller pair lack a cutting edge. Rabbits differ from hares in having shorter ears and legs, producing naked young, and in other respects. Rabbits and hares have a split upper lip, which gave rise to the term "harelip."

□ Ungulates and subungulates

It is now necessary to introduce two orders of mammals, **Perissodactyla** and **Artiodactyla**, known as ungulates. They are mostly large herbivores, and all walk on the tips of their toes (Fig. 9-23, deer), which are protected by hooves (modified nails or claws). Modern ungulates have no more than four toes on each foot and some, like horses, have only one. Ancestral ungulates probably had five toes. Reduction in the number of toes is illustrated by the familiar example of the horse, the little Eocene ancestors of which had four toes in front and three behind. With successive mutations the number was reduced to one.

The teeth of modern ungulates are also characteristic. They are high crowned, and their grinding surfaces exhibit crosswise ridges separated by deep grooves specialized for grinding grasses. There is little morphological difference between premolars and molars. Ungulates lack a clavicle and as a result probably have more freedom of neck movements in grazing. They are the only mammals with horns, although several models come without them.

In addition to ungulates, three other living orders are thought to have perhaps radiated from early ungulate stock—**Proboscidea, Hyracoidea,** and **Sirenia**. They are called subungulates.

□ Perissodactyla

Modern perissodactyls occur in three families: horses, tapirs, and rhinoceros. They walk on the hoofed tips of one, three, or occasionally four toes and are distinguished by the fact that the body weight is borne chiefly on a single digit (mesaxonic foot, Fig. 9-25). Perissodactyls are referred to as odd-toed ungulates. However, tapirs and some rhinoceros have four toes on the forelimb.

□ Artiodactyla

Artiodactyls are ungulates in which the weight of the body is supported by two toes. This is known as a paraxonic foot (Figs. 9-24 and 9-25). Living artiodactyls have an even number of toes, but at least one extinct artiodactyl had five toes on the forelimb. Artiodactyls include pigs, hippopotamuses, peccaries, cows, camels, llamas, deer, antelope, giraffes, goats, and sheep. With the

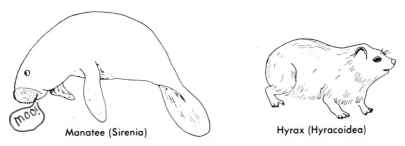

Fig. 3-31. Two subungulates: a sea cow (manatee) and a coney *(Hyrax).* A third subungulate, the mastodon, is illustrated in Fig. 11-6.

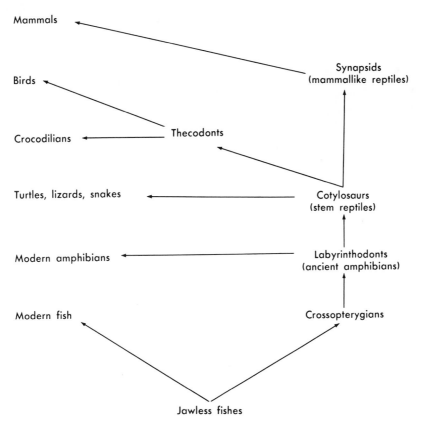

Fig. 3-32. Mainstreams of vertebrate evolution based on the monophyletic theory of the origin of tetrapods.

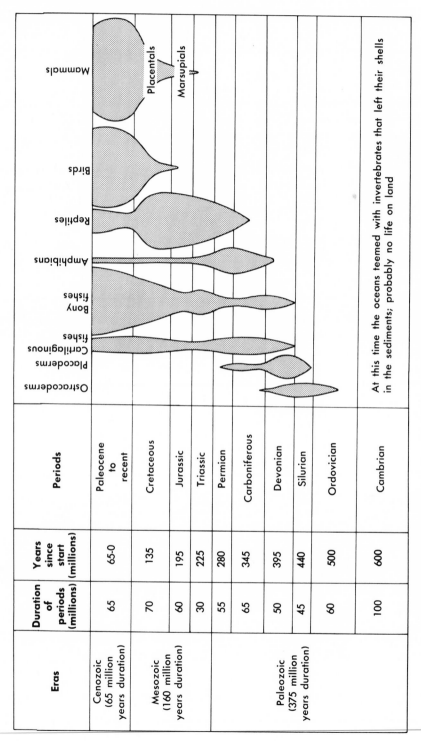

Fig. 3-33. Range and relative abundance of vertebrates through time. (Modified from Colbert.[4])

exception of pigs, hippopotamuses, and peccaries, they have stomachs divided into not fewer than three chambers, sometimes four (Fig. 11-15). They bolt their food, which then passes to the rumen, the first segment of the stomach. Such animals are called **ruminants.** At their leisure, they force undigested food balls back up the esophagus and masticate this cud more thoroughly.

☐ **Proboscidea**

Proboscidea is an order of subungulates that includes elephants, mastodons, and their relatives. They have a proboscis, or trunk, composed of a greatly drawn out upper lip, which is accompanied in its overgrowth by the nostrils (Fig. 11-6, mastodon). They have scanty hair and thick, wrinkled skin. The incisor teeth of one or both jaws are elongated to form tusks, canine teeth are absent, and molars are very large grinders, as in ungulates. Proboscideans are bulky animals, and the limbs are almost vertical pillars of bone and muscle. They have five toes that end in hooves, and on the back of each toe is an elastic pad that bears much of the body weight.

☐ **Hyracoidea**

Hyracoidea is an order of subungulates containing a single genus, the little *Hyrax*, popularly known as the coney. In some traits coneys resemble rodents and hares (Fig. 3-31). They have short ears, they hunch the body when at rest, the upper lip is split (harelip), and the incisor teeth grow continuously. Although they are plantigrade, indications are of a closer kinship to ungulates than to other mammals. All digits except one end in small, flat hooves, fingers have been reduced to four and toes to three, the middle toe being the longest, and they have ungulatelike molar teeth.

☐ **Sirenia**

Sirenians, or sea cows, are the manatees and dugongs. They are stout, clumsy-ap-

pearing, freshwater or marine mammals with an overgrown, wrinkled, almost pathetic-looking snout covered with scattered, coarse bristles (Fig. 3-31). The rest of the body is naked except for a few scattered hairs. The forelimbs are paddles, but within them the tetrapod complement of bones is intact. Hind limbs are absent, but there are internal skeletal vestiges of them and of the pelvic girdle. The tail in some is flattened horizontally like that of whales. They are considered by authorities the most aberrant descendants of very primitive ungulate stock.

☐ **Mainstreams of vertebrate evolution**

The main lines of evolution from ancient jawless fishes to modern vertebrates may pass through the groups indicated in Fig. 3-32. Notice that the lines lead through very ancient groups all of which are now extinct. Fig. 3-33 shows the relative abundance of the vertebrate classes through time.

■ **VARIATION, ISOLATION, AND SPECIATION**

In this chapter we have had a glimpse of the variety of vertebrate life. But variation is not restricted to differences between species, or even between isolated populations of a single species. Variation occurs within populations, and even among littermates. Indeed, no two products of sexual reproduction are identical. The number of kinds of gametes or of homozygous genotypes possible with n pairs of genes is 2^n. The number of different kinds of genotypes possible is 3^n. Even if there were only 100 pairs of genes, forty-eight digits would be needed for writing the number of possible genotypes. And there are thousands and thousands of pairs of genes. In addition, there may be as many as 400 mutational sites on a single gene. Therefore, the number of possible combinations of hereditary traits staggers the imagination. For example, the gastrohepatic artery in a population of *Felis cattus* is sometimes 20 mm long, sometimes 1 mm long, or the gastric and hepatic arteries may arise independently, and there is no gastrohepatic artery. The number of variations that actually occurs in the arterial channels alone in any interbreeding population is such that one could identify a single shark, or a single human being, from his arterial channels alone. In practice, each of us does better than that! We identify a single human being simply by looking at the face or at a print made from a fingertip. We could, of course, recognize an individual organism by looking at the sequence of nucleotides in its DNA. Genetic variation, coupled with geographical isolation, seems to be the matrix out of which new species evolve.

What happens when a small group of individuals on the periphery of a population wanders sufficiently far from the other members that they establish a new colony out of contact with the original population? Such a geographically isolated group may have one of three fates. If the individuals are not partially preadapted to the newer environment, they may become extinct. Or they may mate with nearby populations of another, closely related species, establishing a zone of hybridization. Or as a result of chance mutations affecting one population

and by recombinations and selection pressures, a steady genetic divergence may occur, and the two populations may ultimately become reproductively isolated. When reproductive isolation has occurred, a new species will have been produced. The length of time required for the origin of new species by geographical isolation depends on many factors, but the final emergence of new species seems almost inevitable.

Although new animal species may arise other than by geographical isolation, the origin of new species by natural selection *within an existing interbreeding population* is no longer an acceptable concept to most students of evolution.

Not only have new species been *appearing* at different rates during different epochs of the earth's history, depending in part on changes that produce extreme modifications of climate and geography, but species are also *disappearing* at varying rates. Since the start of the twentieth century fifty animal species are known to have become extinct. Many more species have come and gone than exist today. Because of the continued rise and demise of species, the biota on earth has been changing since life first appeared on this planet.

☐ Chapter summary

1. The vertebrate classes are Agnatha, Placodermi, Chondrichthyes, Osteichthyes, Amphibia, Reptilia, Aves, and Mammalia. Major subdivisions of these classes are given on pp. 436 to 439.
2. Fishes include the first four classes listed above. Agnathans have no jaws. The remaining vertebrates are gnathostomes.
3. Living agnathans (cyclostomes) are fishes with eel-like bodies that lack jaws, paired appendages, typical vertebrae and scales, and have a round, suckerlike mouth. Extinct agnathans (ostracoderms) are the oldest vertebrates known. They were armored fishes.
4. Placoderms are extinct armored gnathostomes. They include arthrodires and acanthodians.
5. Chondrichthyes (elasmobranchs and holocephalans) have cartilaginous skeletons, five to seven naked gill slits (except Holocephali), and bony placoid scales. They were more abundant in ancient times. In many respects they illustrate the basic architectural plan of vertebrate construction.
6. Osteichthyes are lobe-finned or ray-finned bony fishes. They have ganoid, cycloid, or ctenoid scales and a bony operculum. The crossopterygian *Latimeria* and ganoids are the oldest living fishes in the class. Teleosts are the most modern. Most lobe-fins have internal nares, and the air sacs are often used as lungs.
7. Amphibians probably came from rhipidistian crossopterygians (monophyletic theory). They include anurans and urodeles, with glandular skin devoid of scales, and apodans. Most amphibians lay eggs in water and have a gill-bearing larval stage. A few are viviparous. Neotenic amphibians show incomplete metamorphosis.
8. Reptiles, birds, and mammals are amniotes. Their embryos have an amnion, chorion, and allantois as well as a yolk sac. Fishes and amphibians are anamniotes.
9. Reptiles are amniotes with epidermal and dermal scales and claws but lacking feathers and hair. They typically lay large eggs surrounded by a shell (cleidoic eggs). Some are viviparous.
10. Birds are endothermic descendants of archosaurs and have feathers. Reptilian scales persist on the legs and feet, which bear claws. The anterior paired appendages are modified for flight in carinates. In ratites, forelimbs are usually vestigial or absent. The oldest birds were Archaeornithes. They had long tails and teeth.
11. Mammals are descendants of therapsid reptiles. They have hair. Prototheria (monotremes) lay reptilian eggs, their mammary glands are primitive, and they have a cloaca. Theria give birth to living young. Marsupials use the yolk sac as a placenta. The remaining orders have a chorioallantoic placenta.
12. Main lines of vertebrate evolution probably pass through groups indicated in Fig. 3-32.
13. Genetic variation coupled with geographical isolation seem to account for most new species.

LITERATURE CITED AND SELECTED READINGS
Comprehensive

1. Alexander, R. M.: The chordates, London, 1975, Cambridge University Press.
2. Bardack, D., and Zangerl, R.: First fossil lamprey: a record from the Pennsylvanian of Illinois, Science **162**:1265, 1968.
3. Carter, G. S.: Structure and habit in vertebrate evolution, Seattle, 1967, University of Washington Press.
4. Colbert, E. H.: Evolution of the vertebrates, ed. 2, New York, 1969, John Wiley & Sons, Inc.
5. Colbert, E. H.: Wandering lands and animals, New York, 1973, E. P. Dutton & Co., Inc.
6. Hull, D. L.: Certainty and circularity in evolutionary taxonomy, Evolution **20**:174, 1976.
7. Jarvik, E.: The oldest tetrapods and their forerunners, Scientific Monthly **80**:141, 1955.
8. Ørvig, T., editor: Current problems of lower vertebrate phylogeny, Stockholm, 1968, Almqvist & Wiksell Förlag AB.
9. Romer, A. S.: Vertebrate paleontology, ed. 3, Chicago, 1966, University of Chicago Press.
10. Schmalhausen, I. I.: The origin of terrestrial vertebrates (translated from the Russian by Leon Kelso), New York, 1968, Academic Press, Inc.
11. Stahl, B. J.: Vertebrate history: problems in evolution, New York, 1974, McGraw-Hill Book Co.

Fishes

12. Brodal, A., and Fänge, R., editors: The biology of *Myxine*, Oslo, 1963, Universitetsforlaget (Norway).
13. Daniel, J. F.: The elasmobranch fishes, Berkeley, 1934, University of California Press.
14. Gilbert, P. W., Mathewson, R. F., and Rall, D. P., editors: Sharks, skates, and rays, Baltimore, 1967, The Johns Hopkins University Press.
15. Hardisty, M. W., and Potter, I. C., editors: Biology of lampreys, 2 vols., New York, 1972, Academic Press, Inc.
16. Thompson, K. S.: The biology of the lobe-finned fishes, Biological Review **44**:91, 1969.

Amphibians

17. Parsons, T. S., and Williams, E. E.: The relationships of the modern Amphibia: a reexamination, Quarterly Review of Biology **38**:26, 1963.
18. Taylor, E. H.: The caecilians of the world, Lawrence, Kans., 1968, The University Press of Kansas.

Reptiles and birds

19. Bakker, R. T.: Dinosaur renaissance, Scientific American **232**(4):58, 1975.
20. Carroll, R. L.: Problems of the origin of reptiles, Biological Reviews **44**:393, 1969.
21. Colbert, E. H.: Dinosaurs, London, 1962, The Hutchinson Publishing Group, Ltd.
22. de Beer, G. R.: The evolution of ratites, Bulletin of the British Museum (Natural History) Zoology **4**:59, 1956.
23. Romer, A. S.: Osteology of the reptiles, Chicago, 1956, University of Chicago Press.

Mammals

24. Andersen, H. T., editor: Biology of marine mammals, New York, 1969, Academic Press, Inc.
25. Burrell, H.: The platypus, Sydney, 1927, Angus and Robertson Publishers (reprinted in 1974 by Rigby Ltd., Sydney).
26. Griffiths, M.: Echidnas, Oxford, 1968, Pergamon Press Ltd.
27. Hopson, J. A., and Crompton, A. W.: Origin of mammals. In Dobzhansky, T., Hecht, M. K., and Steere, W. C., editors: Evolutionary biology, New York, 1969, Appleton-Century-Crofts.
28. Hunsaker, D., editor: The biology of marsupials, New York, 1977, Academic Press, Inc.
29. Kermack, D. M., and Kermack, K. A., editors: Early mammals, Zoological Journal of the Linnean Society, vol. 50, supplement 1, New York, 1971, Academic Press, Inc.
30. Luckett, W. P., and Szalay, F. S., editors: Phylogeny of the primates: a multidisciplinary approach, New York, 1975, Plenum Publishing Corp.
31. Simpson, G. G.: Horses: the story of the horse family in the modern world and through sixty million years of history, New York, 1951, Oxford University Press, Inc.
32. Slijper, E. J.: Whales, New York, 1962, Basic Books, Inc., Publishers.
33. Tyndale-Biscoe, C. H.: Life of marsupials, London, 1973, Edward Arnold (Publishers) Ltd.

Symposia in American Zoologist

Evolution and relationships of the amphibia, **5**:263, 1965.
Recent advances in the biology of sharks, **17**(2): 287-515, 1977.

4

Viviparity, the germ layers, and extraembryonic membranes

In this chapter we will look briefly at some varieties of viviparity. We will take a minilook at vertebrate blastulae and gastrulation and see how the coelom forms. The major structures that form from mesoderm, ectoderm, and endoderm will be summarized (details are given in appropriate chapters throughout the book). Finally, we will meet the major extraembryonic membranes and see what they do for the unhatched or unborn vertebrate.

■ OVIPARITY AND VIVIPARITY
□ Oviparity

Animals that lay or spawn their eggs are said to be **oviparous**. The egg, when laid, contains sufficient nourishment in the form of yolk and sometimes albumen, to support development of the zygote into a free-living, self-nourishing organism. The baby chick develops from a yolk-laden egg and therefore has sufficient nourishment to be hatched fully formed. The frog, on the other hand, arises from an egg with less yolk and hatches in a larval state. When there is very little yolk, as in the egg of the amphioxus, the free-living, self-nourishing state must be achieved very quickly after the egg is deposited. Accordingly, the amphioxus hatches into an externally ciliated, free-swimming embryo 8 to 15 hours after fertilization, at which time the notochord is a mere ridge in the roof of the primitive gut and there are no gill slits.

□ Viviparity

In many vertebrates the egg is retained within the mother's body during embryonic development and living young are delivered. Such species are said to be **viviparous**. The relationship between mother and embryos varies from one in which the mother provides protection and little else, as in viviparous sharks, to one in which the embryos are dependent on the mother for all nourishment, oxygen, and for carrying away waste products of metabolism, as in viviparous mammals. Many intermediate degrees of dependency have evolved. The term **ovoviviparity** has been coined to designate the condition in which protection, oxygen, and little else is provided, and the term **euviviparity** is used when the embryo cannot develop without nourishment being constantly provided by the mother.

Viviparity in one degree or another has evolved in every class of jawed vertebrates except birds. It seems to have developed independently at least fifteen times in teleost fishes, at least ten times in lizards, and at least six times in snakes. It is thought to have occurred in the extinct *Ichthyosaurus*. There are viviparous urodeles and anurans that give birth to fully metamorphosed young in terrestrial environments, and there are viviparous aquatic apodans. Among cartilaginous fishes, at least ten families of sharks and four families of rays are viviparous.

The dogfish shark, *Squalus acanthias*, is an example of an ovoviviparous organism. The pups are nourished entirely by the yolk from their own yolk sac (Fig. 4-9), but they receive oxygen from greatly enlarged and highly tortuous blood vessels in the uterine lining of the mother. The pups may be removed from the mother 2 to 3 months before birth (the gestation period is 20 to 22 months) and will complete their development, utilizing the nourishment from their yolk sac, as long as they are confined in finger bowls or plastic tubes containing oxygenated sea water.

In viviparous teleosts the egg may be fertilized and the young may develop in the ovarian follicle, there having been no ovulation. The young of *Gambusia* develop in this way. Or the embryos may develop in the ovarian cavity (Fig. 14-33). Among adaptations of the embryo for development in these locations are enlargement of the embryo's pericardial sac, which lies in contact with the maternal tissues of the ovary and absorbs necessary substances; enlargement

of the embryonic gut, which lies in contact with maternal tissues; and villuslike projections of the embryonic rectum that protrude through the vent of the embryo into the surrounding nutrient-rich medium. In some teleosts the young develop for awhile in the follicular chamber and exhibit an enlarged pericardial sac, then later pass into the ovarian cavity and develop absorptive villi. Sometimes developing eggs or larvae are ingested by other larvae.

The maternal tissues, under the influence of hormones, exhibit adaptations for viviparity. The wall of the ovarian follicle occupied by teleost embryos may develop vascular folds or villi, and these may even project into the mouth or opercular cavity of the embryos. In *Dasyatis americana*, a euviviparous stingray, the gravid uterus is lined with villi 2 to 3 cm long, which produce a copious secretion that nourishes the embryo. Secretions of uterine glands provide nourishment for unimplanted blastocysts of mammals, and perhaps for implanted blastocysts throughout pregnancy in perissodactyls and artiodactyls. **Histotrophic (embryotrophic) nutrition** is the term applied to nutrition by *glandular secretions* from maternal tissues, as contrasted with nutrition by substances exchanged via a placenta.

☐ Internal and external fertilization

In viviparous vertebrates, fertilization takes place within the body of the female. Fertilization is also internal whenever eggs are covered by an impenetrable shell before being extruded.

In oviparous fishes, frogs, and toads external fertilization is the rule. This is possible only because mating takes place in water and very large numbers of sperm and eggs are shed. In apodans and urodeles, however, fertilization is usually internal even though the eggs are subsequently laid. The male urodele deposits a sac of sperm embedded in jelly (**spermatophore,**

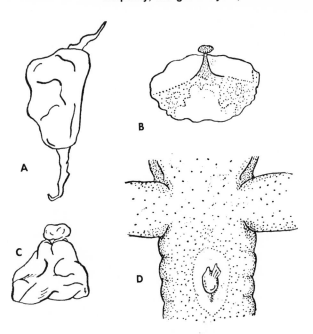

Fig. 4-1. Spermatophores. **A,** *Necturus.* **B,** *Notophthalmus.* **C,** *Ambystoma.*
D, Cloaca of female *Desmognathus* containing spermatophore. (Redrawn from Bishop.[4])

Fig. 4-1) in the immediate vicinity of a female during the mating ritual. The spermatophore is seized by the cloaca of the female or is placed in the cloaca by the male. The sperm then escape from the jelly and migrate up the female reproductive tract to the eggs. Packaging sperm in jelly makes it possible for those few urodeles that live far from water to convey sperm to a female despite lack of a copulatory organ.

□ **Natural selection at work**

When eggs are retained within the female or when they are protected by a parent after being laid, the number of eggs produced is small, and the mortality is low. When, however, eggs are laid outside the body and are left unprotected, mortality prior to hatching is high. A naturally high mortality rate is counterbalanced by production of large numbers of eggs.

It is conceivable that an oviparous species that does not conceal or guard its eggs might tend, as a result of mutations, to produce fewer and fewer eggs. Because of the probable destruction of a percentage of eggs, the species would become more rare. The mutations resulting in smaller numbers of eggs, unaccompanied by other mutations providing for protection of the few eggs produced, would finally result in the destruction of the species. **Natural selection** is a term introduced by Darwin to indicate that natural phenomena inexorably determine which organisms will survive (p. 431). To natural selection may be attributed the following facts: there are very few vertebrates that lay small numbers of eggs in unsheltered locations; there are many vertebrates that lay small numbers of eggs in sheltered or guarded locations; and there are many that lay large numbers of eggs (up to 60 million a season in some fishes) in unguarded locations.

■ **CLEAVAGE AND THE BLASTULA**

Early cell divisions of the zygote are referred to as **segmentation,** or **cleavage.**

As a result, the zygote becomes subdivided into smaller and smaller cells that form (usually) a hollow sphere or **blastula.** Each cell of the blastula is a **blastomere** and the cavity is the **blastocoel** (Figs. 4-2 and 4-3).

When there is very little yolk in an egg, the blastomeres are approximately of equal size (Fig. 4-2, blastula). When cell division is impeded by a moderate amount of yolk, as in amphibian eggs, the cells at this pole divide more slowly and are larger (Fig. 4-3). When there is a massive yolk, as in the eggs of reptiles and birds, segmentation is confined to the animal pole and results in a **blastoderm** (Fig. 4-4) perched like a skullcap upon the massive unsegmented yolk. The embryo develops from this blastoderm.

The eggs of mammals exhibit an animal-vegetal polarity, just like the eggs of reptiles, even though mammalian eggs above monotremes have almost no yolk. The first cleavage division divides the egg into two blastomeres, one representing the animal pole, the other, the vegetal pole (Fig. 4-5, cleavage). The descendants of the latter rapidly become a nutritional membrane, the **trophoblast,** for absorbing nourishment from the uterine fluids. The animal pole cells become the blastoderm, better known in mammals as the **inner cell mass** and, later, the **embryonic disc** (Fig. 4-5). Because of the cystlike nature of the mammalian blastula, it is called a blastocyst.

As a result of cleavage, all vertebrates, including man, pass through a stage of development known as the blastula. During

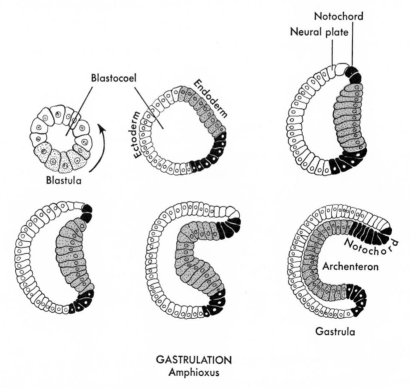

GASTRULATION
Amphioxus

Fig. 4-2. Blastula and gastrulation in an amphioxus, sagittal section. Cells in black surround the blastopore (entrance to archenteron) and are the most active, mitotically. Red indicates presumptive endoderm.

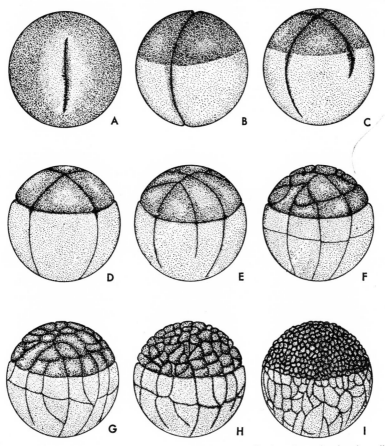

Fig. 4-3. Cleavage and the blastula of amphibian egg. *Dark cells,* animal pole cells containing little yolk; *light cells,* vegetal pole cells containing much yolk. (After Eycleshymer; from Nelsen.[8])

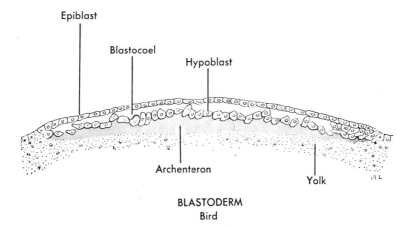

Fig. 4-4. Blastoderm at animal pole of an egg containing a massive yolk, sagittal section. The hypoblast is the roof of the archenteron. Head end is to the left.

this stage major organ-forming areas are being established in preparation for gastrulation.

■ GASTRULATION

Gastrulation is the process whereby presumptive (that is, future) endoderm, mesoderm, and notochordal cells of the blastula migrate to the interior of the embryo. Rapid cell proliferation provides cells for these migratory movements.

In an amphioxus presumptive endoderm folds into the blastocoel (Fig. 4-2). This produces the earliest gut (**archenteron**). The entrance to the archenteron is the **blastopore**. Soon, presumptive notochord cells flow inward to become the temporary roof of the archenteron (Fig. 4-2, gastrula). Then, presumptive mesoderm cells pass inward to

lie in the roof of the gut at either side of the notochord. At this stage the amphioxus embryo consists of an outer tube of ectoderm and an inner tube mostly of endoderm surrounding the archenteron. The roof of the archenteron is future notochord and mesoderm (Fig. 4-6, A). Cells proliferated from the rim (lips) of the blastopore result in elongation of the embryo, now a **gastrula**.

In vertebrates, gastrulation is affected by the presence of yolk. Amphibians manage to tuck the unwieldy yolk inside the embryo, where it becomes the floor of the archenteron. This is accomplished partly when the small cells of the animal pole grow downward over the large cells of the vegetal pole, a process called **epiboly**. Yolk cells then constitute the archenteric floor, and the notochord is temporarily in its roof.

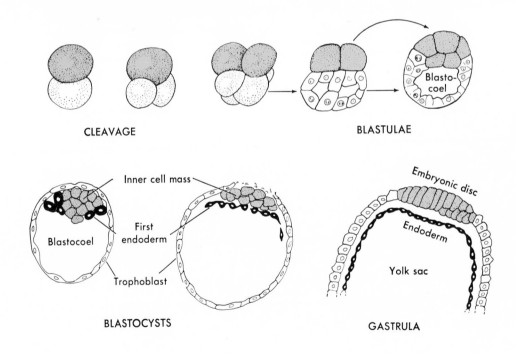

Mammal

Fig. 4-5. Cleavage, blastula, and gastrulation in a mammal. At cleavage (upper left) the red cell at the animal pole gives rise to the embryo. The other cell gives rise to the trophoblast. The endoderm (hypoblast) comes from the inner cell mass. The yolk sac contains no yolk.

Vertebrates with massively yolked eggs have a more difficult problem incorporating the yolk into the body of the developing embryo and, in fact, they cannot. The small cells of the blastoderm cannot grow downward around the massive yolk. Instead, the blastoderm splits into an upper sheet of cells (**epiblast,** Fig. 4-4), which gives rise to the epidermis and neural tube, and a lower sheet (**hypoblast**), which is the future endoderm. The yolk is the temporary floor of the archenteron. Later, the hypoblast grows downward around the yolk to form a yolk sac.

Mammalian gastrulation, too, commences with formation of an epiblast and hypoblast from the inner cell mass. Then the hypoblast (endoderm) grows downward around a yolk that is not there (Fig. 4-5)! During gastrulation in all vertebrates the notochord is temporarily in the roof of the archenteron (Fig. 4-7).

■ MESODERM FORMATION AND THE COELOM

In the amphioxus the first mesoderm arises from a pair of mesodermal folds located in the dorsolateral wall of the archenteron (Fig. 4-6, *A*). These bands fold upward, pinch off, and form a series of hollow **mesodermal (coelomic) pouches** (Fig. 4-6, *B*). About the time that two pouches have appeared, the embryo hatches into a free-swimming larva. As the larva elongates, additional pouches form.

After the mesodermal pouches are established, they grow ventrad, pushing between ectoderm and endoderm (Fig. 4-6, *C*). They finally meet underneath the gut and fuse, establishing a temporary ventral mesentery. The outer wall of each pouch lies against the ectoderm and is called **somatic mesoderm.** Together with the ectoderm it forms the **somatopleure,** which is the early **body wall.** The inner wall of

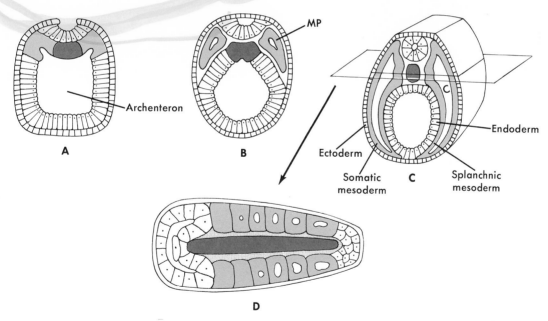

Fig. 4-6. Mesoderm and coelom formation in an amphioxus. **A,** Longitudinal bands of mesoderm (light red) lie in the dorsolateral wall of the archenteron lateral to the notochord (dark red). **B,** Mesodermal pouches (MP) have formed. **C,** The pouches have grown ventrad between ectoderm and endoderm to form a coelom, **C.** **D,** Early larva in frontal section showing segmented coelom.

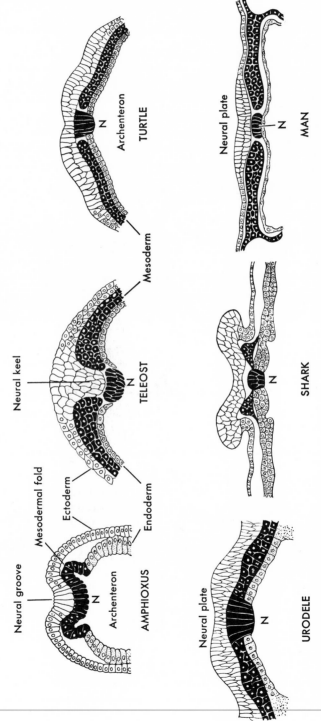

Fig. 4-7. Cross sections of dorsum of representative chordate embryos of three germ layers. Mesoderm and notochord (N) are black.

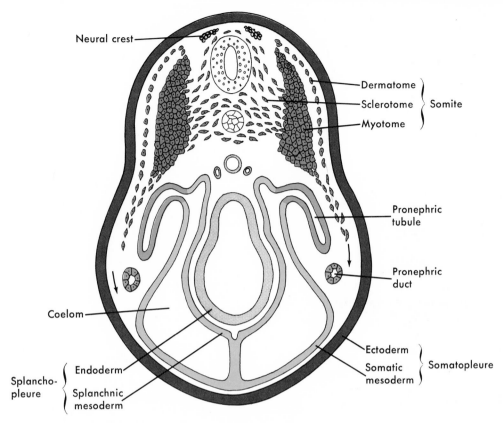

Fig. 4-8. Vertebrate embryo in cross section showing fate of mesoderm. *Dark red,* mesodermal somite (dorsal mesoderm) and its derivatives; *medium red,* derivatives of intermediate mesoderm; *light red,* lateral mesoderm. Sclerotome cells are streaming toward notochord and neural tube to form a vertebra; myotomal cells are streaming into lateral body wall (arrows) to form myotomal muscle; dermatome will form dermis of skin of back.

each pouch lies against the endoderm and is called **splanchnic mesoderm.** Together with endoderm it constitutes the **splanchnopleure,** which becomes chiefly the digestive tube of the trunk. The cavity between somatic and splanchnic mesoderm is the coelom (Fig. 4-6, *C*).

In the amphioxus the coelom is segmented for awhile because of its origin from a series of pouches. Later, the walls between pouches rupture, establishing a single long coelom on each side. Still later, the ventral mesentery ruptures, and the left

and right coelomic cavities become confluent underneath the gut.

Vertebrates do not go through the process of forming mesoderm as outpocketings of the archenteron. Instead, the first mesoderm migrates forward from the blastopore, pushing between ectoderm and endoderm. These migrating cells form a metameric series of **mesodermal somites** alongside the notochord and neural tube (Figs. 1-12 and 4-8). The somites are homologous with the mesodermal pouches of the amphioxus and constitute collectively **dorsal mesoderm.**

Lateral mesoderm, consisting of somatic and splanchnic mesoderm separated by the coelom, also forms (Fig. 4-8), but it is unsegmented. Between the somites and lateral mesoderm is **intermediate mesoderm.**

■ FATE OF THE MESODERM
□ Mesodermal somites

The somites constitute collectively **dorsal mesoderm.** They are aligned beside the neural tube the entire length of the trunk and tail, and they form in the head in varying numbers. Typical somites exhibit three regions—**dermatome, sclerotome,** and **myotome** (Fig. 4-8). Myotomes contribute mesenchyme that gives rise to skeletal muscles except those of the visceral arches. Muscles derived from myotomes are known as **myotomal muscles.**

Dermatomes contribute mesenchyme that gives rise chiefly to the connective tissues, blood vessels, and skeleton of the body wall matic mesoderm and, in the head, neural crests.) Sclerotomes give rise to the vertebral column and to the proximal portion of each rib. (Sclerotomes of head somites contribute little to the skull.)

□ Lateral mesoderm

Lateral mesoderm is confined to the trunk and consists of somatic and splanchnic mesoderm (Fig. 4-8). Somatic mesoderm gives rise chiefly to the connective tissues, blood vessels, and skeleton of the body wall and limbs. The dermis of the body wall is its outermost product, and the parietal peritoneum is its innermost product. Splanchnic mesoderm gives rise to smooth muscles and connective tissue of the digestive tract and its outpocketings and to the heart. The visceral peritoneum is its outermost derivative. Special regions of the peritoneum contribute adrenal cortical (interrenal) tissue, germinal epithelium, and, in some species, muellerian ducts.

□ Intermediate mesoderm

Intermediate (**nephrogenic**) mesoderm consists of a long ribbon of mostly unsegmented mesoderm extending the length of the trunk and lying between the somites and the lateral mesoderm (Fig. 4-8). It gives rise to kidney tubules and to most of the ducts associated with kidneys and gonads.

■ FATE OF THE ECTODERM

A median invagination of the ectoderm of the embryonic head, the **stomodeum,** becomes the anteriormost part of the oral cavity (Fig. 1-1). This part of the oral cavity is therefore ectodermal. Stomodeal ectoderm gives rise to the enamel of the teeth and to some of the more anterior glands of the oral cavity. An evagination from the roof of the stomodeum, **Rathke's pouch,** becomes the adenohypophysis (Fig. 17-5).

An ectodermal invagination similar to the stomodeum develops in relation to the cloaca. This is the **proctodeum** (Fig. 1-1). It gives rise to the terminal lining of the cloaca in lower vertebrates and to the anal canal, the segment of the digestive tract just caudal to the rectum, in mammals.

Ectoderm gives rise to the epidermis of the skin and all its derivatives, including glands that lie in the dermis but open onto the surface by ducts. The nervous system and the pia mater and arachnoid meninges (in amphibians, at least), are ectodermal. The olfactory epithelium, retina and lens of the eye, membranous labyrinth, neuromast system, and taste buds in the skin or stomodeal area are ectodermal.

All neural crest contributions (Fig. 4-8) are ectodermal. In the head these include structures usually thought of as mesodermal —dermis, subcutaneous connective tissue, teeth, most of the pharyngeal skeleton, parts of the neurocranium, and the branchiomeric musculature. Neural crest derivatives outside of the head include chromaffin tissue everywhere, adrenal medulla, and all pigment cells wherever located. Neural crests also give rise to certain neurons and neurilemmal sheath cells. The foregoing derivatives illustrate the diverse potentialities and nonspecific character of ectoderm. Mesenchyme of ectodermal origin is called **ectomesenchyme.**

■ FATE OF THE ENDODERM

Endoderm gives rise to the epithelium of the entire alimentary canal between the stomodeum and proctodeum. Any taste buds or oral glands behind the stomodeal area are endodermal. Since pharyngeal pouches arise as outpocketings of endoderm, the derivatives of the pouches—thymus, parathyroids, ultimobranchial glands, auditory tube, middle ear cavity, and crypts associated with mammalian tonsils—have endodermal components. Any midventral evagination of the embryonic pharynx such as thyroid, lungs or swim bladders, and their ducts, when present, are lined with endoderm or have endodermal components. Caudal to the pharynx the endoderm evaginates to form liver, gallbladder, pancreas, and various gastric and intestinal ceca. Most urinary bladders and the urinogenital sinus of mammals are lined with endoderm, since these are derived from the cloaca. Any other structures that arise as evaginations of the endodermal tube have endodermal components.

■ EXTRAEMBRYONIC MEMBRANES

Most embryonic vertebrates are provided with special membranes that extend beyond the body. These extraembryonic membranes arise early in ontogeny and perform important services for the embryo until hatching or birth. The chief extraembryonic membranes are the **yolk sac, amnion, chorion,** and **allantois.** The last three are found only in reptiles, birds, and mammals. The yolk sac is the most primitive.

□ Yolk sac (Figs. 4-9 and 4-10)

The yolk sac surrounds the yolk. It empties into the midgut and is usually lined by endoderm, although not in bony fishes. The sac is highly vascular and its vessels (**vitelline arteries and veins**) connect with circulatory channels within the embryo. Yolk particles in the sac are usually digested by enzymes secreted by the lining of the sac and then transported to the embryo by vitelline veins. (In sharks, yolk also enters the intestine directly from the sac, pro-

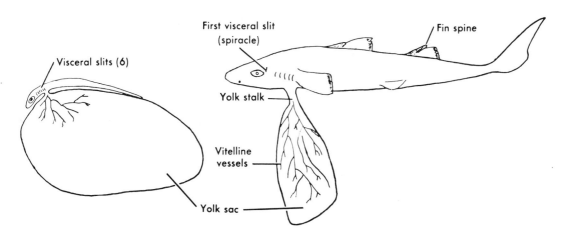

Fig. 4-9. Dogfish embryos with yolk sacs, removed from the uterus.

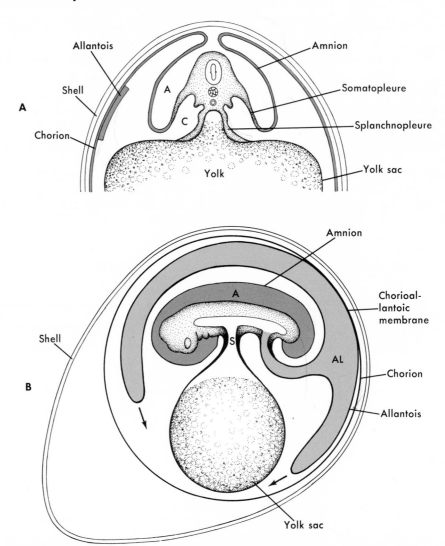

Fig. 4-10. Extraembryonic membranes of an egg-laying amniote. **A,** Cross section of embryo showing amnion and chorion arising from amniotic folds of somatopleure.
The position of the allantois, when fully differentiated, is indicated on left of drawing.
B, Relationships of allantois, which is growing in direction of arrows. *A,* Amniotic cavity; *AL,* allantoic cavity; *C,* intraembryonic coelom; *S,* yolk stalk.

pelled by rapidly beating cilia that line the sac and stalk.[12]) The yolk sac therefore grows smaller as the embryo grows larger. As it shrivels, it slowly disappears through the ventral body wall. The intracoelomic remnant of the yolk sac is finally incorpo-

rated into the wall of the midgut, or it may remain as a small diverticulum.

In embryonic sharks a large diverticulum of the yolk sac develops within the coelom close to where the sac opens into the duodenum. In pups ready to be born and with the yolk sac almost completely retracted, the diverticulum is easily demonstrated. It is distended with yolk, which has also spilled over into the spiral intestine. This remaining yolk serves to nourish the newborn pup for several days until it is able to obtain food from the environment.

Despite the fact that there is no yolk in typical mammalian eggs, embryonic mammals develop a yolk sac (Fig. 4-5)—a reminder of their genetic relationship with egg-laying reptiles. In man a vestige of the yolk sac (Meckel's diverticulum) remains in about 2% of the adult population. Its average position is on the ileum, 1 meter before the ileocolic valve, and its average length is about 5 cm.

Since the yolk sac in viviparous vertebrates is highly vascularized and lies close to the maternal tissues, it often serves as a membrane for absorbing oxygen from the parent. After the yolk in the sac is depleted, or if there is none, it may also absorb nourishment. When functioning in either capacity, it constitutes a **yolk sac placenta.**

□ **Amnion and chorion**

Embryos developing from eggs exposed to the air would be subject to dessication if they had no protection. Compensating for this in reptiles, birds, and mammals are two almost watertight saccular membranes, the amnion and chorion. These come into existence simultaneously when delicate upfoldings of the somatopleure meet above the embryo (Fig. 4-10, A).

The **amnion** is a sac filled with salty fluid that surrounds the embryo. The source of the fluid is chiefly metabolic water—water produced during cellular respiration when oxygen from the air unites with hydrogen from nutrients. The fluid is secreted by the amnion into the amniotic sac. (Turtle eggs laid in moist sand absorb considerable water from the environment, but this appears to be unusual.) Amniotic fluid buffers the fetus against mechanical injury. Man, like all other amniotes, spends the first 9 months in amniotic fluid.

The **chorion** forms a much larger **chorionic sac** surrounding the amniotic sac (Fig. 4-11, pig) and lying in more or less intimate relationship with the eggshell or the lining of the maternal uterus (Fig. 4-10, A).

□ **Allantois**

The allantois is an extraembryonic membrane that arises from the embryonic cloaca as a midventral evagination (Fig. 4-10, B). Typically, it grows until it comes in contact with the chorion to form a **chorioallantoic membrane.**

In oviparous and ovoviviparous reptiles, in birds, and in monotremes the chorioallantoic membrane comes in contact with the inner surface of the porous eggshell (Fig. 4-10, B). There it serves as a respiratory organ for exchange of oxygen and carbon dioxide between fetal organism and environment. In mammals above marsupials the chorioallantoic membrane comes in contact with the lining of the mother's uterus, since there is no eggshell. Here, too, the membrane serves as a respiratory organ, but it performs two additional functions. It serves as a site for transfer of nutrients from mother to young, and transfer of metabolic wastes from young to mother. Thus the chorioallantoic membrane in fetal mammals constitutes a **chorioallantoic placenta.**

In some mammals, including cats, rabbits, and man, the allantoic evagination grows only part way out the umbilical cord, then dwindles to terminate as a blind sac. However, its vessels, the **allantoic (umbilical) arteries and veins,** continue toward and vascularize the chorion, which is the fetal part of the placenta.

The base of the allantois—the part closest to the cloaca—becomes the urinary bladder (Fig. 13-29). The part of the allantois between the bladder and the umbilicus may remain after birth as a **middle umbilical ligament (urachus).** The portion outside the

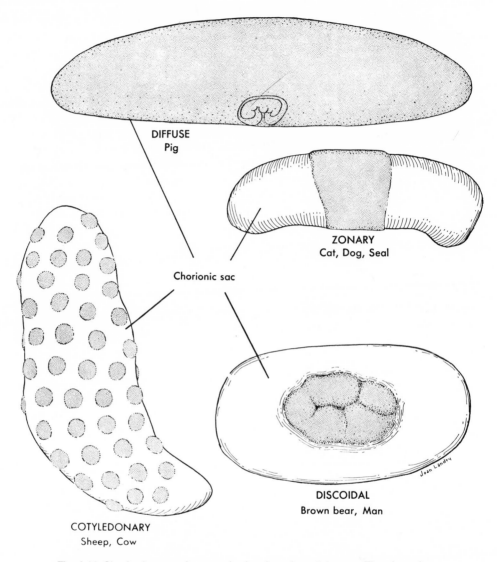

DIFFUSE
Pig

ZONARY
Cat, Dog, Seal

Chorionic sac

DISCOIDAL
Brown bear, Man

COTYLEDONARY
Sheep, Cow

Fig. 4-11. Chorionic sacs of mammals showing placental areas. The pig embryo surrounded by the amnion can be seen through the chorionic sac.

body is discarded at hatching or birth. It is likely that the allantois is the old amphibian urinary bladder that has acquired an added role during fetal life.

□ **Placentas**

The term placenta in a broad sense refers to any region in a viviparous species where parental and embryonic tissues of any kind are closely apposed, and which serves as a site for exchange of necessary substances between parent and embryo. In this sense, placentas are exhibited by some viviparous fishes that develop in an ovarian follicle. In a more restricted sense, a placenta is an organ composed of (1) the modified region of an extraembryonic membrane (yolk sac, choriovitelline membrane, chorioallantoic membrane, or chorion alone) that lies in intimate association with the maternal uterine lining and (2) the associated modified lining of the maternal uterus.

The yolk sac frequently serves as a placenta. In viviparous anamniotes—vertebrates lacking amnion, chorion, and allantois—the yolk sac may lie against the lining of the uterus and constitute the fetal part of a **simple yolk sac placenta.** It serves in this capacity in many sharks and rays. In euviviparous reptiles and the lowest placental mammals (marsupials), the yolk sac lies against the chorion and the chorion is in intimate association with the uterine lining. This is a **choriovitelline placenta.** It is more modern than a simple yolk sac placenta. Mammals above marsupials have a **chorioallantoic placenta** (see p. 87).

The intimacy of the anatomical relationship between fetal and maternal tissues in mammals varies greatly. In cats, pigs, cattle, and some other mammals the fetal placenta lies in simple contact with the uterine lining, and, at birth the fetal membranes simply peel away from the uterus without any bleeding of the uterus. This is said to be a **contact (nondeciduous) placenta.** In more intimate relationships chorionic villi, which are fingerlike outgrowths of the chorionic sac, become more or less rooted into the uterine lining, or even dangle into uterine blood sinuses. When the fetal part of such a placenta disengages from the uterus at parturition (birth), the invaded portion of the uterine lining (decidua) is shed, and some bleeding occurs. This is said to be a **deciduous placenta.** In any case, the extraembryonic membranes along with any sloughed uterine tissue are delivered as the **afterbirth.**

Chorionic villi (indicating functional placental regions) are distributed on the chorionic sac (Fig. 4-11) in isolated patches (**cotyledonary placenta**), in a band encircling the chorion (**zonary placenta**), in a single, large, discoidal area (**discoidal placenta**), or diffusely over the entire surface of the chorion (**diffuse placenta**). The diffuse placenta is thought to be more primitive. Placentas of mammals, at least, are a source of hormones that assist in maintaining pregnancy.

☐ Chapter summary

1. Oviparous vertebrates deposit eggs that contain enough nourishment to support development of a free-living, self-nourishing organism.

2. The amphioxus deposits small eggs with very little yolk that hatch quickly into immature, free-living, ciliated larvae with poorly developed organ systems.

3. Viviparous vertebrates develop in the body of the female parent. In ovoviviparous species the eggs are laden with yolk, and the mother provides oxygen, protection, and little else. In euviviparous species the eggs are practically devoid of yolk, and nourishment must be continually supplied by the parent.

4. In viviparity special highly vascularized embryonic tissues (pericardial sac, gills, villi, opercular lining, extraembryonic membranes, and other tissues) are sites for absorption of necessary substances.

5. Viviparity is found in every vertebrate class except birds.

6. Fertilization is internal in viviparous species, in urodeles and apodans, and in species that cover the egg with a shell.

7. As a result of fertilization and cleavage, a blastula with a blastocoel is formed. A blastoderm forms in yolk-laden eggs. An inner cell mass forms in the eggs of placental mammals.

8. Gastrulation is characterized by notogenesis and formation of an archenteron.

9. Mesoderm consists of segmental dorsal mesoderm (mesodermal somites), intermediate (nephrogenic) mesoderm, and lateral mesoderm.

10. Mesodermal somites consist of sclerotome, dermatome, and myotome. These give rise respectively to vertebrae and ribs, dermis of the back, and myotomal muscles.

11. Intermediate mesoderm gives rise to kidney tubules and to most of the ducts of the kidneys and gonads.

12. Lateral mesoderm consists of somatic and splanchnic layers. Somatic mesoderm plus ectoderm becomes somatopleure (body wall); splanchnic mesoderm plus endoderm gives rise chiefly to splanchnopleure, which becomes the digestive tube.

13. The coelom is the cavity between somatic and splanchnic mesoderm. It is initially segmented in the amphioxus, unsegmented in vertebrates.

14. Ectoderm contributes epidermis and its derivatives, nervous system and many special sense organs, and the derivatives of the stomodeum and proctodeum. Ectodermal neural crests contribute much of the neurocranium and pharyngeal arch derivatives (visceral skeleton and branchiomeric muscle), chromaffin tissue, pigment cells, adrenal medulla, and other miscellaneous tissues.

15. Endoderm gives rise to the epithelium of the digestive tract and its evaginations, to taste buds and oral glands behind the stomodeum, to the epithelial derivatives of the pharyngeal pouches and pharyngeal floor,

to the urinary bladder in tetrapods, and to most of the cloaca and its derivatives.

16. The chief extraembryonic membranes are yolk sac, amnion, chorion, and allantois.

17. A yolk sac develops in all vertebrates even though no yolk is present. Vitelline vessels transport digested yolk, when present, to the embryo. The yolk sac sometimes serves as a placenta, either alone (simple yolk sac placenta) or in association with the chorion (choriovitelline placenta).

18. The amnion and chorion are concentric sacs surrounding the amniote fetus. Amniotic fluid helps prevent desiccation of fetuses and mechanical injury. The chorion is applied to the eggshell or to the lining of the maternal uterus. In the latter case it constitutes part of a placenta.

19. The allantois is a midventral evagination of the embryonic cloaca. It comes in contact with the chorion to form a chorioallantoic membrane lying close to the eggshell or to the uterine lining of the mother. In the latter case it is part of a chorioallantoic placenta. The base of the allantois becomes the urinary bladder in amniotes.

20. The most common vertebrate placentas with respect to fetal membrane components are chorioallantoic, choriovitelline, and simple yolk sac placentas. The latter are the most primitive.

21. Placentas may be contact or deciduate. Villi may be arranged in a zonary, cotyledonary, discoidal, or diffuse pattern.

LITERATURE CITED AND SELECTED READINGS

1. Amoroso, E. C.: Comparative anatomy of the placenta, Annals of the New York Academy of Sciences **75**:855, 1959.
2. Arey, L. B.: Developmental anatomy, ed. 7, Philadelphia, 1965, W. B. Saunders Co.
3. Balinsky, B. I.: An introduction to embryology, ed. 3, Philadelphia, 1970, W. B. Saunders Co.
4. Bishop, S. C.: The salamanders of New York, New York State Museum Bulletin, no. 324, Albany, 1941.
5. DeHaan, R. L., and Ursprung, H., editors: Organogenesis, New York, 1965, Holt, Rinehart and Winston, Inc.
6. Hamilton, W. J., Boyd, J. D., and Mossman, H. W.: Human embryology, Baltimore, 1962, The Williams & Wilkins Co.
7. Matthews, L. H.: The evolution of viviparity in vertebrates, Memoirs of the Society for Endocrinology, no. 4, New York, 1955, Cambridge University Press.
8. Nelsen, O. E.: Comparative embryology of the vertebrates, New York, 1953, McGraw-Hill Book Co.
9. Patten, B. M.: Human embryology, ed. 3, New York, 1968, McGraw-Hill Book Co.
10. Patten, B. M.: Early embryology of the chick, ed. 5, New York, 1971, McGraw-Hill Book Co.
11. Smith, C. C., and Fretwell, S. D.: The optimal balance between size and number of offspring, American Naturalist **108**:499, 1975.
12. TeWinkel, L. E.: Observations on later phases of embryonic nutrition in *Squalus acanthias*, Journal of Morphology **73**:177, 1943.

5

Skin

In this chapter we will look at the basic structure of vertebrate skin. We will examine the epidermis of fishes and then see how the epidermis has been modified for life in air. We will look also at some of the remarkable structures that form from epidermis. Next we will look at the dermis and discover that the dermis of early vertebrates was loaded with bone, that bone is still present in much modern skin, and that the absence of bone is a modification of a basic pattern. Finally, we will look briefly at skin pigment and some of the things that skin does to help vertebrates survive.

Skin of the eft
The epidermis
    Epidermis of fishes and aquatic amphibians
    Epidermal glands of terrestrial vertebrates
    Stratum corneum of terrestrial vertebrates
The dermis
    Bone-forming potential of dermis
    Bony dermis of fishes
    Dermal ossification in tetrapods
Integumentary pigments
Some functions of skin

The skin of all vertebrates is built in accordance with a basic blueprint. It consists of a multilayered **epidermis,** derived from ectoderm, and a **dermis,** derived chiefly from mesoderm. Modifications of the epidermis and dermis involve (1) the relative number and complexity of skin glands, (2) the extent of differentiation and specialization of the most superficial layer (stratum corneum) of the epidermis, and (3) the extent to which bone develops in the dermis. The skin of an amphioxus exhibits epidermis and dermis, but the epidermis is only one cell thick (Fig. 5-1).

92

■ SKIN OF THE EFT

As an introduction to study of the skin, we will examine briefly a skin unencumbered by scales, feathers, or hair—the skin of an eft (immature, land stage of the urodele *Notophthalmus*). It illustrates generalized vertebrate skin with one notable difference: the absence of any ossification in the dermis.

The epidermis is a stratified epithelium (Fig. 5-2). The columnar cells in the deepest (basal) layer are constantly undergoing mitosis and replacing those lost from the surface. Proliferation from the basal layer causes older cells to be pushed outward. As they approach the surface, they become flattened (squamous) and synthesize **keratin,** a scleroprotein that is insoluble in water. When keratin is synthesized in a cell, the cell is said to be **keratinized,** or **cornified,** and it dies. Thus the outermost layer **(stratum corneum)** of the epidermis is made up of flattened, dead, cornified (keratinized) cells. The stratum corneum is constantly being shed in patches and replaced.

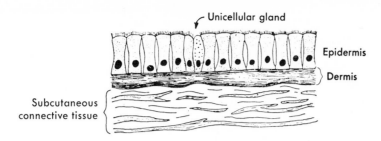

Fig. 5-1. Skin of a young amphioxus.

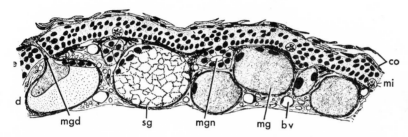

Fig. 5-2. Skin of *Notophthalmus*, land stage (eft). **bv,** Blood vessel; **co,** cells of stratum corneum; **d,** dermis; **e,** epidermis; **mg,** mucous gland; **mgd,** duct of discharging mucous gland; **mgn,** new epidermal gland invading the dermis; **mi,** mitotic figure in the basal layer of the epidermis; **sg,** exhausted mucous gland.

The integumentary glands develop from epidermis and are multicellular and saclike. Most of them secrete mucus. As the glands grow, they invade the dermis where there is room for expansion and where they are close to capillary beds. In some regions of the body the glands are so numerous that they constitute the chief bulk of the dermis. Each gland is drained by a duct lined with ectoderm.

There is a positive correlation in vertebrates between a terrestrial existence, relatively inactive mucous glands, and the presence of a stratum corneum. This is dramatically illustrated in the life cycle of *Notophthalmus.* The larvae of this species live in water and have many mucous glands and no stratum corneum. When the larvae metamorphose into efts and assume life on land, the skin glands become quiescent, and a stratum corneum forms and persists as long as the eft lives on land. As the eft approaches sexual maturity and migrates back to water, the mucous glands again become active, and the stratum corneum is shed and does not reappear.

The dermis underlies the epidermis and is thicker. It is made up chiefly of collagenous connective tissue, blood vessels, tiny nerves, lymphatics, the bases of epidermal glands, and pigment cells.

■ THE EPIDERMIS

The epidermis of most fishes and aquatic amphibians has many skin glands, chiefly mucous, and few keratinized cells on the surface. Entirely terrestrial vertebrates, on the contrary, have few mucous glands and a thick layer of keratinized cells that constitute a prominent stratum corneum (Fig. 5-3). These conditions minimize the loss of water through the skin. The epidermis has no blood vessels and is nourished by capillaries in the dermis.

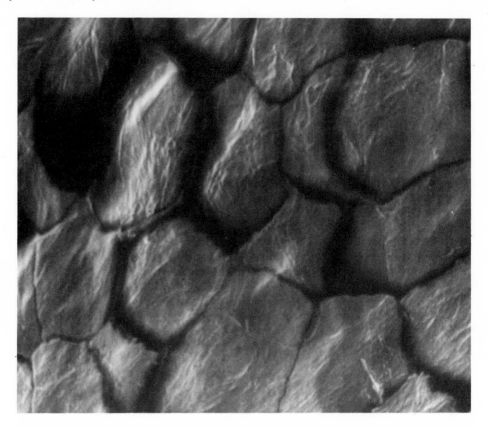

Fig. 5-3. Keratinized (cornified) cells of the stratum corneum on the surface of human skin. (Courtesy D. T. Rovee and B. Farber, Johnson & Johnson Research, New Brunswick, N.J.)

□ Epidermis of fishes and aquatic amphibians

Unicellular epidermal glands abound in cyclostomes (Fig. 5-4) and are common in jawed fishes (Figs. 5-5 and 5-6). These cells form mucous granules while still close to the basal layer of the epidermis. When they reach the surface, the cells rupture, and mucus exudes onto the skin. In addition to unicellular glands, most fishes and aquatic amphibians have multicellular skin glands of two general varieties, mucous and granular. The latter secrete proteins, some of which are toxic, hence protective. The multicellular glands are simple sacs. They may be confined to the epidermis (Fig. 5-6), or they may grow until they push into the underlying dermis (Figs. 5-2 and 5-7). Mu-

cous glands are especially advantageous to aquatic species that spend time out of water. The mucus keeps the skin moist in air, enabling it to continue to serve as a respiratory surface; also, when the animal is captured, mucus is exuded in such abundance that the animal becomes very slimy, slippery, and difficult to restrain. The lungfish *Protopterus* secretes a cocoon of mucus that it inhabits during the annual dry season, and some teleost mothers secrete a nutritious mucus that is eaten by the young offspring.

Some of the multicellular glands of fishes, especially deep-water teleosts, have become light-emitting organs called **photophores.** Like other skin glands, photophores arise in the epidermis and invade the der-

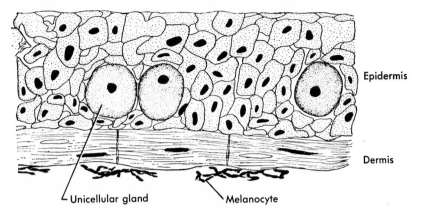

Fig. 5-4. Skin of a larval cyclostome. The dermis contains very dense collagenous fibers.

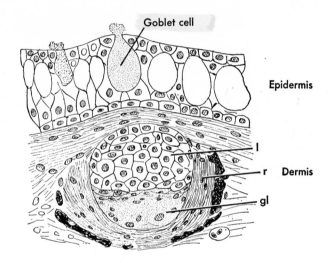

Fig. 5-5. Skin and light organ (photophore) of a luminous fish. **gl,** Luminous cells; **l,** lens cells; **r,** reflector cells (absent in some species) surrounded by pigment cells. The goblet cells are unicellular glands.

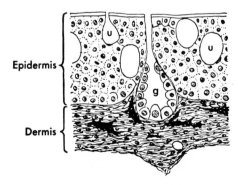

Fig. 5-6. Glandular skin of the dipnoan *Protopterus.* Multicellular, **g,** and unicellular, **u,** glands in the epidermis. Pigment cells are seen in the dermis.

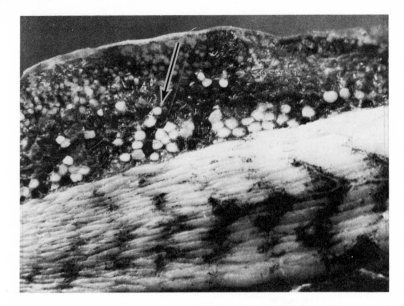

Fig. 5-7. Undersurface of skin from the tail of *Necturus,* showing mucous glands (arrow) projecting into subcutaneous tissue just external to muscle.

mis. In one variety (Fig. 5-5) the basal part of the gland consists of luminous cells, and the more superficial part consists of mucous cells that serve as a magnifying lens. Surrounding the base of the photophore are a blood sinus and a heavy concentration of pigment cells. The light emitted by photophores is not intense and may be of many hues. Among other functions, the light serves for species and sex recognition and sometimes aids in concealment by countershading.

The stratum corneum in fishes and predominantly aquatic amphibians is either thin or absent. This makes it possible for fishes and aquatic amphibians to have many functional unicellular glands because there is no layer of dead cells to prevent the secretions from exuding onto the surface. However, localized areas of the skin sometimes develop cornified structures, although these are relatively few. Conical horny epidermal spines and teeth develop in the buccal funnel and on the tongue of cyclostomes (Fig. 5-8). These rasping structures are pe-

riodically shed. Tadpoles have horny jaws, lips, and teeth that enable them to feed by rasping on vegetation during the larval stage. They are shed at metamorphosis. Callouslike caps develop on the toes of aquatic urodeles subjected to buffeting in mountain streams. These are probably protective.

☐ **Epidermal glands of terrestrial vertebrates**

Skin glands in terrestrial vertebrates are fewer but more diverse than in aquatic species. Mucous glands still exist, but the predominant variety are granular (serous) glands (Fig. 5-9) that secrete an irritating or even highly toxic alkaloid.

Reptiles exhibit a greater diversity of skin glands than amphibians. Most of the secretions are holocrine, that is, the cells of the gland constitute the secretion; and many of them are pheromones (substances secreted into the environment by an organism that have an effect on the behavior or physiology of other members of the population). Some

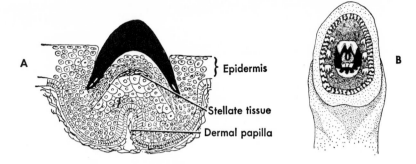

Fig. 5-8. Lamprey teeth and buccal funnel. **A,** Cornified tooth (black), a product of the epidermis. The stellate tissue may be the beginning of a replacement tooth. **B,** Buccal funnel. The teeth are shown as white against a dark background.

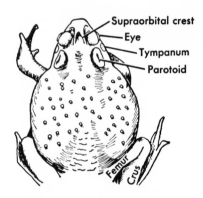

Fig. 5-9. Warty skin of toad *(Bufo)*. A serous gland, the parotoid, is seen behind the eye.

glands aid in sloughing of the skin. In lizards as many as four different sets encircle the vent, and in male lizards glands on the medial side of each hind leg secrete a substance that hardens on the skin to form temporary spines that help to restrain the female during copulation. Musk turtles exude a yellowish fluid from two glands on each side of the body just below the carapace, and a longitudinal row of glands of doubtful function occurs along the back in crocodilians.

Birds have the fewest integumentary glands. One is the uropygial gland, which causes a prominent swelling at the rump. It is best developed in aquatic birds. The oily secretion that exudes from the uropygial gland is transferred to the feathers during preening. Small oil glands in some birds line the outer ear canal and surround the exit from the cloaca. These are the only integumentary glands that have been described in birds.

Mammals have a wide variety of skin glands, but all seem to be variations of two major groups, sudoriferous (sweat) and sebaceous (oil) glands. Mammary glands appear to be modified sebaceous glands.

Mammary glands arise in both sexes from a pair of elevated ribbons of ectoderm, called **milk lines,** which extend along the ventrolateral body wall of the fetus from the axilla to the groin (Fig. 5-10). Patches of undifferentiated mammary tissue develop along the milk lines, invade the dermis (Fig. 5-11), and then spread under it in the superficial fascia. As development progresses, a nipple forms above each patch. Further development of mammary tissue usually occurs in a circular patch beneath each nipple.

As the female mammal approaches sexual maturity (adolescence in primates), rising titers of female sex hormones cause the juvenile duct system to spread and branch. During pregnancy a battery of hormones causes the formation of saclike secreting terminals (alveoli) at the ends of the duct system. In pregnant cats and many other

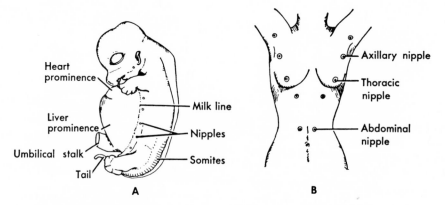

Fig. 5-10. A, Milk line and nipples in a 20-mm pig embryo. **B,** Supernumerary nipples in man.

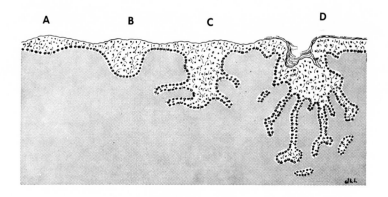

MAMMARY GLAND

Morphogenesis

Fig. 5-11. Successive stages of mammary gland development. **A,** Equivalent to a human embryo at 6 weeks; **B,** equivalent to a human embryo at 9 weeks and to mouse embryo at 16 days; **C,** intermediate stage; **D,** at birth. Gray area represents dermis.

mammals, the patches of mammary tissue on each side spread toward each other and unite to form two long, thick masses of considerable weight.

The distribution and number of mammary glands and nipples vary with the species. A single pair of thoracic nipples occurs in apes and man. Bats also have thoracic nipples. Insectivores and some lemurs have one pair of thoracic and one pair of inguinal nipples. Flying lemurs and marmosets have a single pair in the armpit (axillary nipples). In Ce-

tacea nipples occur near the groin (inguinal nipples), and the baby porpoise or whale holds onto the nipple as the mother swims about in the sea. Male lemurs have a nipple on each shoulder, and nutrias have four on the back so that the babies are able to ride along on the mother's back above water while nursing. In pigs, dogs, edentates, and many other mammals, a series of axillary, thoracic, abdominal, and inguinal nipples is scattered all along the milk line. Supernumerary nipples may occur in any mammal,

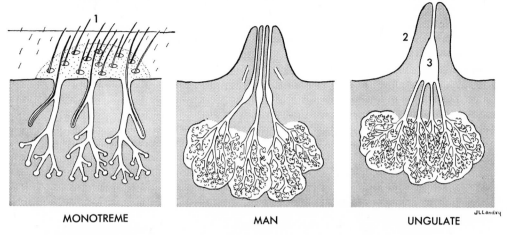

Fig. 5-12. Mammary glands, ducts, and nipples. The monotreme lacks nipples, and the glands resemble modified sweat glands. **1,** Hairs. **2,** False teat. **3,** Cistern.

including man (Fig. 5-10, *B*). In general, there are sufficient nipples for the number of young in a litter and these are in locations appropriate to the habits of the species.

In true teats (Fig. 5-12, man) all ducts open at the tip of the nipple as a result of the elevation of the duct-bearing area during development. In false teats (Fig. 5-12, ungulate) the skin around the duct openings becomes elevated, and all ducts empty into one cistern. Suckling young are not concerned about the terminology as long as the teats are provident.

Monotremes do not develop typical mammary glands or nipples. Instead, in both sexes modified sweat glands produce a nutritious secretion, which is lapped off a convenient tuft of hairs by the young (Fig. 5-12, monotreme). Teats would probably be useless in the duckbill, since it appears doubtful whether the young, hindered by horny beaks and lacking muscular cheeks and lips, could nurse. Except during lactation, the teats of the opossum are hygienically stored in depressions within the skin.

Sudoriferous (sweat) glands are long, slender, coiled tubes of epidermal cells extending deep into the dermis (Fig. 5-13).

Their secretions ooze onto the surface of the skin through tortuous channels that perforate the stratum corneum and open as pores. Sweat glands are widespread among mammals, but they may be absent, as in scaly anteaters and marine mammals. They may occur only on the soles of the feet (cats and mice) or on the toes, lips, ears, back, or head. Man, covered with less hair, has the greatest number of sweat glands per square inch of body surface. The **ciliary glands,** which open into the hair follicles of the eyelashes and along the margins of the eyelids, are modified sudoriferous glands.

Sebaceous glands secrete an oily exudate, sebum, into the hair follicles (Fig. 5-13). The oil lubricates the skin. Fur (including human hair) glistens after brushing because of the oil on the hairs. Usually several glands open in association with one follicle, but in some areas they open directly onto the surface of the skin. Marine mammals are practically devoid of hair and do not have sebaceous glands.

In the outer ear canal of mammals modified sebaceous glands (**ceruminous glands**) secrete cerumen, a waxy grease. The hairs and the wax trap insects that may otherwise wander deep into the canal and touch the highly sensitive eardrum. **Tarsal** (meibomian) **glands** secrete oil onto the conjunctiva of the eyeball. They are embedded in a dense connective tissue plate, the tarsus, in each eyelid.

Mammals have a variety of scent glands (sebaceous and sudoriferous), and most of the scents are pheromones. They carry a message to other members of the species or to the other sex. One species, man, takes the pheromone from musk glands on the abdomen of male musk deer, adds other odorants to it, and dabs it on his ear lobes. (It is then called perfume, but it is still performing the function of a pheromone.) Occasionally, pheromones have been converted into defensive weapons, such as the spray from the anal glands of skunks. Scent glands develop on the feet in goats and rhinoceros, and calluslike growths on the feet of horses appear to be remnants of similar glands. Kangaroo rats have sebaceous scent glands along the middorsal line in the most exposed area of the arched back. Grisons (South American rodents) emit such a pungent odor at all times that it would be unthinkable to remain in a closed room with these animals for more than a few minutes. Many of the odors of the mammalian zoo are caused by scent glands, not by unhygienic conditions in the pens and cages.

Male elephants have a temporal gland that swells during the breeding season and secretes a sticky, brown fluid. At this time

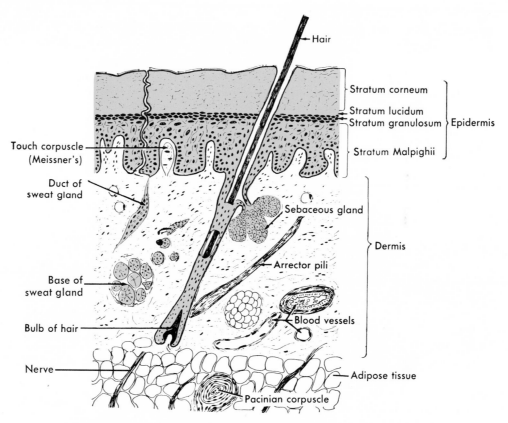

Fig. 5-13. Mammalian skin, with hair follicle extending deep into the dermis. Supporting the constituents of the dermis are dense bundles of collagenous connective tissue. Epidermal derivatives are red.

the male elephant is very dangerous. A gland above the eye of the peccary looks like a navel and secretes a watery fluid. The male lemur in some species has a hardened patch of spiny skin on the forearm, under which lies a gland the size of an almond. Bats have many glands in the skin of the face and head. All of the foregoing appear to be modifications either of sebaceous or sudoriferous glands.

☐ **Stratum corneum of terrestrial vertebrates**

Terrestrial vertebrates have a stratum corneum that varies in thickness among species and on different parts of the body. The most generalized cornified structures

are epidermal scales. Specializations include claws, nails, hoofs, feathers, hair, and horns.

EPIDERMAL SCALES

Epidermal scales are regular overlapping thickenings of the stratum corneum found only in amniotes. Reptiles are almost completely covered with them (Fig. 5-14), and in these animals epidermal scales reach an evolutionary peak. In birds and mammals epidermal scales are confined to restricted regions of the body such as legs and feet, the base of the beak, and the tail, as in rodents. Armadillos (Fig. 5-15) and pangolins (Fig. 5-24) are exceptions in that they are almost completely covered with epidermal scales.

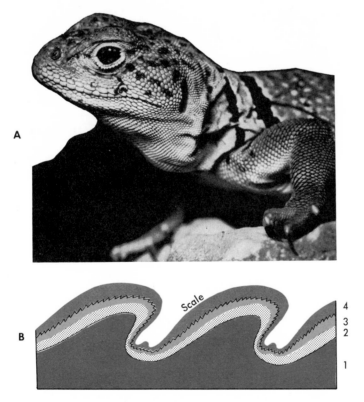

Fig. 5-14. Squamate scales. **A,** Collared lizard. **B,** Diagrammatic section of skin of a snake or lizard. **1,** Dermis; **2,** actively mitotic layer of epidermis; **3,** newly cornified layer of epidermis; **4,** older cornified layer to be shed at next molt. (**A** courtesy F. W. Schmidt, naturalist photographer, La Marque, Tex.)

The scales on the head of snakes exhibit a characteristic number and arrangement for the species and are often named for underlying bones (Fig. 5-16). The large quadrilateral scales that come in contact with the ground are **scutes**. The broad scales on the surface of a turtle's shell are identified by name (Fig. 5-17), but their arrangement does not coincide with that of the under-

lying bones of the shell. Warty or horny outgrowths replace scales in some parts of the body, particularly in lizards (Figs. 5-18 and 16-9).

Lizards and snakes have a double layer of stratum corneum, the inner layer having the same structure as the entire epidermis of amphibians (Fig. 5-14, *B*). The outer layer is shed periodically. In lizards it flakes off in large patches, but in snakes the outer layer of the entire body including the transparent covering (spectacle) of the eye is shed in one piece.

The stratum corneum is impervious to water. The double layer in lizards and snakes and the tough corneum of turtles and crocodilians result in very slow loss of water in dry air. This, coupled with the cleidoic egg (p. 442), enabled reptiles to completely emancipate themselves from water and spend their entire life on dry land.

CLAWS, NAILS, AND HOOFS

Claws, nails, and hoofs are modifications of the stratum corneum at the ends of digits. Claws first appeared in reptiles, and have persisted in birds and most mammals. (The "claws" of the African clawed toad are not comparable to reptilian claws.) Claws evolved into nails in primates and into hoofs

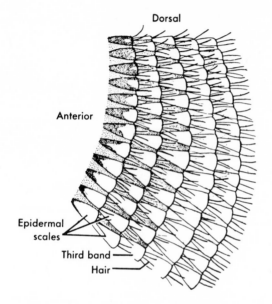

Fig. 5-15. Skin from an armadillo, showing epidermal scales and hair.

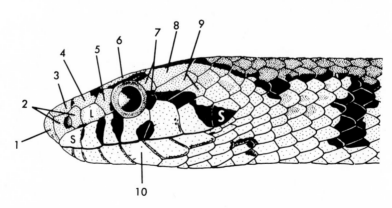

Fig. 5-16. Epidermal scales on head of milk snake *Lampropeltis triangulum*.
1, Rostral; **2,** nasal; **3,** internasal; **4,** prefrontal; **5,** preocular; **6,** supraocular; **7,** postoculars; **8,** parietal; **9,** anterior temporals; **10,** fifth of seven infralabials; **L,** loreal; **S,** first and last supralabial. (Courtesy Kenneth L. Williams.)

in ungulates. Claws, nails, and hoofs have the same basic structure. They consist of two curved parts, a horny dorsal plate, the **unguis,** and a softer ventral plate, the **subunguis** (Fig. 5-19). The two plates wrap partially around the terminal phalanx, which is usually pointed when associated with a claw, blunt when associated with a hoof or nail. A still softer calluslike, cornified pad, the **cuneus** (called the "frog" by horsemen), is frequently present in ungulates, partially surrounded by the subunguis. The thick, hard unguis of a hoof is U- or V-shaped, and since it consists of dead cells, a shoe can be nailed into it. The unguis of a nail has become flattened, and the subunguis is much reduced. As a result, nails cover only the dorsal surfaces of digits; but if permitted to grow nails become clawlike.

Although claws in birds are often thought of as associated only with the feet (Fig. 5-20), sharp claws are frequently borne at the end of one or two of the digits of the wings (ostriches, geese, some swifts, and others). Young hoatzins use the claws on the wings

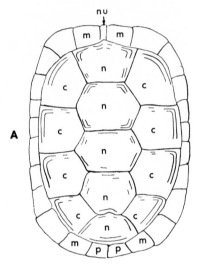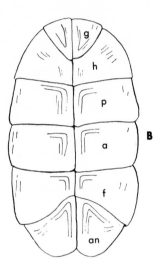

Fig. 5-17. Epidermal scales of the turtle *Chrysemys.* **A,** Carapace. **B,** Plastron. On the carapace the scutes shown are, **c,** costals; **m,** marginals all around the periphery, including the nuchal **(nu),** and pygals **(p); n,** neurals. On the plastron the scutes shown are, **g,** gulars; **h,** humerals; **p,** pectorals; **a,** abdominals; **f,** femorals; **an,** anals.

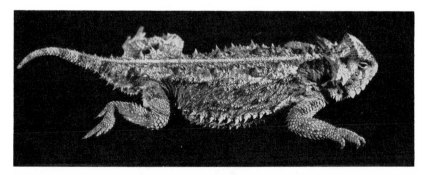

Fig. 5-18. Warty and spiny skin of horned toad, a lizard. (Courtesy Carolina Biological Supply Co., Burlington, N.C.)

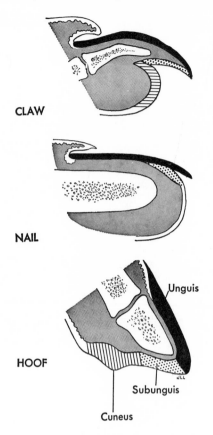

Fig. 5-19. Claw, nail, hoof, and terminal phalanx diagrammed in sagittal section.

Fig. 5-20. Claw on the middle toe of a great blue heron. (From Atwood, W. H.: Comparative anatomy, ed. 2, St. Louis, 1955, The C. V. Mosby Co.)

for climbing about on the bark of trees. *Archaeopteryx* had three claws on each wing. Only reptilian claws are shed. In other animals, claws, nails, and hoofs must be worn down by friction.

FEATHERS

Feathers are remarkably complicated cornified outgrowths of the epidermis. There are three morphological varieties: contour feathers, down feathers (plumules), and hairlike feathers (filoplumes).

Contour feathers (Fig. 5-21, *A*) are the conspicuous feathers that give the bird its contour, or general shape. A typical contour feather consists of a horny, hollow **shaft** and two flattened **vanes**. The base of the shaft devoid of vanes is the **calamus (quill)**. The vane-bearing segment of the shaft is the **rachis**. Each vane consists of a row of **barbs**, which in turn have **barbules**. The latter have **hooklets**, which interlock with the barbules of adjacent barbs and stiffen the vane (Fig. 5-21, *B*, *C*). When a contour feather is ruffled, the barbules have become unhooked. Preening rehooks the hooklets, returning the feather to its tailored state. During preening the oily secretion of the uropygial gland is applied to the barbs. On the smaller contour feathers of the wings (covert feathers), the lower barbules lack hooklets, and this region of the feather is fluffy. In ostriches and some other birds all feathers are fluffy, since hooklets are absent.

Arising from a notch (**superior umbilicus**) on the shaft of a contour feather at the base of the rachis is an **afterfeather**. Usually the aftershaft is much smaller than the main shaft, but in some birds (emu and cassowary) the afterfeather is of the same length. This results in a double feather.

Although contour feathers completely clothe most of the body, the follicles from which they grow are disposed in feather tracts, or **pterylae** (Fig. 5-22). A few birds, including ostriches and penguins, lack feather tracts.

Inserting on the walls of the feather follicles in the dermis are erector muscles (**arrectores plumarum**), which, along with extrinsic integumentary muscles, enable a bird to fluff its feathers.

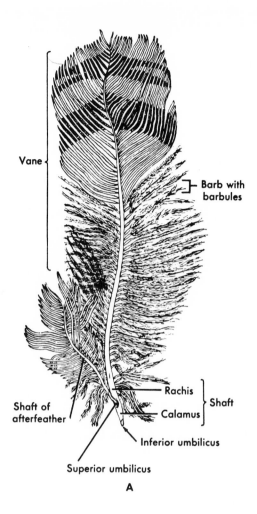

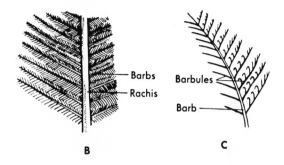

Fig. 5-21. **A,** Contour feather from a grouse. **B,** Section of feather showing barbs coming off the rachis. **C,** Each barb has barbules, and the barbules have hooklets. (**B** and **C** from Atwood, W. H.: Comparative anatomy, ed. 2, St. Louis, 1955, The C. V. Mosby Co.)

Down feathers are small, fluffy feathers lying underneath and between contour feathers. Very young birds lack contour feathers and are covered with down. Down feathers may be ancestral to contour feathers. They have a short calamus, with a crown of barbs arising from the free end. Hooklets are lacking. Eiderdown, used in pillows, is the down feathers of the eider duck.

Filoplumes are hairlike feathers familiar to housewives who singe hens before cooking them. Filoplumes consist chiefly of a threadlike shaft. They are usually scattered throughout the skin among the contour

feathers. They are the very long colorful feathers of a peacock.

Development of feathers. Feathers arise from feather follicles lined by epidermis (Fig. 5-23). The feather primordium (**pinfeather**) is covered by an epidermal feather sheath and contains a core of dermis, the dermal papilla. The mitotic layer of epidermis at the base of the follicle proliferates tall columns of epidermal cells that push toward the tip of the growing feather just under the epidermis. These columns separate from one another, cornify, and develop into barbs with barbules. When the feather sheath splits open, the fluffy barbs stretch out of their cramped quarters, and the quill elongates.

When a feather is full grown, the dermal papilla in the shaft dies and becomes the

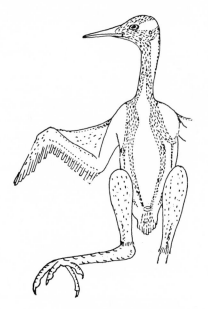

Fig. 5-22. Feather tracts.

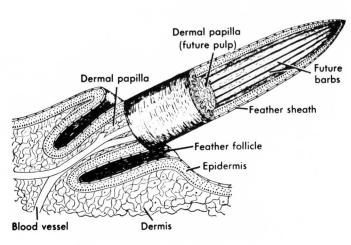

Fig. 5-23. Developing feather (pinfeather).

pulp. The living basal portion of the papilla withdraws from the base of the shaft leaving an opening, the **inferior umbilicus.** Feathers that arise during successive molts develop from reactivated dermal papillae that have already given rise to feathers. In many birds—perhaps in all—feathers of the old generation are passively pushed out of the follicles by incoming feathers.

HAIR

Hairs are much like feathers but are far less complex. They may form a dense, furry covering over the entire body, or there may be only one or two bristles on the upper lip, as in some whales. Where fur is dense, there are usually short, fine hairs (underfur) as well as long, coarse ones. When sufficiently dense, hair has an insulating effect.

Hairs are sensitive tactile organs. The root of each is surrounded by a basketlike network of sensory nerve endings, and displacement of the root initiates a train of sensory impulses to the brain. Disturb a single hair on the back of your hand and note

the sensation evoked. Vibrissae (stiff long whiskers on the face of many mammals) perform this role exclusively.

Hairs grow in isolated groups of two to a dozen or more, and linearly between scales when the latter are present (Fig. 5-15). Some monkeys have groups of three, apes have groups of five, and man has groups of three to five. A glance at the side of your own hand, near the base of the thumb, will reveal the linear arrangement of hairs in that location. The presence of hairs between scales, or in what appear to be interscalar locations when no scales are present, seems to have been an adaptation for the receipt of tactile stimuli in a skin covered with insensitive scales.

The phylogenetic origin of hair is not known, and existing theories are speculative.

Hairs grow out of hair follicles that project deep into the dermis (Fig. 5-13). Each hair consists of a long shaft, a root, and a bulb. The **shaft** lies free within the follicle and projects above the surface of the skin. It is composed of layers of keratinized cells numbering about 500,000 per linear inch. The hair may contain a **medulla** (equivalent to the pulp of a feather) composed chiefly of air spaces. The **root** of the hair is that portion deep within the follicle where the hair has not yet separated from the surrounding epidermal cells of the follicular wall. Here the cells are becoming cornified and are dying. The **bulb** is a swelling at the base of the follicle containing the dermal papilla. It is an area of rapid mitosis, which is constantly contributing new cells that make the hair longer.

Inserting on the wall of each hair follicle in the dermis is a tiny smooth muscle, the **arrector pili,** or "elevator of the hair" (Fig. 5-13). When the arrectores pilorum contract, the hairs are drawn toward a vertical position and the skin around the base of each hair is pulled into a tiny mound, causing (in man) "gooseflesh," or "chill bumps." Elevation of the hairs in many mammals causes the animal to look ferocious. It also increases the insulating ability of the fur.

The scales of pangolins (Fig. 5-24) and, perhaps, the horns of rhinoceros (Fig. 5-25) may be evolutionary products of agglutinated hairs. Other modifications of hair include bristles, spines like those of spiny anteaters, and porcupine quills.

Development of hairs. Hair follicles first develop as cylindrical ingrowths of the epidermis into the dermis (Fig. 5-26). Beneath the epidermal ingrowth and indenting its base, a dermal papilla organizes. With continued proliferation of epidermal cells, the hair primordium grows deeper and deeper into the dermis, nourished by vessels within the papilla. When the bulb at the base of the primordium is sufficiently differentiated, cornified cells start to appear, and a hair shaft begins to rise out of the follicle.

HORNS

Most artiodactyls of the bovine family (cattle, antelope, sheep, goats) have true horns. These consist of a core of dermal bone covered by a horny epidermal sheath.

Fig. 5-24. Pangolin showing epidermal scales. (From Flower and Lydekker: Introduction to the study of mammals, London, A. & C. Black, Ltd.)

The sheath is the actual horn, and when removed it is hollow (Fig. 5-25). It is a product of the stratum corneum, and in bovines it is never shed. Horns usually occur in both sexes, although not always. Polled cattle have lost their horns by selective breeding. Male pronghorn "antelopes" also have true horns that are branched, the horny part, but not the bony core, is shed annually.

The horns of a rhinoceros differ structurally from bovine horns. They have no bony core and are composed of agglutinated keratinized hairlike epidermal fibers that form a solid structure perched on a roughened area of the nasal bone (from which rhinoceros derive their name). Some African rhinoceros have two horns, one behind the other. They occur on both sexes and are not shed.

The so-called horns of giraffes and the antlers of deer are not horns but dermal bone attached to the skull. They are covered initially (antlers) or permanently (giraffes) with normal skin containing velvety hair.

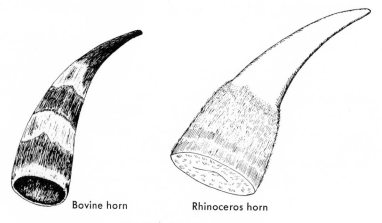

Bovine horn   Rhinoceros horn

Fig. 5-25. Mammalian horns.

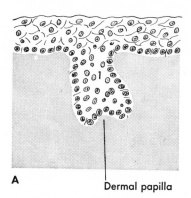

A

Dermal papilla

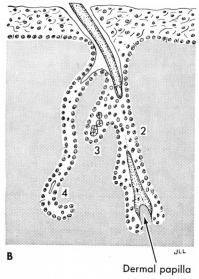

B

HAIR
Morphogenesis

Dermal papilla

Fig. 5-26. Successive stages in the development of hair and associated glands. **1,** Initial epidermal ingrowth into dermis; **2,** hair follicle; **3,** developing sebaceous gland; **4,** sweat gland. Gray area, dermis.

OTHER CORNIFIED STRUCTURES OF AMNIOTES

Rattlesnake **rattles** are rings of horny stratum corneum that remain attached to the tail after each molt (Fig. 5-27). **Beaks** are covered with a horny sheath, and roosters' **combs** are covered with a thick, warty stratum corneum. Monkeys and apes sit on thick **ischial callosities,** and camels kneel on **knee pads. Tori** are epidermal pads that most mammals other than ungulates walk on

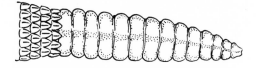

Fig. 5-27. Rattles from a rattlesnake.

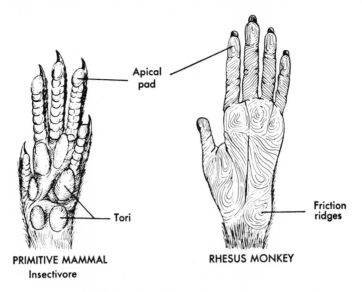

Apical
pad

Friction
ridges

Tori

PRIMITIVE MAMMAL
Insectivore

RHESUS MONKEY

Fig. 5-28. Tori and friction ridges.

Fig. 5-29. Sheets of baleen (whalebone) removed from the oral cavity of a whalebone whale. The sheets may be from 2 to 12 feet long. (Courtesy General Biological Supply House, Inc., Chicago, Ill.)

(Fig. 5-28). Cats "pussyfoot" by retracting their claws and walking stealthily on tori. At the ends of digits tori are called **apical pads. Corns** and **calluses** are temporary thickenings of the stratum corneum that develop where the skin has been subjected to unusual friction.

Toothless whales have great frayed horny sheets of oral epithelium called **baleen** hanging from the roof of the mouth (Fig. 5-29). As many as 370 of these sheets have been counted in one whale. The apparatus serves as a massive strainer of food. They are among the many cornified structures that develop from the skin.

■ THE DERMIS

The basic component of dermis, whether fish or man, is collagenous connective tissue. Collagen is a proteinaceous fibril demonstrable by electron microscopy, which aggregates with other collagen fibrils to form dense bundles of collagenous connective tissue visible by light microscopy.

Reticular and elastic tissue bundles also form in the dermis. Capillaries supply the dermis and, by diffusion, the epidermis, since the latter lacks vessels. Dermal papillae become exceptionally vascular because much nourishment and oxygen are needed to maintain the rapid mitosis taking place in the basal layer of the epidermis at that location. A variety of bulbous (encapsulated) general sense receptors are found in the dermis, particularly in birds and mammals (Fig. 16-16). Nerves invade the dermis, lymphatics enter, and pigment cells migrate in from neural crests, multiply, and become established close to the epidermis. Smooth muscle, including erectors of feathers and hairs, form in birds and mammals, and mesenchyme cells become adipose for storage of fat. Most important phylogenetically, bone is deposited in the dermis of many vertebrates.

Bone-forming potential of dermis

The dermis has an ancient and persistent potential to form bone. The earliest vertebrates, ostracoderms, had so much dermal bone that they are called armored fishes. Fishes that succeeded ostracoderms—placoderms, then ganoids—continued to develop much bone in their skin. Even the most modern fishes with a few exceptions develop bone in their skin, but the bone has been becoming thinner with time. Many modern tetrapods develop dermal bone. In fact, the only vertebrate class that does not have any living members with dermal bone is birds.

The deposit of collagen is the first major step in bone formation. Once this is accomplished—and it is accomplished in all vertebrate skin—it is only necessary that bone-forming cells differentiate from the mesenchyme and deposit hydroxyapatite crystals. The result is dermal bone. By reason of its intramembranous origin, dermal bone is a variety of membrane bone (Chapter 6).

Absence of bone in the dermis seems not to be primitive. The inherent primitive potential of dermis is to become ossified, given the appropriate stimulus. Animals lacking bone in the dermis have the necessary collagenous matrix but fail to deposit hydroxyapatite crystals. Cyclostome dermis, for instance, is exceptionally tough because the collagen bundles are very densely packed, but there is no bone. It is as though the dermis had been prepared for a shower of hydroxyapatite crystals that never materialized! Dense collagen in the dermis of mammals, when tanned, makes leather.

Bony dermis of fishes

Ancient dermal armor varied in detail, but a basic structural pattern is evident. It consisted of lamellar bone, spongy (vascular) bone, and dentin. On the surface there was often a layer of an enamellike material (Fig. 5-30). The armor was disposed either as broad, flat plates or as smaller bony scales. In either case, it covered the entire body in the oldest species (Fig. 2-1). Presumably, this armor was protective, but it may also have served as a storage site for calcium and phosphates.

Eventually, the large bony plates on trunk and tail gave way to smaller and thinner bony scales, whereas on the head and pectoral region they evidently contributed to the bony skull and pectoral girdle. Dermal scales can be conveniently classified into cosmoid, ganoid, placoid, and modern bony scales. The latter are of two kinds, cycloid and ctenoid.

Cosmoid scales were found on early lobe-finned fishes and resembled ancient armor. They were precursors of ganoid, placoid, and modern bony scales and are not found on any fish alive today. The dentin in these scales is sometimes called **cosmine**, which accounts for the name cosmoid.

Ganoid scales were dominant when ganoid fishes were at their zenith. They almost disappeared when the burgeoning teleosts

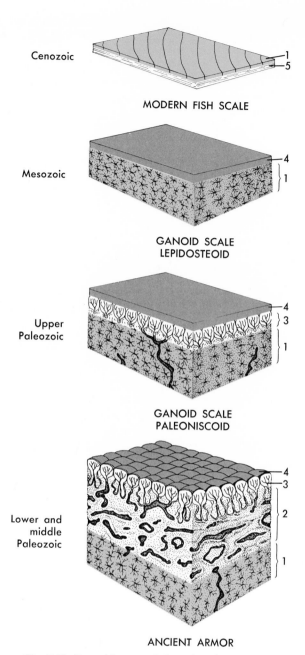

Cenozoic

MODERN FISH SCALE

Mesozoic

**GANOID SCALE
LEPIDOSTEOID**

Upper
Paleozoic

**GANOID SCALE
PALEONISCOID**

Lower and
middle
Paleozoic

ANCIENT ARMOR

Fig. 5-30. Dermal bone and dermal scales through the ages, diagrammatic. **1,** Lamellar bone; **2,** spongy bone; **3,** dentin; **4,** enameloid of one kind or another, including ganoin; **5,** fibrous plate characteristic of modern fish scales. The surface elevations in ancient armor are "denticles" consisting of layers **3** and **4.** The lamellar bone in modern fish scales is acellular. Cosmoid scales (not illustrated) resemble ancient armor.

took over as the chief bony fishes. Genuine, old-fashioned ganoid scales, called paleoniscoid (Fig. 5-30), are found only on *Polypterus* and *Calamoichthyes* and on *Latimeria*, the sole surviving crossopterygian. All other surviving ganoids have lost either the dentin (lepidosteoid scale of garfish, Fig. 5-30) or the ganoin. The scales of gars (Fig. 5-31) and of *Polypterus* completely cover the body. The remaining ganoids—how much longer will they survive?—have ganoid scales on the cephalic part of the trunk and modern flexible cycloid scales elsewhere.

Placoid scales are found on elasmobranchs and are similar to the denticles of armored fishes. The difference is that instead of forming a flat denticle the dentin and enamel form a spine that protrudes through typical epidermis (Fig. 5-32). The basal plate embedded in the dermis is thin lamellar bone. These scales, or denticles, become teeth at the edge of the jaws.

Cycloid scales are characteristic of teleosts and modern lobe-finned fishes. **Ctenoid scales** occur only on teleosts (Fig. 5-33). These modern scales seem to be lepidosteoid scales that have lost their ganoin so that all that remains is a thin layer of acellular lamellar bone underlaid by a fibrous plate of dense collagen (Fig. 5-30, modern scale). The overlying epidermis becomes quite thin and may even wear off. Cyclostomes, catfishes, eels, and some other recent fishes have lost the ability to form scales. Nevertheless, scale anlagen form transitorily in teleost embryos.

☐ **Dermal ossification in tetrapods**

When the first tetrapods lumbered onto land they carried with them a thin version of the cosmoid scales of their lobe-finned ancestors. Some labyrinthodonts and even some cotylosaurs had large bony scales. Others had minute scales. Bony dermal scales, often called **osteoderms** in tetrapods, continue to form in some amphibians, most

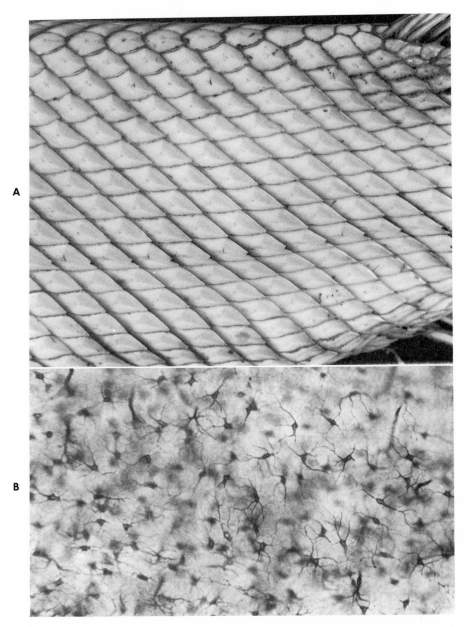

Fig. 5-31. Ganoid scales. **A,** Scales on trunk of a garfish. The tail is at the right. **B,** Canaliculi and lacunae within a single scale (microscopic).

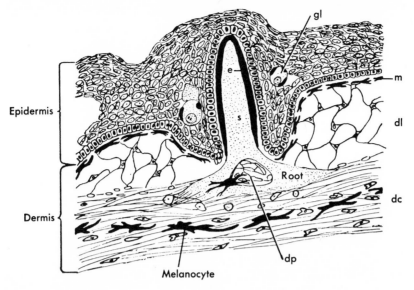

Fig. 5-32. Developing placoid scale in embryonic skin of shark. Dermis consists of compact layer, **dc,** and loose layer, **dl; dp,** dermal pulp within root and spine; **e,** enamel; **gl,** unicellular epidermal gland; **m,** melanocyte; **s,** spine, composed of dentine and enamel.

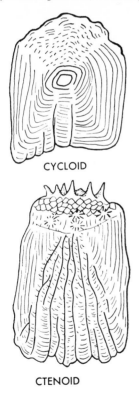

CYCLOID

CTENOID

Fig. 5-33. Modern flexible fish scales. The upper edges are caudal free borders.

reptiles, and a few mammals. Only among birds are there no dermal scales. Their loss was certainly no disadvantage in flight.

Apodans and some tropical toads have dermal scales. In apodans the scales are microscopic between the furrows of the skin (an apodan is illustrated in Fig. 3-14), and macroscopic (1 to 2 mm in diameter, Fig. 5-34, *B*) within the furrows. Crocodilians have large oval osteoderms in the dermis, especially along the back (Fig. 5-35) where they are often associated with epidermal crests that give the animal an awesome appearance. A few lizards have similar but smaller bony scales.

Turtles are truly armored vertebrates! The armor, or shell, consists of large bony plates that meet in immovable sutures (Fig. 5-36). The arched carapace and ventral, flattened plastron are united by bony lateral bridges that must be sawed through to expose the internal organs. Leatherback turtles have leathery shells because the dense collagenous tissue of the dermis does not become ossified.

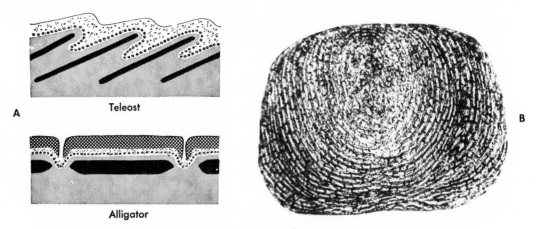

Fig. 5-34. A, Site of dermal scales (black) in a teleost and of osteoderms in an alligator. Gray represents dermis. **B,** A single bony scale from an apodan.

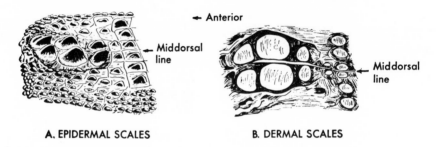

Fig. 5-35. Alligator skin. **A,** from dorsum of neck, showing epidermal scales. **B,** Same section turned upside down to show osteoderms (dermal scales) embedded in skin beneath the tall crests. The middorsal line may be used as a reference point.

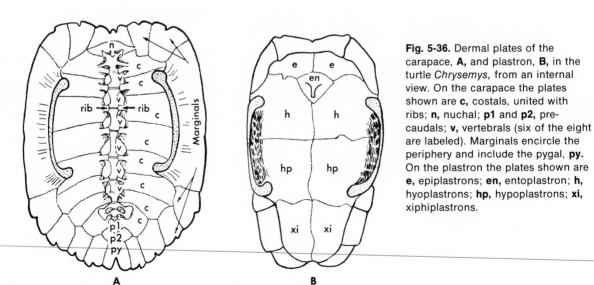

Fig. 5-36. Dermal plates of the carapace, **A,** and plastron, **B,** in the turtle *Chrysemys,* from an internal view. On the carapace the plates shown are **c,** costals, united with ribs; **n,** nuchal; **p1** and **p2,** pre-caudals; **v,** vertebrals (six of the eight are labeled). Marginals encircle the periphery and include the pygal, **py.** On the plastron the plates shown are **e,** epiplastrons; **en,** entoplastron; **h,** hyoplastrons; **hp,** hypoplastrons; **xi,** xiphiplastrons.

Among mammals, only armadillos have dermal armor, and it lies beneath epidermal scales (Fig. 5-15). Dermal scales in mammals and, perhaps, in some other tetrapods may well represent a reacquisition of bone that was once lost, rather than being a vestige of the past. Nevertheless, an earlier statement is worth repeating: The dermis of vertebrates has an ancient and persistent *potential* to form bone.

■ INTEGUMENTARY PIGMENTS

Chromatophores are cells that contain pigment granules. They have permanent processes extending from the cell, and, although located at many deep sites including meninges and even within striated muscles, they are most common in the skin, where they account for skin color. They arise from cells that migrate from neural crests.

Chromatophores are identified by the color of their granules. **Melanophores** contain melanin granules (melanosomes), which are varying shades of brown. **Xanthophores** contain yellow, and **erythrophores** contain red, granules. (The two latter are sometimes called lipophores because the granules are soluble in lipid solvents.) **Iridophores** contain a prismatic substance, guanine, which reflects and disperses light, producing silvery or iridescent skin.

Dermal chromatophores (chromatophores in the dermis) are responsible for rapid color changes (**physiological color changes**), such as are seen in chameleons. Physiological color changes occur only in ectotherms and are induced reflexly by neurotransmitters (p. 355) and by hormones, such as intermedin and melatonin. The color change results from dispersal of granules into the processes of pigment cells, or aggregation of granules to a position close to the nuclei. Dispersal of granules forms a blanket that masks underlying pigments; aggregation exposes them. Since not all varieties of chromatophores respond alike

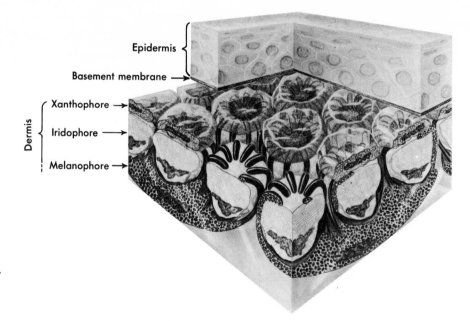

Fig. 5-37. Dermal chromatophore unit responsible for physiological color changes in anurans. Adaptation to a dark background is illustrated, with processes of melanophores overlying iridophores. (Modified from Bagnara, J. T., Taylor, J. D., and Hadley, M. E.: The Journal of Cell Biology **38**:67, 1968.)

to the same stimulus, various color combinations result. A functional association of dermal chromatophores is illustrated in Fig. 5-37.

Birds and mammals have epidermal as well as dermal pigment cells, but their skin cannot change color reflexly because the granules within the cells cannot aggregate and disperse. These animals change color only as pigment granules are synthesized in response to long-term stimuli such as exposure to sunlight ("getting a tan," for example), or when hairs, feathers, or epidermis is shed and replaced with new hairs, feathers, or epidermis with different color combinations or densities. These are **morphological color changes.**

The pigment seen in hair and feathers is not in pigment cells. Electron microscopy has shown that pigment cells in feather and hair follicles actively *inject* pigment into the epidermal cells that are being added to the growing feather or hair. Hairs receive only melanin pigment, and hair color is attributable to the specific distribution and density of the pigment and to the number of air vacuoles in the medulla of the hairs. Gray and white hair are the result of large numbers of air vacuoles and little melanin. Feathers receive brown, yellow, and red pigments.

Despite what appear to be blue feathers, vertebrates have no blue granules. When viewed under a microscope by *transmitted* light, the "blue" feather is seen to be brown, the color of the melanin granules beneath the prismatic layer. The blue color observed in *reflected* light is a dispersion phenomenon, like the blue of the sky. A red feather is red even when viewed in transmitted light because of red pigment. The iridescence of feathers is also a dispersion phenomenon.

■ SOME FUNCTIONS OF SKIN

Dermal armor provides protection from attack and, since it is usually heaviest in the head, it protects the brain and special sense organs from mechanical injury. Even in the absence of armor a leathery dermis provides some protection against penetration of foreign objects. Skin glands secrete obnoxious or poisonous substances that ward off enemies. If the enemy lives through the first encounter, it may not seek another. Some glands keep skin moist and the conjunctiva of the eye free of irritants. Integumentary pigments provide protective coloration and, in naked skin, absorb excess solar radiation. The bristling coat of an angry mammal and the ruffled plumage of a frightened bird make them ominous. Claws, nails, horns, spines, barbs, needles—all confer advantages in the struggle for existence.

Temperature regulation is effected largely by the skin. Fur and feathers insulate against heat and cold, sweat cools through evaporation, and dilation of blood vessels within the dermis increases heat loss by radiation. When heat conservation is necessary, the vessels constrict. Fat deposits insulate deep tissues from icy waters or frigid air, or provide energy stores for hibernation, migration, or periods of seasonal famine.

The skin assists in maintaining homeostasis. Bony scales are reservoirs for calcium and phosphate storage. Chloride-secreting glands and sweat glands excrete salts and water. Heavy layers of stratum corneum conserve water. Aquatic amphibians excrete carbon dioxide through the skin. Mammary glands provide nourishment to newborn mammals. Adhesive pads and claws assist in climbing, and feathers provide an airfoil for aerial locomotion. The distribution of pigment and the pheromonal secretions of scent glands often signal species and sex of the bearer. Nerve endings in skin alert vertebrates against inimical forces. Brood pouches under the skin house developing young. Vitamin D is synthesized in some skin. You will probably be able to think of other roles that the integument performs.

☐ Chapter summary

1. The integument of vertebrates consists of epidermis and dermis. The epidermis is a stratified epithelium and gives rise to skin glands. In terrestrial forms it has a surface layer (stratum corneum) of dead, keratinized (cornified) cells.

2. Major constituents of the dermis are blood vessels, general sense organs, nerves, lymphatics, the bases of multicellular epidermal glands, pigment cells, and an abundance of collagenous connective tissue.

3. The skin of fishes is characterized by bony or fibrous scales in the dermis, many unicellular mucous glands in the epidermis, some multicellular glands, and a poorly differentiated stratum corneum or none.

4. The skin of modern amphibians has many multicellular mucous and granular glands. The stratum corneum is thin or absent in aquatic forms, thick and sometimes warty in terrestrial species. Apodans and a few tropical toads have dermal scales.

5. The skin of reptiles has a thick stratum corneum that gives rise to a variety of cornified appendages, such as epidermal scales, claws, horny protuberances, spikes, and rattles. Integumentary glands are few but diverse. Bony scales (osteoderms) occur in crocodilians and some lizards, and dermal plates form the shell of turtles.

6. The thin skin of birds is characterized by feathers (contour, down, and filoplumes) derived from stratum corneum. Epidermal scales occur chiefly on the legs and head. Contour feathers develop from feather tracts (pterylae). Pinfeathers are developing feathers. There are no dermal scales and few skin glands.

7. The skin of mammals is characterized by hair derived from stratum corneum. Modifications of hair include spines, quills, bristles, and vibrissae. The stratum corneum is thick and exhibits localized epidermal scales. Multicellular epidermal glands are variants of sebaceous and sudoriferous types. Bony dermal plates occur in armadillos.

8. There is a correlation between terrestrial life, few mucous glands, and a thick stratum corneum. The latter conserves water.

9. The stratum corneum of amniotes gives rise to epidermal scales, claws, nails, hoofs, horn, spines, rattles, beak coverings, feathers, hairs, quills, ischial callosities, tori, and baleen. Epidermal scales are especially characteristic of reptiles but are present in restricted sites on birds and mammals.

10. Claws consist of an unguis and subunguis. They are characteristic of reptiles, birds, and mammals. Nails and hoofs are modified claws.

11. True horns are found in bovines and pronghorn antelopes. They consist of a horny epidermal cap that covers a bony core. Only pronghorns are shed. Rhinoceros horns consist of agglutinated hairlike fibers.

12. Skin glands are mostly unicellular and mucus-producing in fishes, multicellular and abundant in amphibians, few in reptiles and birds, and numerous and diversified in mammals. All skin glands are epidermal in origin.

13. Photophores are modified multicellular epidermal glands of fishes. They emit light.

14. The dermis has always been a potential site for deposit of bone. This potential was expressed primitively as dermal armor. Subsequent armor took the form of scales with less bone—cosmoid at first, then ganoid, and finally ctenoid and cycloid. Placoid scales are also derived from cosmoid scales. In tetrapods the bone-forming potential of dermis is expressed as bony scales or plates in the dermis of apodans, tropical toads, crocodilians, lizards, turtles, and armadillos.

15. Cosmoid scales are thick bony scales derived from dermal armor and found in early lobe-finned fishes. They do not occur today.

16. Placoid scales are characteristic of elasmobranchs. They have a basal plate composed of lamellar bone and a spine composed of dentin and enamel. They become teeth on the jaws.

17. Ganoid scales (paleoniscoid and lepidosteoid) composed of bone covered by ganoin were once predominant among ray-finned fishes. They are found today only in Chondrostei and Holostei.

18. Cycloid and ctenoid scales are characteristic of teleosts and modern lungfish. They consist of a very thin layer of acellular bone, underlaid by a flexible fibrous plate of collagen.

19. Living cyclostomes and a few gnathostome fishes have lost all dermal scales.

20. Chromatophores are cells that contain pigment granules. They originate from neural crests. Xanthophores, erythrophores, melanophores, and iridophores are dermal chromatophores. Their pigment granules in ectotherms may be aggregated or dispersed reflexly, resulting in physiological color changes. Melanocytes are melanin-containing pigment cells, the granules of which cannot be aggregated. They occur in the dermis and epidermis of all vertebrate classes.

21. Among functions of the integument are protection, temperature regulation, nutrition for young, maintenance of homeostasis, respiration, receipt of sensory stimuli, production of pheromones, and others.

LITERATURE CITED AND SELECTED READINGS

1. Fox, D. L.: Coloration of mammals. In Gray, P., editor: The encyclopedia of the biological sciences, New York, 1961, Reinhold Publishing Corp.
2. Lillie, F. R.: On the development of feathers, Biological Reviews 17:247, 1942.
3. Ling, J. K.: Adaptive functions of vertebrate molting cycles, American Zoologist 12:77, 1972.
4. Lucas, A. M., and Stettenheim, P. R.: Avian anatomy, Integument, Agriculture Handbook 362, 2 parts, Washington, D.C., undated, U.S. Government Printing Office.
5. Maderson, P. F. A.: When? Why? and How? Some speculations on the evolution of the vertebrate integument, American Zoologist 12:159, 1972.
6. Modell, W.: Horns and antlers, Scientific American 220(4):114, 1969.
7. Montagna, W., and Lobitz, W. C., editors: The epidermis, New York, 1964, Academic Press, Inc.
8. Moss, M. L.: The origin of vertebrate calcified tissues. In Ørvig, T., editor: Current problems in lower vertebrate phylogeny, New York, 1968, Interscience-Wiley.
9. Ørvig, T.: The dermal skeleton: general considerations. In Ørvig, T., editor: Current problems in lower vertebrate phylogeny, New York, 1968, Interscience-Wiley.
10. Parakkal, P. F., and Alexander, N. J.: Keratinization: a survey of vertebrate epithelia, New York, 1972, Academic Press, Inc.
11. Quay, W. B.: Integument and the environment: glandular composition, function, and evolution, American Zoologist 12:95, 1972.
12. Taylor, J. D., and Bagnara, J. T.: Dermal chromatophores, American Zoologist 12:43, 1972.

Symposium in American Zoologist

The vertebrate integument, 12:12, 1972.

6

Mineralized tissues: an introduction to the skeleton

In this chapter we will examine the nature of mineralized connective tissues, meet some common varieties, see how these are formed, and look briefly at their phylogeny. We will discover that the skeleton has a role in homeostasis, and we will find out why it is necessary that a skeleton undergo continual remodeling.

The skeleton of vertebrates is composed of hardened (mineralized) connective tissue. For the most part, this tissue is **bone.** However, three other mineralized tissues are part of a vertebrate skeleton. These are **dentin** (a kind of bone), **cartilage** (often a precursor of bone), and **enamel** or enamel-like (enameloid) substances.

All connective tissues, whether mineralized or not, are produced by highly specialized cells that secrete around themselves extracellular materials that characterize them. **Osteoblasts** produce bone, **chondroblasts** produce cartilage, **odontoblasts** produce dentin, and **ameloblasts** produce enamel. These four "differentiated" (specialized) cell types arise from undifferentiated (unspecialized) connective tissue cells called scleroblasts (Fig. 6-1). Scleroblasts, in turn, arise from even less differentiated embryonic cells, which, given the proper stimulus could equally well give rise to muscle, blood cells, or certain other tissue. These embryonic, highly undifferentiated cells are mesenchymal cells, and an aggregation of them is referred to as **mesenchyme.** The adult body retains islands of mesenchyme that have the potential to repair or replace bone and other connective tissues throughout life.

■ BONE

Bone is composed of a matrix consisting of bundles of **collagen** (a proteinaceous fibril), the spaces between which have been impregnated with **hydroxyapatite crystals** composed of calcium, phosphate, and hydroxyl ions $[3Ca_3(PO_4)_2 \cdot Ca(OH)_2]$. A **cementing substance** composed of water and mucopolysaccharides binds the crystals to the collagen bundles. The inorganic crystals are deposited under the influence of osteoblasts, which ultimately become trapped in tiny pools of fluid called **lacunae** (Fig. 6-2). Thereafter, the osteoblasts are bone cells, or osteocytes. The process of bone forma-

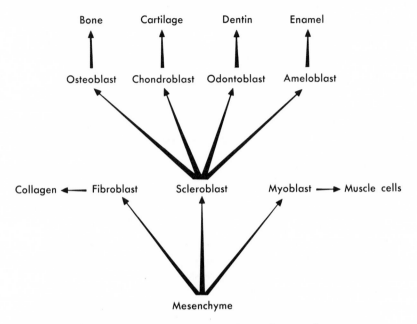

Fig. 6-1. Some differentiated products of mesenchyme.

Fig. 6-2. Section of one osteon (haversian system) in compact bone to show lacunae (black) and canaliculi (canals radiating from the lacunae). (From Bevelander, G., and Ramaley, J. A.: Essentials of histology, ed. 7, St. Louis, 1974, The C. V. Mosby Co.)

tion is known as **osteogenesis.** In addition to osteocytes, lacunae contain dissolved salts. Some of these will be converted into crystals, and some result from the dissolu- tion of crystals, since salt deposit and with- drawal are constantly taking place. Inter- connecting the lacunae are tiny, fluid-filled canals, or **canaliculi.**

In **osteon bone** the collagen bundles are deposited in concentric layers, or lamellae, surrounding an artery and vein. This results in haversian systems, or **osteons** (Fig. 6-2). In **surface bone** the collagen bundles are deposited as flat lamellae instead of concentrically, and there are no haversian systems. Osteon and surface bone are **compact bone,** also called **lamellar bone.** The interior of many bones is spongy because of bony trabeculae that separate spaces filled with bone marrow. This is **spongy, or cancellous, bone.**

In some bones the osteoblasts retreat from the region where they are producing bone, there are no lacunae, and the canaliculi are obliterated. This is **acellular bone,** sometimes called aspidin. The flexible scales of modern fish contain acellular bone.

■ DENTIN

Dentin has the same constituents as bone, and it will be referred to as bone in this chapter. The odontoblasts that deposit dentin are not trapped in lacunae. Instead, they retreat as they deposit it and are located at the inner border of dentin (Fig. 6-3). The canaliculi are delicate canals called **dentinal tubules** that extend from the odontoblasts all the way to the surface of the dentin. The tubules contain processes of the odontoblasts. Dentin forms only in the outer layer of the dermis just beneath the epidermis, and it is frequently coated on its surface by

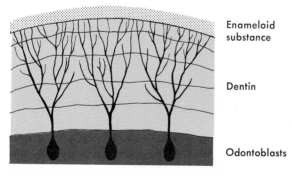

Enameloid substance

Dentin

Odontoblasts

Fig. 6-3. Dentin covered with enameloid. In dentin, canaliculi are called dentinal tubules.

enamel or enameloid substances (Fig. 5-30, ancient armor). Some enamels are produced by ameloblasts evidently derived from ectoderm. Others may possibly be a very hard variety of dentin.

■ PRESKELETAL BLASTEMAS, MEMBRANE BONE, AND REPLACEMENT BONE

A blastema is any aggregation of embryonic mesenchymal cells, which, given the appropriate stimulus, differentiate into some tissue such as muscle, bone, or cartilage. Before bone or cartilage can be deposited, a preskeletal mesenchymal blastema must develop. The mesenchymal cells that contribute to preskeletal blastemas are chiefly of mesodermal origin. However, mesenchyme that gives rise to preskeletal (and premuscular) blastemas in the head and pharyngeal arches arises from neural crests, which are ectoderm. Mesenchyme of neural crest origin is called **ectomesenchyme,** or **mesectoderm.**

Once the blastema of the future bone or cartilage has aggregated, some mesenchymal cells become fibroblasts and secrete collagen. Other mesenchymal cells become osteoblasts or chondroblasts and secrete bone or cartilage. Of course, the capacity to deposit either one depends in part on inheritance of appropriate enzyme systems.

Bone deposited directly in a blastema is **membrane bone.** The process of membrane bone formation is known as **intramembranous ossification.** It gives rise to certain bones of the lower jaw, skull, and pectoral girdle; to dentin and other bone that arises in the skin; to vertebrae in a few vertebrates (teleosts, urodeles, apodans); and to bones in a few other locations. Periosteal bone (bone deposited by the periosteal membrane) is membrane bone.

Some bone is deposited in preexisting cartilage, and this is called **replacement bone.** The cartilage is first removed, then bone is deposited where cartilage previously existed. The process is known as **endochondral ossification.** The processes of endochondral and intramembranous ossification are the same in that they consist of impregnation of a collagenous matrix with hydroxyapatite crystals. However, in endochondral ossification, cartilage must be removed before bone may be deposited.

■ DERMAL BONE

Bone that forms from mesenchyme within the dermis of the skin is called **dermal bone.** It arises by intramembranous ossification and may be dentin, spongy bone, or lamellar bone, depending on the species and location. All three contributed to the dermal armor of ancient fishes (Fig. 5-30). Somewhere in geological time and along phylogenetic pathways leading from armored fishes, some of the collagen that was secreted by fibroblasts in the skin evidently began to be exported to membranes beneath the skin, where it formed a matrix on which osteoblasts deposited membrane bone. Many membrane bones appear to have such a phylogenetic origin and to represent bones that, at one time, formed *in,* rather than *under,* the dermis. They include some of the bones alongside of and above the brain, some bones of the lower jaw, and certain bones of the pectoral girdles. Only those membrane bones that arise phylogenetically or ontogenetically from skin should be called dermal bones.

■ CARTILAGE

Cartilage resembles bone in that the cells (chondrocytes) lie in pools of fluid surrounded by a collagenous interstitial matrix. The matrix of cartilage, however, contains a sulfated mucopolysaccharide. Unlike bone, cartilage has no canaliculi demonstrable by light microscopy, and no blood vessels penetrate it except those en route to other organs. Therefore the cells are supplied by diffusion.

Cartilage is formed within a prechondral mesenchymal blastema by the deposit of chondroitin sulfate. The process is known as **chondrogenesis.** Once cartilage has formed, it may remain throughout life, or it may be resorbed and replaced by bone. The latter is more likely, since cartilage in vertebrates, especially hyaline cartilage, appears to be primarily an embryonic or juvenile tissue. As long as cartilage is present, endochondral bone formation may continue and the skeleton is resilient and resistant to fracture.

Hyaline cartilage is a translucent cartilage found in many locations. In vertebrate embryos it is abundant, constituting a temporary skeleton to be replaced later by bone. Cartilage with thick, dense collagenous bundles in the interstitial matrix is **fibrocartilage.** The intervertebral discs of mammals are fibrocartilage. **Elastic cartilage** contains elastic fibers. In mammals it occurs in the pinna of the ear, in the walls of the outer ear canal, in the epiglottis, and elsewhere. Cartilage may have calcium salts deposited within the interstitial substance. This **calcified cartilage** is often mistaken for bone. The jaws of sharks contain much calcified cartilage.

Cartilaginous fishes appear to have lost the genetic code necessary for ossification except for formation of dentin in scales and teeth.

■ CALCIUM REGULATION AND SKELETAL REMODELING

Bone and cartilage not only serve as mechanical supports for the vertebrate body but, with scales and teeth, constitute important storage places for calcium and other mineral salts. They therefore participate in maintaining homeostasis. Calcium is constantly being deposited and withdrawn in response to rising or falling levels of calcium in the blood serum. When serum calcium levels are rising, calcium is deposited in bone or cartilage, along with

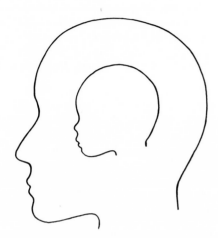

Fig. 6-4. Comparative size of skull of newborn and 21-year-old human being.

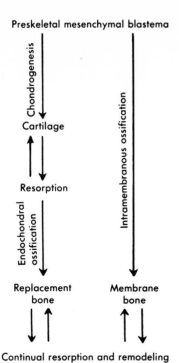

Fig. 6-5. Steps in osteogenesis and remodeling.

inorganic phosphate, as hydroxyapatite crystals. When serum calcium levels are falling, calcium is withdrawn from the skeleton and other depots by dissolution of the crystals.

In addition to bone resorption in response to lowered blood calcium levels, cartilage and bone are constantly being resorbed for another reason. Consider the skull of a newborn human being. Can one envision any way this skull can become that of a 21-year-old human being (Fig. 6-4) unless cartilage and bone are being constantly resorbed on the inner surface and added to on the outer surface day by day as the individual grows? A 21-year-old human brain would never fit into the cranial cavity of a newborn baby! Skeletal remodeling is a necessary and continuous process. Although it is a more active process in growing skeletons, it is characteristic of all (Fig. 6-5).

Parathyroid hormone withdraws calcium from bone and its other storage places. Calcitonin, a hormone of the ultimobranchial, thyroid, and parathyroid glands protects the calcium of bone, preventing bone from being excessively resorbed. The precise relationship of these hormones to one another in the remodeling process, especially in growing skeletons and in lower vertebrates, remains to be discovered.

■ TENDONS, LIGAMENTS, AND JOINTS

Tendons and ligaments are made of thick, closely packed bundles of collagen between which are connective tissue cells called fibroblasts. **Tendons** connect muscles with bone; when a muscle contracts, the pull on the bone is exerted through the tendon. In accordance with this mechanical requirement the collagen is in parallel bundles and the fibroblasts lie in rows between them. **Ligaments** connect bone to bone, and the collagen bundles have a less regular arrangement. Ligaments and tendons that become flat and very wide are sometimes called **aponeuroses.** The term ligament is sometimes applied to fibrous connective tissue membranes or cords that hold visceral organs in place as, for example, the falciform ligament and round ligament of the ovary. However, these are not skeletal structures.

In some species tendons and ligaments normally become mineralized in one or more locations. Turkeys, for example, have ossified tendons in their legs. Ornithischian dinosaurs had ossified tendons millions of years ago. **Sesamoid bones** or cartilages (so named because they resemble sesamoid seeds) are nodules of bone or cartilage that form in tendons or ligaments. Best known is the patella, or kneecap, which is endochondral in some species, intramembranous in others.

A **joint** is the site where two bones (or cartilages) meet. If the joint is movable in one or more planes, it is a **diarthrosis.** The articular surfaces of the bones in a movable joint are usually covered with a layer of hyaline cartilage, which is readily replaceable with wear. Ligamentous bands hold the bones together and form a fluid-filled sac, or **bursa,** around the joint. The lubricatory **synovial fluid** is secreted by a synovial membrane that lines the bursa.

In some joints movement is not possible, but you can see the suture between the bones. This joint is a **synarthrosis.** The suture between the frontal and parietal bones (Fig. 8-38) is an example of a synarthrosis. Sometimes bones meet and become united by collagen and hydroxyapatite crystals that obliterate the suture. This condition is said to be an **ankylosis.** The premaxillary and maxillary bones of the human embryo ankylose and, as a result, the premaxillary cannot be distinguished as a separate bone in adults.

An **amphiarthrosis (symphysis)** is a joint in the midline, in which two bones are separated by fibrocartilage and in which movement is severely limited. Such joints are found in the vertebral column and at the symphysis pubis. The latter joint becomes a bit more movable by hormonal dissolution of the fibrocartilage shortly before labor begins in female mammals.

■ HETEROTOPIC ELEMENTS

In addition to the usual cartilages and bones comprising the axial and appendicular skeletons, miscellaneous heterotopic elements develop. A heterotopic bone, as the term is employed here, refers to a bone that develops in an aberrant location in one vertebrate or another, either by endochondral or intramembranous ossification, and frequently without a phylogenetic precursor. Heterotopic bones, and also cartilages, are especially common in mammals. They are usually missing from routine skeletal preparations. Among heterotopic bones are the **os cordis** in the interventricular septum of the heart of ungulates, and the **baculum (os priapi** or **os penis)** embedded between the spongy bodies in the penis of bats, rodents, marsupials, carnivores, insectivores, bovines, and lower primates (Fig. 6-6). The baculum reaches a length of nearly 60 cm in walruses. An **os clitoridis** is embedded in the female penis (clitoris) in otters, several rodents, rabbits, and numerous other female mammals.

Osseous or cartilaginous tissue forms in the wall of the gizzard in some doves. The syrinx of birds often develops an internal skeletal element, the **pessulus.** At least one species of bat has bone in the tongue. Bone develops in the gular pouch of a South American lizard, in the muscular diaphragm of camels, and in the upper eyelid of crocodilians (**adlacrimal,** or **palpebral, bone**). A similar plate of connective tissue, the **tarsus,** develops in man. A **rostral bone** develops in the snout of a number of mammals including pigs, and a **cloacal bone** in the ventral wall of the cloaca of some lizards. **Epipubic (marsupial)** bones are embedded in the ventral body wall of some monotremes and marsupials.

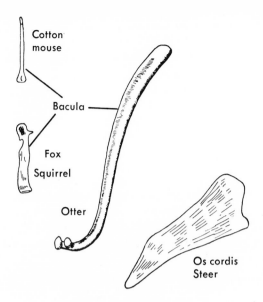

Fig. 6-6. Heterotopic bones.

tebrates are more often calcium carbonate than calcium phosphate. Cartilage is found among invertebrates, including squids, some gastropods, and protochordates. The earliest vertebrates, therefore, had access to genetic information for skeletal building, and bone, dentin, cartilage, and enameloid were all well developed in Ordovician ostracoderms. For that reason, we cannot say that any one of these is phylogenetically older. The chief contribution of vertebrates seems to be the capacity of ectoderm and mesoderm to *interact* during ontogeny to produce a wide variety of ectodermal and mesodermal mineralized tissues in a variety of locations in a single individual. This tendency is more pronounced in fishes.

■ REGIONAL COMPONENTS OF THE SKELETON

The skeleton may be divided regionally into the following parts:

Axial skeleton
    Notochord and vertebral column
    Ribs and sternum
    Skull and visceral skeleton
Appendicular skeleton
    Pectoral and pelvic girdles
    Skeleton of paired fins or limbs
    Skeleton of median fins of fishes
Heterotopic elements

■ MINERALIZED TISSUES AND THE INVERTEBRATES

Mineralized tissues are not unique among vertebrates. In fact, two-thirds of the living species of animals that contain mineralized tissues are invertebrates.[4] The matrix is collagen and it goes back as far as the sponges; but the inorganic crystals in inver-

☐ Chapter summary

1. The chief mineralized tissues are bone (from osteoblasts), dentin (from odontoblasts), cartilage (from chondroblasts), and enamel (from ameloblasts). These cell types arise from mesenchyme of mesodermal or neural crest origin.
2. Bone consists of collagen, inorganic salts, cementing substance, and osteocytes that usually occupy canaliculi interconnected by lacunae.
3. Bone is lamellar or spongy (compact or cancellous). Lamellar bone is osteon bone or surface bone. Acellular bone lacks canaliculi and lacunae.
4. Membrane bone forms directly in mesenchyme. Endochondral bone forms in pre-existing cartilage.
5. Dermal bone arise ontogenetically or phylogenetically from skin. Dermal bones are membrane bones.
6. Dentin is a form of bone with peripheral odontoblasts, long canaliculi, and no lacunae. It is confined to the skin and teeth.
7. Cartilage consists of a matrix of collagen and chondroitin sulfate, and of chondrocytes. It may be hyaline, fibrocartilage, or elastic and is sometimes calcified.
8. Bone and cartilage are constantly being deposited and resorbed, thereby maintaining homeostasis and orderly growth.
9. Heterotopic elements are miscellaneous bones or cartilages of membranous origin that usually lack phylogenetic precursors.
10. Tendons attach muscles to bone. Ligaments connect bone to bone. Both may be calcified.
11. Joints are diarthroses, synarthroses, or am- phiarthroses (symphyses). Bursas are ligamentous and contain synovial fluid if the joint is movable.
12. Many invertebrates have mineralized tissues consisting of collagen, inorganic salts, and cementing substances. Among vertebrates cartilage, bone, dentin, and enamel seem equally old.
13. The skeleton may be subdivided into axial and appendicular parts.

LITERATURE CITED AND SELECTED READINGS

1. Biltz, R. M., and Pellagrino, E. D.: The chemical anatomy of bone. 1. A comparative study of bone composition in sixteen vertebrates, Journal of Bone and Joint Surgery **51A:**456, 1969.
2. Burt, W. H.: Bacula of North American mammals, University of Michigan Museum of Zoology Miscellaneous Publications **113:**1-76, 1970.
3. Friant, M.: Vue d'ensemble sur l'evolution du "cartilage de Meckel" de quelques groupes de Mammifères, Acta Zoologica **47:**67, 1966.
4. Hall, B. K.: Evolutionary consequences of skeletal differentiation, American Zoologist **15:**329, 1975.
5. Halstead, L. B.: Calcified tissue in the earliest vertebrates, Calcified Tissue Research **3:**107, 1969.
6. Hancox, N. M.: Biology of bone, Cambridge, 1972, Cambridge University Press.
7. Jollie, M. C.: Some developmental aspects of the head skeleton of the 35-37 mm *Squalus acanthias* foetus, Journal of Morphology **133:**17, 1971.
8. Moss, M. L.: The origin of vertebrate calcified tissues. In Ørvig, T., editor: Current problems of lower vertebrate phylogeny, New York, 1968, John Wiley & Sons, Inc.
9. Moss, M. L.: Skeletal tissues in sharks, American Zoologist **17:**335, 1977.
10. Ørvig, T.: The dermal skeleton: general considerations. In Ørvig, T., editor: Current problems of lower vertebrate phylogeny, New York, 1968, Interscience-Wiley.

7

Vertebrae, ribs, and sternum

In this chapter we will look at the structure of representative vertebral columns
from fishes to man, consider some of the morphological differences in terms
of their adaptive value, and then look briefly at vertebrate ribs and sterna.

The vertebral column and ribs, along with
the sternum of tetrapods, constitute the
major axial skeleton of vertebrates behind
the skull. The vertebral column replaces the
embryonic notochord in position, and ribs
and sternum form in the lateral and ventral
body wall.

■ VERTEBRAL COLUMN

The vertebral column is a series of cartilaginous or bony vertebrae that extends
from the skull to the tip of the tail. More
than one morphological variety of vertebrae
is found in every column. In fishes, trunk
vertebrae differ from those of the tail, and
in tetrapods, trunk vertebrae are further
modified when associated with ribs (**thoracic vertebrae**), with the pelvic girdle (**sacral
vertebrae**), or in the neck (**cervical vertebrae**). Snakes have the longest vertebral
columns, with up to 500 vertebrae. Caecilians have up to 250, and some urodeles
have as many as 100.

□ Basic structure of vertebrae

Most vertebrae have a centrum, or body,
one or two arches, and certain processes
(Fig. 7-1). **Centra** lie immediately ventral to
the neural tube. The most primitive centra
are concave at both ends (amphicelous), and
the space between them contains notochordal tissue (Figs. 7-1, *A* and *D*, and 7-2,
A). Most fishes have this type. So do less
specialized urodeles such as *Necturus*, as
well as caecilians and primitive lizards. The
notochord in these animals usually extends
the length of the vertebral column, although
it is constricted within centra. More specialized centra are concave at one end or
the other, or flat on both ends (Fig. 7-2, *B* to
D), and in the long, flexible neck of birds
they are saddle-shaped at the ends (heterocelous). In such species the notochord may
be completely gone in adults, or a vestige
may remain in the intercentra or intervertebral discs between centra (Fig. 7-2, *D*). In
crocodilians and man this remnant is the
pulpy nucleus.

Perched on a centrum is a **neural arch**
(Fig. 7-1). Successive neural arches and
their interconnecting ligaments enclose a
long **vertebral canal** that houses the spinal
cord. Inverted beneath centra in the tail
may be a **hemal arch,** or chevron bone, as
it is called in amniotes (Figs. 7-1, *C* and *D*,
and 7-3). Fishes, urodeles, most reptiles,
some birds, and many long-tailed mammals,
including cats, have hemal arches or chev-

127

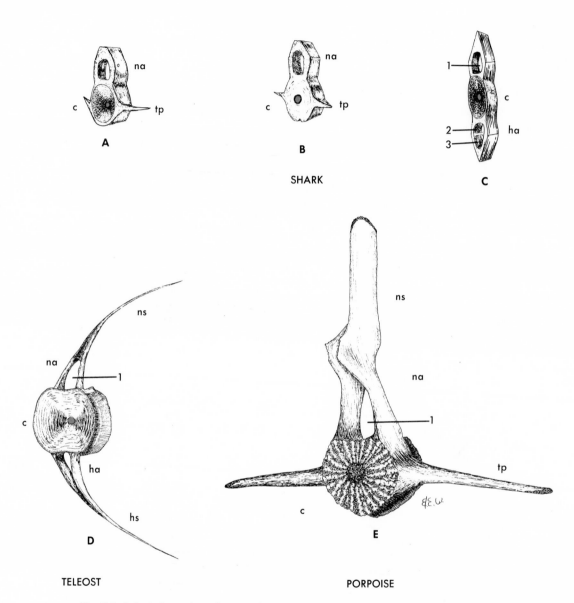

SHARK

TELEOST   PORPOISE

Fig. 7-1. Selected vertebrae from cephalic view. **A,** Trunk vertebra of shark.
B, Cross section of trunk vertebra near middle of centrum. **C,** Tail vertebra of shark.
D, Tail vertebra of teleost. **E,** Lumbar vertebra of porpoise. The fish vertebrae are
amphicelous, the porpoise vertebra is acelous. **c,** Centrum; **ha,** hemal arch;
hs, hemal spine; **na,** neural arch; **ns,** neural spine; **tp,** transverse process.
1, Vertebral canal housing spinal cord; **2** and **3,** canals for caudal artery and
caudal vein, respectively. *Red,* site of notochord.

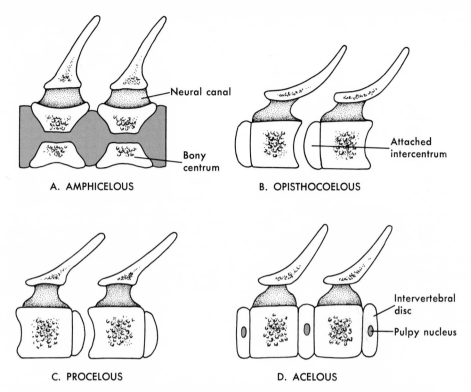

Fig. 7-2. Vertebral types based on shape of articular surfaces of the centra. Midsagittal sections, cephalic end to left. Color indicates notochordal tissue. Amphicelous vertebrae are found in fishes, primitive urodeles, caecilians, and lizards. Opisthocoelous vertebrae are found in salamanders. Procelous vertebrae are found in anurans and modern reptiles. Acelous vertebrae are found in mammals.

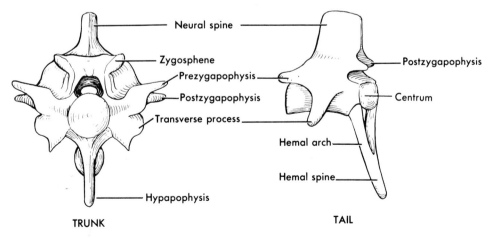

Fig. 7-3. Python vertebrae, cephalic and lateral views. The hemal arch in amniotes is also called a chevron bone.

ron bones in the tail. Chevron bones are usually lost in the preparation of mammalian skeletons.

Vertebral processes are projections from arches and centra. Some provide for rigidity of the column, some prevent excessive torsion, some articulate with ribs, and some serve to attach muscles. **Transverse processes** are the most common. They extend laterad from the base of the neural arch or from the centrum and separate the epaxial and hypaxial muscles (Fig. 1-2). Among the many processes of tetrapods are **diapophyses** and **parapophyses,** which articulate with the heads of ribs (Fig. 7-18), and **prezygapophyses** and **postzygapophyses** that articulate with one another and limit the flexion and torsion of the column (Fig. 7-3).

A typical vertebra arises from mesenchymal cells that stream out of mesodermal somites, surround the notochord and neural tube, and produce the blastema for a future vertebra (Fig. 4-8). Next, some of the blastemal cells (chondroblasts) deposit a cartilaginous centrum and neural arch within the blastema. This results in a cartilaginous vertebra. Later the cartilage is removed and osteoblasts deposit bone where cartilage previously existed. By this process a

vertebra is formed consisting of replacement bone. Teleosts, apodans, and urodeles have a modification of this process. In these vertebrates, bone is deposited directly in the mesenchymal blastema without an intermediate cartilaginous stage, and the vertebrae are membrane bone instead of replacement bone. Cartilaginous fishes, too, show a modification. Cartilaginous vertebrae are deposited and never replaced by bone.

In some species, in the formation of centra, chondroblasts not only deposit cartilage *around* the notochord (perichordal cartilage) but they penetrate the notochord sheath and deposit cartilage *within* the sheath and *within* the notochord. This results in **chordal cartilage** (Fig. 7-4). The centra of many lower vertebrates, including adult urodeles and apodans, contain much chordal cartilage that never ossifies, as well as perichordal membrane bone. Anuran centra, on the other hand, contain replacement bone around, as well as within, the chord. Variations such as these may be of significance in the study of phylogenetic relationships of anurans on the one hand and urodeles and apodans on the other.

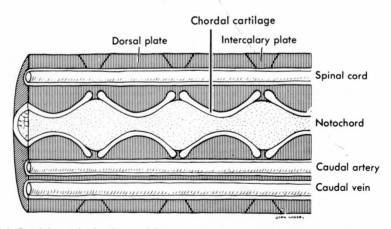

Fig. 7-4. Caudal vertebral column of *Squalus,* sagittal section. The notochord is constricted within each centrum. Chordal cartilage is the notochord sheath impregnated by calcified cartilage. The perichordal cartilage is hyaline cartilage (cross-hatched).

Vertebral columns of fishes

The columns of cartilaginous fishes are not typical fish columns. Neural arches consist of cartilaginous dorsal plates between which intercalary plates complete the enclosure of the vertebral canal (Fig. 7-4). Hemal arches have similar components. In some holocephalans chordal cartilage converts the notochord into a column consisting of calcified cartilaginous rings that are much more numerous than the body segments (Fig. 7-5), and the notochord does not show constrictions and expansions.

Teleosts have well-ossified amphicelous vertebrae (Fig. 7-1, *D*). The notochord persists within each centrum, though usually much constricted. It is prominent between centra, and the notochord sheath forms strong intervertebral ligaments. A neural

arch is associated with each centrum except at the end of the tail, and hemal arches develop in the tail. The neural spines are often greatly elongated, and successive ones may unite in the posterior region of the trunk and in the tail to form a delicate bony rod paralleling the vertebral column. In lower teleosts the neural spines may be surmounted by pointed **supraneural bones.** A variety of processes protrude from the centra and arches and articulate with similar processes on adjacent vertebrae. For the most part, the processes are not comparable with those of tetrapods.

Chondrosteans and modern lungfishes have incomplete centra (Fig. 7-6). The notochord is unconstricted the length of the body, and cartilage deposited within the notochord sheath provides rigidity. Lying against the notochord in each body segment are paired cartilages (basidorsal, interdorsal, basiventral, interventral). It is not known whether this is an arrested embryonic state or an extreme specialization. In another ganoid fish, *Amia*, a centrum and intercentrum surround the notochord in each body segment (Fig. 7-7). Only the centrum bears neural and hemal arches. Some fishes, including sharks, normally have two centra and two sets of arches in each body segment, especially in the tail. This condition is known as *diplospondyly*.

Agnathans have a strange vertebral column if, indeed, they may be said to have

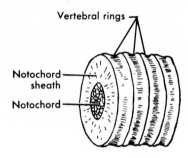

Fig. 7-5. Calcified cartilaginous notochordal rings of some holocephalans. The rings are more numerous than the somites. The neural and hemal arches are not illustrated.

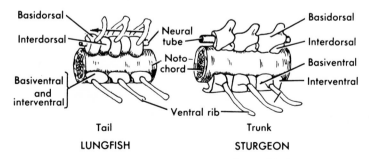

Fig. 7-6. Vertebral components in adult lungfish *(Neoceratodus)* and sturgeon. Arrow in tail of lungfish indicates canal occupied by a longitudinal ligament.

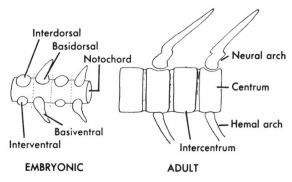

Fig. 7-7. Vertebrae in tail of *Amia*. A centrum and intercentrum develop in each body segment. Basidorsal and basiventral cartilages contribute to each centrum, and interdorsal and interventral cartilages contribute to each intercentrum.

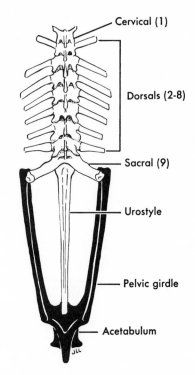

Fig. 7-8. Vertebral column and pelvic girdle (black) of an anuran. The transverse processes include short ribs. The pelvic girdle is braced against the sacral vertebra.

one. The only skeletal elements associated with the notochord are paired lateral neural cartilages (Fig. 1-5). Caudally, the lateral neural cartilages of lampreys fuse to form a single cartilaginous plate perforated by foramina for spinal nerves. Some lampreys have two pairs of cartilages in each body segment, and in hagfishes lateral neural cartilages are limited to the tail. These cartilages may be vestigial vertebrae, primitive vertebrae, or may bear no phylogenetic relationship to vertebrae. Whether one calls them vertebrae or not depends on one's preferred definition of a vertebra. What is yours?

□ **Vertebral columns of tetrapods**

The advent of terrestrial life was accompanied by adaptive changes in the vertebral column. Tetrapod limbs push against the earth, and the pressure is transmitted via an expanded pelvic girdle to the caudalmost trunk vertebrae. These became modified and are called **sacral** (Fig. 7-8). Land vertebrates also developed an increasingly flexible neck. This was accomplished in part by greatly reducing the length of the ribs associated with the anteriormost trunk vertebrae. These are called **cervical.** The remaining trunk vertebrae are **dorsals** (Fig. 7-8).

Amphibians were the first tetrapods to exhibit these modifications, and their vertebral column consists of cervicals (1), dorsals (variable number), sacral (1), and caudals (variable number), the latter being absent in anurans.

CERVICAL REGION

Amphibians have a single cervical vertebra and, therefore, little independent movement of the head. The number became higher in reptiles and still higher in birds. The flexibility of the head as an independent movable part increased proportionately. Mammals have about as many cervical vertebrae as do reptiles (Table 7-1). The cervical region in birds is unusually flexible. Many birds are capable of turning their heads almost backward, of lowering them

Table 7-1. Number of vertebrae in selected tetrapods

	Cervical	Thoracic	Dorsals	Lumbar	Sacral	Caudal
Anura	1		7		1	Urostyle
Salamander	1		10		1	24
Lizard (*Lacerta*)	8		22		2	Numerous
Painted turtle	8		10		2	25 to 30
Alligator	9*	10*		5 to 6	2	34 to 40
Pigeon	12 to 14	5		6	2	15
Mammals	6 to 9	9 to 25		5 to 8	2 to 5†	3 to 50
Horse	7	18 to 20		6	5	15 to 21
Opposum	7	13		6	2	19 to 35
Hamster	7	13		6	4	13 to 14
Sheep	7	13		6 to 7	4	16 to 18
Cat	7	13		7	3	18 to 25
Dog	7	12 to 13		7	3	19 to 23
Rabbit	7	12		7	4	16+
Man	7	12		5	5	3 to 5
Bat	7	11		5	5	9
Sperm whale	7	11		8	0	24

*Or 8 C and 11 T, depending on the definition of thoracic.
†Except cetaceans with none.

considerably below the level of their feet, and of bending their necks into sinuous curves. The specialization is especially useful in feeding and is made possible by heterocelous vertebrae, the caudal ends of which are saddle shaped, with a convexity in the right-left axis and a concavity in the dorsoventral axis. The cephalic end of the next centrum is shaped to accommodate this configuration. The number of cervical vertebrae varies in birds; the long neck of the swan has twenty-five.

In reptiles, birds, and mammals the first two cervical vertebrae are modified to permit movements of the head in several directions. The first one (**atlas**) is ringlike (Fig. 7-9), since most of its centrum has been detached. The atlas articulates with the skull in a condyloid joint that provides a cradle in which the skull may rock, as in nodding "yes." The detached centrum of the atlas is attached to the second vertebra (**axis**) as an **odontoid process** that projects forward to rest on the floor of the atlas (Fig.

7-10). In mammals the skull and atlas pivot as a unit on the odontoid process, as in shaking the head "no." In some reptiles and in an occasional mammal, an additional neural arch, the **proatlas,** is interposed between atlas and skull (Fig. 7-11). It is membrane bone and fills a dorsal gap between the skull and atlas. In the lizard *Lacerta* the gap remains membranous.

Mammals almost always have seven cervical vertebrae (Table 7-1). This is as true in the stubby, rigid neck of whales as in the neck of the tallest giraffe. The only exceptions are edentates with six, eight, or nine and manatees with six. In moles several cervical vertebrae ankylose, perhaps strengthening the neck for burrowing. In cetaceans and armadillos there is no external evidence of a neck, and all cervical vertebrae are shortened and more or less fused together.

Cervical vertebrae in birds and mammals have a **transverse foramen** (Fig. 7-9). Successive foramina provide a **vertebrarterial**

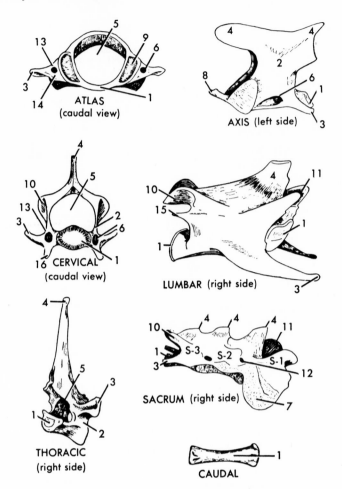

Fig. 7-9. Cat vertebrae. **1,** Centrum; **2,** pedicle; **3,** transverse process; **4,** neural spine; **5,** vertebral canal; **6,** transverse foramen; **7,** site of articulation with ilium; **8,** odontoid process; **9,** articular facet for axis; **10,** postzygapophysis; **11,** prezygapophysis; **12,** intervertebral foramen; **13,** diapophysis; **14,** parapophysis; **15,** accessory process; **16,** vestige of a cervical rib with two heads, one fused with a diapophysis, the other with a parapophysis to form a transverse foramen; **S-1, S-2, S-3,** sacral vertebrae.

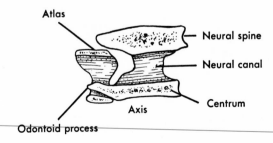

Fig. 7-10. First two cervical vertebrae of a cat, sagittal sections. Much of the centrum of the atlas has become attached to the axis as an odontoid process.

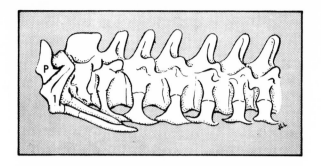

Fig. 7-11. The eight cervical vertebrae, eight cervical ribs, and proatlas, **P,** of an alligator, left lateral view. **1,** Atlas and attached first rib. Immediately behind the first rib is the rib of the axis. The ribs are fused to transverse processes.

canal that houses the vertebral artery and vein.

There is survival value in having a neck under some conditions and not in others. To explore this idea, one might ask what disadvantage a neck could be to some animal —a long-legged, fisheating wading bird, for example—if mutations were to shorten its neck? On the other hand, what disadvantage might it be to a whale to *have* a neck? Whales have seven cervical vertebrae but no external neck. Plesiosaurs lived in the oceans and had a long neck. A neck may be an aid or an impediment in the struggle for existence, depending on the habitat and habits of the animal. We must attribute the difference between whales and plesiosaurs to natural selection.

DORSAL REGION

The vertebrae between cervicals and sacrals are called dorsals when they all bear similar ribs, which may be long or short. These vertebrae are pretty much alike. However, the ribs of crocodilians, lizards, birds, and mammals are confined to the anterior region of the trunk. The vertebrae that bear the long ribs in these instances are called **thoracic,** and the rest are lumbars (Fig. 7-12). The latter bear greatly reduced ribs or none.

SACRUM AND SYNSACRUM

Sacral vertebrae bear short, stout transverse processes that brace the pelvic girdle and hind limbs against the vertebral column (Fig. 7-8). Amphibians have only one sacral

vertebra, and living reptiles and most birds have two. Most mammals have three to five. When there are more than one the sacral vertebrae usually ankylose to form a single bony complex, the **sacrum** (Fig. 7-13). No sacrum forms in whales, a condition correlated with the absence of hind limbs and pelvic girdle.

In birds the last thoracic vertebrae, all lumbars, the two sacrals (three in the ostrich), and the first few caudals unite to form one adult bone, the synsacrum (Figs. 7-14 and 7-15). The latter, in turn, becomes more or less fused with the pelvic girdle. The transverse processes of the vertebrae of the synsacrum may be seen from a ventral view (Fig. 7-15, *B*). The specific number of vertebrae incorporated in the synsacrum varies; thirteen to eighteen or more is common. The synsacrum provides a rigid framework for the two-legged stance of birds. The thoracic vertebrae anterior to the synsacrum also unite more or less completely so there is little flexibility in the avian column behind the neck.

The sacrals in armadillos fuse with a variable number of caudal vertebrae to form a synsacrum of up to thirteen vertebrae. Extensive fusion also occurs among cervical vertebrae in these armored mammals. Fusion involves neural and hemal arches as well as centra.

CAUDAL REGION

Primitively, caudal vertebrae in tetrapods may have numbered fifty or more, but in modern tetrapods the number is much re-

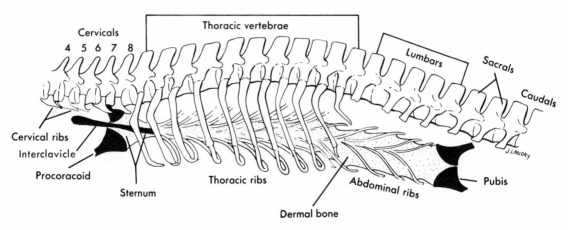

Fig. 7-12. Vertebrae and ribs of an alligator.

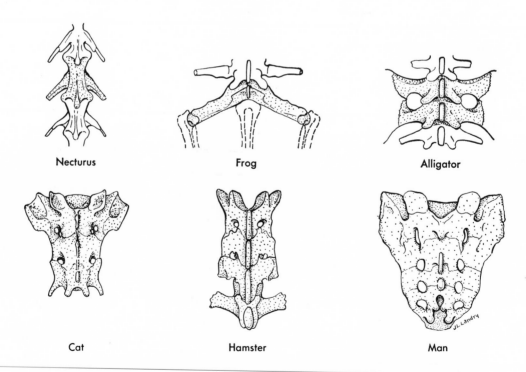

Fig. 7-13. Sacral vertebrae (stippled) of selected vertebrates, dorsal views. They have ankylosed to form a sacrum in the amniotes illustrated.

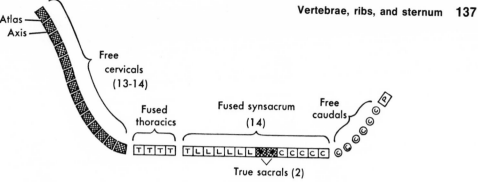

Fig. 7-14. Vertebral column of pigeon, diagrammatical. **T,** Thoracic; **L,** lumbar; **C,** caudal; **P,** pygostyle, composed of four fused vertebrae.

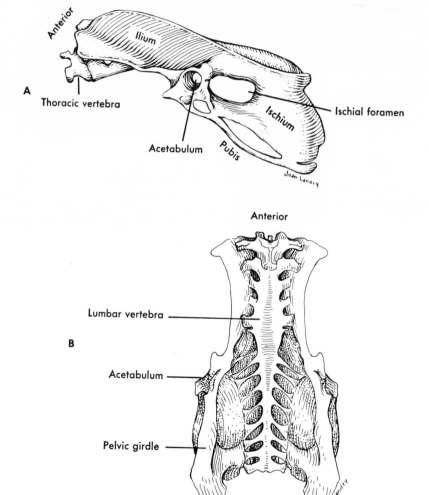

Fig. 7-15. Synsacrum and pelvic girdle of the guinea hen. **A,** Left lateral view. **B,** Ventral view.

Fig. 7-16. Complete set of tail vertebrae from a hamster, left lateral view.

duced. Toward the end of the tail all arches and processes become progressively shorter until finally the last caudals consist of small cylindrical centra only (Fig. 7-16).

Anurans have a unique terminal segment of the column, the urostyle (Fig. 7-8). In some species the cranial end of the urostyle exhibits one or more coalesced centra, vestigial arches, transverse processes, and nerve foramina. The urostyle develops from a continuous perichordal cartilage that surrounds the larval notochord in the tail. It is apparently an unsegmented vertebral column homologous with the separate caudal vertebrae of earlier anurans.

Birds and mammals still have remnants of an ancestral reptilian tail, and some birds retain more tail vertebrae than do some mammals. The number of postsacral vertebrae in a pigeon is fifteen. The half dozen or so immediately behind the synsacrum are free; the remaining four or five are fused together to form a **pygostyle**, the skeleton within the stumpy tail (uropygium). The pygostyle develops as independent cartilaginous centra. The three to five caudal vertebrae in apes and man are called **coccygeal**, since two or more usually ankylose during life to form a **coccyx**.

Many lizards, when captured by the tail, break it proximal to the point of capture and scurry away. Such **autotomy** is implemented by a zone of soft tissue that divides each caudal centrum into cephalic and posterior halves, the location being at the level of a myoseptum. At this location the break occurs and the tail begins to regenerate.

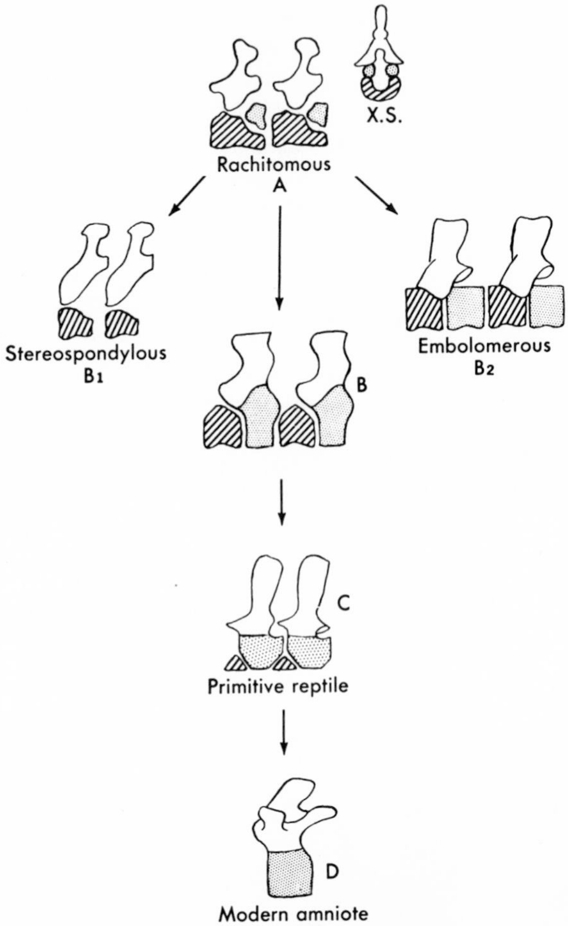

Fig. 7-17. Modifications of tetrapod vertebrae leading to modern amniotes. The rachitomous type (shown also in cross section, X.S.) occurred in crossopterygians and in the earliest amphibians. **B** is from a labyrinthodont thought to be in the reptile line. **B₁** and **B₂** are from other labyrinthodonts. *Diagonal lines,* hypocentrum; *stippled,* pleurocentrum. An interpretation of the phylogeny of amphibian vertebrae will be found in Schmalhausen.[7]

□ Evolution of vertebrae

The vertebral column of the earliest tetrapods did not consist of one vertebra per body segment, as in most tetrapods today. The "vertebra" of crossopterygians and of the earliest amphibians (Fig. 7-17, *A*) consisted of a **hypocentrum (intercentrum),** a large anterior, median, wedge-shaped element that was incomplete dorsally, and two **pleurocentra** (smaller, intersegmental, posterodorsal elements). A vertebra of this kind is **rachitomous.** All later tetrapod vertebrae are probably modifications of the rachitomous type. Successive changes leading to modern amniotes appear to have been characterized by progressive increase in size of the pleurocentrum. The rachitomous vertebral type was also modified in other directions (Fig. 7-17, B_1 and B_2). At present authorities are not sure whether the centrum of modern amphibians represents a hypocentrum or pleurocentrum.

Even today in modern tetrapods each vertebra commences development at several loci surrounding the spinal cord and notochord. As the separate anlagen enlarge, they may remain independent or unite. The typical adult vertebra is, therefore, a **composite** structure. The findings of embryology, comparative anatomy, and paleontology all point to the same conclusion: An adult vertebra composed of a single centrum and neural arch for each body segment is a specialized condition. Primitively, several skeletal elements were associated with each body segment.

■ RIBS

Ribs are long or short, cartilaginous or bony myosepta articulating medially with vertebrae and extending into the body wall. *Polypterus* and a few teleosts have two pairs of ribs for each centrum of the trunk (Fig. 7-18, *A*). A **dorsal rib** passes laterad into a horizontal septum between epaxial and hypaxial muscles, and a **ventral rib** arches ventrad in the body wall just external to the parietal peritoneum. Dorsal and ventral ribs may be primitive. Most teleosts have ventral ribs only, and these are characteristic "fish" ribs. Sharks and some other fishes develop dorsal ribs only. Agnathans have no ribs. This may be correlated with absence of centra. In the tail the paired ventral ribs frequently meet underneath the centrum to form hemal arches (Fig. 7-1, *D*).

Tetrapod ribs usually articulate with vertebrae in movable joints. They are cartilaginous or bony and are deposited in myosepta. Snakes have an extended series commencing immediately behind the head and extending into the tail. The long cervical and anterior thoracic ribs of the cobra (Fig. 7-19), when rotated outward, cause the neck to "spread" characteristically.

In early tetrapods, ribs articulated with every vertebra from the atlas to the end of the trunk (Fig. 7-20). But as tetrapods developed better limbs for locomotion on land, long ribs became confined to the cephalic

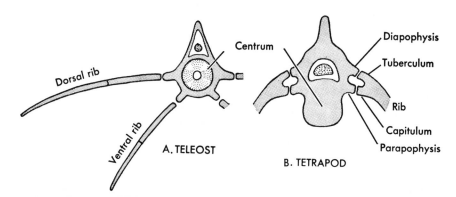

Fig. 7-18. Relationship of teleost ribs, **A,** and bicipital (two-headed) tetrapod rib, **B,** to vertebral column.

part of the trunk (that is, to the thorax) where they often assisted in pulmonary respiration. The other ribs did not disappear entirely, however. They usually became shortened and fused with transverse processes (Fig. 7-11). All frog ribs are this kind (Fig. 7-8). It can be verified that these are ribs by studying their embryonic development. Remnants of abdominal ribs (**gastralia**) have persisted in the ventral body wall of crocodilians and a few lizards (Fig. 7-12). Extra ribs in the neck or trunk are frequent anomalies in many vertebrates, including man.

Most thoracic ribs are composed of a dorsal segment (**vertebral rib**) and a ventral

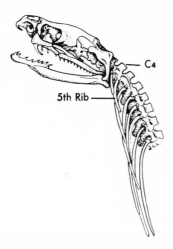

segment (**sternal rib**). The latter may be ossified, as in birds, or remain as **costal cartilages,** as in mammals. Sternal ribs and costal cartilages usually connect, directly or indirectly, with the sternum. Floating ribs lack this connection. The thoracic ribs of modern birds and some lizards have flat **uncinate processes** that overlap the next rib (Fig. 7-21). All the ribs of turtles (as well as the neural arches of the dorsal, sacral, and first caudal vertebrae) are fused with the carapace (Fig. 5-36, *A*).

A typical tetrapod rib is bicipital; that is, it exhibits two heads (Fig. 7-18, *B*). The dorsal head (**tuberculum**) articulates with the diapophysis of a vertebra; the ventral head (**capitulum**) articulates with the parapophysis or with a centrum.

The part of each rib proximal to a centrum arises from somitic mesenchyme.* The distal part usually arises from the somatopleure. Bony ribs usually are replacement bone.

■ STERNUM

The sternum is strictly a tetrapod structure and primarily an amniote structure. There was no sternum in the early amphibians, and only anurans have a well differentiated one today (Fig. 9-4). It is absent in caecilians and poorly developed or absent in urodeles. In *Necturus,* for example (Fig. 7-22), it consists solely of scattered centers

Fig. 7-19. Cervical ribs of a cobra, or "spreading adder." The first three ribs, associated with the atlas, axis, and third vertebra, are short and hidden by the jaws. **C4,** Neural spine of the fourth cervical vertebra.

*Mesenchyme from mesodermal somites.

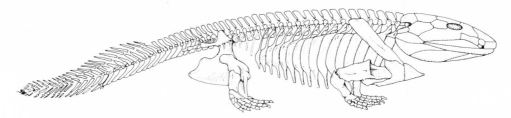

Fig. 7-20. Skeleton of *Ichthyostega,* the oldest known tetrapod. These amphibians were a little less than 1 meter in length. (After Jarvik, E.: Scientific Monthly **80**:152, March, 1955.)

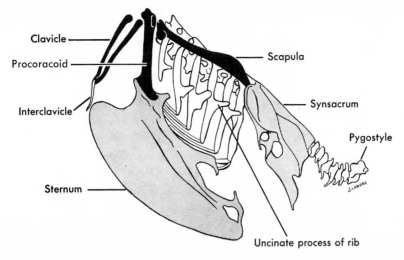

Fig. 7-21. Skeleton of trunk, tail, and pectoral girdle of a pigeon. Bones of the pectoral girdle are black. The ventral keel of the sternum is the carina.

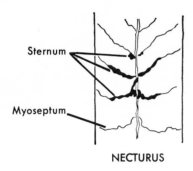

Fig. 7-22. The "sternum" of a necturus.

NECTURUS

of chondrification in myosepta of the pectoral region. In salamanders it is a simple little cartilaginous plate, at best. It may be that the sternum of amphibians is not homologous with that of amniotes.

In amniotes the sternum is a plate of cartilage and replacement bone that articulates with the pectoral girdle anteriorly and with a variable number of ribs (Fig. 7-23). In birds other than ratites the sternum has a midventral keel, or **carina,** on which the massive flight muscles insert (Fig. 7-21).

The elongated sternum of mammals is composed of segments, or sternebrae, except in cetaceans and sirenians (Fig. 7-23, monkey). The anteriormost sternebra is the **manubrium.** The posteriormost is the **xiphisternum,** which bears a cartilaginous or bony **xiphoid process.**

The amniote sternum arises as paired mesenchymal bars that later unite and undergo chondrogenesis, and in many mammals presternal and suprasternal blastemas also develop (Fig. 7-24). The median center con-

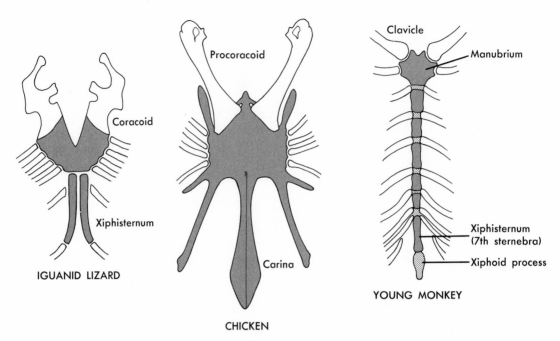

Fig. 7-23. Sternum (light gray) of a reptile, bird, and mammal. Stipple in the young monkey indicates cartilage. The sternum articulates anteriorly with the pectoral girdle (coracoid, procoracoid, clavicle) and laterally with ribs.

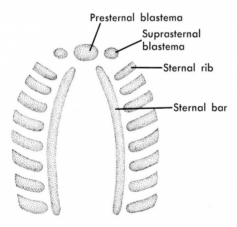

Fig. 7-24. Mesenchymal blastemas that contribute to the amniote sternum. Presternal and suprasternal blastemas develop in mammals only. The ventral ends of developing ribs are also shown.

tributes to the manubrium, and the paired (suprasternal) centers sometimes do. In a few mammals (insectivores, edentates, rodents, and a few others) the suprasternal centers give rise to independent **suprasternal ossicles** that lie between the clavicle and the manubrium. This sometimes happens in man, but the condition remains undiscovered unless there is an occasion to x-ray the sternoclavicular joint.

One might speculate that the presternal and suprasternal centers may be vestiges of the median interclavicle and paired coracoids of the pectoral girdle of reptiles.

□ CHAPTER SUMMARY

1. Vertebrae consist of centra, arches, and processes.

2. Fishes have trunk and tail vertebrae. In tetrapods, trunk vertebrae are divided into cervical, dorsal, and sacral. Dorsal vertebrae are divided into thoracic and lumbar when long curved ribs are limited to the anterior trunk region.

3. One cervical vertebra develops in amphibians. The number is higher in reptiles, still higher in birds. Mammals have seven. The first two cervicals in amniotes are atlas and axis. A proatlas occurs in many reptiles and in an occasional mammal.

4. Thoracic vertebrae are associated with long ribs and are usually related functionally to respiratory movements of the body wall.

5. Sacral vertebrae bear stout processes that brace the hind limbs and pelvic girdle against the vertebral column. Amphibians have one, reptiles and birds have two, and mammals have two to five.

6. The sacrals in amniotes usually unite to form a sacrum. The sacrum of birds unites with adjacent lumbar and caudal vertebrae to form a synsacrum.

7. Caudal vertebrae frequently bear hemal arches or chevron bones. They are reduced to archless centra near the end of the column. The anuran urostyle is probably a series of unsegmented caudal vertebrae. The terminal caudals in birds form a pygostyle. In man the caudals form a coccyx.

8. The notochord is prominent within the column of adult fishes. It becomes reduced but persists in many amphibians and reptiles. It disappears in adult birds and mammals except for the pulpy nucleus.

9. Cyclostomes have lateral neural cartilages but no centra.

10. The centra of Chondrichthyes consist of chordal and perichordal cartilage and a core of notochord. Neural arches consist of dorsal and intercalary plates. Diplospondyly occurs in the tail.

11. Primitive vertebrae are amphicelous. They occur in most fishes, primitive amphibians, and primitive lizards. Anurans and modern reptiles usually have procelous vertebrae, higher urodeles have opisthocelous, birds have heterocelous, and most mammals have acelous, vertebrae.

12. Transverse processes are present in most vertebrates. Tetrapod processes include zygapophyses, diapophyses, and parapophyses. Supraneural bones are often associated with neural arches in fishes.

13. Vertebrae and the proximal parts of ribs arise from somitic mesenchyme. Most bony centra are endochondral but in teleosts, apodans, and urodeles they are intramembranous.

14. A vertebra consisting of centrum and neural arch is probably specialized. Primitively, a hypocentrum and two pleurocentra formed in each body segment. Intercentra or intervertebral discs are common from fish to man.

15. Fishes have dorsal ribs, ventral ribs, or both. Agnatha lack ribs.

16. In recent tetrapods long ribs are mostly confined to the trunk. Most thoracic ribs reach

143

the sternum but some may be floating. Cervical, lumbar, and sacral ribs are usually short and fused with vertebral processes. All ribs are short and fused with vertebrae in amphibians.

17. Gastralia are abdominal ribs found in some ancient and modern reptiles.
18. A sternum is limited to tetrapods. It is absent in caecilians, snakes, legless lizards, and turtles and is poorly developed in urodeles. It has a carina in flying birds and is usually segmented in mammals.
19. The sternum of amniotes arises as a pair of sternal bars that later unite.
20. Suprasternal ossicles in mammals may contribute to the sternum or persist as separate bones.

LITERATURE CITED AND SELECTED READINGS

1. Alexander, R. M.: Functional design in fishes, London, 1967, Hutchinson University Library.
2. Evans, F. G.: The morphology and functional evolution of the atlas-axis complex from fish to mammals, Annals of the New York Academy of Sciences 39:29, 1939.
3. Goodrich, E. S.: Studies on the structure and development of vertebrates, London, 1930, The McMillan Co., Ltd. (Reprinted by Dover Publications, Inc., New York, 1958.)
4. Hoffstetter, R., and Gasc, J.-P.: Vertebrae and ribs of modern reptiles. In Gans, C., Bellairs, A. d'A., and Parsons, T. S., editors: Biology of the reptilia, vol. 1, New York, 1969, Academic Press, Inc.
5. Wake, D. B., and Lawson, R.: Development and adult morphology of the vertebral column in the Plethodontid salamander *Eurycea bislineata*, with comments on vertebral evolution in the amphibia, Journal of Morphology 139:251, 1973.
6. Williams E. E.: Gadow's arcualia and the development of tetrapod vertebrae, Quarterly Review of Biology 34:1, 1959.
7. Schmalhausen, I. I.: The origin of terrestrial vertebrates, New York, 1968, Academic Press, Inc.

8

Skull and visceral skeleton

In this chapter we will see that vertebrates from fishes to man build their skulls out of three components: a cartilaginous braincase, ancient dermal armor, and contributions from the branchial skeleton. We will see how these components are assembled in cartilaginous and bony skulls and note the new functions achieved by the old branchial skeleton when vertebrates took up life on land.

The word "skull" is seldom misunderstood by the layman. To him, it is the bony structure that Hamlet held in his hand and gazed at dolefully as he spoke his now famous words, "Alas, poor Yorick." To the morphologist, however, the term poses a problem because of the intimate relationship in fishes between the skeleton that protects the brain and that of the jaws and branchial arches. The latter has been inherited in modified form by higher vertebrates including, alas, poor Yorick. For

this reason, the comparative anatomist may avoid the term "skull" and refer instead to (1) the **neurocranium**, or primary braincase; (2) the **dermatocranium**; and (3) the **visceral skeleton**, or splanchnocranium. In this chapter, for practical purposes, "skull" will mean what the layman probably thinks it means, but without the lower jaw. We can then classify the parts of the cranial skeleton of vertebrates as follows.

Skull
   Neurocranium
   Dermatocranium
Visceral skeleton
   Embryonic upper jaw cartilage and its replacement bones
   Embryonic lower jaw cartilage and its replacement and investing bones
   Skeleton of the branchial arches

The embryonic upper jaw cartilage is really visceral skeleton because it is part of the first visceral arch (Fig. 8-1). However, in bony vertebrates it becomes incorporated into the skull.

■ NEUROCRANIUM

The neurocranium (sometimes called endocranium or primary braincase), is that part

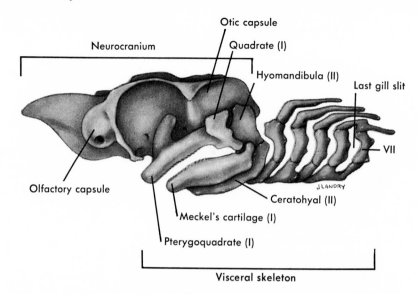

Fig. 8-1. Skull and visceral skeleton of the shark *Squalus acanthias.* **I, II,** and **VII,** Skeleton of first, second, and seventh pharyngeal arches. Labial cartilages, gill rakers, and gill rays are omitted.

of the skull that (1) protects the brain and certain special sense organs, (2) arises as cartilage, and (3) is subsequently partly or wholly replaced by bone except in cartilaginous fishes. The neurocranium in all vertebrates develops in accordance with the basic pattern described below.

□ **The cartilaginous stage**
PARACHORDAL AND PRECHORDAL CARTILAGES

The neurocranium commences as a pair of parachordal and prechordal cartilages (Fig. 8-2, *A*) underneath the brain. Parachordal cartilages parallel the anterior end of the notochord beneath the midbrain and hindbrain. Prechordal cartilages (also called trabeculae cranii) develop anterior to the notochord underneath the forebrain. The parachordal cartilages expand across the midline toward each other and unite. In the process, the notochord and parachordal cartilages are incorporated into a single, broad, cartilaginous **basal plate.** The prechordal cartilages likewise expand and unite across

the midline at their anterior ends to form an **ethmoid plate.**

SENSE CAPSULES

While parachordal and prechordal cartilages are forming, cartilage also appears in two other locations: (1) as an **olfactory capsule** partially surrounding the olfactory epithelium and (2) as an **otic capsule** completely surrounding the otocyst, which is the developing inner ear (Fig. 8-2, *A* and *B*). The olfactory capsules are incomplete anteriorly, since water (in fishes) or air (in tetrapods) must have access to the olfactory epithelium. The walls of the olfactory and otic capsules are perforated by foramina that transmit nerves and vascular channels.

An **optic capsule** forms around the retina but it is not the orbit, or skeletal socket, in which the eyeball lies. It is the **sclerotic coat** of the eyeball. Although this capsule is fibrous in mammals, cartilaginous or bony plates very often form a **scleral ring** within the sclerotic coat (Fig. 8-3). The ring helps to maintain the shape of the eyeball. This is

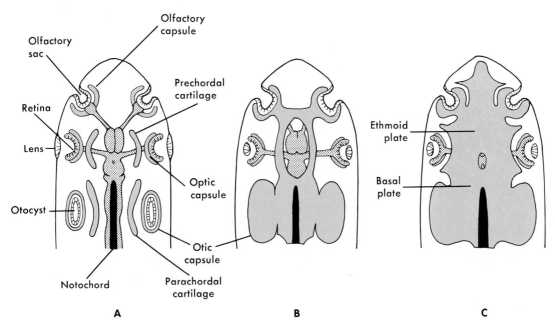

Olfactory capsule
Olfactory sac
Prechordal cartilage
Retina
Lens
Optic capsule
Otocyst
Ethmoid plate
Basal plate
Notochord
Parachordal cartilage
Otic capsule

A   B   C

Fig. 8-2. Early stages in development of a cartilaginous neurocranium, as seen from a ventral view. In **A** the notochord is seen underlying the midbrain and hindbrain. In **B** the notochord has been incorporated into the caudal floor of the neurocranium (basal plate). In **C** a cartilaginous floor has been completed beneath the entire brain. The optic capsule will later become the sclerotic coat of the eyeball.

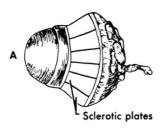

Sclerotic plates

Fig. 8-3. Ossicles of the sclerotic coat of the eye. **A,** Owl's eye, showing sclerotic plates in place. **B,** Scleral ring of overlapping ossicles dissected from a lizard's eye (after Gugg[7]).

an ancient condition, having been present in crossopterygians and extinct amphibians and reptiles. Because the optic capsule does not fuse with the neurocranium, the eyeball is free to move independently of the skull. Therefore, the sclerotic coat is not conventionally considered part of the neurocranium.

COMPLETION OF FLOOR, WALLS, AND ROOF

The expanding ethmoid plate fuses anteriorly with the olfactory capsules, and the expanding basal plate fuses with the otic capsules that lie lateral to the hindbrain. The ethmoid and basal plates also expand toward one another until they meet to form a floor on which the brain rests (Fig. 8-2, *C*). Further development of the cartilaginous neurocranium involves construction of car-

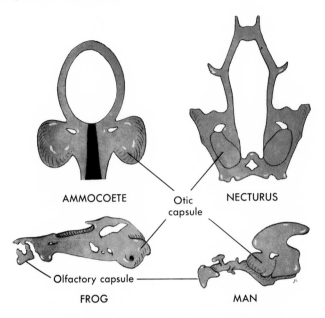

AMMOCOETE   Otic   NECTURUS
                  capsule

                  Olfactory capsule
FROG   MAN

Fig. 8-4. Cartilaginous neurocrania from selected embryonic, larval, or immature vertebrates. Dorsal view of ammocoete and *Necturus;* lateral view of frog and man.

tilaginous walls alongside the brain and, in lower forms, a cartilaginous roof over the brain. The cranial nerves and blood vessels are already present by this time, and the cartilage is deposited in a manner that leaves foramina for these structures. The largest is the foramen magnum in the rear wall of the neurocranium.

In cartilaginous and lower bony fishes the brain is completely covered by a cartilaginous roof. But in teleosts and tetrapods the brain is never completely roofed over by cartilage.

The preceding pattern of development recurs throughout the vertebrate series and produces a cartilaginous neurocranium that protects much of the embryonic brain, the olfactory epithelia, and the inner ear (Fig. 8-4). The blastema that gives rise to the cartilaginous neurocranium is a contribution chiefly of ectomesenchyme with sclerotomal contributions in the occipital region.

ADULT CARTILAGINOUS NEUROCRANIA

Cartilaginous fishes retain a cartilaginous neurocranium, sometimes called a chondrocranium, throughout life. The neurocranium in these fishes (Fig. 8-1) completely encloses the brain, and the otic and olfactory capsules are fused into it along with the notochord. Dorsally, an endolymphatic fossa is perforated by endolymphatic and perilymphatic ducts and there is an opening, the pineal (epiphyseal) foramen, that is occupied by the pineal body in life (Fig. 8-5).

Among bony fishes, lungfishes and most ganoids retain a highly cartilaginous neurocranium throughout life. In order to see it, the membrane bones that overlie it must be stripped away (Fig. 8-9).

In cyclostomes the several cartilaginous components of the embryonic neurocranium remain in adults as more or less independent cartilages (Fig. 8-20). An olfactory capsule (median, protecting the median

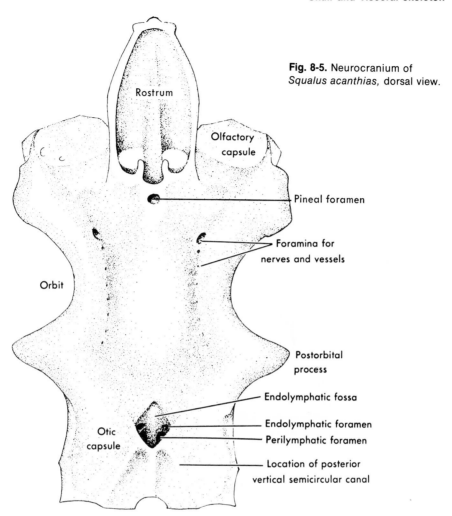

Fig. 8-5. Neurocranium of *Squalus acanthias,* dorsal view.

Rostrum

Olfactory capsule

Pineal foramen

Foramina for nerves and vessels

Orbit

Postorbital process

Endolymphatic fossa

Endolymphatic foramen

Perilymphatic foramen

Otic capsule

Location of posterior vertical semicircular canal

olfactory sac), otic capsules, a basal plate, a notochord (not fused with the basal plate), and other cartilages not homologizable with those of gnathostomes can be identified. The roof above the brain remains fibrous.

Neurocranial ossification centers

In bony vertebrates the embryonic cartilaginous neurocranium is mostly replaced by replacement bone. The process of endochondral ossification occurs more or less simultaneously at numerous separate ossification centers (Fig. 8-6, *B*). Although the specific number of such centers varies in different species, four regions are universally involved. These regions—occipital, sphenoid, ethmoid, and otic—will be discussed next.

OCCIPITAL CENTERS

The cartilage surrounding the foramen magnum may be replaced by as many as four bones. One or more endochondral ossification centers ventral to the foramen magnum produce a **basioccipital bone** underlying the hindbrain (Fig. 8-7). Centers in the lateral

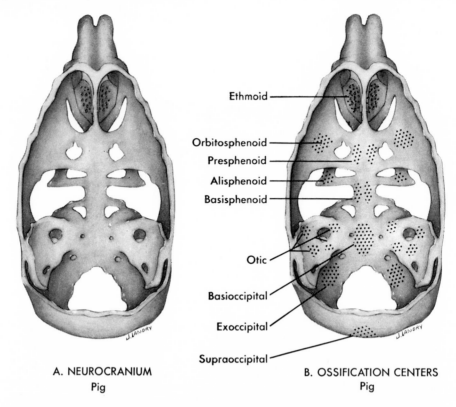

A. NEUROCRANIUM
Pig

B. OSSIFICATION CENTERS
Pig

Fig. 8-6. A, Cartilaginous neurocranium of fetal pig. The structure is complete as shown, there being no cartilage above the brain. **B,** Ossification centers in typical mammalian cartilaginous neurocranium, based on fetal pig. The otic centers are multiple centers in the otic capsule. The ethmoid centers are interspersed among the olfactory foramina. (The alisphenoid center is in the pterygoquadrate cartilage.)

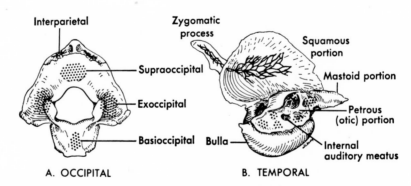

A. OCCIPITAL

B. TEMPORAL

Fig. 8-7. Endochondral ossification centers (dots) and intramembranous ossification centers (black networks) superimposed on the occipital and right temporal bones of an adult cat. **A,** Caudal view. **B,** Medial view. The bulla arises from new cartilage not associated with the earlier neurocranium. The mastoid portion is an outgrowth of the petrous portion.

walls of the foramen magnum produce two **exoccipital bones.** Above the foramen a **supraoccipital bone** may develop. In mammals all four occipital elements usually fuse to form a single **occipital bone.** In modern amphibians, one or more of these may remain cartilaginous, although they were bony in stem amphibians.

The neurocranium of tetrapods articulates with the first vertebra via one or two **occipital condyles.** Stem amphibians had a single condyle borne chiefly on the basioccipital bone. Living reptiles and birds still have a single condyle. Modern amphibians and mammals diverged from the early tetrapod condition with two condyles, one on each exoccipital.

SPHENOID CENTERS

The embryonic cartilaginous neurocranium underlying the midbrain and pituitary

gland ossifies to form a **basisphenoid bone** (anterior to the basioccipital) and a **presphenoid.** Thus a bony platform consisting of occipital and sphenoid bones underlies the brain. The side walls above the basisphenoid and presphenoid ossification centers form other sphenoid elements (**orbitosphenoid, pleurosphenoid,** and others*), and these may remain separate or unite with the basisphenoids and presphenoids to form a single adult **sphenoid bone** with "wings" (Fig. 8-8). The pituitary rests in the sella turcica of the basisphenoid region. Since the embryonic neurocranium is usually incomplete dorsally, no replacement bones lie above the brain.

*The alisphenoid of mammals evidently forms in the pterygoquadrate cartilage rather than in the neurocranium.

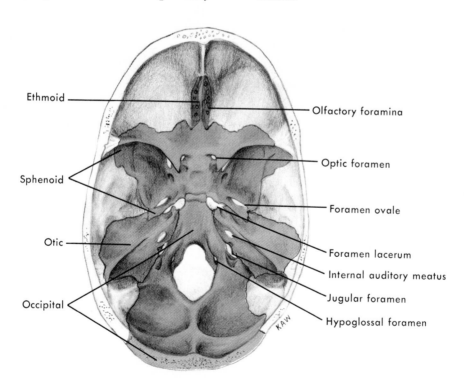

Fig. 8-8. Bony neurocranium (red) of human skull. The calvarium (roof) of the skull has been sawed off and view is looking down into skull from above. Major endochondral ossification centers are labeled at left.

ETHMOID CENTERS

The ethmoid region lies immediately anterior to the sphenoid and includes the ethmoid plate and olfactory capsules. Of the four major ossification centers in the cartilaginous neurocranium (occipital, sphenoid, ethmoid, otic), the ethmoid more than the others tends to remain cartilaginous. Ossification centers in this region become the **cribriform plate** of the ethmoid, perforated by olfactory foramina, and several of the conchae, or turbinal bones (**ethmoturbinals**) in the nasal passageways of crocodilians, birds, and mammals. **Mesethmoid** bones ossify in some mammals and contribute to the otherwise cartilaginous median nasal septum. In anurans the **sphenethmoid** is the sole bone arising in the sphenoid and ethmoid regions.

OTIC CENTERS

The cartilaginous otic capsule is replaced in lower vertebrates by several bones with such names as **prootic, opisthotic,** and **epiotic.** One or more of these may unite with adjacent replacement or membrane bones. For example, in frogs and most reptiles the opisthotics fuse with the exoccipitals and in birds and mammals the prootic, opisthotic, and epiotics all unite to form a single **periotic,** or **petrosal bone.** The petrosal, in turn, may unite with the squamosal to form a **temporal** bone (Fig. 8-7, *B*). Six ossification centers have been described in the otic capsule of a human fetus.

■ **DERMATOCRANIUM**

In ganoid fishes the superficial membrane bones of the head are sculptured in appearance, scalelike, and located in the dermis (Fig. 8-9). They are often continuous with the typical scales of the trunk. The "cheek plates" of a gar (Fig. 8-27, *A*) are either "skull bones" or "dermal scales," whichever you prefer to call them. These scalelike cranial bones of ganoids are part of their dermal armor and are descendants of the der-

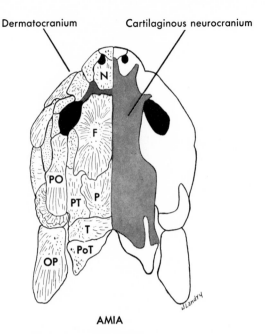

Fig. 8-9. Skull of *Amia calva,* dorsal view. Dermal bones have been removed on the right side to reveal underlying cartilaginous neurocranium.
F, Frontal bones; **N,** nasal; **OP,** operculum; **P,** parietal; **PO,** postorbital; **PoT,** posttemporal; **PT,** pterotic; **T,** tabular. The bones anterior to the nasals are ethmoids. Premaxillae are not visible in this view.

mal armor of the earliest vertebrates. Because they are dermal in location they lie superficial to the neurocranium (Fig. 8-9). In later bony vertebrates these dermal bones of the head no longer formed *within* the skin, but *under* it, where, along with the neurocranium, or primary braincase, they became part of the skull. Collectively, these dermal bones of the skull constitute the dermatocranium.

For convenience, the dermatocranium will be discussed under the following topics: (1) bones that form a roof over the brain and contribute to the lateral walls of the skull, (2) bones of the upper jaw, (3) bones of the palates, and (4) opercular bones.

Teleost skulls are highly specialized, and the dermal bones of their skulls are mostly

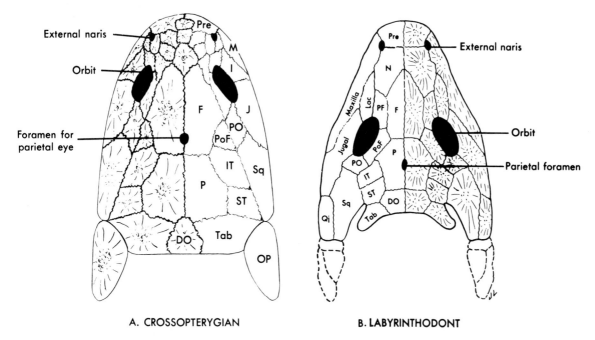

A. CROSSOPTERYGIAN   B. LABYRINTHODONT

Fig. 8-10. Early dermal bone patterns from which tetrapod dermatocrania probably evolved. **A,** Skull of the crossopterygian fish *Eusthenopteron.* Note midline bones and small, scalelike bones in the rostral region. The location of the parietal foramen is in dispute. **B,** Skull of a Carboniferous labyrinthodont, representing the primitive tetrapod condition. Broken lines indicate deleted opercular bones. **DO,** Dermoccipital (postparietal); **F,** frontal; **I,** infraorbital; **IT,** intertemporal; **J,** jugal; **Lac,** lacrimal; **M,** maxilla; **N,** nasal; **OP,** opercular; **P,** parietal; **PF,** prefrontal; **PO,** postorbital; **PoF,** postfrontal; **Pre,** premaxilla; **Sq,** squamosal; **ST,** supratemporal; **Qj,** quadratojugal; **Tab,** tabular. (Modified from numerous sources.)

not homologizable with those of older fishes or tetrapods. For that reason they will be mentioned only occasionally.

☐ **Roofing bones**

The basic pattern for roofing bones is diagrammed in Fig. 8-10. In crossopterygians a series of paired and unpaired bones extended along the middorsal line from the nares to the occiput, overlying the olfactory capsules and brain. In labyrinthodonts the unpaired bones were lost and a series of paired bones—**nasals, frontals, parietals, dermoccipitals (postparietals)**—resulted. In the midline between the two frontal or parietal bones was a parietal foramen. It is still present in many fishes, amphibians, and reptiles and houses the third eye.

Forming the walls of the orbit in the basic pattern were **lacrimal, prefrontal, postfrontal, postorbital,** and **infraorbital** bones (of which the **jugal** is one). At the posterior angle of the skull were **intertemporal, supratemporal, tabular, squamosal,** and **quadratojugal** bones. The quadratojugal disappeared as an independent bone in lizards, snakes, and mammals.

Roofing bones overlie the neurocranium when the latter is complete above the brain (Fig. 8-9). When the neurocranium is incomplete dorsally, soft spots (fontanels) can be felt in the head until the membranes under the skin have ossified (Fig. 8-11). In mammals a small **bregmatic bone** sometimes ossifies in the fontanel at the junction of the coronal and sagittal sutures (Fig.

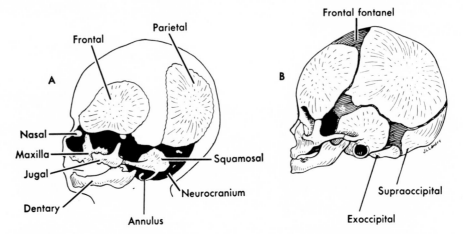

Fig. 8-11. Two stages in the development of the human skull. **A,** Intramembranous ossification is under way. The neurocranium (black) is incomplete lateral to and above the brain. **B,** Intramembranous ossification has progressed, but "soft spots" (fontanels) remain where there is no cartilage or bone. The exoccipital and supraoccipital are of endochondral origin.

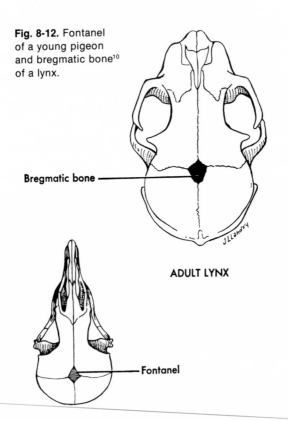

Fig. 8-12. Fontanel of a young pigeon and bregmatic bone[10] of a lynx.

ADULT LYNX

YOUNG PIGEON

8-12). It is sometimes present in man, and Paracelsus called it the antiepileptic bone because he thought it served as a "pop-up" valve!

□ **Upper jaw bones**

The first upper jaw skeleton that a vertebrate embryo develops is the **pterygoquadrate (palatoquadrate) cartilage** on each side. It is part of the visceral skeleton and the only upper jaw that cartilaginous fishes develop (Fig. 8-1). In bony vertebrates this cartilage becomes covered (ensheathed) by dermal bones that "lock it into" the skull. These dermal bones, the **premaxillae** and **maxillae,** comprise the adult upper jaw of most bony vertebrates and usually bear teeth. In birds the premaxillae are elongated and are part of the beak. Premaxillae are not seen in adult human skulls because they unite with the maxillae very early in development.

Garfishes have a long series of scalelike bones that have been named maxillaries (Fig. 8-27). Bones called maxillae in teleosts may be toothless, reduced, or not in the up-

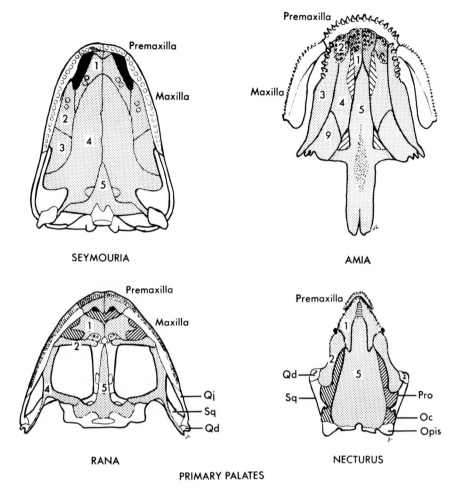

SEYMOURIA

AMIA

RANA

NECTURUS

PRIMARY PALATES

Fig. 8-13. Primary palates of a primitive reptile *(Seymouria),* a bony fish *(Amia),* and two amphibians *(Rana* and *Necturus).* Cartilage is indicated by diagonal lines; internal nares are black, and palatal bones are stippled. **1,** Vomer; **2,** palatine (in *Necturus,* palatopterygoid); **3,** ectopterygoid; **4,** endopterygoid; **5,** parasphenoid; **9,** epipterygoid (of endochondral origin). **Oc,** Cartilaginous portion of otic capsule; **Opis,** opisthotic; **Pro,** prootic; **Qd,** quadrate; **Qj,** quadratojugal; **Sq,** squamosal.

per jaw (Fig. 8-27, *B*). They may not be homologous with the maxillae of older fishes and tetrapods.

□ **Palatal bones**

The floor on which the brain rests is at the same time the roof of the oral cavity in fishes and amphibians. This part of the skull is the **primary palate.** In sharks it is cartilaginous.

In bony vertebrates membrane bones form in this location (Figs. 8-13 and 8-14). These are **vomers** (beneath the olfactory capsules), **palatines, endopterygoids** and **ectopterygoids** (beneath the pterygoquadrate cartilages) and a median **parasphenoid** (beneath the sphenoid region of the neurocranium). Primitively, teeth formed on all these bones, and many persist today in lower vertebrates.

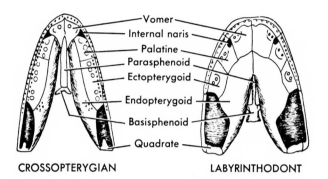

Fig. 8-14. Primary palates of a crossopterygian and labyrinthodont. Note similarity of structure. The basisphenoid and quadrate are not part of the palate.

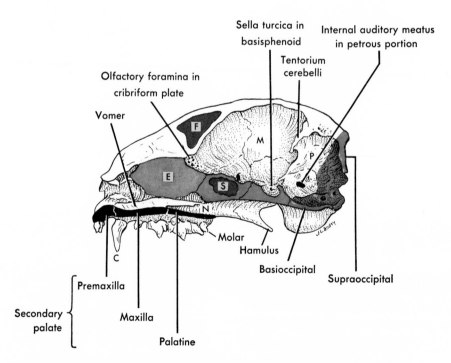

Fig. 8-15. Sagittal section, cat skull, showing bony part of secondary palate in black. **C,** Canine tooth; **E,** mesethmoid (perpendicular plate of ethmoid) in nasal septum; **F,** frontal sinus in frontal bone; **M,** middle cranial fossa housing cerebral hemispheres; **N,** nasal passageway; **P,** posterior cranial fossa housing cerebellum; **S,** sphenoidal sinus in presphenoid bone. Light gray designates ethmoid, sphenoid, and occipital components of the neurocranium.

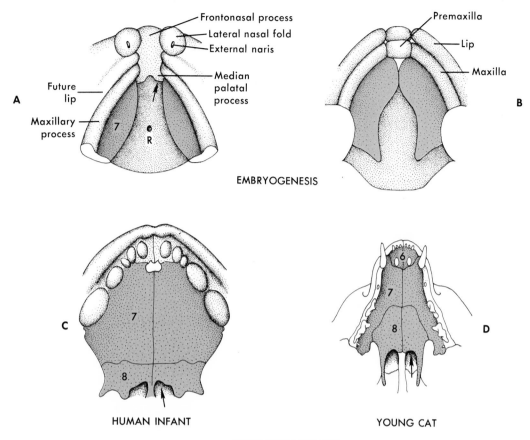

EMBRYOGENESIS

SECONDARY PALATES

Fig. 8-16. A to **C,** Formation of secondary palate in man. **D,** Secondary palate of young cat for comparison. Arrows indicate nasal passageways. **6,** Palatine process of premaxilla; **7,** palatine process of maxilla; **8,** palatine process of palatine bone. In **A** (fetus approximately 18 weeks old) the palatine processes of the maxillae are growing toward the midline, forming a secondary roof (red) in the oral cavity. **R,** Rathke's pouch in primary roof. In **B** the palatine processes of the maxillae have met anteriorly. In **C** the palate is complete. Failure of palatine processes to meet in midline results in a cleft palate.

In birds and mammals and some reptiles a **secondary (false) palate** develops. It is a horizontal partition separating the primitive oral cavity into nasal and oral passageways (Figs. 11-10, *B,* and 8-15). Membrane bones in the secondary palate usually include **palatal processes** of premaxillae, maxillae, and palatines. In crocodilians the pterygoid bone also contributes.

Palatal processes arise as horizontal shelves of the foregoing bones, which grow toward one another in the roof of the embryonic oral cavity (Fig. 8-16, *A* and *B*). Failure of these processes to meet results in a **cleft palate.** This is a normal condition in turtles and birds, a congenital abnormality in mammals.

☐ **Opercular bones**

The operculum is a fold of the hyoid arch that extends back over the gill slits in holocephalans and bony fishes. In the latter it is stiffened by dermal **opercular bones.** In many bony fishes a series of **branchiostegal rays** forms in a ventral flap of the operculum, but there are no vestiges of opercular bones in tetrapods.

■ VISCERAL SKELETON

The visceral skeleton, or splanchnocranium, is the skeleton of the pharyngeal arches. In fishes, therefore, it is the skeleton of the jaws and gill arches. In tetrapods this skeleton has become modified to perform new functions on land.

The blastemas that give rise to the visceral skeleton come from neural crests and they first secrete cartilage. Later, the cartilage may be partly or wholly replaced by bone. In the first arch only, much of it is ensheathed by dermal bone. We will look first at a shark, in which no bone forms and in which we can see the visceral skeleton in its primitive capacity, that of supporting the jaws and gills.

□ Fishes
SHARKS

The visceral skeleton of a shark consists of seven sets of paired cartilages in the seven

visceral arches (Fig. 8-1) and a series of midventral cartilages, **basihyal** and **basibranchials,** in the pharyngeal floor (Fig. 8-17). All seven pairs conform to a basic pattern, but the mandibular and some of the hyoid cartilages are modified for feeding (Fig. 8-18).

The skeleton of the mandibular, or first visceral, arch consists of two cartilages on each side, a **pterygoquadrate** (also called **palatoquadrate**) **cartilage** dorsally and **Meckel's cartilage** ventrally (Fig. 8-18, C). The left and right pterygoquadrates meet in the middorsal line to form the skeleton of the upper jaw, and the left and right Meckel's cartilages meet ventrally to form the skeleton of the lower jaw. Delicate labial cartilages embedded in a position corresponding to lips articulate with the upper and lower jaw cartilages at the corners of the mouth, but they have no known phylogenetic significance.

The skeleton of the hyoid arch consists of paired **hyomandibular** cartilages dorsally and **ceratohyals** laterally (Fig. 8-18, B). The latter bear a demibranch and articulate with the basihyal cartilage. In embryos the basihyal is paired. The cartilages of the re-

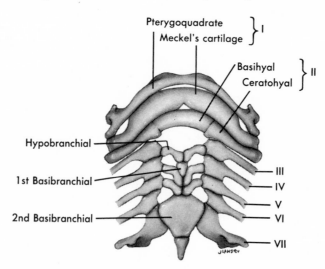

Fig. 8-17. Visceral skeleton of *Squalus acanthias,* ventral view. **III** to **VII,** Ceratobranchial cartilages of the third to seventh pharyngeal arches.

maining visceral arches are essentially alike (Fig. 8-18, *A*), and all but the last pair support gills. Hypobranchials, unless absent, articulate ventrally with a basibranchial.

At the corner of the mouth Meckel's cartilage, the pterygoquadrate cartilage, and the hyomandibular cartilage articulate in a movable joint that participates in operation of the jaws (Fig. 8-1). The hyomandibula is bound by ligaments to the otic capsule, and it therefore suspends the jaws and branchial skeleton from the skull. This is **hyostylic jaw suspension.**

BONY FISHES

The visceral skeleton of bony fishes resembles that of sharks except that bone

is added (Fig. 8-19). The caudal ends of the cartilaginous pterygoquadrate cartilage undergo endochondral ossification to be replaced by **quadrate bones.** The remainder is replaced by **palatine,*** **epipterygoid,** and **metapterygoid bones,** and these endochondral bones contribute to the palate. Meckel's cartilage makes little contribution to the lower jaw. The posterior tip becomes an **articular bone** (Fig. 8-27, *B*). (When this bone incorporates some dermal bone contributions, it is called **derm-articular.**) The remainder of Meckel's cartilage becomes invested by dermal bones, chiefly dentary and

---

*The palatine of bony fishes is not homologous with the bone of the same name in tetrapods.

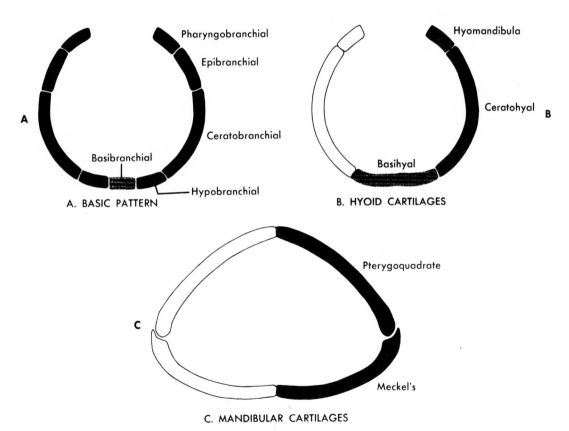

A. BASIC PATTERN

B. HYOID CARTILAGES

C. MANDIBULAR CARTILAGES

Fig. 8-18. Skeletal components of a typical branchial arch, **A,** and modifications in the hyoid and mandibular arches of *Squalus acanthias,* **B** and **C.** Midventral elements in the pharyngeal floor are shown in white on black.

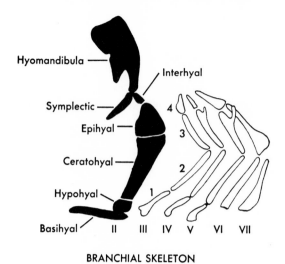

Hyomandibula

Interhyal

Symplectic

Epihyal

Ceratohyal

Hypohyal

Basihyal

4

3

2

1

II   III   IV   V   VI   VII

BRANCHIAL SKELETON

Salmon

Fig. 8-19. Visceral skeleton of a salmon, upper and lower jaws removed. Hyoid cartilages are in black. The basihyal is unpaired. **1** to **4**, Hypobranchial, ceratobranchial, epibranchial, and pharyngobranchial elements of the third arch.

angular in modern fishes (Fig. 8-27), and it either persists as a cartilaginous core within the lower jaw or disappears. In some fishes a segment of Meckel's cartilage at the chin is replaced by a **mentomeckelian bone.**

The hyoid skeleton of bony fishes undergoes extensive ossification. Most common replacement bones are the **hyomandibulas** (which articulate with the otic capsules), **symplectics** (which articulate with the quadrates of the upper jaw), **interhyals, epihyals, ceratohyals,** and **hypohyals** (Fig. 8-19). A median **basihyal** develops in the pharyngeal floor. These separate elements enable the hyoid arch to perform key roles in the specialized movements of food taking and respiration. When a teleost feeds, the jaws move forward and backward in a telescopic fashion independently of the rest of the skull. The hyoid arch participates in such movements via the articulation of the symplectic with the quadrate.

CYCLOSTOMES

The visceral skeleton of cyclostomes (Fig. 8-20) is quite unlike that of jawed fishes. For example, *Myxine* has no identifiable pterygoquadrate or Meckel's cartilages. It does have a **"dental plate,"** or **lingual cartilage** that forms a V-shaped trough in the floor of the oral cavity, and, beneath this, an immovable basal plate to which the muscles of the dental plate are attached. There is evidence that these may be derived from the first visceral arch[3]; if so, the rasping tongue-like structure of which they are a part may be considered a type of lower jaw. With regard to an upper jaw, a careful study of the visceral skeleton of a hagfish led to the conclusion that a rudimentary upper jaw is fused with the neurocranium.[1] The rest of the visceral skeleton of cyclostomes consists of cartilages of unknown homology, including a basketlike cartilaginous framework immediately under the skin surrounding the gill slits.

JAW SUSPENSION IN FISHES

The jaw-hyoid complex of fishes requires bracing against some support so it may function effectively, and the nearest one is the neurocranium. Students of the anatomy of *Squalus* are familiar with the bracing design in that animal. The hyomandibular cartilage is braced against the otic capsule, and the jaws are braced against the hyomandibula. This is a fairly recent arrangement and is seen also in modern bony fishes. Technically, the condition is known as **hyostyly** (hyostylic jaw suspension). A more primitive condition is seen in some older sharks in which the jaws and hyoid are both braced directly against the braincase, thus providing a double brace, a condition known as **amphistyly.** Still another variant is seen in lungfishes and chimaeras, in which the hyomandibula plays no role in bracing the jaws. This condition is one of "self-bracing" of the jaws, or **autostyly.** More sophisticated terminologies of the relationships between

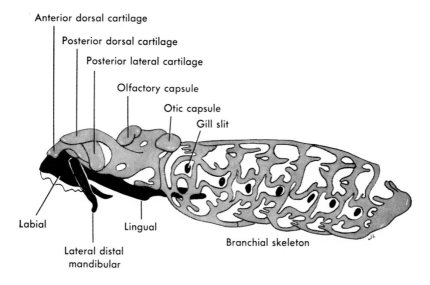

Anterior dorsal cartilage
Posterior dorsal cartilage
Posterior lateral cartilage
Olfactory capsule
Otic capsule
Gill slit
Labial
Lingual
Lateral distal
mandibular
Branchial skeleton

NEUROCRANIUM and VISCERAL SKELETON
Lamprey

Fig. 8-20. Neurocranium and visceral skeleton of a lamprey. Black elements may represent vestiges of jaws. Olfactory capsule is a midline structure, otic capsules are paired. The lingual cartilage is also called basal plate cartilage.

jaws, hyoid arch, and otic capsule are employed by specialists.

□ **Tetrapods**

With pulmonary respiration and life on land, the ancestral visceral skeleton, so necessary in gill-bearing vertebrates, underwent profound adaptive modifications. Some previously functional parts were deleted, and those that persisted perform new and, sometimes, surprising functions.

Not only have changes in the visceral skeleton taken place during the *evolution* of tetrapods; they occur also during the *ontogeny* of every gill-bearing amphibian that undergoes complete metamorphosis. For example, larval frogs have six pairs of visceral cartilages, and the last four (III to VI) support gills. These branchial cartilages unite ventrally in a hypobranchial plate (Fig. 8-21, *A*). During metamorphosis (Fig. 8-21, *B* and *C*) visceral cartilages V and VI regress and disappear, the hypobranchial

plate enlarges and, along with the first basibranchial, becomes incorporated into a broad plate (body of the hyoid) in the buccal and pharyngeal floor. The ceratohyal cartilage (arch II) is reduced to a slender anterior horn (cornu) of the hyoid, and the cartilage of arch IV becomes a posterior horn. Other changes take place with the result that a visceral skeleton initially adapted for branchial respiration becomes converted, in the span of a few short days, to one suitable for life on land. Perennibranchiate amphibians, on the other hand, retain a branchial skeleton throughout life (Fig. 8-22). In the paragraphs that follow, we will examine modifications of the visceral skeleton of tetrapods.

PTERYGOQUADRATE CARTILAGE
(PALATOQUADRATE CARTILAGE)

Pterygoquadrate cartilages are the embryonic upper jaw cartilages. In amphibians, reptiles, and birds their posterior end,

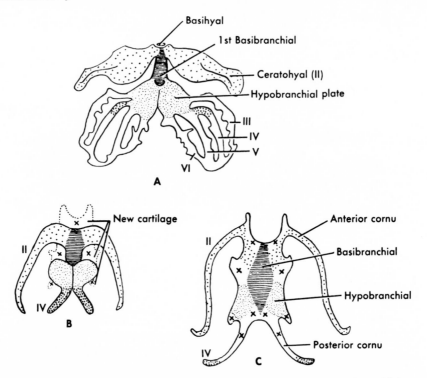

Fig. 8-21. Metamorphosis of visceral skeleton of a frog, jaws omitted. **A,** Branchial skeleton of larva. **B,** Condition in late metamorphosis. **C,** Hyoid of a young frog. Coarse and fine stipple and cross hatching indicate homologous areas. **x,** Cartilage added at metamorphosis. **II** to **VI,** Skeleton of second through sixth pharyngeal arches.

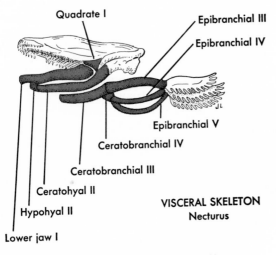

VISCERAL SKELETON
Necturus

Fig. 8-22. Skull and visceral skeleton of *Necturus*. **I** to **V,** Skeleton of the five pharyngeal arches. Some derivatives of the pterygoquadrate cartilage are in the palate.

which is the quadrate region, usually undergoes endochondral ossification to become the **quadrate bone** at the caudal angle of the skull (Fig. 8-28, *B*). There it forms a joint with the articular bone of the lower jaw. Thus, in tetrapods, as in sharks, the caudal ends of Meckel's cartilages (now ossified) rock against the caudal ends of the pterygoquadrate cartilages (also ossified). In mammals the same joint exists; but the quadrate bone has become surrounded by the middle ear cavity, has separated from the rest of the pterygoquadrate cartilage, and has become the **incus** of the middle ear. The evolutionary transition from a jaw bone to ear ossicle was gradual. The intermediate steps are seen in mammallike reptiles.

The anterior part of the pterygoquadrate cartilage becomes ensheathed by dermal

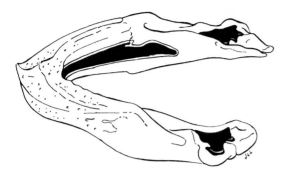

Fig. 8-23. Mandible of an adult sea turtle, from the left and above, showing core of Meckel's cartilage (black) ensheathed by membrane bone.

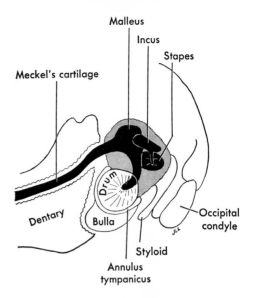

Fig. 8-24. Fate of the posterior tip of Meckel's cartilage in mammals. It becomes surrounded by the developing middle ear cavity (gray) and develops into a malleus.

bones including the premaxilla and maxilla and by some of the membrane bones of the palate. As a result, this part of the visceral skeleton is "locked into" the skull. In perennibranchiate amphibians some of it remains cartilaginous and contributes to the palate. It has been claimed that the bony ring (**annulus tympanicus**) to which the tympanic membrane of anurans is attached is a derivative of the pterygoquadrate cartilage.*

MECKEL'S CARTILAGE

Parts of the embryonic Meckel's cartilage become replacement bone, parts remain cartilaginous, and much of it is ensheathed by dermal bones (Figs. 8-23 and 8-34). (Dry, preserved mandibles of reptiles are hollow because Meckel's cartilage has been removed.) In adult birds and mammals Meckel's cartilage fails to grow beyond the embryonic state and few or no remnants remain within the adult mandible.

The posterior tip of Meckel's cartilage is not ensheathed. Instead, below mammals it becomes the **articular bone** of the lower jaw, which articulates with the quadrate of the upper jaw. In anurans the articular portion remains unossified.

In mammals the articular portion projects into the middle ear cavity and later separates from the rest of Meckel's cartilage to become the **malleus,** a middle ear ossicle (Fig. 8-24). The malleus still forms a joint with the quadrate (incus), but in the middle ear instead of at the end of the jaw.

An ossification center sometimes develops in Meckel's cartilage on either side of the mandibular symphysis, giving rise to a **mentomeckelian bone.** In some species, however, the mentomeckelian is a membrane bone.

HYOMANDIBULAR CARTILAGE
(COLUMELLA, OR STAPES)

It will be recalled that the hyomandibula of sharks is interposed between the quadrate region of the upper jaw and the otic capsule containing the inner ear. Studies have shown that the hyomandibula of tetrapod embryos ossifies to become part of the **columella,** or **stapes,** of the middle ear (Fig. 16-8). It still articulates with the otic capsule but now transmits sound waves. Further-

*If so, the annulus of anurans is not homologous with that of mammals, since the latter is membrane bone.

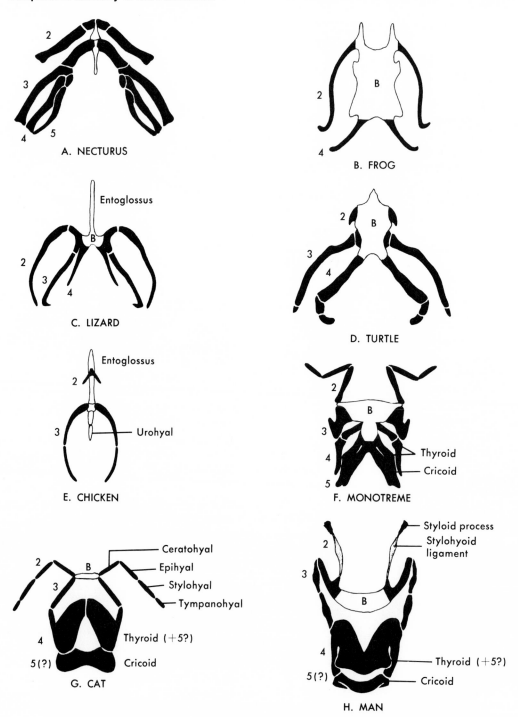

Fig. 8-25. Skeletal derivatives of the second through fifth pharyngeal arches in selected tetrapods. **B,** Body of hyoid. **2** to **5,** Derivatives of arches 2 through 5. The projections from the body in **B** to **H** are the horns of the hyoid.

more, in mammals it exhibits its ancient relationship, extending between the otic capsule and the quadrate bone (incus).

HYOID APPARATUS

The term hyoid apparatus is used here to designate the skeletal derivatives of the hyoid arch other than the columella or stapes, and the derivatives of the more caudal visceral arches other than those contributing to the larynx. The hyoid apparatus—or simply hyoid—consists of a median plate, the **body** of the hyoid, derived from basihyal and basibranchial elements, and two or more **horns,** or cornua (Fig. 8-25, *B* to *H*). The anterior horns arise from arch II and are homologous with the ceratohyals of fishes. The more caudal horns arise from arches III and, frequently, IV.

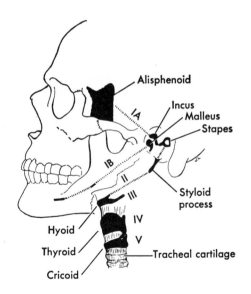

VISCERAL SKELETON

Fig. 8-26. Visceral skeleton of man. **IA,** Broken line connects derivatives of pterygoquadrate cartilage; **IB,** broken line connects vestiges and derivatives of Meckel's cartilage; **II,** stylohyoid ligament; **III to V,** derivatives of third, fourth, and fifth arches. **III,** Greater (posterior) horn of hyoid bone (illustrated also in Fig. 12-10).

In lizards and birds the body of the hyoid is narrow, and an elongated process extends forward into the tongue as an **entoglossal bone** (Fig. 8-25, *C* and *E*). In some male lizards (anoles and related genera) a similar process extends caudad into the gular pouch, or dewlap. The process is flexible, and when bent like a bow, as during mating display in sunlight, the dewlap protrudes far out from the underside of the neck, where its many blood vessels and pigment cells produce a bright coloration. In snakes the entire branchial skeleton is vestigial.

The hyoid of mammals has two horns, an anterior pair from arch II and a posterior pair from arch III. In cats the anterior horns are longer (**greater horns**) and are composed of four segments (Fig. 8-25, *G*). The dorsalmost, or **tympanohyal,** ends in a notch in the tympanic bulla. In man (Figs. 8-25, *H*, and 8-26) the anterior horns are shorter (**lesser horns**), a **stylohyoid ligament** represents the middle segment, and the **styloid process** of the temporal bone (Figs. 8-26 and 8-38) is equivalent to the tympanohyal. In rabbits, too, the anterior horn is shorter, and a slender **stylohyal bone** embedded in the tendon of the stylohyoideus minor muscle close to the skull is equivalent to the tympanohyal of cats. These are examples of the kind of variations found among mammals.

Branchiomeric and hypobranchial muscles approach the hyoid from many directions and insert on it. Because of this it is part of the buccopharyngeal pressure pump in anurans. In amniotes, tongue muscles and muscles used in swallowing attach to it.

LARYNGEAL SKELETON

Nearly all tetrapods have cricoid and arytenoid elements (Fig. 12-9), and crocodilians and mammals have thyroid elements as well (Figs. 8-25 and 12-10). Thyroid cartilages arise from the mesenchyme of arch IV and perhaps V; cricoid and arytenoid cartilages may be products of arch V. Since the caudal end of the visceral arch

Table 8-1. Skeletal derivatives of pharyngeal arches in sharks and *approximate* homologues in bony vertebrates

Arch	Shark	Teleost	Necturus	Frog	Reptile and bird	Mammal
I	Meckel's cartilage	Articular*	Articular	Articular, Mentomeckelian†	Articular	Malleus
	Pterygoquadrate	Quadrate, Epipterygoid, Metapterygoid	Quadrate, Palatal cartilage	Quadrate	Quadrate, Epipterygoid	Incus, Alisphenoid
				Annulus tympanicus (?)		
II	Hyomandibula	Hyomandibula	Rudimentary	Columella (stapes)		
		Symplectic, Interhyal, Epihyal, Ceratohyal, Hypohyal		Styloid process in mammals, Anterior horn of hyoid		
	Ceratohyal		Ceratohyal			
	Basihyal	Basihyal	Hypohyals	Body of hyoid, Entoglossus in reptiles and birds		
III	Pharyngobranchial, Epibranchial, Ceratobranchial, Hypobranchial	Pharyngobranchial, Epibranchial, Ceratobranchial, Hypobranchial	Epibranchial, Ceratobranchial	Body of hyoid	2nd horn of hyoid	
IV	Branchial skeleton	Branchial skeleton	Branchial skeleton	Last horn and body of hyoid, Laryngeal cartilages (?)	Thyroid cartilages	
V	Branchial skeleton			Laryngeal cartilages (?) (precise homologies unknown)		
VI	Branchial skeleton			Not present		
VII	Branchial skeleton					

*Sometimes part of derm-articular.
†Of intramembranous origin in some species.

series has been subject to reduction during evolution, it is not surprising that problems are encountered in relating laryngeal cartilages to specific arches. Also, there may be instances in which the laryngeal cartilages are not visceral arch derivatives.

Perspective

It is evident that the visceral skeleton is an ancient mechanism associated at an earlier time with feeding and branchial respiration. In tetrapods it has been modified for transmission of sound (malleus, incus, stapes), for attachment of tongue and other muscles, and for support of the larynx. These adaptations illustrate how mutations alter ancient structures for new functions. Some of the skeletal derivatives of the visceral arches in representative living vertebrates are listed in Table 8-1.

EVOLUTIONARY TRENDS

Teleosts

Teleost skulls have become highly specialized. They are laterally compressed, arched, vaultlike (tropibasic) as contrasted with the broad, flat platybasic skulls of more primitive fishes (Fig. 8-27). The neurocranium is incomplete above the brain, and the remainder is well ossified. The dermal bones no longer resemble scales and have sunk deeper into the head. Strict homologies between the many dermal bones of teleosts and those of tetrapods cannot be established. The eyeball muscles occupy two longitudinal cavities called myodomes in the posterior cranial floor, rather than arising on the orbital wall, as in other vertebrates

Transition to a tetrapod skull

It is highly probable that labyrinthodonts arose from a crossopterygian type of fish. Some of the evidence comes from the many similarities between the skulls of these two major groups (Figs. 8-10 and 8-14). The skulls of both were flat, and the many

dermal bones were similar in arrangement and sculptured by the imprint of the overlying skin. The neurocranium was extensively ossified except for the olfactory capsules, and a single occipital condyle was borne chiefly on the basioccipital bone. Internal nares pierced the primary palate, and, farther back, palatal vacuities separated the endopterygoid and parasphenoid bones.

In the transition from fish to tetrapod, the gill chambers disappeared and the operculum was lost. As a result, the labyrinthodont skull became shorter from the eyes back. The hyomandibula no longer participated in jaw movements, and it was freed to conduct sound from the newly acquired eardrum to the otic capsule. Meanwhile, labyrinthodonts developed a longer "facial" area by increasing the length of the bones between the nares and orbits, and this may have been associated with new methods of getting food into the mouth. The branchial skeleton became modified for new functions. As time passed, labyrinthodonts and their descendants, which were the later amphibians and amniotes, slowly lost more fishlike features.

Modern amphibians

The skulls of modern amphibians have become extensively modified, although they remain markedly platybasic (Fig. 8-28). Two occipital condyles are now present, one on each exoccipital bone. Fewer membrane bones form, and ossification has been reduced in the cartilaginous neurocranium. Apodan skulls undergo considerably more endochondral ossification than others.

Reduction of the dermatocranium has resulted in loss of lacrimal (present in *Ranodon*), postfrontal, postorbital, supratemporal, tabular, and postparietal bones, and, in anurans, prefrontals, and has exposed the bones of the otic capsule to dorsal view. Palatal vacuities have increased in

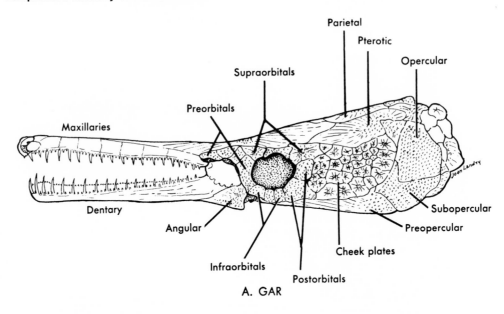

A. GAR

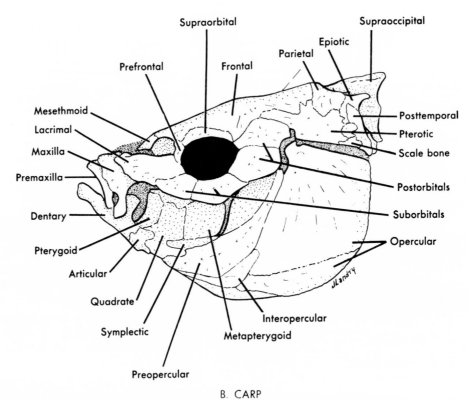

B. CARP

Fig. 8-27. Skull of modern fish (carp) and of more ancient fish (gar) for contrast. Note the scalelike nature of many of the dermal bones in the gar and the series of maxillary bones. In the carp the dark stipple represents unossified cartilage.

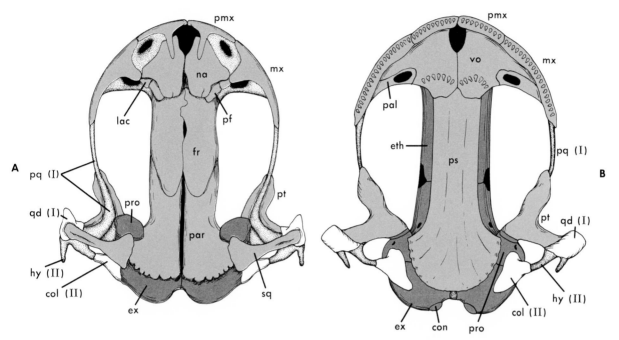

Fig. 8-28. Skull of *Ranodon,* a primitive urodele in the family Hynobiidae. **A,** Dorsal view; **B,** palatal view. Light red indicates dermal bones; dark red, neurocranial bones; stipple, cartilage. **col,** Columella; **con,** occipital condyle; **eth,** sphenethmoid; **ex,** exoccipital; **fr,** frontal; **hy,** dorsal process of hyoid; **lac,** lacrimal; **mx,** maxilla; **na,** nasal; **pal,** palatine; **par,** parietal; **pf,** prefrontal; **pmx,** premaxilla; **pq,** pterygoquadrate cartilage; **pro,** prootic; **ps,** parasphenoid; **pt,** pterygoid; **qd,** quadrate; **sq,** squamosal; **vo,** vomer. I and II indicate origin from first or second visceral arch. (After Schmalhausen,[13] courtesy Academic Press, Inc. From Lebedkina.)

size, with the result that the eyeballs in anurans may be retracted into the oral cavity. In the process, ectopterygoids have been lost, the palatines have become reduced to transverse splinters bracing the upper jaws against the braincase anteriorly, and the pterygoids have been reduced to slender bipartite bones bracing the upper jaws posteriorly (Fig. 8-28, *Rana*). Dermal bones of the mandible have been reduced to a slender dentary and an angulosplenial (Fig. 8-34, frog).

Much of the neurocranium remains cartilaginous throughout life. The only replacement bones in the neurocranium of most anurans and urodeles are two exoccipitals, prootics, and a sphenethmoid. Much of the pterygoquadrate cartilage persists in the palate, either exposed or invested. The quadrate is sometimes unossified.

The skulls of perennibranchiate amphibians are not representative amphibian skulls. Bones that usually ensheath the olfactory epithelium and pterygoquadrate cartilages fail to develop, except premaxillae. Therefore they have no nasals, sphenethmoids, maxillae, jugals, or quadratojugals. The pterygoquadrate cartilages remain in a larval, unossified state in the lateral palatal roof, and the columella remains rudimentary. The evolution of the hyomandibula into a columella has been described by the Russian scientist Schmalhausen with specific reference to two primitive, living urodeles, *Hynobius* and *Ranodon*.[13]

☐ Reptiles

Among generalized features transmitted from labyrinthodonts to cotylosaurs and modern reptiles are the single occipital condyle, numerous membrane bones, extensive ossification of the neurocranium, and, in many lizards, a parietal foramen.

Among specializations in reptilian skulls are temporal fossae (Fig. 8-29), a secondary palate, kinetism in the quadrate bone, turbinal elements in the nasal canals, increased prominence of the dentary in synapsids, and mandibular vacuities in archosaurs. The neurocranium has become tropibasic and has fontanels dorsally (i.e., it has openings) and is therefore incomplete. Most of these mutations have been transmitted to birds, mammals, or both.

TEMPORAL FOSSAE

One of the major structural changes that has overtaken reptiles is development of one or two fossae bounded by bony arches in the temporal region (Fig. 8-29). The fossae provide "working room" for the stout muscles that arise on the temporal region of the skull and insert on the lower jaw. Stem reptiles did not have temporal fossae, and so their skulls are **anapsid.** Today, only turtles lack temporal fossae, and they are placed, with misgivings, in the subclass Anapsida, along with cotylosaurs.

Synapsida, the extinct mammallike reptiles, developed a single temporal fossa bounded by the postorbital, squamosal, and jugal bones, the last two forming an underlying **zygomatic arch.** This **synapsid skull** was transmitted to mammals.

Some reptiles developed a **superior** and an **inferior fossa.** The inferior fossa corresponds in location to the fossa of synapsids. When two fossae are present, there are two lateral temporal arches; hence the term **diapsid skull.** The superior temporal arch formed by the postorbital and squamosal bones lies between the superior and inferior fossae. The outer ear canal passes under the

superior arch to terminate at the tympanic membrane recessed within the inferior temporal fossa. The first birds arose from diapsid reptiles and they, like crocodilians, have diapsid skulls.

If current interpretations are correct, lizards and snakes have modified diapsid skulls, lizards having lost the inferior arch and snakes both arches. Neither lizards nor

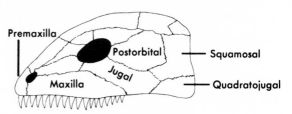

A. ANAPSID (stem reptile)

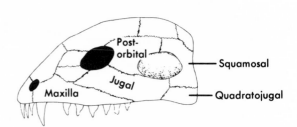

B. SYNAPSID (mammal stock)

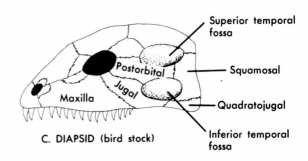

C. DIAPSID (bird stock)

Fig. 8-29. Temporal fossae in reptiles leading to birds and mammals. The squamosal and postorbital bone in the diapsid skull form the superior temporal arch. The squamosal and jugals form the zygomatic arch in the synapsid skull.

snakes have jugal and quadratojugal bones.

Ichthyosaurs and plesiosaurs had one dorsally located temporal fossa, which may or may not have been equivalent to the superior temporal fossa of diapsids. (The postorbital bone met the squamosal below the

fossa.) This **euryapsid** condition no longer exists.

The temporal region of a turtle skull is an enigma. It has no temporal fossae, which suggests a primitive condition with complete roofing. Yet there seems to have been

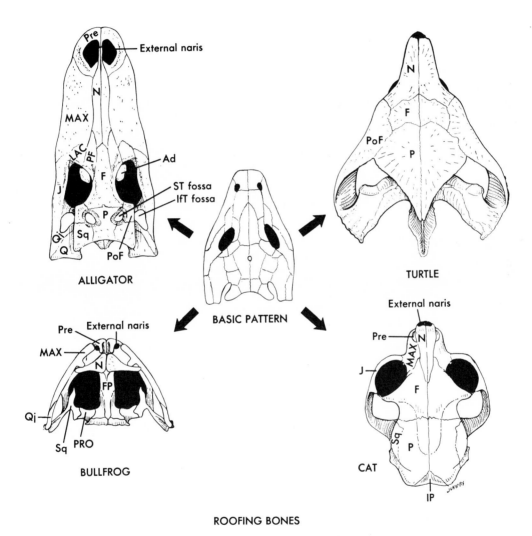

ROOFING BONES

Fig. 8-30. Roofing area and associated bones in selected tetrapods, dorsal views. The basic pattern represents a labyrinthodont. The turtle is an alligator snapping turtle, *Macroclemys temminckii.* **F,** Frontal; **LAC,** lacrimal; **N,** nasal; **P,** parietal; **PF,** prefrontal; **PoF,** postfrontal; **Pre,** premaxilla; **Sq,** squamosal; **ST,** supratemporal fossa; **Qj,** quadratojugal; **Ad,** adlacrimal; **FP,** frontoparietal; **IfT,** infratemporal fossa; **IP,** interparietal; **J,** jugal; **MAX,** maxilla; **PRO,** prootic; **Q,** quadrate. As a study aid you may wish to color homologous bones on the different skulls with the same colors.

considerable excavation at the rear. Supratemporal, tabular, and postparietal bones are missing, the postorbital is included in the postfrontal, and parietal and squamosal bones have receded, leaving a wide gap in the temporal region (Fig. 8-30). Therefore, it cannot be said with certainty that the turtle skull is a true anapsid condition.

PALATES

Development of a secondary palate occurred first in reptiles. In crocodilians it is exceptionally long because of the very long facial region (Fig. 8-31). Palatal processes of the premaxilla, maxilla, palatine, and pterygoid bones all unite in crocodilians to form a completely bony secondary palate,

with the internal nares far to the rear. In most other reptiles the palatal processes of these bones do not meet, and the secondary palate is incomplete. Varying degrees of completeness in turtles are illustrated in Fig. 8-32.

The primary palatal bones of snakes and some lizards, the quadrate bones, and sometimes certain bones of the upper jaw are movable as independent units, so that some snakes can open their mouth wide enough to swallow objects larger than their own head (Fig. 8-33, *A* and *B*). The condition is known as cranial **kinesis.** Kinetism was present in crossopterygians and in early amphibians leading to reptiles. It has been transmitted to birds (Fig. 8-33, *C*) but has been lost

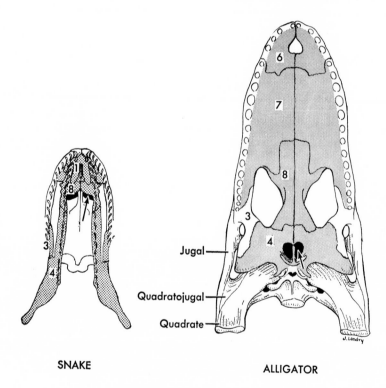

SNAKE   ALLIGATOR

SECONDARY PALATES

Fig. 8-31. Secondary palates (gray) of two reptiles. Compare location of internal nares (arrows) in alligator with position in snake *(Natrix)* and turtles (Fig. 8-32). **1,** Vomer (of primary palate); **3,** ectopterygoid; **4,** pterygoid; **6,** palatine process of premaxilla; **7,** palatine process of maxilla; **8,** palatine process of palatine.

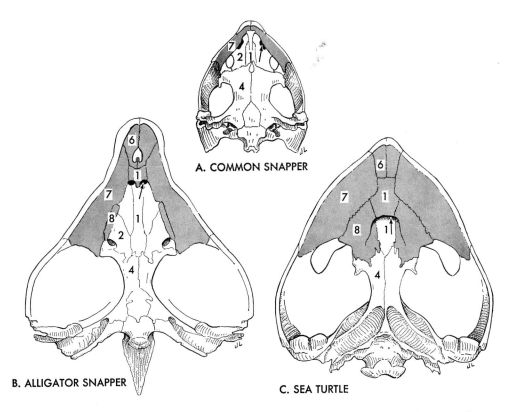

A. COMMON SNAPPER

B. ALLIGATOR SNAPPER

C. SEA TURTLE

Fig. 8-32. Species differences in the secondary palate (gray) of turtles. **A,** *Chelydra serpentina.*
B, *Macroclemys temminckii.* **C,** *Lepidochelys olivacea* Ridley, the posterior part of the
quadrate and the squamosal and supraoccipital regions omitted. In **A** only the maxilla,
7, participates in formation of the secondary palate. In **B** the premaxillae, **6,** also participate,
and the palatines, **8,** make a small contribution. In **C** all three bones make contributions,
as does the vomer also. **1,** Vomer; **2,** palatine bone of primary palate; **4,** pterygoid; **6,** palatine
process of premaxilla; **7,** palatine process of maxilla; **8,** palatine process of palatine bone.
Arrows indicate position of internal nares. You may wish to color homologous bones
with the same colors.

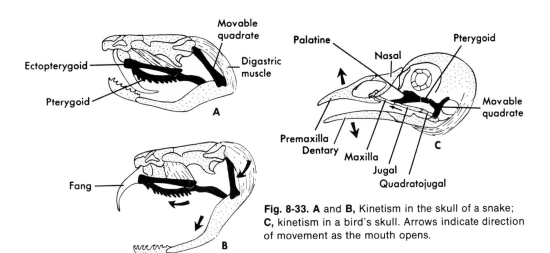

Fig. 8-33. A and **B,** Kinetism in the skull of a snake;
C, kinetism in a bird's skull. Arrows indicate direction
of movement as the mouth opens.

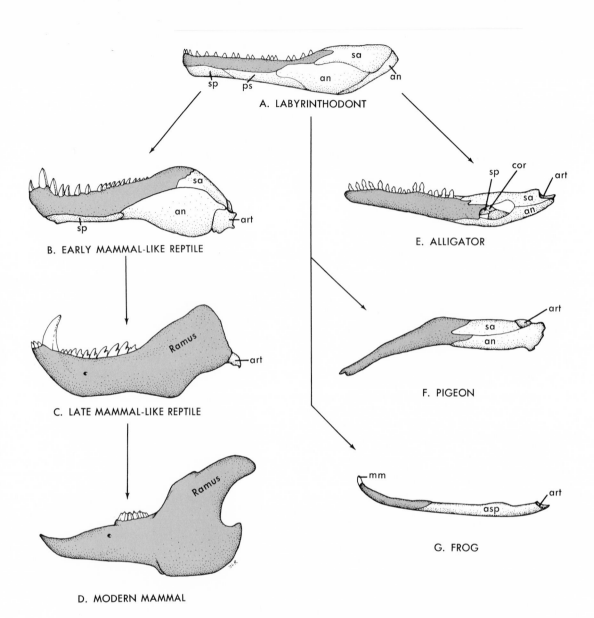

Fig. 8-34. The mandibles of tetrapods. **A** to **D,** Probable evolutionary stages leading to modern mammals. The dentary (red) became increasingly larger, whereas other bones were reduced and finally lost. Arrows indicate phylogenetic pathways. **E** to **G,** Lower jaws of three modern tetrapods for comparison with basic pattern. All are dermal bones except the mentomeckelian and articular. **an,** Angular; **art,** articular (cartilage in frog); **asp,** angulosplenial; **cor,** coronoid; **mm,** mentomeckelian; **ps,** postsplenial; **sa,** surangular; **sp,** splenial. **B,** Pelycosaur; **C,** late therapsid; **D,** rabbit. Left lateral views.

Table 8-2. Skulls of early tetrapods contrasted with those of modern amphibians and reptiles with reference to a few selected characteristics

	Early tetrapods	Modern reptiles	Modern amphibians
Neurocranium	Well ossified	Well ossified	Mostly cartilage
	One condyle	One condyle	Two condyles
	Platybasic	Tropibasic	Platybasic
Primary palate	Complete complement of dermal bones	Relatively complete	Fewer
	Parasphenoid small	Small	Large in urodeles
	Vacuity small	Small	Large in anurans
	Internal nares lateral	Medial	Lateral
Secondary palate	None	Partial or complete	None
Dermal roofing bones	Complete complement	Some reduction	Extensive reduction
Parietal foramen	Present	Present in some	Confined to larvae
Marginal bones	Complete complement	Usually complete	Fewer
Bones ensheathing Meckel's cartilage	Numerous	Numerous	Fewer

in modern amphibians and mammals. The movable quadrate in reptilian precursors of mammals may have facilitated the transition of the quadrate from a bone of the upper jaw to a bone of the middle ear.

INCREASED IMPORTANCE OF THE
DENTARY BONE

In fossil reptiles leading to mammals the dentary became increasingly prominent. A ramus developed on the dentary and extended upward toward the temporal region (Fig. 8-34, C). It served for the insertion of the increasingly massive lower jaw muscles. Meanwhile, the other bones of the mandible became reduced, presaging the loss in mammals of all mandibular bones except the two dentaries (Fig. 8-34, D).

In Table 8-2 the skulls of early tetrapods and modern amphibians and reptiles are contrasted with respect to a few traits. The primitive pattern is more nearly expressed in modern reptiles than in modern amphibians.

□ **Birds**

The bird skull is reptilian in structure. There is a reptilian complement of dermal bones, an incomplete secondary palate, a single occipital condyle, reptilian vacuities and fossae, and kinetism (Fig. 8-33, C). Also, the neurocranium is well ossified and incomplete dorsally, and there is a single ossicle, the columella, in the middle ear.

Modifications of the reptilian condition are associated partly with flight, altered feeding habits, and increased brain size. The dermal bones are very thin, and the sutures are obliterated except in ratites. Although the skull is diapsid, the bony arch between the superior and inferior fossae has been lost along with other bones, which decreases the flying load. The premaxillae and dentaries (and sometimes the maxillae and nasals) are elongated to form a beak, which is used in feeding. Two turbinal cartilages (instead of one, as in crocodilians) are formed. The much larger brain of birds results in a vaulted skull, the frontal and parietal bones arching downward alongside the brain. The parietal foramen has been lost, and the lacrimal bones are pierced by a nasolacrimal duct that drains fluid from the surface of the eyeball into the nasal cavity.

□ Mammals

Mammalian skulls are derived from synapsid reptiles. The single temporal fossa is bounded ventrally by a zygomatic arch, although the arch is very thin or incomplete in some insectivores. There are two occipital condyles inherited from synapsids, and the secondary palate is complete. The skull has become increasingly domed as the cerebral hemispheres have expanded. As a result, the frontal and parietal bones are vaulted and extend downward beside the brain.

Unique in mammals was the capture of the articular and quadrate bones by the middle ear cavity, which provided mammals with three ear ossicles—**malleus** (articular), **incus** (quadrate), and **stapes** (columella or hyomandibula). Removal of the quadrate and articular from the jaws and expansion of the dentary during synapsid evolution resulted in the dentary articulating with the squamosal, or squamous portion of the temporal bone, an innovation among vertebrates.

DERMATOCRANIUM

The dermatocranium is usually represented in mammals by paired premaxillae, maxillae, nasals, lacrimals, jugals (more often called malars or zygomatics in mammals), and squamosals; and by unpaired frontals, parietals, and interparietals, all of which are paired in embryos. The squamosal may be an independent bone, as in rabbits, or it may become part of the temporal, as in cats and man. *Homo erectus* had an interparietal, and it is still present in some populations of Mongolians. Because it was first described in Inca Indians, it is sometimes called the Inca bone (Fig. 8-35).

Fig. 8-35. Inca bone in a human skull from the Aleutian Islands. **P**, Parietal; **O**, occipital. (Courtesy William S. Laughlin.)

The frontal bone often houses an air-filled sinus (Fig. 8-15). In sheep and goats that butt heads as part of a mating ritual, the frontal sinus extends into the horns, which are outgrowths of the frontal bones. The males come together at speeds up to 35 miles per hour and the presence of the sinus, with its bony internal braces, causes the compression wave to be shunted away from the brain via the bones of the skull to the vertebral column.[11]

Of the primary palatal elements inherited from reptiles, the vomer, now unpaired, lies at the base of the nasal septum and the palatines lie in the lateral wall of the nasopharynx, where they contribute to the orbit. The pterygoids are reduced to small wing-like processes of the sphenoid complex, where they form part of the lateral wall of the nasopharynx medial to the pterygoid fossa (Fig. 8-36). The parasphenoid and ectopterygoids have been lost.

The secondary palate consists of two parts. The "hard," or bony, palate consists of palatine processes of the premaxilla, maxilla, and palatine bones (Fig. 8-16, *D*).

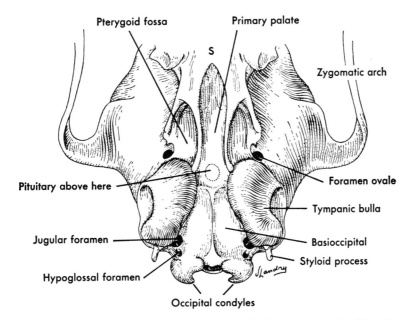

Pterygoid fossa

Primary palate

S

Zygomatic arch

Pituitary above here

Foramen ovale

Tympanic bulla

Jugular foramen

Basioccipital

Styloid process

Hypoglossal foramen

Occipital condyles

Fig. 8-36. Hamster skull, caudal part, ventral view. **S,** Secondary palate. The primary palate is the roof of the nasopharynx.

Behind the bony palate is the soft palate, so named because bone does not form in it.

NEUROCRANIUM

No neurocranium forms above the brain. Occipitals form a caudal wall and floor for the hindbrain; sphenoids form chiefly a floor for most of the rest of the brainstem; and ethmoids, which ossify from the olfactory capsule, form a floor for the olfactory bulbs, house the olfactory epithelium, and transmit the olfactory nerve fibers. The otic capsule becomes a single periotic, or petrosal, which lies mostly beneath the brain because of the great expansion of the mammalian cerebral hemispheres. The endochondral components of a typical mammalian skull are diagrammed in Fig. 8-37.

TEMPORAL BONE COMPLEX

The temporal complex (Figs. 8-38 and 8-39) consists of numerous components of

intramembranous and endochondral origin. The **squamous portion** represents the squamosal bone of lower vertebrates. The **petrosal (petrous) component** is the ossified otic capsule. The **tympanic portion,** new in mammals, surrounds the middle ear to form a tympanic bulla (Fig. 8-36). Associated with the tympanic portion is a bony ring, the **annulus tympanicus** (Fig. 8-24), derived (according to evidence from embryonic opossums) from the angular bone of reptiles. The tympanic membrane is attached to it. A **mastoid portion** of endochondral origin is new in mammals.

Although the tympanic and petrous components are separate bones in some mammals, they frequently unite to form a **petrotympanic bone,** as in rabbits. The petrotympanic may unite with the squamosal to form a **temporal bone,** as in cats and man. A **styloid process** from the hyoid arch may coalesce with the temporal bone ventrally (Fig. 8-38).

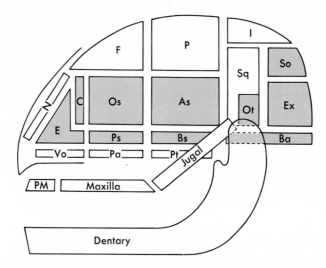

Fig. 8-37. Chief endochondral (red) and dermatocranial (white) components in a typical mammalian skull. The vomer, palatine, and pterygoid are parts of the primary palate. The premaxilla and maxilla contribute horizontal processes to the secondary palate. The dentary is a membrane bone of the visceral skeleton. The alisphenoid is said to be derived from the pterygoquadrate cartilage or, occasionally, to be intramembranous in origin.

As, Pleurosphenoid (alisphenoid)
Ba, Basioccipital
Bs, Basisphenoid
C, Cribriform plate of ethmoid
E, Ethmoid, perpendicular plate
Ex, Exoccipital
F, Frontal

I, Interparietal
N, Nasal
Os, Orbitosphenoid
Ot, Otic (petrous)
P, Parietal
Pa, Palatine

PM, Premaxilla
Ps, Presphenoid
Pt, Pterygoid
So, Supraoccipital
Sq, Squamosal
Vo, Vomer

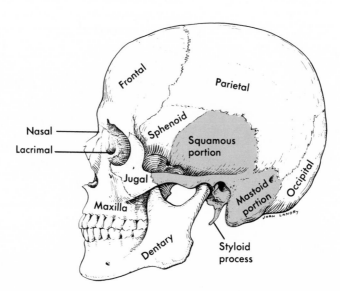

Fig. 8-38. Skull of modern man. The temporal bone is red.

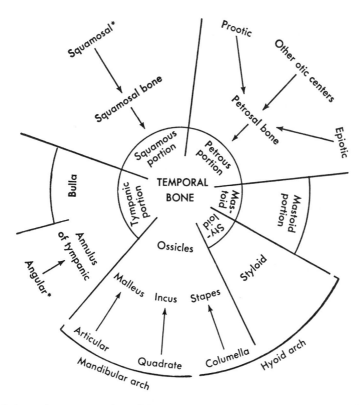

Fig. 8-39. Schematic representation of the multiple nature of the temporal bone of mammals. Note reduction in number of separate elements from the condition in reptiles (outer circle) to mammals (other circles). The two dermal elements have asterisks. The mastoid portion and tympanic bulla are mammalian innovations. The ossicles are within the temporal bone but not part of it.

☐ Reduction in number of bones during phylogeny

The number of individual bones, especially membrane bones, has tended to be reduced during phylogeny. With reference to Fig. 3-32, any group at the end of an arrow has fewer bones in the skull than the group preceding it in the phylogenetic line. Labyrinthodonts had fewer than crossopterygians, cotylosaurs had fewer than labyrinthodonts, modern reptiles have fewer than cotylosaurs, and mammals have fewer than mammallike reptiles. Modern amphibians have fewer than their ancestors, the labyrinthodonts. This generalization does not mean that *modern* reptiles have fewer bones than

modern amphibians. In fact, reptiles have more. But then, modern reptiles were not derived from modern amphibians.

The reduction is a result of fusion of adjacent ossification centers to form composite bones and, sometimes, reduction in the number of ossification centers. In frogs the frontal and parietal bones are represented by a single **frontoparietal.** In man the left and right frontal bones of the fetus (Fig. 8-11) unite at about the eighth year of life to form a single bone. Reduction in the number of membrane bones in the mandible (Fig. 8-34 and Table 8-3) illustrates this trend.

Membrane bones frequently unite with adjacent replacement bones, thereby giving

Table 8-3. Reduction in number of dermal bones investing Meckel's cartilage when early vertebrates are contrasted with later ones

Fishes			Tetrapods				
			Primitive	Modern			
Primitive	Crossop-terygians	Modern	Labyrintho-donts	Reptiles and birds	Amphibians	Mammals	
Dentary	Dentary	Dentary†	Dentary	Dentary	Dentary	Dentary
Angular	Angular	Angular‡	Angular	Angular	Angular¶	
Surangular	Surangular		Surangular	Surangular		
Infradentary*	Splenial		Splenial	Splenial	Splenial¶	
Infradentary	Coronoid		Coronoid	Coronoid		
Infradentary	Prearticular	Derm-articu-lar§	Prearticular			
Infradentary			Intercoro-noid			
Infradentary			Precoronoid			
Infradentary			Postsplenial			

Primitive forms had a greater number of bones than modern ones. Reptiles have retained more of the primitive elements than other modern tetrapods.

*Variable number.
†Dentary incorporates mentomeckelian of endochondral origin in some teleosts.
‡May be absent. Sometimes named surangular.
§May include articular of cartilage origin.
¶Sometimes incorporated in an angulosplenial.

rise to a single bone with a dual history. Postfrontal and supratemporal bones may unite with replacement bones of the otic capsule to form a **sphenotic** and **pterotic bone,** respectively. The squamosal in mammals unites with otic and other elements to contribute to a **temporal bone.** The mammalian interparietal may unite with the supraoccipital. Unions such as these result in reducing the number of individual bones in the skulls of more recent tetrapods.

□ Chapter summary

1. Sharks have a neurocranium constructed of prechordal and parachordal cartilages, notochord, and cartilaginous olfactory and otic capsules knit solidly together and completed by cartilaginous walls and a roof above the brain.

2. Sharks have a visceral skeleton consisting of pterygoquadrate and Meckel's cartilages, hyoid cartilages, and a series of additional branchial cartilages, all braced against the otic capsule by the hyomandibula (hyostylic jaw suspension).

3. The cranial skeleton of cyclostomes consists of loosely articulated cartilages and a branchial basket. The latter is not readily homologizable with the visceral skeleton of other vertebrates. There are no recognizable pterygoquadrate and Meckel's cartilages.

4. The visceral skeleton of bony fishes is replaced most commonly by the following bones: In the **pterygoquadrate cartilage**—palatine, metapterygoid, epipterygoid, and quadrate; in **Meckel's cartilage**—articular; in the **hyoid arch**—symplectic, interhyal, epihyal, ceratohyal, hypohyal, basihyal. Other centers form in the basibranchials and gill arches.

5. Jaw suspension in fishes is mostly hyostylic, autostylic, or amphistylic.

6. Bony vertebrates have a neurocranium, dermatocranium, and visceral skeleton. The neurocranium is cartilaginous initially but is later partly or wholly replaced by bone. Except in lower vertebrates it is usually incomplete above the brain. The visceral skeleton is initially cartilaginous but is partly replaced by bone, and in the lower jaw it is partly invested by membrane bone. The dermatocranium consists of bones of the skull that arise by intramembranous ossification.

7. The chief centers of ossification in the neurocranium are occipital, sphenoid, ethmoid, and otic. Common replacement bones are exoccipital, basioccipital, supraoccipital; presphenoid, basisphenoid, orbitosphenoid, laterosphenoid (pleurosphenoid); mesethmoid, cribriform plate, ethmoturbinals; prootic, epiotic, opisthotic, and petrosal. In higher forms they may be reduced in number by loss or fusion. Scleral rings often form in the optic capsule (sclera of eyeball).

8. Dermatocranial bones are numerous, superficial, and scalelike in generalized fishes, smooth and more deeply situated in modern fishes and tetrapods.

9. The chief dermal bones of tetrapod skulls include (a) **roofing bones**—nasal, frontal, parietal, interparietal, intertemporal, supratemporal, squamosal, tabular, lacrimal, prefrontal, postfrontal, postorbital, infraorbital (jugal), quadratojugal; (b) **primary palatal bones**—vomer, parasphenoid, palatine, endopterygoid, ectopterygoid; (c) **upper jaw bones**—premaxilla, maxilla; and (d) opercular bones.

10. The visceral skeleton of tetrapods has been modified for life on land, including sound transmission, attachment of tongue muscles, and support of the larynx.

11. The chief centers of endochondral ossifica-

tion in the visceral skeleton of tetrapods are (a) in the **pterygoquadrate cartilage**—epipterygoid (alisphenoid in mammals), quadrate (incus in mammals); (b) in **Meckel's cartilage**—articular (malleus in mammals), mentomeckelian; and (c) in the hyoid and successive visceral arches—hyomandibula (columella or stapes in tetrapods), body and horns of hyoid, styloid process (mammals), laryngeal skeleton.

12. Occipital condyles in ancient tetrapods were single. The condition was transmitted to modern reptiles and birds. Modern amphibians, synapsid reptiles, and mammals have two occipital condyles.

13. Modern amphibians have lost many membrane bones, and much of the neurocranium remains unossified. Apodan skulls have diverged least among modern amphibians.

14. Modern reptiles retain extensive ossification of the neurocranium, a single occipital condyle, numerous membrane bones (fewer in snakes and lizards), and a parietal foramen in lizards. Among specializations are temporal fossae except in turtles, a partial secondary palate (complete in crocodilians), mandibular vacuities and turbinal bones in crocodilians, cranial kinetism in snakes and lizards, and increased prominence of the dentary in synapsids.

15. Temporal fossae result in diapsid (two-arch) and synapsid (one-arch) skulls, and variants. The former were transmitted to birds, the latter to mammals.

16. Bird skulls are diapsid, essentially reptilian, and often kinetic. The dermal bones are thin, and sutures have been obliterated except in ratites. The jaws and facial bones have elongated to form a beak. The skull is highly domed to accomodate the expanded brain.

17. Mammalian skulls are synapsid and domed because of expansion of the cerebral hemispheres. The number of separate bones has been reduced. There are three ear ossicles, a complete secondary palate, a mastoid region, and, frequently, a tympanic bulla. The dentary is the sole bone of the mandible. It articulates with the squamosal.

18. The cranial skeleton is chiefly ectomesenchymal, although sclerotomes contribute partly to the neurocranium.

LITERATURE CITED AND SELECTED READINGS

1. Ayers, H., and Jackson, C. M.: Morphology of the Myxinoidei. I. Skeleton and musculature, Journal of Morphology **17**:185, 1901.
2. Bock, W. J.: Kinetics of the avian skull, Journal of Morphology **114**:1, 1964.
3. Brodal, A., and Fänge, R., editors: The biology of *Myxine*, Oslo, 1963, Norway Universitetsforlaget.
4. De Beer, G. R.: The development of the vertebrate skull, Oxford, 1937, The Clarendon Press.
5. Frazzetta, T. H.: Adaptive problems and possibilities in the temporal fenestration of tetrapod skulls, Journal of Morphology **125**:145, 1968.
6. Gregory, J. T.: The jaws of the cretaceous toothed birds, *Ichthyornis* and *Hesperornis*, The Condor **54**:73, 1952.
7. Gugg, W.: Der Skleralring der plagiotremen Reptilien, Zoologische Jahrbücher, part 2, **65**:339, 1939.
8. Harrington, R. W., Jr.: The osteocranium of the American cyprinid fish, *Notropis bifrenatus*, with an annotated synonymy of teleost skull bones, Copeia, no. 4, p. 267, 1955.
9. Jollie, M. T.: The head skeleton of the lizard, Acta Zoologica **41**:1, 1960.
10. Manville, R. C.: Bregmatic bones in North American lynx, Science **130**:1254, 1959.
11. Reed, C. A., and Schaffer, W.: Evolutionary implications of cranial morphology in the sheep and goats, American Zoologist **6**:565, 1966.
12. Ørvig, T.: The latero-sensory component of the dermal skeleton in lower vertebrates and its phyletic significance, Zoologica Scripta **1**:139, 1972.
13. Schmalhausen, I. I.: The origin of terrestrial vertebrates, New York, 1968, Academic Press, Inc.
14. Stahl, B. J.: Vertebrate history: problems in evolution, New York, 1974, McGraw-Hill Book Co.
15. Watson, D. M. S.: The evolution of the mammalian ear, Evolution **7**:159, 1953.

9

Appendicular skeleton

In this chapter we will examine the skeletal components of the girdles, fins, and limbs. We will see that fin skeletons vary but that all tetrapod limbs are built in accordance with a single basic blueprint. Finally, we will examine some of the theories of how paired fins and limbs may have originated.

The appendicular skeleton includes the pectoral and pelvic girdles and the skeleton of the fins and limbs. Agnathans, caecilians, snakes, and some lizards have no appendicular skeleton, and it has been much reduced in certain other vertebrates, as we will see shortly.

■ PECTORAL GIRDLES

The pectoral girdle is embedded in the body wall at the anterior end of the trunk. There it serves as a brace for the anterior appendages, since the anterior fin or limb articulates with the pectoral girdle in a glenoid fossa on the scapula. In bony vertebrates the girdle consists of membrane bones and replacement bones. The membrane bones represent dermal armor. Replacement bones form from the underlying cartilaginous endoskeleton. Early fishes had three replacement bones—**coracoid, scapula,** and **suprascapula**—and a series of dermal bones, including **clavicle, cleithrum, supracleithrum,** and **posttemporal.** Fig. 9-1 shows subsequent phylogenetic trends. In later bony fishes (Fig. 9-1, ganoid fish) there was a tendency to reduce the number and size of the replacement bones, whereas in tetrapods there was a tendency to reduce the number of dermal bones.

☐ Fishes

The pectoral girdles of living bony fishes (Figs. 9-2 and 9-11) have a reduced coracoid and scapula (often united in teleosts to form a coracoscapula), a large prominent cleithrum, and a supracleithrum. An old-fashioned posttemporal bone still connects the supracleithrum to the skull. When a clavicle is present it is much reduced, but some ganoid fishes have acquired an extra couple of dermal bones (Fig. 9-2, postcleithra).

Cartilaginous fishes lack the ability to deposit bone anywhere other than in their skin. As a result, they have no dermal bone contributions to their girdle, and their coracoid, scapula, and suprascapula remain cartilaginous throughout life (Fig. 9-3).

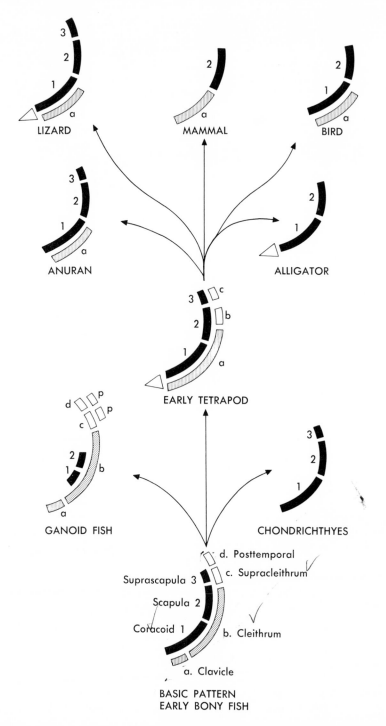

Fig. 9-1. Pectoral girdle in selected phylogenetic lines. Cartilage and replacement bones are black. Triangles represent interclavicle. **1,** In alligator and bird, procoracoid; **p,** postcleithrum. Only one-half of each girdle is illustrated, and relationships have been distorted when necessary to emphasize homologies.

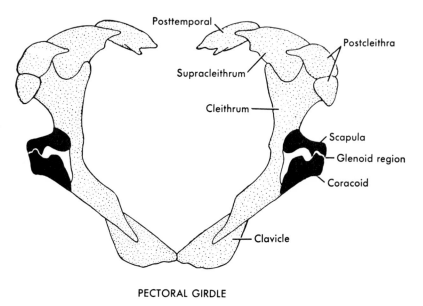

PECTORAL GIRDLE

Fig. 9-2. Pectoral girdle of the ganoid fish *Polypterus.* Dermal bones are stippled, replacement bones are black.

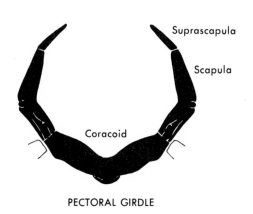

PECTORAL GIRDLE

Fig. 9-3. Cartilaginous pectoral girdle of the shark *Squalus,* anterior view.

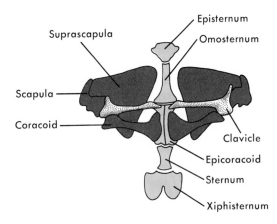

Fig. 9-4. Sternum (red) and pectoral girdle of frog, ventral view. Replacement bones of girdle are black.

□ **Tetrapods**

The pectoral girdles of early tetrapods were quite similar to those of their crossopterygian ancestors (Fig. 9-1). A new midventral dermal bone, the **interclavicle**, appeared, and the coracoids began to brace the forelimbs against the newly acquired sternum (Figs. 7-23, lizard, chicken, and 9-4). This was essential if the limbs were to push more and more forcefully against the earth for support and locomotion. Later, the clavicle also expanded until it assisted, or even replaced, the coracoid in bracing the forelimbs (Figs. 7-23, monkey, and 9-4). Meanwhile, the cleithrum and supracleithrum were lost. The clavicle has also been reduced or lost in crocodilians, legless lizards, and some mammals. Urodeles have

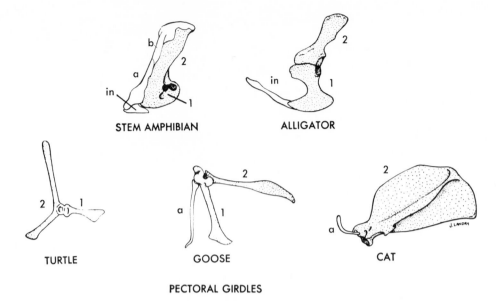

STEM AMPHIBIAN   ALLIGATOR

TURTLE   GOOSE   CAT

PECTORAL GIRDLES

Fig. 9-5. Left half of the pectoral girdle of selected tetrapods, lateral views. Replacement bones are stippled. **1,** Coracoid (procoracoid in alligator and birds); **2,** scapula. **a,** Clavicle; **b,** cleithrum; **in,** interclavicle. In turtles the clavicle and interclavicle are fused with the shell.

lost all dermal bones of the girdle. The entire girdle is missing in snakes.

Monotremes have a reptilian pectoral girdle, but above monotremes only a clavicle and scapula remain, and sometimes the clavicle is absent. Cetaceans, ungulates, subungulates, and some carnivores have no clavicle. In other carnivores, such as cats, the clavicle has been reduced to a slender bony splinter that fails to reach either the sternum or the scapula. The scapula, on the other hand, is always present in mammals. It has a broad, flat surface divided by a scapular spine into supraspinous and infraspinous fossae for the origin of the extensive muscles that insert on the humerus. The pectoral girdles of a few tetrapods are illustrated in Fig. 9-5.

■ **PELVIC GIRDLES**

Pelvic girdles brace the posterior paired appendages just as pectoral girdles brace

the anterior ones. Unlike pectoral girdles, however, they have no dermal components.

□ **Fishes**

The pelvic girdle in most fishes consists of two cartilaginous or bony **pelvic (ischiopubic) plates** that articulate with the pelvic fins (Fig. 9-6, herring). The plates usually meet medially in a symphysis. In sharks (Figs. 9-6 and 9-7) and lungfishes two embryonic plates unite to form a single adult plate. In teleosts with a short trunk the pelvic girdle lies immediately behind the pectoral and is often attached to the latter.

□ **Tetrapods**

Tetrapod embryos, like fishes, form a pair of cartilaginous pelvic plates. Each plate ossifies at two centers to form a **pubic bone (pubis)** and an **ischium** (Fig. 9-8). Dorsally on each side an additional cartilage becomes the **ilium.** At the junction of the three

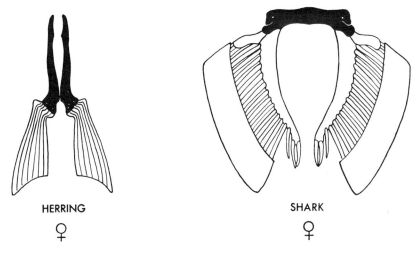

HERRING   SHARK
♀   ♀

Fig. 9-6. Pelvic plates, or girdles (black), of a bony and cartilaginous fish.

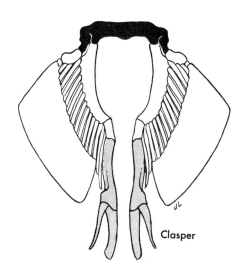

Clasper

Fig. 9-7. Pelvic girdle (black) and fin of a male shark, showing basal fin cartilages modified as claspers (gray). Compare female shark, Fig. 9-6.

components is the acetabulum, a socket accommodating the head of the femur.

The femurs push against the ilia and the ilia push against the sacral vertebrae (via transverse processes or sacral ribs), so that the sacral region of the vertebral column receives part of the force exerted against the trunk as a result of standing with four feet on the ground. Midventrally, the left and right pubic bones meet in a pubic symphysis, the ischia meet in an ischial symphysis, and these symphyses receive some of the force. In bipedal animals, and in those that are modified for jumping, aerial flight, swimming, and other specialized activities, the three bones of the girdle are adaptively modified in size, shape, or relationship. A squatting and jumping frog, for instance, has a different set of vectors affecting the pelvic girdle than does a turtle, whale, or kangaroo, and this is expressed in the anatomy of the girdle. An analysis of these different vectors is in the purview of biomechanics.

In frogs (Fig. 7-8) the ilia are greatly elongated and extend from the sacral vertebra to the end of the urostyle, where they meet the ischia and pubic bones and where the acetabula are located. The joint between ilium and sacral rib (sacroiliac joint) is freely movable, and there is motion in the joint when a frog pushes off and stretches out on a jump. In many other vertebrates the ilia are firmly ankylosed to the sacral vertebrae and there is no movement in the joint.

In reptiles each ilium is braced against *two* sacral vertebrae, and the ilia become

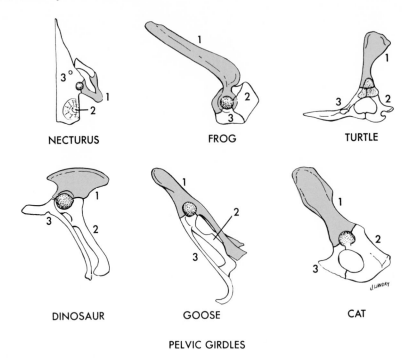

NECTURUS   FROG   TURTLE

DINOSAUR   GOOSE   CAT

PELVIC GIRDLES

Fig. 9-8. Left halves of the pelvic girdles of selected tetrapods. Left lateral views except of *Necturus,* which is a ventral view. **1,** Ilium, red; **2,** ischium; **3,** pubis. The acetabulum is stippled. In *Necturus* the ischium is an ossification center in the cartilaginous pubic plate, and the ilium is shown with a sacral rib attached dorsally.

broad for the attachment of vertebral as well as hind limb muscles. Both of these sets of muscles are necessary in a four-legged animal that keeps its trunk elevated above the ground.

In birds the ilium and ischium are greatly expanded to accommodate the musculature for bipedalism (Fig. 7-15, *A*), and the girdle is braced against lumbar as well as sacral vertebrae (Fig. 7-15, *B*). The pubic bones, on the contrary, are reduced to long splinters (Fig. 9-8, goose). The absence of a pubic symphysis provides a larger pelvic outlet for laying the massive eggs.

In mammals the ilium, ischium, and pubis unite early in postnatal life to form an **innominate bone** (hip bone, or **os coxae**) on each side. The two innominates comprise the pelvic girdle and are ankylosed with the sacrum dorsally. Occasionally an **acetabular (cotyloid) bone** ossifies in the wall of the acetabulum. Marsupials and monotremes have two small **epipubic (marsupial)** bones articulating with the pubic bones and extending forward in the abdominal wall. These support the abdominal pouch in which baby marsupials are transported.

Because of the pubic and ischial symphyses ventrally, and because the ilia are united with the sacrum dorsally, the caudal end of the coelom is surrounded by a bony ring, the **pelvis.** The pelvis contains the caudal end of the digestive and urinogenital systems. The inferior border of the ring is the pelvic outlet through which eggs or living young must pass in females. In limbless tetrapods the pelvic girdle has been either reduced or lost altogether (Fig. 9-22, *C*).

In mammals the ligaments uniting the

pubic and ischial bones ventrally at their symphyses are softened during pregnancy by hormones. This facilitates expansion of the pelvic outlet for delivery of fetuses. In mice 6 days pregnant, the gap between the two pubic bones at the pubic symphysis has been shown by x-ray films to be only 0.25 mm. Thirteen days later, on the day of birth, the gap has widened to 5.6 mm.

■ FINS

All jawed fishes have pectoral and pelvic fins except eels, which lack pelvics. Fins, both paired and median, serve primarily as steering devices and rudders rather than for propulsion except in fishes with unusually rigid bodies.

There has been divergence in the structure of fins in different phylogenetic lines. The chief varieties today are **lobed fins, fin fold fins,** and **ray fins** (Fig. 9-9). Flexible filaments called fin rays produced in the dermis of the fin stiffen all fins to their edges. In cartilaginous fishes the rays are fibrous (ceratotrichia); in bony fishes they are modified dermal scales (lepidotrichia) that are lined up end to end.

Lobed fins are found in sarcopterygians. Early lobed fins had an axial skeleton (Fig. 9-10, *A*) consisting of jointed cartilages or bones (**axials**) with anterior (preaxial) and posterior (postaxial) **radials** extending from the axis. Fins like these are still present in the lungfish *Neoceratodus*. Such a fin is said to be biserial because of its skeletal arrangement. Fin fold fins (Fig. 9-9) are found in cartilaginous fishes. Early ones had what appears to be a modification of the biserial fin (Fig. 9-10, *B*). Modern sharks have a fin skeleton consisting of one to five **basal cartilages** and one or several rows of **radials** (Fig. 9-10, *C*). Ray-finned fishes have tended to lose proximal components of the fin skeleton, and modern teleosts (Fig. 9-11) have lost axials, basals, and some have even lost the radials. As a result, fin rays in modern fishes extend all the way from the girdle to

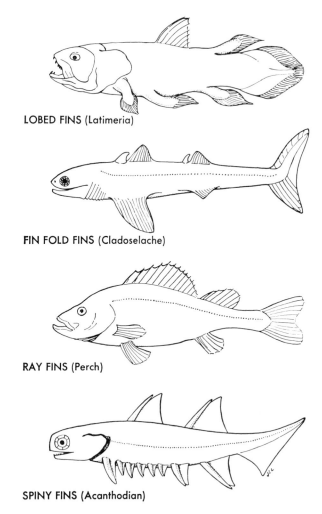

LOBED FINS (Latimeria)

FIN FOLD FINS (Cladoselache)

RAY FINS (Perch)

SPINY FINS (Acanthodian)

Fig. 9-9. External appearances of some fins.

the free edge of the fin. Spiny fins (Fig. 9-9) were characteristic of acanthodian placoderms. Because they are so ancient they provide a basis for speculation concerning the origin of paired fins (p. 205).

The skeleton of the pelvic fins of male cartilaginous fishes has been modified to form an intromittent organ (**clasper**) for conducting sperm to the female uterus (Fig. 9-7).

Most fishes have median fins that help to prevent torque and yaw during swimming. There may be one or two dorsal fins, an anal

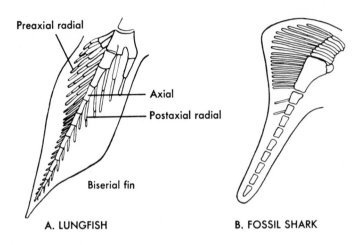

Preaxial radial

Axial

Postaxial radial

Biserial fin

A. LUNGFISH

B. FOSSIL SHARK

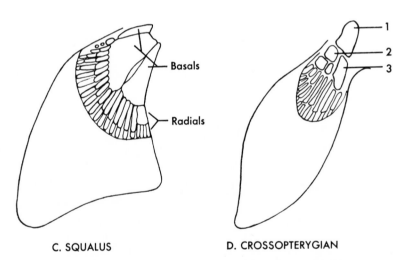

Basals

Radials

1

2

3

C. SQUALUS

D. CROSSOPTERYGIAN

Fig. 9-10. Selected fin skeletons, reoriented. Fin rays are not shown. **A,** Biserial pelvic fin of *Neoceratodus,* a living lobefinned fish. **B** and **C,** Pectoral fins of the Paleozoic shark *Cladodus* and the modern shark *Squalus.* **D,** Pectoral fin of an ancient crossopterygian, *Eusthenopteron.* **1** to **3** are equivalent by position to the humerus, ulna, and radius of a tetrapod limb.

fin behind the anus or vent, and caudal fins. The anal fin may be lost in bottom dwellers, and in males of viviparous teleost species it is sometimes modified as an intromittent organ, the **gonopodium.** In egg-laying teleost species it is sometimes modified as an **ovipositor.** The skeleton of median fins usually consists of radials and rays.

■ **LIMBS**

Commencing with amphibians vertebrates typically have four legs. However, some have lost one or both pairs and in others one pair is modified as arms, wings, or paddles. By employing limbs, tetrapods swim, crawl, walk, run, hop, jump, dig, climb, glide, or fly from place to place and thus avoid enemies, seek food, and find a mate.

Tetrapod limbs, anterior and posterior, consist of five segments (Table 9-1 and Fig. 9-12). The skeleton within homologous segments is remarkably similar despite differences in outward appearance. For the most part, adaptive modifications affect chiefly the distal end of the appendage.

A few tetrapods have completely lost one or both pairs of limbs. Lacking limbs altogether are caecilians, most snakes, and snakelike lizards. Having forelimbs only are sirens (urodeles in the family Sirenidae), the lizard *Bipes* (= *Chirotes*), and manatees and dugongs. A few lizards and a few ratites have hind limbs only, and pythons and boa constrictors have only vestiges of hind limbs. Cetaceans have lost all external manifestations of hind limbs, but vestiges sometimes remain embedded within the body wall. In many instances in which a limb is absent, an embryonic limb bud appears transitorily but fails to develop. Loss of limbs in terrestrial vertebrates is correlated with elongation of the body, which confers certain defensive advantages.[4]

The limbs of early tetrapods were short and the first segment extended straight

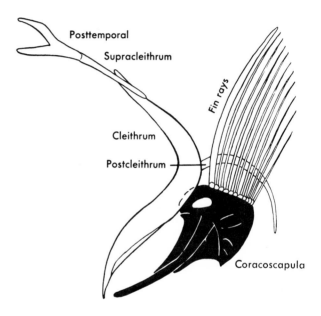

MODERN FIN SKELETON

Fig. 9-11. Pectoral girdle and fin skeleton of ribbonfish, a teleost. Replacement bone is black.

Table 9-1. Homologous segments in anterior and posterior limbs of tetrapods

Anterior limb		Posterior limb	
Name of segment	Skeleton	Name of segment	Skeleton
Upper arm (brachium)	Humerus	Thigh (femur)	Femur
Forearm (antebrachium)	Radius and ulna	Shank (crus)	Tibia and fibula
Wrist (carpus) ⎫	⎧Carpals	Ankle (tarsus) ⎫	⎧Tarsals
Palm (metacarpus) ⎬ Manus	⎨Metacarpals	Instep (metatarsus) ⎬ Pes	⎨Metatarsals
Digits ⎭	⎩Phalanges	Digits ⎭	⎩Phalanges

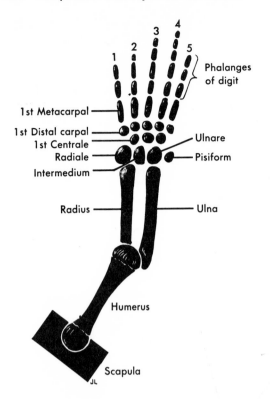

Fig. 9-12. Basic pattern of a right anterior limb, viewed from above, palm down. **1** to **5**, First to fifth digits.

out from the body. This posture persists among many lower tetrapods, but in most reptiles and in mammals there has been a rotation of the appendages toward the body so the long axis of the humerus and femur more nearly parallels the vertebral column. To a marked degree the knee is directed cephalad and the elbow caudad (Fig. 9-22, A). Limbs oriented in this fashion are much better shock absorbers. They also permit greater leverage between axial skeleton and appendage, which increases speed and agility. Such reorientation was an essential step toward bipedalism in reptiles.

□ Arm and forearm

The humerus is the bone of the upper arm. The similarity of the humeri of all tetrapods is more striking than their differ-ences (Fig. 9-13). Variations in length, diameter, and shape are adaptive modifications. The odd humerus of the mole, for example (Fig. 9-14), has expansions for the insertion of massive muscles for digging.

The radius and ulna are bones of the forearm. The radius is a preaxial (anterior) bone articulating proximally with the humerus and ulna and distally with wrist bones on the thumb side of the hand. The radius bears most of the body weight. The ulna is the longer, postaxial bone articulating proximally with the humerus and radius and distally with wrist bones on the side opposite the thumb. The ulna sometimes fuses with the radius, or it may fail to develop (Fig. 9-13, frog, bat).

Extending between the radius and ulna is an interosseous ligament that sometimes ossifies. In man the ligament is incomplete and flexible. Place your forearm and hand on a flat surface and turn your palm down (pronation); then turn your hand over (supination). Few animals can accomplish this movement because of the interosseous ligament. In frogs the ligament is so ossified that adults appear to exhibit only one bone —the radioulna.

□ Manus: the hand

Wrist, palm, and digits constitute a functional unit—the hand, or **manus.** The manus is surprisingly uniform throughout the tetrapod series.

The skeleton of the wrist is the most stable of the regions of the manus. In generalized hands it consists of three rows of carpal bones (Fig. 9-12). (1) The proximal row has a radial carpal (**radiale**) at the base of the radius, an ulnar carpal (**ulnare**) at the base of the ulna, and an **intermedium** between the two. At the ulnar end of the proximal row in most reptiles and mammals is the **pisiform,** a sesamoid bone. (2) The middle row of carpals in a generalized hand consists of three central carpals (**centralia**). (3) The distal row is composed of five

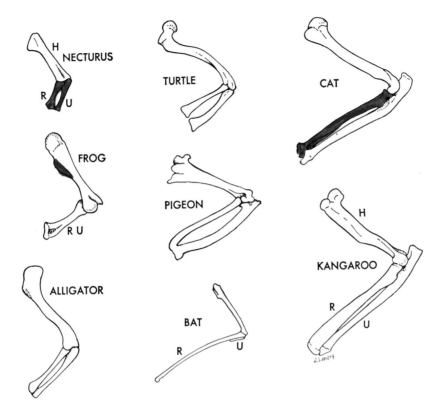

Fig. 9-13. Humerus, radius, and ulna of the left forelimb, lateral views. **H,** Humerus; **R,** radius; **U,** ulna. In the frog the radius and ulna have united to form a radioulna, **RU.** In the bat the ulna is vestigial.

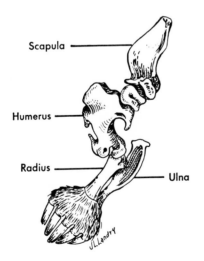

Fig. 9-14. Right anterior limb of a mole, which has been modified for digging. This is a medial view! The palms of a mole turn outward from the body.

distal carpals numbered 1 to 5 commencing on the thumb (radial) side. Table 9-2 lists the names of the carpal bones.

The metacarpals constitute the skeleton of the palm. Primitively, there were probably as many distal carpals and metacarpals as there were digits.

Each digit consists of a linear series of phalanges. The generalized phalangeal formula commencing with the thumb is usually given as 2-3-4-5-3, the formula in generalized reptiles.

Modifications of the manus with few exceptions involve *reduction of the number of bones by evolutionary loss or fusion.* A less common modification is the *disproportionate lengthening or shortening of some of the*

Table 9-2. Synonymy of carpal bones

Terms preferred by comparative anatomists	Nomina anatomica*	Anglicized names and synonyms
Radiale	Os scaphoideum	Scaphoid, navicular
Intermedium	Os lunatum	Lunate, lunar, semilunar
Ulnare	Os triquetrum	Triquetral, cuneiform
Pisiform	Os pisiforme	Pisiform, ulnar sesamoid
Centralia (0 to 3)	Os centrale	Central carpal(s)
Distal carpal 1	Os trapezium	Trapezium, greater multangular
Distal carpal 2	Os trapezoideum	Trapezoid, lesser multangular
Distal carpal 3	Os capitatum	Capitate, magnum
Distal carpal 4 ⎫ Distal carpal 5 ⎭	Os hamatum	Hamate, unciform, uncinate

*Terms approved by the Eighth International Congress of Anatomists at Wiesbaden in 1965.[8]

bones. Least common is an *increase in the number of phalanges.*

Among the first elements of the manus to show loss or fusion were the central carpals. These have been reduced in number or lost altogether in most modern tetrapods. Fusion of distal carpals 4 and 5 is common and results in a **hamate bone.** Digits may be lost, whereupon the corresponding metacarpals become rudimentary or lost.

In modern amphibians, at least one digit has been lost and one metacarpal has been reduced (Fig. 9-15, frog). However, a vestige of the missing finger or its corresponding metacarpal remains in several species of frogs (*Rana catesbeiana* and *R. temporaria*). Members of the urodele genus *Amphiuma* have one to three digits.

Several carpals have been lost by fusion or deletion in amphibians. In frogs and *Necturus,* for example, the embryonic intermedium and ulnare fuse, and there are only one central and three distal carpals.

Since *Necturus,* along with other modern amphibians, has only four fingers, the first digit, or thumb, may be missing. In accordance with this interpretation, the three distal carpals have been considered to be, commencing on the radial side (Fig. 9-15), carpal 2, carpal 3, and hamate (fused carpals 4 + 5). However, the fifth finger rather than the thumb may be missing.[6] The three distal carpals might then represent a prepollex (an extra bone that sometimes occurs near the thumb, or pollex), carpals 1 + 2, and carpals 3 + 4. (Carpal 5 would be missing.) The hypothesis is based on the fact that the bone labeled *P* in Fig. 9-15 has no muscle connecting it with a finger, and the muscle from the first and second digits attaches to the second of the three distal carpals. (The second carpal often shows double ossification centers.) This is the kind of reasoning that is employed in determining some homologies.

Reptiles, insectivores, and primates tend to remain pentadactyl and to have five metacarpals and a nearly full complement of carpals, except for centralia (Fig. 9-16, man, turtle). In alligators, however, the wrist has been reduced to five bones (Fig. 9-16, alligator). Retention of an adult centrale is characteristic of numerous mammals, including most monkeys. It may lie in the distal row of carpals, as in rabbits, or it may fuse with the radiale and intermedium to form a bone of triple origin (scapholunar), as in cats. Occasionally, a human fetus has a centrale.

ADAPTIVE MODIFICATIONS OF THE MANUS

Among major modifications of the hand are those for flight, life in the ocean, swift-

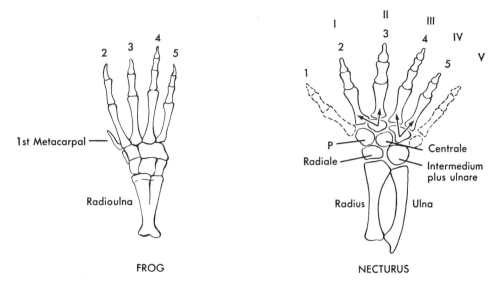

Fig. 9-15. Hands of *Rana catesbeiana* and *Necturus,* dorsal views. Which finger is missing in *Necturus?* Arabic numerals suggest that in a necturus the thumb, **1,** is missing and the little finger, **5,** is present. Roman numerals suggest that the thumb, **I,** is present and the little finger, **V,** and last distal carpal and metacarpal are missing. Arrows indicate existing muscle attachments. Broken lines represent nonexisting elements, one of which has been lost.
P, Distal carpal number two or prepollex, depending on interpretation.

footedness, and grasping. These will be discussed briefly.

Adaptations for flight. In most birds the hand has little independent role in propulsion in air but, being at the end of an airfoil, it has an aerodynamic effect. A tetrapod hand on a bird would provide considerable drag. A bird's hand does not do so because there has been a loss and fusion of bones, which has reduced the hand to a rigid, tapering structure (Fig. 9-17). Despite this, most of the basic components of tetrapod hands are identifiable in embryos. Two carpals (radiale and ulnare) form in the proximal row and three in the distal row. As development progresses the three distal carpals unite with the three metacarpals to form a rigid **carpometacarpus.** Three fingers are usually present, and the number of phalanges has been reduced. (Terns often develop four embryonic digits, but only three persist.) The fingers often bear claws that are used in nonaerial locomotion. In birds

with high-speed wings (up to 50 strokes per second in hummingbirds) the hand is long in proportion to the arm, whereas in soaring birds the hand is proportionately short.[7]

Contrary to the condition in birds, the skeleton of the hand of pterosaurs (Fig. 3-17) and bats is the chief skeleton of the wing membrane, or **patagium.** Pterosaurs had four fingers (Fig. 9-18, *B*). Three were normal and bore claws. The fourth was embedded in the patagium and consisted of four enormously elongated phalanges that made this finger as long as the entire body. The associated metacarpal was not elongated but it was much enlarged. Bats (Fig. 9-18, *A*) have five fingers. The thumb is normal and bears a claw. The other four fingers are elongated and are associated with four greatly elongated metacarpals. The four metacarpals and fingers constitute the skeleton of the patagium. The three proximal carpals are united in a single bone. Movement of the hand is responsible for takeoff and true

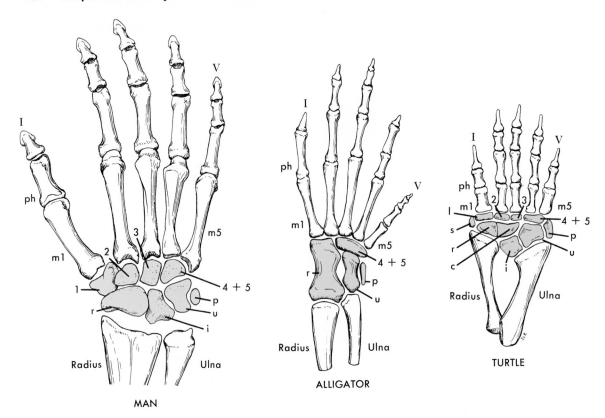

Fig. 9-16. Right manus of man, alligator, and turtle; dorsal views. **c,** Centrale; **i,** intermedium; **m1** and **m5,** first and fifth metacarpals; **p,** pisiform; **ph,** proximal phalanx; **r,** radiale; **s,** radial sesamoid; **u,** ulnare; **1** to **5,** distal carpals; **I** and **V,** first and fifth digits. The alligator has an additional carpal that cannot be seen from this view. Wrist bones are red.

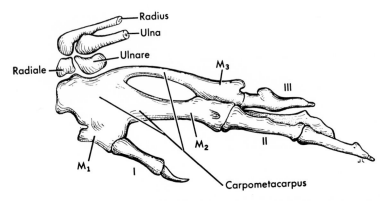

Fig. 9-17. Left manus of a bird. **I** to **III,** Digits; **M₁** to **M₃,** metacarpals fused at their bases with three carpals to form a carpometacarpus.

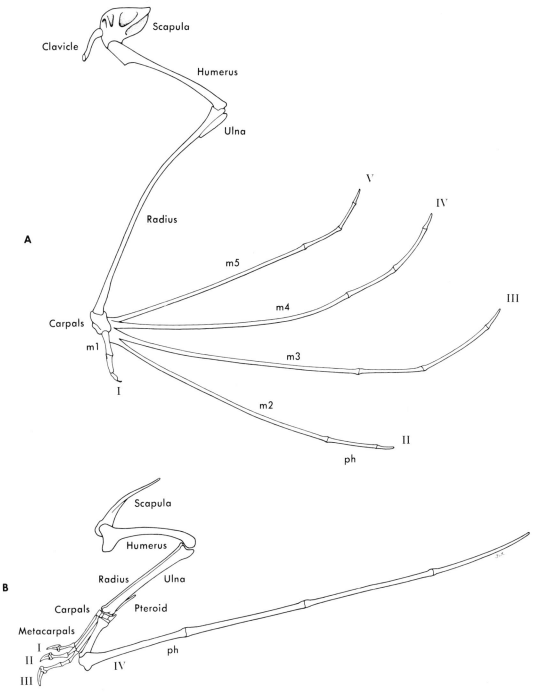

Fig. 9-18. Pectoral girdle and limb of two flying vertebrates. **A,** Bat, right wing; **B,** Jurassic pterosaur, left wing. **m1** to **m5,** First through fifth metacarpals; **ph,** proximal phalanx; **I** to **V,** digits.

Ichthyosaurus

Fig. 9-19. Jurassic and Cretaceous ocean-dwelling reptile, ranging up to 3 meters in length. (From Colbert, E. H.: Evolution of the vertebrates, ed. 2, New York, 1969, John Wiley & Sons, Inc.)

flight in bats. No one has ever seen a pterosaur take off!

Flying lemurs have a patagium, but it is less well developed and the fingers, although embedded in it, are not elongated. Flying lemurs soar but are not capable of true flight. The formation of a patagium in such unrelated animals as pterosaurs, bats, lemurs, and even flying lizards, demonstrates instances of evolutionary convergence (p. 63).

Adaptations for life in the ocean. The hands of ichthyosaurs (Figs. 9-19 and 9-20), plesiosaurs, some sea turtles, penguins, cetaceans, sirenians, seals, and seal lions have become paddlelike flippers. They are flattened, short, and stout, and in several groups the number of phalanges has greatly increased. In some ichthyosaurs there were as many as twenty-six phalanges per digit, or over a hundred in a single hand. Dolphins show the same modification (Fig. 9-20). Within the flippers of most of the other swimmers, however, the bones conform

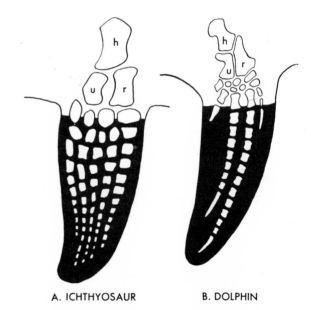

A. ICHTHYOSAUR   B. DOLPHIN

Fig. 9-20. Convergent evolution in anterior limbs. **A,** Extinct, water-dwelling reptile. **B,** Water-dwelling mammal. **h,** Humerus; **r,** radius; **u,** ulna.

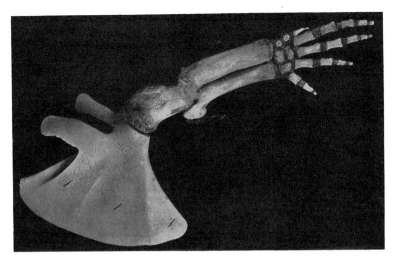

Fig. 9-21. Forelimb and pectoral girdle of a beaked whale. A remarkable resemblance to basic pattern remains, despite the fact that the limb has become paddlelike. (Courtesy American Museum of Natural History, New York, N.Y.)

closely to the primitive tetrapod pattern (Fig. 9-21). Some aquatic mammals have lost all traces of hind limbs (Fig. 9-22, *C*).

Adaptations for swift-footedness. Mammals with pentadactyl hands and feet are usually plantigrade, that is, flat-footed. The palm, wrist, and digits of the hand and the metatarsals, ankle, and digits of the foot all tend to rest more or less on the ground (Fig. 9-23, monkey). This is a primitive stance among tetrapods. Insectivores, monkeys, apes, man, bears, and some other animals walk in this manner. Bipedal mammals with plantigrade feet have a "metatarsal arch," or "instep," that not only distributes body weight over four solid bases (the heel and "ball" of each foot) but also provides "spring" for walking or running. Du Brul[2] has compared this four-pedestal architectural pattern to that of the Eiffel Tower.

Mammals in which only the first digit has been reduced or lost (rabbits, rodents, many carnivores, and others) tend to be digitigrade, that is, bear their weight on digital arches with wrist (and ankle) elevated (Fig. 9-23, dog). Digitigrade animals usually run faster than plantigrade species. They also walk more silently and are more agile.

The extreme modification of reducing the number of digits and walking on the tips of the remaining ones is seen in ungulates, or hoofed mammals. These walk on three, two, or even one digit (Fig. 9-23, deer). The claws are thickened to form tough hoofs, which bear the weight of the body. The metacarpals corresponding to missing digits are likewise reduced in size or lost, and those metacarpals that remain are elongated and often united (Fig. 9-24). Although there are many swift-footed digitigrade mammals, hoofed animals, as a group, are the most fleet-footed. Their feet are specialized for running through grassy meadows and plains, but they also function well in climbing over rocky crags. However, the specialization has made their fingers and toes practically useless for anything else.

Sequential evolutionary steps leading to the most specialized unguligrade stance may be illustrated by placing the fingers and palm flat (prone) on a tabletop, with the forearm perpendicular to the surface. This

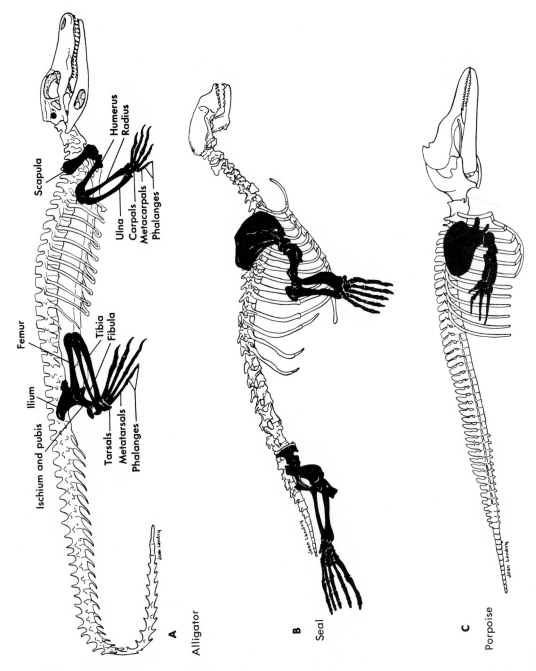

Fig. 9-22. A, Skeleton of a land-dwelling amniote. **B** and **C,** Skeletal adaptations for life in the water. Appendicular skeleton is shown in black. **B** is a "wriggling seal" (*Phoca*). In the seal and porpoise the hand is a paddle in which the phalanges are embedded.

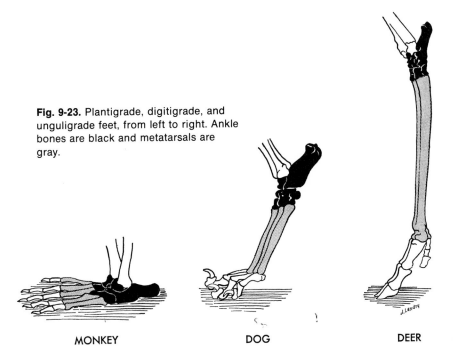

Fig. 9-23. Plantigrade, digitigrade, and unguligrade feet, from left to right. Ankle bones are black and metatarsals are gray.

MONKEY   DOG   DEER

represents roughly the plantigrade position. Raising the palm off the table while keeping the fingers flat on the table illustrates roughly the digitigrade position. Unguligrade conditions may be illustrated by placing only the fingertips on the table and then raising the thumb, the little finger, the second finger, and finally the fourth finger, leaving only the third finger to bear the body weight, as in modern horses. Those fingers that fail to reach the table represent digits that have been successively reduced or lost in ungulates.

The horse underwent these successive changes commencing with the early *Eohippus*, which had four digits on the manus, and culminating in the modern *Equus*, which has a single digit. Despite extreme specialization of the manus of the modern horse, the proximal row of carpals (Fig. 9-25) is intact, and the distal row lacks only the first carpal. With the loss of digits I, II, IV, and V, metacarpals 1 and 5 have been lost, and 2 and 4 have been reduced to splinters.

Metacarpal 3, associated with digit III, has elongated.

Evolution among the ungulates seems to have progressed along two independent lines. In the line leading to artiodactyls, the weight of the body tended to be distributed equally between digits III and IV. Thus arose the "cloven" hoof (Fig. 9-26). Such a foot is said to be **paraxonic** because the body weight is borne on two parallel axes (Fig. 9-25, camel). Artiodactyls of today have an even number of digits. In the evolutionary line leading to perissodactyls, the body weight increasingly tended to be borne on digit III, the middle digit. This is a **mesaxonic** foot (Fig. 9-25, horse). Most perissodactyls have an odd number of digits, although some have four. It is the mesaxonic foot and not the number of digits that defines the perissodactyls.

Adaptations for grasping. Many mammals are able to flex the hand at the joint between the palm and the base of the fingers. Rodents, for example, sit on their haunches and

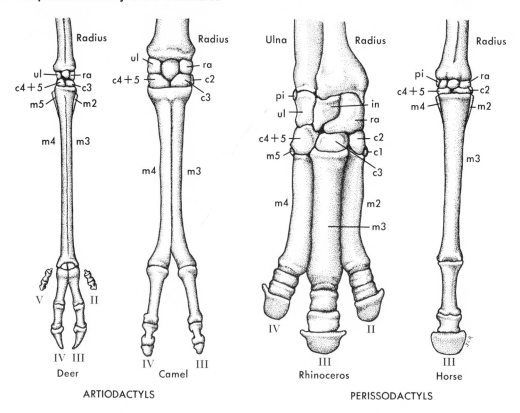

Fig. 9-24. Right manus of several ungulates as seen from in front. **c2** to **c5,** Distal carpals 2 to 5; **in,** intermedium; **m2** to **m5,** metacarpals 2 to 5; **pi,** pisiform; **ra,** radiale; **ul,** ulnare; **II** to **V,** digits.

nibble on food held between *two hands,* which are flexed in this manner and *which face one another.* A further specialization is the ability to wrap the fingers (but not the thumb) around an object such as a pencil so that it is held securely in *one hand.* This is accomplished by flexing the fingers at each interphalangeal joint. Only primates can do that.

Another step in the evolution of the mammalian hand was development of an opposable thumb—one that can be made to touch the tips of each of the other digits. This was accomplished by formation of a saddle joint at the base of the thumb where it meets the palm, by setting the thumb at increasingly wider and wider angles to the

index finger, and by the evolution of strong adductor pollicis (thumb) muscles. True opposability is found in Old World monkeys, but even there the hand does not have the full range of functional capability that has evolved in man. Neither New World monkeys nor anthropoid apes have a perfectly opposable thumb. With such a hand, man was able to fashion increasingly sophisticated instruments, commencing with rocks chipped by design and continuing to the electronic computer. As John Napier[13] has said, "The implements of early man were as good (or as bad) as the hands that made them." Of course, evolution of the brain was an essential concomitant; still it seems impossible that any species lacking a prehen-

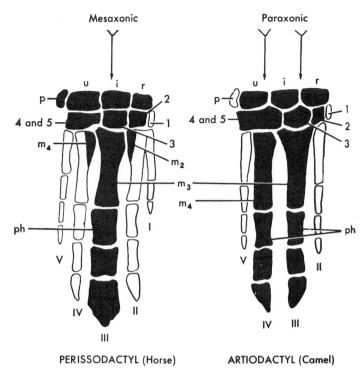

Fig. 9-25. Mesaxonic and paraxonic manus of representative ungulates, showing bones lost (white) and retained (black) and indicating distribution of body weight through wrist and digits. The horse and camel are used as specific examples, and the number of bony elements are correct for these animals. **i,** Intermedium; m_2 to m_4, second, third, and fourth metacarpals; **p,** pisiform; **ph,** first phalanx; **r,** radiale; **u,** ulnare; **1** to **5,** distal carpals; **I** to **V,** digits.

Table 9-3. Comparison of skeletal elements of manus and pes

Manus*	Pes, with synonyms	
Radiale	Tibiale	Talus or astragalus†
Intermedium	Intermedium	
Ulnare	Fibulare	Calcaneus
Pisiform		
Centralia (0 to 3)	Centralia (0 to 3)	Navicular
Distal carpal 1	Distal tarsal 1	Entocuneiform
Distal carpal 2	Distal tarsal 2	Mesocuneiform
Distal carpal 3	Distal tarsal 3	Ectocuneiform
Distal carpal 4 ⎱ Hamate	Distal tarsal 4 ⎱	Cuboid
Distal carpal 5 ⎰	Distal tarsal 5 ⎰	
Metacarpals (1 to 5)	Metatarsals (1 to 5)	
Digits (I to V)	Digits (I to V)	

*For synonyms, see Table 9-2.
†Often incorporates the intermedium.

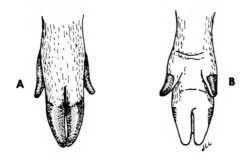

Fig. 9-26. Foot of fetal pig. There is a cleft between digits. Thus the hoof is "cloven." **A,** As seen from in front; **B,** as seen from behind.

sile hand, even if it had the brain, could have evolved so sophisticated an existence.

☐ **Posterior limbs**

The bones of the hind limbs are comparable, segment by segment, to those of the forelimbs, except that the ankle lacks an equivalent of the pisiform bone (Table 9-3). A sesamoid bone, the **patella,** or kneecap, develops in birds and mammals.

Variations among femurs, tibias, and fibulas are of the same nature as variations of

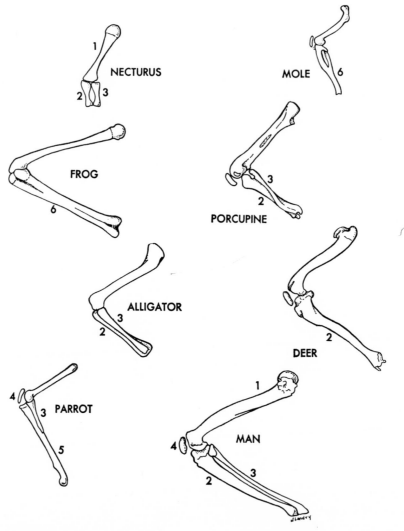

Fig. 9-27. Left thigh and shank bones of representative tetrapods, lateral views.
1, Femur; **2,** tibia; **3,** fibula; **4,** patella; **5,** tibiotarsus; **6,** tibiofibula.

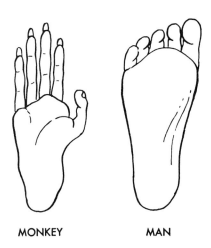

MONKEY   MAN

Fig. 9-28. The partially opposable big toe of an Old World monkey and the nonopposable toe of man.

equivalent bones of the forelimb. The fibula may unite partially or completely with the tibia (Fig. 9-27, mole and frog), it may be reduced to a splinter (parrot), or it may be absent, as in ungulates (deer). In birds the tibia fuses with the proximal row of tarsals to form a **tibiotarsus,** and the metatarsals fuse with the distal row of tarsals to form a **tarsometatarsus.** A joint between the tibiotarsus and tarsometatarsus permits flexion at this joint. Reptiles have a similar intratarsal joint.

Digits of the foot may be more or less numerous than digits of the manus in the same animal. Most amphibians have four fingers and five toes. Tapirs, on the contrary, have four fingers and three toes. Birds have four, three, or two functional toes. The big toe (hallux) is opposable in many primates, but not in man (Fig. 9-28). The generalized phalangeal formula for the foot is thought to be 2-3-4-5-4, the formula of generalized reptiles such as *Sphenodon*.

■ ORIGIN OF FINS

What was the nature of the first paired appendages of vertebrates? This is one of the more perplexing questions encountered in a study of vertebrate origins. Most of the systems and organs exhibited by the verte-

brate body commenced in antiquity as relatively simple structures that through successive mutations became more and more modified. Presumably paired appendages underwent similar changes. Our lack of specific knowledge of the nature of the first appendages stems from a lack of early vertebrate fossils. Numerous very ancient fish have been discovered, but were these actually ancestral to modern forms, or were they offshoots from the main evolutionary line? May we rely on them for evidence of the nature of ancestral appendages? These are questions to which we cannot give an answer at present. Three highly speculative hypotheses will be mentioned.

The **fin fold hypothesis** derives paired fins from a continuous fold of the lateral body wall that became interrupted in the middle of the trunk, leaving later fishes with pectoral and pelvic fins like those in *Cladoselache* (Fig. 9-9). There is little evidence for this beyond the metapleural fold of the amphioxus, which is probably irrelevant. Historically interesting is the **gill arch hypothesis** of Gegenbaur, according to which the girdles may be a modification of the skeleton of the last two gill arches and the fin skeleton an expansion of gill rays. The location of the pectoral girdle immediately behind the last gill in some fishes and its resemblance to gill cartilages prompted this hypothesis. A third guess is the **fin spine hypothesis.** In some acanthodians (Fig. 9-9) pectoral and pelvic appendages were members of a series of spiny appendages that extended the length of the trunk and perhaps served as stabilizers. Associated with each spine was a fleshy web. If all the appendages in the series were lost except an anterior and posterior pair, this would have produced paired pectoral and pelvic fins.[5]

No known prechordate or protochordate has structures that conceivably could have given rise to vertebrate fins, even permitting free reign to the imagination. Reliable clues to the origin of fins are probably hidden forever in the obscurity of time.

■ ORIGIN OF LIMBS

Although the problem of the origin of paired fins may never be satisfactorily resolved, this is not true of the origin of tetrapod limbs. All evidence points to the origin of tetrapods from fishes, and so the paired fins of some ancient fish must have been the precursors of tetrapod limbs. The question then arises: Did the skeleton of the fin of any known Devonian fish evince the potentiality of becoming a limb? For an answer we turn to the lobe-finned crossopterygians, which resembled the first tetrapods in many respects.

The skeleton in the basal lobe of some early crossopterygian fins bears a striking resemblance to that in a tetrapod limb (Fig. 9-29). In crossopterygians a single bone (we will call it a "humerus") articulates proximally with the scapula and distally with a pair of bones (we will call them "radius" and "ulna"). Loss of the fin rays and relatively minor modifications of the fin distal to the radius and ulna could have produced the first tetrapod limb. Meanwhile, the girdle

of the labyrinthodonts remained fishlike.

It is possible that crossopterygian fins were used as props for resting on the water bottom, without bearing any appreciable weight.[3] Minor modifications would have permitted "walking" on the muddy or sandy bottoms close to shore. Several hundred living species of fishes do this, including the Australian lungfish. Some living fishes move several feet inland nightly. Some climb inclined planes with little fins that have a remarkable resemblance to hands.

The pressures that drove vertebrates onto the land must, of necessity, be conjectural. It may be that there were fewer predators on land, that there was less competition for food, or simply that food was abundant on land. Or perhaps it was simply a manifestation of the tendency of organisms to invade a contiguous environment whenever nothing prevents the invasion. Whatever the explanation, it seems that it was almost inevitable that a limb more suitable for life on land would evolve sooner or later from the fin of a fish.

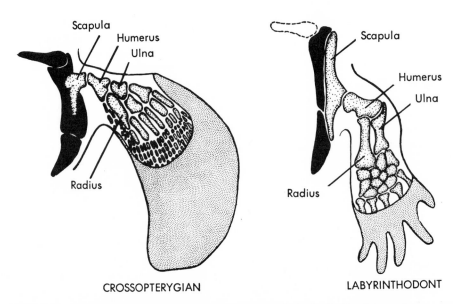

CROSSOPTERYGIAN

LABYRINTHODONT

Fig. 9-29. Pectoral fin of crossopterygian fish and forelimb of primitive tetrapod, oriented to show similarity of skeleton. Dermal bones are black and replacement bones or cartilage are stippled. Dotted line represents missing girdle bone. The "ulna" and "radius" of the crossopterygian are enlarged radials.

☐ Chapter summary

1. The appendicular skeleton includes the pectoral and pelvic girdles and the skeleton of the fins and limbs.

2. Pectoral girdles are all built in accordance with a single architectural pattern. Some components belong to the primary endoskeleton and consist of cartilage or replacement bone. Others are membrane bones derived from dermal armor.

3. Coracoids, scapulas, and suprascapulas arise as cartilages and may be replaced by bone. Coracoids are absent above monotremes. A scapula is almost universally present.

4. Clavicles, cleithra, supracleithra, and posttemporals arise by intramembranous ossification. Cleithra and supracleithra are confined to fishes. Clavicles occur in most vertebrate classes and are best developed in tetrapods. A midventral interclavicle of membranous origin is found in reptiles, birds, and monotremes.

5. The pelvic girdle of fishes consists of two pelvic plates. These usually meet ventrally in a symphysis and articulate laterally with a pelvic fin. The plates unite to form a median ischiopubic bar in sharks and lungfishes.

6. In tetrapods two ossification centers arise in each pelvic plate and form ischial and pubic bones. A third bone, the ilium, braces the girdle against the vertebral column. There is no dermal armor in the pelvic girdle.

7. The pelvic girdle and sacrum of tetrapods enclose a pelvic cavity.

8. Excepting sesamoids, bones distal to the girdles arise by endochondral ossification.

9. Paired fins are of three types: (a) lobed fins (in Sarcopterygii), having a central jointed axis and a series of preaxial and postaxial radials; (b) fin fold fins (in Chondrichthyes), with a proximal row of basals and distal rows of radials; and (c) ray fins (in Actinopterygii), reduced to a few short radials or none and supported chiefly by fin rays. Median fins have a skeleton of radials and rays.

10. Tetrapod limbs have a humerus or femur, radius and ulna or tibia and fibula, and a manus (carpals, metacarpals, and phalanges) or pes (tarsals, metatarsals, and phalanges).

11. Structural modifications of the manus and pes involve reduction in the number of bones. Digits have been reduced to four or fewer in modern amphibians, three in birds, and as few as one in some ungulates. Reduction of digits is usually accompanied by reduction of associated carpals and metacarpals, or tarsals and metatarsals.

12. Other structural modifications include disproportionate lengthening or shortening of bones and, infrequently, increase in number of phalanges.

13. The most striking modifications are found in flying tetrapods (pterosaurs, birds, bats), in water-adapted reptiles and mammals (ichthyosaurs, plesiosaurs, cetaceans, sirenians), and in hoofed mammals (ungulates). A few species lack anterior appendages, posterior appendages, or both.

14. Ungulate feet are hoofed and either mesaxonic (perissodactyls) or paraxonic (artiodactyls).

15. Three theories of the origin of fins are the fin fold, gill arch, and fin spine theories.

16. Tetrapod limbs probably arose from lobed fins resembling those of crossopterygians.

LITERATURE CITED AND SELECTED READINGS

1. Darnell, R. M.: Nocturnal terrestrial habits of the tropical gobioid fish *Gobiomorus dormitor*, with remarks on its ecology, Copeia, no. 3, p. 237, 1955.
2. Du Brul, E. L.: Biomechanics of the body, BSCS Pamphlet, no. 5, Boston, 1965, D. C. Heath & Co.
3. Eaton, T. H.: Origin of tetrapod limbs, American Midland Naturalist **46**:245, 1951.
4. Gans, C.: Tetrapod limblessness: evolution and functional corollaries, American Zoologist **15**:455, 1975.
5. Gregory, W. K., and Raven, H. C.: Studies on the origin and early evolution of paired fins and limbs, Annals of the New York Academy of Sciences **42**:273, 1941.
6. Harris, J. P., Jr.: The skeleton of the arm of *Necturus*, Field and Laboratory **20**:78, 1952.
7. Hildebrand, M.: Analysis of vertebrate structure, New York, 1974, John Wiley & Sons, Inc.
8. Nomina anatomica, ed. 3, Amsterdam, 1966, Excerpta Medica Foundation.
9. Nursall, J. R.: Swimming and the origin of paired appendages, American Zoologist **2**:127, 1962.
10. Romer, A. S.: The early evolution of fishes, Quarterly Review of Biology **21**:33, 1946.
11. Savile, D. B. O.: Gliding and flight in the vertebrates, American Zoologist **2**:161, 1962.
12. Snyder, R. C.: Adaptations for bipedal locomotion of lizards, American Zoologist **2**:191, 1962.
13. Vertebrate structures and functions: readings from Scientific American with introductions by Norman K. Wessells, San Francisco, 1955-1974, W. H. Freeman and Co. Publishers. Includes Gray, J.: How fishes swim; Hildebrand, M.: How animals run; Gans, C.: How snakes move; Napier, J.: The antiquity of human walking; Napier, J.: The evolution of the hand; and Welty, C.: Birds as flying machines.
14. Westoll, T. S.: The lateral fin-fold theory and the pectoral fins of ostracoderms and early fishes. In Westoll, T. S., editor: Studies on fossil vertebrates, London, 1958, University of London Press.

Symposium

Symposium on vertebrate locomotion, American Society of Zoologists **2**:127, 1962.

10

Muscles

In this chapter we will study the skeletal muscles. We will classify them from several viewpoints and then examine them in functional groups, observing the basic architectural pattern and seeing how the pattern was modified as vertebrates became increasingly adapted for life on land.

Muscle is a tissue specialized to do one thing, and to do it well: to shorten when stimulated. Shortening is a result of chemical changes in two muscle proteins, actin and myosin. Shortening of a sufficient number of fibers causes a corresponding shortening and fattening of the entire muscle mass. If the muscle surrounds a lumen, the lumen is compressed. If it extends between two structures, one of these is drawn toward the other. The usual stimulus for muscle contraction is a nerve impulse (Fig. 10-1).

■ KINDS OF MUSCLES

Muscles may be classified according to several criteria. When classified histologically muscles may be striated or smooth. Striated muscles are usually definite organs composed of long multinucleate fibers that have cross striations (Fig. 10-1). Smooth muscle, on the other hand, occurs most often in sheets as part of an organ. Smooth muscle cells are spindle shaped and uninucleate and lack striations (Fig. 10-2). Cardiac muscle is a special type of striated muscle found only in the heart wall.

Muscles may be either **voluntary** or **involuntary** according to whether or not they can be operated at will. Voluntary muscles contract at will unless they are fatigued. This does not mean they may not be operated unintentionally (reflexly) if the body is endangered, as when the skin comes in contact unexpectedly with a pin. Involuntary muscles, on the contrary, are not ordinarily contracted at will, being predominantly under reflex control.

Muscles attached to the skeleton are **skeletal muscles** and are striated and voluntary. Skeletal muscles are disposed in accordance with a single architectural pattern most evident in fishes but recognizable in all higher vertebrates. Evolutionary modifications of

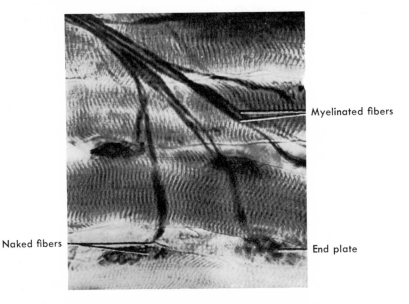

Fig. 10-1. Striated muscle fibers innervated by myelinated nerve fibers that end on motor end-plates. (From Bevelander, G., and Ramaley, J. A.: Essentials of histology, ed. 8, St. Louis, 1978, The C. V. Mosby Co.)

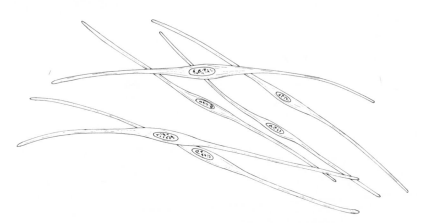

Fig. 10-2. Isolated smooth muscle cells from intestine.

the basic pattern are correlated chiefly with assumption of life on land. Muscles not attached to the skeleton must be referred to as **nonskeletal,** since no other term (smooth, involuntary, etc.) is the exact opposite of the word skeletal. Nevertheless, most nonskeletal muscles are smooth and involuntary.

Using the foregoing criteria, the muscles of the vertebrate body could be classified

according to the following scheme:

Skeletal, striated, voluntary muscles
    Axial
        Body wall and tail
        Hypobranchial and tongue
        Extrinsic eyeball
    Appendicular
    Branchiomeric
    Integumentary

Nonskeletal, smooth, chiefly involuntary
   muscles
     Muscles of tubes, vessels, and hollow
       organs
     Intrinsic eyeball muscles
     Erectors of feathers and hairs
Cardiac muscle
Electric organs

The foregoing categories are not mutually exclusive and no such scheme can be devised. Some branchiomeric muscles have become secondarily associated with the appendages, intrinsic eyeball muscles of reptiles and birds are often striated, and erectors of feathers and hairs are intrinsic integumentary muscles.

It is sometimes convenient or appropriate to classify muscles as somatic or visceral. In general, **somatic muscles** orient the animal (its body, or soma) with respect to the external environment. They enable an animal to go deeper into an environment to pursue food or a mate, and to withdraw from an environment that is hazardous. Therefore, skeletal muscles, other than those of the visceral arches (branchiomeric muscles), are somatic.

The chief role of **visceral muscle** is to maintain an appropriate internal milieu. It is the muscle of hollow organs, vessels, tubes, and ducts; the intrinsic muscle of the eyeball (iris diaphragm, ciliary body, cam-

panula); the erectors of hairs and feathers; and the striated muscles of the jaws and remaining visceral arches. Visceral muscle compresses lumens as in emptying bladders or heart; causes peristalsis, which propels substances within tubes; and serves as sphincter and dilator muscles in such diverse locations as the iris diaphragm, pyloric sphincter, and gill pouches.

If one wishes to classify musculature on the basis of whether it is somatic or visceral, a scheme somewhat like the following would be useful.

Somatic muscles
   Muscles of the body wall and tail
   Hypobranchial and tongue muscles
   Extrinsic eyeball muscles
   Appendicular muscles
Visceral muscles
   Branchiomeric muscles
   Muscles of tubes, vessels, hollow organs
   Intrinsic eyeball muscles
   Erectors of feathers and hairs
   Cardiac muscle

As in the preceding classification, the above categories are not mutually exclusive. A few appendicular muscles are branchiomeric, erectors of feathers and hairs are in the body wall, and electric organs have been omitted because some are somatic and some are visceral.

Table 10-1. Some contrasts between somatic and visceral muscles

Somatic muscles	Visceral muscles except branchiomeric
Striated, skeletal, voluntary	Smooth, nonskeletal, involuntary
Primitively segmented	Unsegmented
Arise primitively from somites (myotomal muscle)	Arise mostly from lateral mesoderm
Mostly in body wall	Mostly in splanchnopleure*
Primarily for orientation in external environment	Regulate internal environment
Innervated directly by spinal nerves and cranial nerves III, IV, VI, XII	Innervated by postganglionic fibers of autonomic nervous system (Fig. 10-2)

*Those in the body wall erect hairs or feathers or constrict blood vessels.

Except for branchiomeric muscles, visceral muscles exhibit few evolutionary changes throughout the vertebrate series and will receive little attention in this chapter. Some contrasts between somatic and visceral muscles are presented in Table 10-1.

■ INTRODUCTION TO SKELETAL MUSCLES
□ Skeletal muscles as organs

A skeletal muscle is an organ just as the stomach is, and the organs constitute a **skeletal muscle system.** Skeletal muscles consist of **muscular** and **tendinous portions.** The tendinous portions are part of the muscle, and are extensions of its tough connective tissue sheath (**muscle fascia,** or **epimysium**) and of the connective tissue within the muscle. The tendinous portions anchor the muscle to its origins and insertions.

The **anatomical origin** of a muscle is the site of attachment that, under most functional conditions, remains fixed; that is, it is not displaced when the muscle contracts. For example, when the biceps muscle of the upper arm contracts, the forearm is drawn toward the upper arm. The origin is, therefore, somewhere above the elbow. The **insertion** of a muscle is the site of attachment that is normally displaced by contraction of the muscle. The biceps referred to above inserts on a bone of the forearm. A muscle may cause displacement of the origin instead of the insertion if the latter becomes immobilized by other muscles. For example, the geniohyoid muscle, which is attached to the hyoid bone and to the lower jaw at the chin, either lowers the jaw or draws the hyoid forward, depending on which one is immobilized at the time.

Muscles with a pronounced bulge in the middle are said to have a **belly.** Digastric muscles have two bellies. Some muscles are straplike, or broad and flat, and have no belly. Broad, flat muscles usually have broad tendons. Such a tendon is an **aponeurosis.** Broad muscles that insert on connective tissue raphes (seams) such as the linea alba or the raphe in the floor of the buccal cavity may compress a cavity and the organs within it.

During dissection of adult mammals, students frequently notice that skeletal muscles vary in size in the two sexes. Androgens, the predominant gonadal hormones of males, cause amino acids to be linked together into polypeptides and proteins. Since muscle is 80% protein, androgen results in statistically demonstrable larger muscles in males. This confers on the male an added advantage during mating, which increases the likelihood of impregnation of all ovulated eggs in a population and therefore the probability of survival of the species.

□ Actions of skeletal muscles

Skeletal muscles may be grouped according to their function. **Flexors** bend and **extensors** straighten a structural complex, such as an arm or leg. **Adductors** draw a part toward the midline; **abductors** cause displacement away from the midline. **Protractors** cause a part, such as the tongue or hyoid, to be thrust forward or outward; **retractors** pull it back. **Levators** raise a part; **depressors** lower it. **Rotators** cause rotation of a part on its axis. **Supinators** are rotators that turn the palm upward, **pronators** make it prone (turn it downwards). **Tensors** make a part such as the eardrum more taut. **Constrictors** compress internal parts. **Sphincters** are constrictors that make an opening smaller; **dilators** have an opposite effect. Most sphincter and dilator muscles are nonskeletal.

It is exceptional for a single skeletal muscle to act independently of other muscles. Instead, muscles nearly always cooperate in functional groups. While one functional group is contracting, another group ("antagonistic muscles") must relax simultaneously, or a stalemate would result. In order that muscles may act cooperatively (synergistically), they must be reflexly directed by the nervous system, just as the musicians in a symphony orchestra must be directed by a conductor. The reflex control of skeletal muscles comes chiefly from the cerebellum, which sends motor impulses to appropriate muscles after receiving incoming information. The information is in the form of feedback over sensory fibers from muscle spindles, tendons, and from the bursas of affected joints. The sensory phenomenon is known as proprioception (Chapter 16).

Names and homologies of skeletal muscles

Skeletal muscles have been named for the direction of their fibers (oblique, rectus), location or position (thoracis, supraspinatus, superficial), number of divisions (triceps), shape (deltoid, teres, serratus), origin and/or insertion (xiphihumeralis, stapedius), action (levator scapulae, risorius), size (major, longissimus), and for still other reasons including a combination of these. Insight into the significance of a muscle's name should aid in recalling more information about a muscle.

Names were given to the muscles of man several hundred years ago, and many of these names are still used. Frequently, these names were applied later to the corresponding muscles of animals, especially tetrapods. But the fact that two muscles in two different vertebrate groups bear the same name is no assurance that the two are homologous, and the less related the animals, the more likely it is that they may not be. One reason is that some muscles have apparently shifted their attachments from time to time, and so similarity of origin and insertion is not a highly reliable criterion for homology. An example is the genioglossus muscle, which inserts on the mammalian tongue. The probable homologue in some birds inserts on the sublingual seed pouch (p. 237). Embryonic origin and nerve supply are considered more reliable criteria; yet these, too, may sometimes lead to false assumptions. On the other hand, homologies between *functional groups* of muscles of vertebrates may be deduced with a much greater degree of reliability.

■ AXIAL MUSCLES

Axial muscles are the skeletal muscles of the trunk and tail. They also extend forward beneath the pharynx as hypobranchial muscles and muscles of the tongue, and some are present in the orbits as extrinsic muscles of the eyeballs.

The outstanding characteristic of these muscles in their primitive condition is their metamerism. This is evident in fishes and aquatic amphibians (Fig. 10-3), in which axial muscles are used for locomotion. Waves of contractions sweep from segment to segment along the length of the body from the head to the tip of the tail, producing the sinuous swimming movements characteristic of fishes (Fig. 1-3). In limbless tetrapods these muscles perform the same function—locomotion. In higher tetrapods, however, some of the metamerism of these muscles became obscured when paired appendages took over responsibility for locomotion on land. Nevertheless, much of the axial musculature remains metameric, even in man.

The axial muscles are segmental because of their embryonic origin: they arise from segmental mesodermal somites (Figs. 10-4 and 15-5). Mesenchyme cells from the myotome of each somite stream into the embryonic body wall and make their way ventrad between ectoderm and parietal peritoneum, meanwhile undergoing repeated cell division (Figs. 4-8 and 10-5). They continue to migrate and divide until they reach the midventral line, where the linea alba develops. These myotomal cells (cells whose lineage can be traced to myotomes) aggregate to form blastemas for the body wall muscles; and because the somites are metameric, the blastemas are metamer-

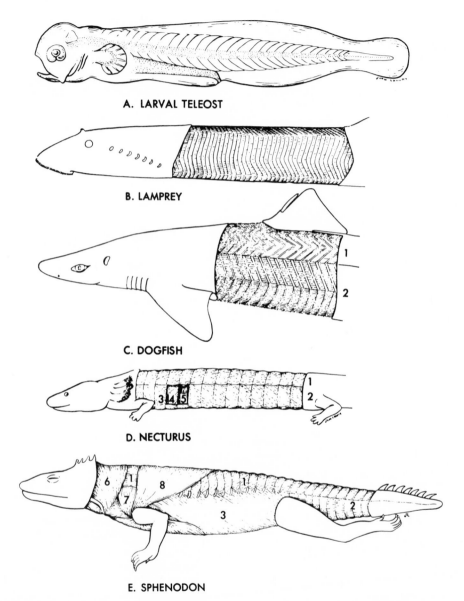

A. LARVAL TELEOST

B. LAMPREY

C. DOGFISH

D. NECTURUS

E. SPHENODON

Fig. 10-3. Trunk muscles of selected vertebrates. **1,** Epaxials (dorsalis trunci); **2,** hypaxials; **3,** external oblique; **4,** internal oblique; **5,** transverse muscle of abdomen. The location of the notochord in the larval teleost is indicated by stipple commencing just behind eye. In *Sphenodon* the appendicular muscles are: **6,** trapezius; **7,** dorsalis scapulae; **8,** latissimus dorsi. (*Sphenodon* modified from Byerly.[6])

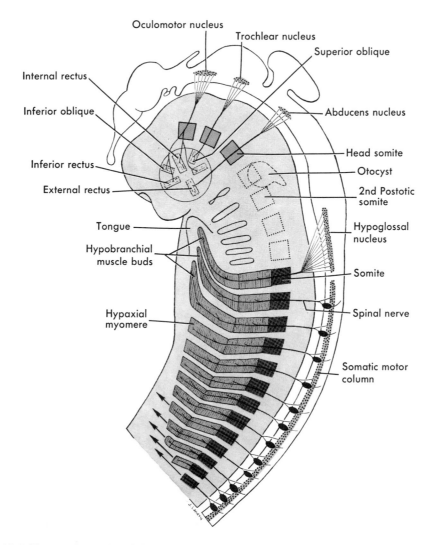

Fig. 10-4. Myotomal muscle origins and innervation in a vertebrate embryo (diagrammatical). The eyeball muscles arise from three preotic head somites associated with the oculomotor, trochlear, and abducens nuclei, respectively. Segmental muscles of the trunk arise from trunk somites and are supplied by corresponding segmental nerves. Hypobranchial musculature migrates forward into the floor of the pharynx accompanied by a nerve supply. Motor fibers innervating myotomal muscle have their cell bodies in the somatic motor column of the cord and brain. The four postotic somites (dotted outlines) make no contribution to the musculature. The central nervous system has been projected above the embryo for clarity.

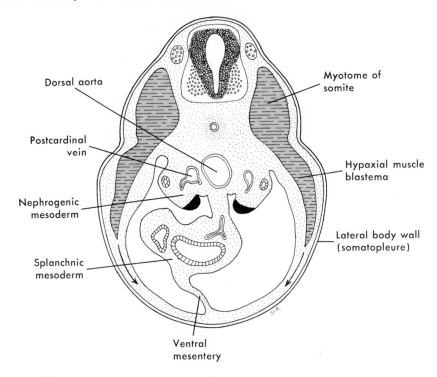

Dorsal aorta

Postcardinal vein

Nephrogenic mesoderm

Splanchnic mesoderm

Myotome of somite

Hypaxial muscle blastema

Lateral body wall (somatopleure)

Ventral mesentery

Fig. 10-5. Cross section of mammalian embryo showing invasion of myotomal cells (red) into lateral body wall to form hypaxial muscles.

ic. The blastemal cells are now **myoblasts,** and several myoblasts unite to form multinucleate striated muscle fibers. As a result, the body wall muscles begin to take shape. The metamerism of the embryonic blastemas is expressed in the adult wherever connective tissue myosepta separate the muscle of one body segment from the next. Because axial muscles form from myotomal cells they are often referred to as **myotomal muscles.**

Myosepta do not form in the abdominal region of higher tetrapods (Fig. 10-3, *E*). As a result the abdominal musculature consists of broad sheets rather than serially repeated myomeres. Nevertheless, these sheets are innervated by as many spinal nerves as there were somites that contributed to the blastema from which the sheet developed.

□ Trunk and tail muscles of fishes

In fishes (Fig. 10-3, *A* to *C*) the axial musculature consists of a series of muscle segments, or myomeres, separated by myosepta in which ribs often ossify.* Myosepta serve as origins and insertions for the segmental muscles. Except in agnathans the myomeres are divided into dorsal and ventral masses by a horizontal septum that extends between the transverse processes of the vertebrae and the skin. Above the septum are **epaxial muscles,** which lie lateral to the neural arches, and below it are **hypaxial muscles** in the lateral body wall. Each myomere is supplied by a spinal nerve from its own body segment, so successive myomeres are innervated by successive spinal nerves (Fig.

---

*The trunk muscles of the freshwater lamprey *Lampetra* are not segmented.

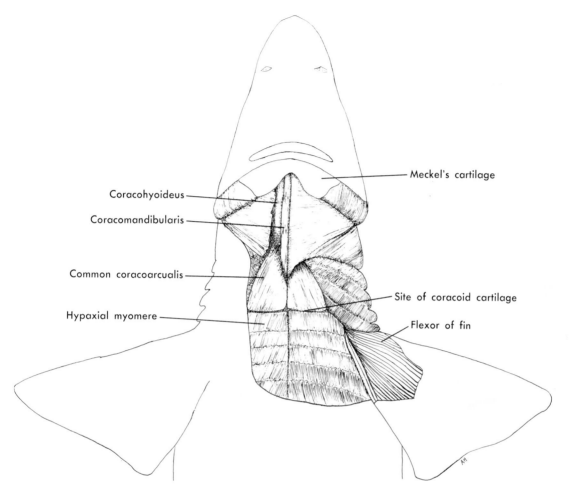

Coracohyoideus

Coracomandibularis

Common coracoarcualis

Hypaxial myomere

Meckel's cartilage

Site of coracoid cartilage

Flexor of fin

Fig. 10-6. Ventral muscles of "neck" of shark. At the left, anterior to the coracoid cartilage, are three hypobranchial muscles. At the right are branchiomerics. The intermandibular is superficial to the coracohyoid and coracomandibular muscles and has been partly removed on the left.

10-4). Middorsal and midventral septa separate the myomeres of the two sides of the body. The midventral septum is a connective tissue raphe, or "seam," the **linea alba.** Absence of epaxial and hypaxial masses in agnatha may be related to the absence of centra and transverse processes.

A thin sheet of **oblique fibers** lies superficial to the main hypaxial mass ventrolaterally. Farther ventrad, near the linea alba, a narrow ribbon of still more superficial fibers

extends cephalocaudad. These frequently form a discrete **rectus abdominis** muscle.

The metamerism of the hypaxial muscles is interrupted where the pectoral and pelvic girdles and fins are built into the body wall. It is also interrupted by the pharynx, since myotomal mesenchyme does not invade the pharyngeal arches. However, hypaxial muscles known as **hypobranchials** form in the pharyngeal floor from myotomal mesenchyme that migrates forward beneath the gills from farther back (Fig. 10-4, hypobranchial muscle buds, and 10-6). The epaxial muscles dorsal to the gills are called **epibranchial muscles.**

□ Trunk and tail muscles of tetrapods

Tetrapods, like fishes, have epaxial and hypaxial masses, and these retain some evidence of their primitive metamerism in even the highest tetrapods. Certain modifications have developed as adaptations to life on land:

1. The epaxial muscles tended to form

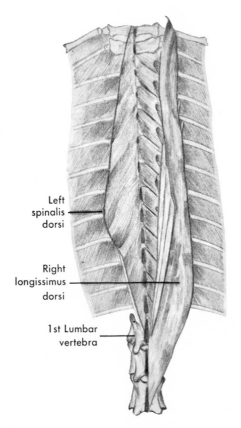

Left spinalis dorsi

Right longissimus dorsi

1st Lumbar vertebra

Fig. 10-7. Two long epaxial bundles of a hamster, dorsal view. The right spinalis dorsi and left longissimus dorsi have been removed. The deeper epaxial muscles seen between the two long bundles connect a transverse process with the neural spine on the second vertebra cephalad and belong to the transversospinalis group.

elongated bundles that extend through many body segments and became increasingly buried under the expanded appendicular muscles required to operate strong tetrapod limbs (Fig. 10-7).

2. The hypaxial muscles of the abdomen tended to lose their myosepta and to form broad sheets (Fig. 10-3, *E*).

3. Orientation of the hypaxial musculature into oblique, rectus, and transverse bundles proceeded to a high degree of specialization in tetrapods.

EPAXIALS

Epaxial muscles of tetrapods lie along the vertebral column dorsal to the transverse processes and lateral to the neural arches. They extend from the base of the skull to the tip of the tail. In urodeles and primitive lizards (Fig. 10-3, *D* and *E*) these epaxial muscles are obviously metameric and are called, collectively, the **dorsalis trunci.** In higher tetrapods the superficial epaxial bundles form long muscles occupying several or many body segments, whereas the deep bundles remain segmental. The longest bundles are named **longissimus, iliocostalis,** and **spinalis.** The shortest bundles, **intervertebrals,** connect one vertebra with the next.

The **longissimus group** (Fig. 10-7) lies on the transverse processes of the vertebrae and is so named because it includes the longest epaxial bundles. Regional subdivisions include the longissimus dorsi in the trunk, longissimus cervicis in the neck, and longissimus capitis, which inserts on the skull. The longissimus muscles continue into the tail as lateral extensors of the tail.

The **iliocostalis group,** located lateral to the longissidmus, arises on the ilium, and the bundles pass forward to insert on the dorsal ends of ribs or on uncinate processes when the latter are present. Extensions occur as far forward as the neck. The iliocostals lie just above where the horizontal septum would be if one were present.

The **spinalis group** (Fig. 10-7 and Table 10-3) includes long, medium, and short bundles lying close to the neural arches and connecting neural spines or transverse processes with neural spines several or many segments anteriorly. They continue into the tail as medial extensors of the tail.

The **intervertebral muscles** have retained the primitive segmental condition of the epaxial musculature. They connect chiefly the transverse processes of one vertebra with the transverse processes of the next anterior vertebra (intertransversarii), successive neural spines (interspinales), successive neural arches (interarcuales), and successive zygapophyses (interarticulares).

The foregoing epaxial muscles and many others occur in amniotes. In mammals most of the epaxial muscles are hidden by the extensive shoulder muscles and by the lumbodorsal fascia.*

HYPAXIALS

The hypaxial muscles of tetrapods may be divided into two chief groups: (1) muscles of the lateral body wall—**oblique, transverse,** and **rectus muscles** and (2) muscles that form longitudinal bands in the roof of the body cavity—**subvertebral (hyposkeletal) muscles.**

Oblique and transverse muscles. In early amphibians and reptiles, long, arching ribs developed in the myosepta throughout the entire length of the trunk. The metamerism may have facilitated locomotion in some species, since appendicular muscles were still weak. Urodeles still have myosepta the length of the trunk, but long ribs no longer form within them. In modern amniotes, excepting snakes and snakelike lizards, myosepta and ribs have become restricted to the anterior part of the trunk, which is thereafter called thorax, and so the abdominal wall muscles lack segmentation. Their

primitive metamerism is revealed, however, by their embryonic origin from successive somites and by their innervation via successive spinal nerves.

In addition to loss of myosepta in the abdominal region, the hypaxial muscles of both thorax and abdomen have become stratified into three layers—**external oblique, internal oblique,** and **transverse** sheets (Fig. 10-3, *D*). In the thorax they are called **external intercostals, internal intercostals,** and **transverse muscle of the thorax.** As their names imply, intercostals occupy the intercostal spaces and the fibers pass from one rib to the next.

One of the oblique layers may split into more than one sheet or, on the contrary, may be reduced or absent. In aquatic urodeles, for example, in which the hypaxial muscles are locomotor, the external oblique sheet of the abdomen is represented by superficial and deep portions. In crocodilians and some lizards all three layers in the abdomen consist of two sheets each. In anurans, on the other hand, the internal oblique layer may be missing. In birds the oblique sheets of the abdomen are greatly reduced, and the transverse sheet may be missing altogether. In mammals the transverse sheet is often incomplete in the thorax. In turtles all layers are almost completely absent. (If you want to make turtle soup it will have to come from the neck, tail, or appendicular muscles!) Chance mutations almost completely eliminated the body wall muscles of turtles, but a turtle couldn't care less. Because of the rigid shell, it could not use them anyway.*

At the deep inguinal ring of male mammals (Fig. 14-20) the spermatic cord gains an investment of muscle from the internal oblique and transverse abdominal sheets. These muscle slips wrap around the spermatic cord for some distance toward the testis and constitute a **cremaster muscle.** It is

---

*The splenius of mammals is an epaxial muscle with no known precursor in reptiles.

*See Lamarckism, p. 432.

better developed in rabbits, rodents, bats, and other species that are able to retract their testes.

External respiration in amniotes with a few exceptions is accomplished with the aid of body wall muscles. When intercostals are employed, their action is supplemented by special costal bundles such as the **scalenus, serratus dorsalis, levatores costarum,** and **transversus costarum.** The scalenus, serratus dorsalis, and levatores costarum may be epaxial derivatives. The transversus costarum is derived from some hypaxial mass.

In mammals diaphragmatic movements are more vital than rib movements in respiration, and mesenchyme that streams from somites in the future neck region invades the embryonic septum transversum to form the mammalian diaphragm. Innervation by ventral rami from the neck indicates a hypaxial origin of the diaphragmatic muscles.

Rectus muscles. The primitive rectus bundles of fishes have evolved into strong rectus muscles in tetrapods. These help support the ventral body wall where ventral ribs are lacking, and also aid in arching the back if the backbone is supple. In mammals the broad, strap-like rectus muscle is called the **rectus abdominis,** since it is confined to the abdominal region in man; however, in most mammals it extends the entire length of the trunk from the anterior end of the sternum to the pelvic girdle. A **pyramidal muscle** associated with the abdominal pouch of marsupials is a slip of the rectus abdominis. Some higher mammals have a vestigial pyramidal muscle.

Subvertebral muscles. A column of longitudinal hypaxial muscles, the subvertebrals, lies immediately underneath and against the transverse processes of the vertebrae in the roof of the body cavity and in the neck. The cavity must be opened to demonstrate most of them. They include the **psoas, iliacus,** and **iliopsoas** in the lumbar region, **longus colli** in the neck, and others. Subvertebrals are less well developed in the

thorax and there are none in the tail. They assist the epaxial muscles in movements of the vertebral column. Subvertebral muscles are sold as tenderloin and filet mignon in butcher shops.

Hypaxial muscles of the tail. From the base of the tail just behind the pelvis, slips of the caudal hypaxial musculature pass cephalad to insert on the pelvic girdle. Farther along on the tail, bundles of hypaxial muscles lie alongside the centra, underneath any transverse processes that are present. These are abductors and flexors of the tail.

FUNCTION OF EPAXIAL AND HYPAXIAL MUSCLES OF TETRAPODS

The short epaxial muscles of tetrapods perform the same function as in fishes, producing side-to-side movements of the vertebral column. Both the short and long bundles arch the back, and the long bundles support the back on land. In turtles and birds the vertebral column of the trunk is rigid, and the associated epaxial muscles are poorly developed, whereas in the neck and tail the column is flexible and the muscles are well developed. The most anterior epaxial bundles attach to the skull and participate in movements of the head. In a hatching chick one of these muscles, the **complexus,** provides the power for cracking the eggshell with the beak.

The hypaxial muscles of aquatic urodeles are used chiefly for swimming, and even terrestrial urodeles use them to assist in locomotion. Limbless tetrapods also use them for locomotion. In the remaining tetrapods the hypaxial musculature is reduced in volume when compared with fishes because of the shift to locomotion by appendages. They support the contents of the abdomen in a muscular sling, participate in external respiration, and (in the case of subvertebral and rectus groups) assist the epaxial muscles in bending the vertebral column. Hypaxial muscles make little contribution to the neck other than the hypobranchial muscles, to be discussed next.

□ Hypobranchial and tongue muscles

Myotomal muscle does not grow down into the pharyngeal arches from somites located above them. However, mesenchyme from postbranchial somites streams forward *beneath* the branchial region to give rise to hypobranchial and, when present, tongue muscles (Fig. 10-4).

In fishes the hypobranchial muscles extend forward from the coracoid and other ventral elements of the pectoral girdle to insert on the mandible, hyoid, and gill cartilages (Fig. 10-6). They strengthen the floor of the pharynx and pericardial cavity and aid the branchiomeric muscles in elevating the floor of the mouth, lowering the jaw, and expanding the gill pouches.

In tetrapods the hypobranchial muscles bear such names as **rectus cervicis, sternohyoid, sternothyroid, thyrohyoid, omohyoid,** and **geniohyoid.** They stabilize the hyoid apparatus and larynx or draw these structures cephalad or caudad, depending on what other muscles are doing at the time.

The tongue of amniotes is essentially a mucosal sac anchored to the hyoid skeleton and stuffed with hypobranchial muscle. The premuscle mesenchyme migrates into the developing tongue from the blastema of the hypobranchial muscles. This explains the fact that, in bats, tongue muscles can extend all the way from the sternum. The chief tongue muscles in mammals are the **hyoglossus, styloglossus, genioglossus,** and **lingualis.** The lingualis is an intrinsic muscle. The tongue of vertebrates below amniotes has no intrinsic musculature.

Because of their derivation from anterior trunk somites, the hypobranchial muscles are supplied by cervical spinal nerves. The motor fibers that innervate them have their central nervous system origin from the same somatic motor column that gives rise to fibers supplying the tongue and the axial muscles of the trunk and tail (Fig. 10-4).

□ Extrinsic eyeball muscles

The muscles that move the eyeballs of elasmobranch fishes arise during embryonic development from three head somites called **preotic somites** because they form cephalad to the otocyst (Fig. 10-4). Head somite I is located where cranial nerve III emerges from the brain. This somite splits to give rise to four eyeball muscles operated by cranial nerve III (Table 10-2). Head somite II is located where cranial nerve IV emerges from the brain. This somite gives rise to the superior oblique eyeball muscle innervated by cranial nerve IV. Head somite III (with contributions from somite II) is lo-

Table 10-2. Chief muscles derived from head somites and their innervation

Head somite pre-otic	Cranial nerve supply	Extrinsic eyeball muscles	Eyelid muscles
I	III (oculomotor)	Superior rectus Inferior rectus* Medial (internal) rectus Inferior oblique	Levator palpebrae superioris
II	IV (trochlear)	Superior oblique	
III (with contributions from somite II)	VI (abducens)	External (lateral) rectus Retractor bulbi	Pyramidalis of eye Quadratus of eye

*In lampreys the inferior rectus is operated by cranial nerve VI.

cated where cranial nerve VI emerges. This somite gives rise to the external (lateral) rectus muscle supplied by cranial nerve VI and to a retractor bulbi, if one is present. In other vertebrates the head somites are evanescent, and the mesenchyme that forms eyeball muscles cannot be traced directly to them.* Nevertheless, the eyeball muscles of all vertebrates are evidently homologous. Vertebrates with vestigial eyeballs have imperfect eyeball muscles.

Many amniotes have muscles inserting on the lids and nictitating membrane (pyramidalis of reptiles and birds, quadratus of birds, levator palpebrae superioris of reptiles and mammals), and lampreys have a corneal muscle that inserts on the cornea and alters its curvature. These, too, are myotomal muscles.

The cell bodies of motor fibers innervating these eyeball muscles are in the somatic motor column of the central nervous system. This column also supplies fibers to all other myotomal muscle and to no other muscle (Fig. 10-4). Therefore it can be said with confidence that eyeball muscles are derivatives of an ancient, segmental body wall musculature.

■ APPENDICULAR MUSCLES

Appendicular muscles move fins or limbs. **Extrinsic** appendicular muscles have their anatomical origins on the axial skeleton or on fascia of the trunk and insert on the girdles and limbs. **Intrinsic** appendicular muscles have their anatomical origin on the girdle or on proximal skeletal elements of the appendage and insert on more distal elements. All extrinsic muscles must be completely severed to permit removal of an appendage from the body; in the process, intrinsic muscles remain intact.

---

*The extrinsic eyeball muscles of an apodan, turtle, and several birds have been described as arising directly from preotic somites, but this has been disputed.

□ Fishes

Paired fins and limbs first appear in embryos as elevated fin folds or limb buds protruding from the lateral body wall (Fig. 10-8). Buds from the developing hypaxial muscles of several body segments grow into the fin folds to attach to skeletal elements at the base of the fin. These invading buds form extensors, located dorsally, and flexors, ventrally (Fig. 10-9). Other hypaxial muscles attach to the girdles.

Appendicular muscles of fishes serve mostly as stabilizers and contribute little to forward locomotion. However, a few fishes swim entirely with their fins. In these species the pelvic fins are far forward, the appendicular muscles are strong, and the body wall and tail muscles are thin. Intrinsic appendicular muscles in fishes are sparse and relatively undifferentiated.

□ Tetrapods

The appendicular muscles of even the lowest tetrapods are far more complicated than those of fishes. This provides the greater leverage required for locomotion on land. Also, jointed appendages require a more "sophisticated" musculature.

The **extrinsic appendicular musculature** is disposed in dorsal and ventral groups. The dorsal group of the forelimbs arises on the fascia of the trunk in lower tetrapods (Fig. 10-10, A, *Necturus*), or on the skull and vertebral column to a point well caudal to the scapula in higher tetrapods (Fig. 10-10, B and C). They converge on the girdle and limb. Muscles of the ventral group arise on ventral skeletal elements (sternum, coracoid, and others) and converge on the limb. Thus the pectoral girdle and limb are joined to the trunk by extrinsic appendicular muscles. The pelvic girdle requires no such *muscular* anchoring because it is usually attached directly to the vertebral column. Therefore, in the posterior limb the volume of extrinsic muscle is relatively small. This accounts for the fact that an agile tetrapod

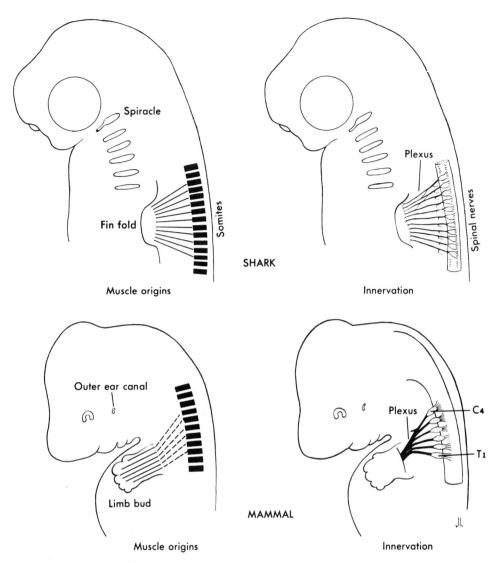

Fig. 10-8. *Above,* Origin of shark appendicular muscles from somites and their innervation by corresponding spinal nerves. *Below,* Probable phylogenetic derivation of appendicular muscles of a mammal from six somites based on their innervation. **C4 and T1,** Dorsal root ganglia of the fourth cervical and first thoracic spinal nerves of the brachial plexus.

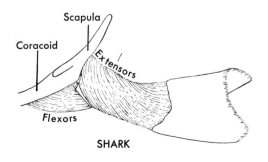

Fig. 10-9. Appendicular muscles of pectoral fin of shark. Extensors are also called levators. Flexors are adductors.

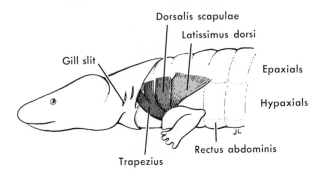

A. NECTURUS

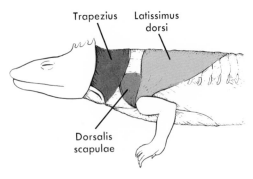

B. SPHENODON

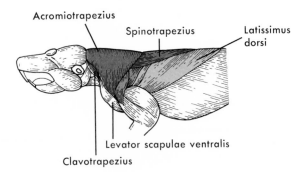

C. HAMSTER

Fig. 10-10. Superficial shoulder muscles of an amphibian, reptile, and mammal, illustrating increased volume and extent of homologous muscles in successively higher tetrapods.

jumps by pushing off with its hind limbs, which are braced against the pelvis; but it lands at least partly on its forelimbs where the extrinsic muscles of the shoulder can absorb much of the jolt. (The hind limbs of digitigrade and unguligrade mammals are also good shock absorbers, however).

Most extrinsic appendicular muscles of tetrapods develop from hypaxial blastemas in the body wall, and they are sometimes referred to as **secondary appendicular muscles** because it was not their original role to operate appendages. However, the rhomboideus is thought to be epaxial (although its innervation does not confirm this); and the trapezius is partly branchiomeric. The chief extrinsic appendicular muscles of the forelimbs of mammals are listed near the bottom of Table 10-3.

The **intrinsic appendicular muscles** of tetrapods do not arise by invasion of the limb bud from hypaxial blastemas as in fishes. Instead, they form from blastemas within the limb bud. These blastemas sometimes spread trunkward to give rise to an occasional extrinsic muscle such as the latissimus dorsi. Muscles that arise from blastemas within the limb bud are **primary appendicular muscles.** Table 10-4 lists the chief intrinsic muscles of the anterior limbs of reptiles and mammals.

The appendicular muscles of reptiles are more numerous, more diversified, and more powerful than those of amphibians, particularly aquatic ones (compare Fig. 10-10, *A* and *B*). The intrinsic muscles become more specialized and thereby provide greater support for the body and increased mobility of the distal segments of the appendage. Reptilian appendicular muscles presage the development of powerful muscles for flight in birds and for digging, flying, swimming, climbing, and general agility in mammals.

The largest and most powerful muscles of birds are the extrinsic muscles of the wings. These are attached to the keeled sternum. In contrast, the intrinsic musculature of the

Table 10-3. Representative myotomal muscles of head, trunk, and forelimbs in mammals

HEAD AND NECK

Eyeball	Tongue	Hypobranchial
Superior oblique	Genioglossus	Geniohyoideus
Inferior oblique	Hyoglossus	Sternohyoideus
Medial rectus	Styloglossus	Sternothyroideus
Lateral rectus	Lingualis	Thyrohyoideus
Superior rectus		Omohyoideus
Inferior rectus		

TRUNK AND TAIL

Epaxial muscles	Hypaxial muscles
Longissimus	Subvertebrals
L. dorsi	Longus colli
L. cervicis	Psoas
L. capitis	Iliacus
Extensor caudae lateralis	Quadratus lumborum
Iliocostales	Oblique group
Spinales	Internal and external intercostals
S. dorsi	Internal and external obliques
S. cervicis	of abdomen
S. capitis	Diaphragm
Transversospinales	Cremaster
Extensor caudae medialis	Transverse group
Intervertebrales	Transversus thoracis
Intertransversarii	Transversus abdominis
Interspinales	Rectus abdominis
Interarcuales	Pyramidalis
Interarticulares	Abductors and flexors of tail

FORELIMB

Extrinsic muscles*	Intrinsic muscles
Levator scapulae†	See Table 10-4
Latissimus dorsi‡	
Rhomboideus†	
Serratus ventralis†	
Pectoral group†	

*The trapezius, sternomastoideus, and cleidomastoideus are probably of branchiomeric origin, hence not myotomal.
†Secondary appendicular muscle.
‡Primary appendicular muscle.

wings is reduced. The musculature of the hind limbs is well developed in support of bipedalism. A specialization for perching is seen in the very long tendons that pass behind the "heel" to insert on the digits. Because of the extreme length of the tendons, the muscles do not have to shorten so much; this reduces the energy expenditure of the perching bird.[5]

The appendicular muscles of mammals are not greatly different from those of reptiles, although many bear different names. Some subdivision of bundles occurred and some bundles were lost. For example, the

Table 10-4. Chief intrinsic muscles of the forelimbs of reptiles and mammals*

MUSCLES OF GIRDLE

Girdle to humerus, proximally
  Deltoideus
  Subcoracoscapularis
    Subscapularis
  Scapulohumeralis
    Teres minor
  Supracoracoideus
    Infraspinatus
    Supraspinatus
  Coracobrachialis
  Teres major (derived from latissimus dorsi)

MUSCLES OF UPPER ARM

Girdle or humerus to proximal end of radius
  or ulna
  Triceps brachii (anconeus)
  Humeroantebrachialis
  Biceps brachii
  Brachialis

MUSCLES OF FOREARM†

Humerus and proximal end of radius and ulna
  to manus
  Extensors and **flexors** of wrist and digits
  Supinators and **pronators** of hand

MUSCLES OF HAND‡

Flexors, extensors, abductors, and **adductors**
  of fingers

*Indentations indicate likely mammalian derivatives.
†These frequently have long tendons of insertion.
‡Reduced in species in which digits are reduced.

pectoral muscles became divided into major, minor, superficial, and deep bundles and are assisted by additional slips (xiphihumeralis and pectoantebrachialis). The rhomboideus has an extra slip, the rhomboideus capitis. The latissimus dorsi evidently split to form a teres major. A few representative somatic muscles of the head, trunk, and forelimbs of mammals are listed in Table 10-3. A recitation of the posterior appendicular muscles would yield little additional insight into the evolution of appendicular muscles.

□ **Innervation of appendicular muscles**

Branches of the ventral rami of a series of spinal nerves grow into the fin folds and limb buds to innervate the muscle within the limb. In fishes these nerves come from the same body segments that contribute myotomal muscle to the appendage (Fig. 10-8). Although in tetrapods the muscle within the appendage cannot be traced directly to myotomes, it seems probable that the number of spinal nerves entering a tetrapod limb is indicative of the number of somites that contributed mesenchyme to the limb phylogenetically.

■ **BRANCHIOMERIC MUSCLES**

Associated with the pharyngeal arches of vertebrates (Fig. 10-11) is a series of striated, skeletal, voluntary, visceral, branchiomeric muscles. The basic architectural pattern of these muscles is illustrated in fishes, in which adductors, constrictors, and levators operate the jaws and successive gill arches (Fig. 10-12, A). In tetrapods they continue to operate the jaws. However, with loss of gills, the more posterior ones have acquired new functions.

That branchiomeric muscles must be classified as visceral is evident from the following facts:

  1. Their position in the wall of the ali-

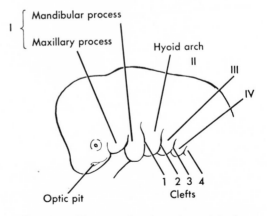

Fig. 10-11. Pharyngeal arches of a vertebrate.

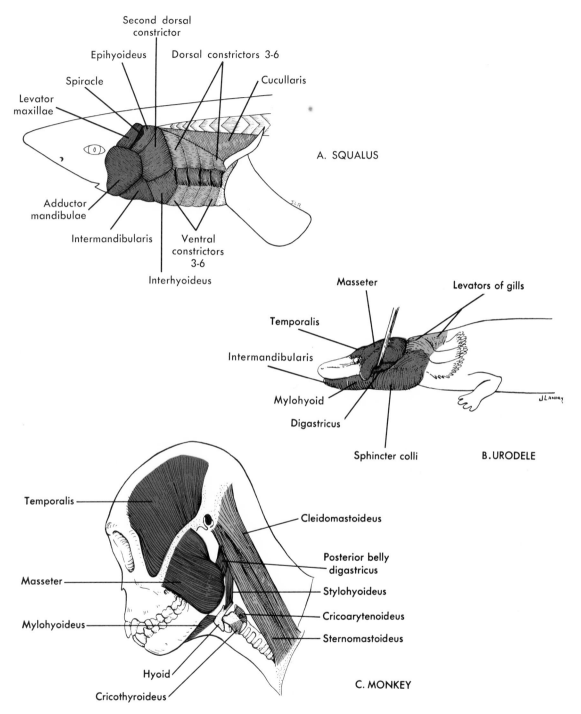

Fig. 10-12. Selected branchiomeric muscles of three vertebrates. The sphincter colli derivatives in mammals are shown in Fig. 10-13. Muscles of the first, second, and successive pharyngeal arches are indicated by three shades of red.

mentary canal identifies them as visceral.

2. The motor nerve fibers innervating them arise in visceral motor columns (Fig. 15-28, SVE) located close to the columns that innervate smooth muscles and glands (Fig. 15-28, GVE).

3. Functionally they are associated with two visceral processes: nutrition and respiration.

The foregoing criteria set branchiomeric muscles apart from somatic (myotomal) muscle.

☐ Muscles of the mandibular arch

In *Squalus* and other fishes (Fig. 10-12, *A*) the branchiomeric muscles of the first pharyngeal arch operate the jaws. A large **adductor of the mandible** and an **intermandibularis** insert on Meckel's cartilage, and a **levator maxillae** inserts on the pterygoquadrate cartilage.

Above fishes, the muscles of the first arch continue to operate the jaws (Fig. 10-12, *B* and *C*). Adductors of the mandible now include **pterygoid muscles,** and the adductor mandibulae is usually divided into a **masseter** and **temporalis.** The intermandibular, now more often called mylohyoid, often gives rise to a separate **digastric.** In mammals an ancient part of the lower jaw musculature continues to insert on the ossified posterior tip of Meckel's cartilage, which is now in the middle ear and called the malleus. The muscle puts tension on the eardrum and is called the **tensor tympani.** Upper jaw muscles become relatively unnecessary when the upper jaw unites with the braincase. However, some of them function in reptiles and birds whose upper jaw or palate are movable. Muscles derived from the first arch are also found in the orbits of some reptiles and birds as **protractors or levators of the eyeball** and **depressors of the lower lid.** All these first arch muscles are innervated by the mandibular division of cranial nerve V.

☐ Muscles of the hyoid arch

Hyoid arch muscles (Fig. 10-12) move the skeleton of the hyoid arch whether it has a gill attached, an operculum, or whether the skeleton becomes the hyoid apparatus in the walls and floor of the pharynx of tetrapods. The **stapedial muscle** of the middle ear cavity of tetrapods is of second arch origin. This muscle attaches to the stapes, which is a derivative of the hyomandibula, and it aids in hearing. It served this function as far back as ancient urodeles,[14] and before that it operated the hyomandibula of some fish. Hyoid arch muscles also assist in moving the lower jaw. For example, the large **depressor mandibulae** of reptiles and the **posterior belly of the digastric** of mammals, when present (Fig. 10-12, *C*), are hyoid arch muscles.

In tetrapods a thin **sphincter colli** muscle lies just under the skin of the neck, somewhat like a collar, which gives it its name (Fig. 10-13, *A* and *B*). In reptiles and birds the sphincter colli spreads upward around the rear of the skull to insert on the skin of the head and is then called **platysma.** In mammals the platysma spreads forward onto the face to become muscles of facial expression, or **mimetic muscles** (Fig. 10-13, *C* to *F*). These and all other muscles derived from the second arch are innervated by cranial nerve VII. It is called the facial nerve because it is widely distributed under the skin of the face in mammals.

☐ Muscles of the third and successive pharyngeal arches

In *Squalus* the branchiomeric muscles of the third through the sixth pharyngeal arches are represented by constrictors located above and below the gill chambers, by levators, including the cucullaris, and by other serial bundles that assist in compressing or expanding the gill pouches (Fig. 10-12, *A*). The muscles of the third arch are innervated by cranial nerve IX, and those of the remaining arches are innervated by

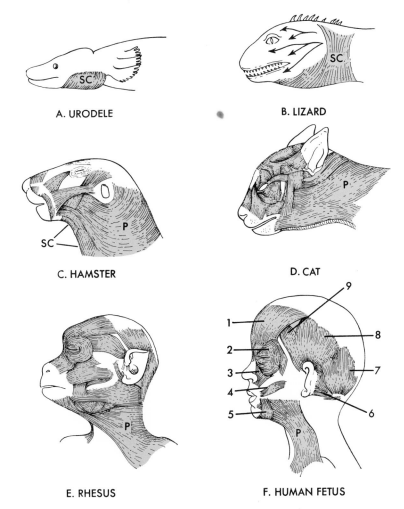

A. URODELE

B. LIZARD

C. HAMSTER

D. CAT

E. RHESUS

F. HUMAN FETUS

Fig. 10-13. Evolution of mammalian mimetic muscles from hyoid arch muscles of lower tetrapods. The sphincter colli, **SC,** spreads onto the neck in reptiles and becomes the platysma, **P,** in mammals. It then spreads forward onto the head and face, as indicated by arrows in **B,** to become muscles of facial expression. Note increasing differentiation of mimetics in the mammals shown. **1,** Frontalis; **2,** orbicularis oculi; **3,** quadratus labii superioris; **4,** risorius; **5,** triangularis; **6,** posterior auricular; **7,** occipital; **8,** superior auricular; **9,** anterior auricular. (**C** after Priddy and Brodie[13]; **E** and **F** after Huber.[9])

cranial nerve X. In bony fishes the branchiomeric musculature caudal to the hyoid arch has become reduced, and the operculum, to a large extent, controls the respiratory stream.

The absence of gills in tetrapods is correlated with reduction of the branchiomeric

muscles caudal to the hyoid arch. The chief branchiomeric muscles of these arches are the **stylopharyngeus** from arch III, used for swallowing, and the intrinsic muscles of the larynx (**thyroarytenoideus, cricoarytenoideus,** and **cricothyroideus**) derived from the fourth and any remaining arches. The

Table 10-5. Chief branchiomeric muscles and their innervation in *Squalus* and in tetrapods

Pharyngeal arch	Pharyngeal skeleton in squalus	Chief branchiomeric muscles		Cranial nerve innervation
		Squalus	Tetrapods*	

I Mandibular arch	Meckel's cartilage	Intermandibularis	Intermandibularis
			  Mylohyoideus (anterior part)
			  Digastricus (anterior part)
				V

		Adductor mandibulae	Adductor mandibulae
			  Masseter
			  Temporalis
			Pterygoidei
			Tensor tympani	

	Pterygoquadrate cartilage	Levator maxillae superioris		

II Hyoid arch	Hyomandibula
	Ceratohyal
	
	Basihyal
		Epihyoideus (dorsal constrictor)
		Interhyoideus (ventral constrictor)
			Stapedius
			Stylohyoideus (anterior part)
			Mylohyoideus (posterior part)
			
			Depressor mandibulae
			Digastricus (posterior part)
			Sphincter colli (gularis)
			  Platysma
			  Mimetics
				VII

III	Gill cartilages	Constrictors and levators	Stylopharyngeus	IX
IV to VI	Gill cartilages	Constrictors and levators	Striated pharyngeal muscles Thyroarytenoideus Cricoarytenoideus Cricothyroideus	X
VII	Seventh visceral cartilages	Cucullaris (probably derived also from dorsal constrictors 3 to 6)	Trapezius Sternomastoideus Cleidomastoideus	Occipitospinal nerves in shark; XI in amniotes

*Indented muscles in this column may be derivatives of the preceding muscle.

trapezius, **cleidomastoid,** and **sternomastoid** muscles of amniotes appear to be derived from the cucullaris on the basis of innervation.

The chief branchiomeric muscles and
their innervations are given in Table 10-5.

■ INTEGUMENTARY MUSCLES

Extrinsic integumentary muscles arise
elsewhere than on the skin, usually on the
skeleton, and insert on the underside of the
dermis. They are striated muscles and, commencing with reptiles, they move the skin.
Intrinsic integumentary muscles lie entirely
within the dermis. They occur in vertebrates with hair and feathers and are mostly
smooth muscles.

In fishes and amphibians, slips of branchiomeric or body wall muscles insert here
and there on the dermis, firmly attaching the
skin to the underlying muscle; however,
the slips cause little movement of the skin.
The **cutaneous pectoris** of anurans is such a
muscle.

Costocutaneous muscles are hypaxial
muscles that erect the ventral scutes of
snakes, thereby providing friction for locomotion. **Patagial muscles** in birds and bats
are slips of pectoral muscles that insert on
the skin of the wing membrane.

Extrinsic integumentary muscles reach an
evolutionary peak in mammals. The trunk is
frequently completely wrapped in a **panniculus carnosus,** also called **cutaneous maximus** (Fig. 10-14), which seems to have been
derived from hypaxial musculature. It helps
animals such as armadillos to roll into a ball
when endangered. In marsupials a sphincter portion of this muscle surrounds the entrance to the marsupial pouch. Horses,
cows, and many other animals use the muscle to shake off flies. It is poorly developed
or absent in primates.

The most notable mammalian integumentary muscles are the **mimetic muscles,**
which evolved from the platysma and
spread onto the face (Fig. 10-13, *C* to *F*).

CAT   MONKEY

Fig. 10-14. Panniculus carnosus (cutaneous
maximus) of a cat and primate. Note difference
in the extent of the muscle in these two animals.

They are best developed in primates. Man
has the largest number of these—about
thirty—some of which are illustrated in Fig.
10-13, *F*. By contracting singly or in groups
they enable primates, and man especially,
to express his emotions without uttering a
word. The **platysma** depresses the corner of
the mouth, as in grief. The **frontalis** raises
the eyebrow. A large **corrugator muscle**
wrinkles the skin of the forehead and also
raises the eyebrow. The **orbicularis oculi**
closes the eye tightly, and a similar muscle
orbiting the mouth, the **orbicularis oris,**
draws the lips together. The **quadratus labii
superioris** dilates the nostril. The **risorius**
draws back the corner of the mouth, as in
grief, despite its name (p. 448). A **zygomat-
icus,** arising on the zygomatic arch and inserting on the corner of the mouth, is the
"smiling" muscle. The **triangularis** pulls
down the corner of the mouth. **Auriculars**
direct the pinna of the ear toward faint
sounds which might spell the difference
between life and death to the hunter or
hunted. Man can scarcely wiggle his ears

with his poorly developed auriculars. The **occipital muscle** draws the scalp backward. One muscle, the **caninus**, elevates the part of the upper lip hiding the spearlike canine tooth. In contracting it, the carnivore is displaying the tooth used for ripping into flesh. When man uses it, he is said to be "sneering."

Intrinsic integumentary muscles are found in birds and mammals. They are mostly smooth muscles that attach to feather follicles (**arrectores plumarum**) and to hair follicles (**arrectores pilorum**). These muscles permit ruffling the feathers or elevating the hairs for insulation or in response to danger. They are innervated by *visceral* motor fibers.

■ ELECTRIC ORGANS

In many fishes certain muscle masses are modified to produce, store, and discharge electricity. In *Torpedo*, the electric ray, an electric organ lies in each pectoral fin near the gills. It is probably of branchiomeric origin, since it is supplied by cranial nerves VII and IX. In *Raia* (a skate) and *Electrophorus* (electric eel) electric organs lie in the tail and are modified hypaxial muscles (Fig. 10-15). The potential produced by these organs in eels amounts to 500 volts and is probably a protective device. Other fishes have electric organs with a lower electric potential, which serve as locating mechanisms or for signaling (communication). Certain neuromasts are receptors for these signals. Electric organs consist of a

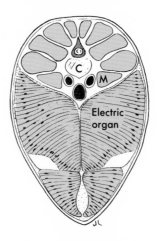

Fig. 10-15. Electric organs in tail of electric eel. Each nucleated horizontal disc (electroplax) is a modified hypaxial muscle cell. **C,** Centrum; **M,** epaxial myomere.

large number of electric discs (up to 20,000 in the tail of one ray) piled in either vertical or horizontal columns.

Each disc (**electroplax**) is a modified multinucleate muscle cell embedded in jellylike extracellular material and surrounded by connective tissue. Nerve endings terminating on each disc induce the discharge. Capillaries develop in the jelly layer.

Electric organs seem to have no systematic distribution among fishes, and the various types probably result from convergent evolution. The electric organ of one teleost (a catfish native to Africa) is said to be a modified skin gland, rather than muscle.

☐ Chapter summary

1. Muscle may be classified as striated, smooth, or cardiac; voluntary or involuntary; skeletal or nonskeletal; somatic (myotomal) or visceral (nonmyotomal).

2. Skeletal muscles are attached to the skeleton. They are striated and voluntary. Excepting branchiomeric muscles, they are somatic and arise from mesodermal somites directly or indirectly.

3. Nonskeletal muscle is found in the walls of hollow viscera, tubes, and vessels; in the eyeball, dermis, and miscellaneous sites. It is typically involuntary, smooth, and visceral.

4. Cardiac muscle is striated visceral muscle constituting the myocardium of the heart.

5. Classified according to their actions, muscles are flexors, extensors, abductors, adductors, protractors, retractors, levators, depressors, rotators (including pronators and supinators), constrictors, dilators, and tensors.

6. Skeletal muscles may be named on the basis of the direction of their fibers, location, number of parts, shape, attachments, action, size, and miscellaneous considerations.

7. Homologies between muscle groups are more easily discernible than homologies between muscles. Embryonic origin and nerve supply are the most reliable criteria.

8. The trunk and tail muscles of fishes and tailed amphibians are locomotor. They are arranged as myomeres separated by myosepta. Myomeres are divided into epaxial and hypaxial masses by a horizontal septum. Loss of myosepta tends to obscure the primitive metamerism.

9. The metamerism of axial muscles is an expression of the metamerism of embryonic somites.

10. Epaxial and hypaxial masses are innervated by dorsal and ventral rami, respectively.

11. With the advent of tetrapod appendages epaxial muscles tended to form long bundles extending over many segments, but the deepest masses (intervertebral muscles) retain their primitive segmentation. In animals with a rigid vertebral column the epaxials are reduced.

12. Hypaxial myomeres in amniotes tend to form large muscle sheets except where long ribs are present.

13. Hypaxial muscles in tetrapods are disposed in transverse, oblique, and rectus layers.

14. The mammalian diaphragm and cremaster muscles are hypaxial muscles that have migrated, accompanied by their spinal nerves.

15. Hypaxial muscles immediately ventral to transverse processes are disposed in long bundles (subvertebral muscles).

16. Hypaxial muscle extends forward under the pharynx as hypobranchial musculature. This contributes the musculature of the amniote tongue.

17. Eyeball muscles arise from three preotic somites in elasmobranchs but develop in situ in higher vertebrates. They are innervated by cranial nerves III, IV and VI in all

vertebrates and must be considered myotomal derivatives in all species.

18. Extrinsic appendicular muscles have anatomical origins on the axial skeleton and insertions on the girdles and limbs. Intrinsic appendicular muscles have no axial skeletal attachments.

19. Appendicular muscles that form from blastemas within the body wall and achieve an attachment to the girdle or limb are secondary appendicular muscles. Most extrinsic appendicular muscles are of this kind.

20. Muscles that organize from blastemas within the limb are primary appendicular muscles. They are the intrinsic and a few extrinsic appendicular muscles of tetrapods.

21. Branchiomeric muscles are striated, skeletal, voluntary, visceral muscles, which arise from ectomesenchyme in the pharyngeal arches.

22. The muscles of the mandibular arch operate the jaws. In mammals a derivative operates the malleus. They are innervated by cranial nerve V.

23. The muscles of the second arch are attached to the hyoid skeleton, lower jaw, and operculum. The sphincter colli spreads over the head to become platysma and facial muscles. Second arch muscle also operates the stapes. The muscles of the second arch are innervated by cranial nerve VII.

24. Muscles of the third and remaining arches operate gills in fishes. In higher vertebrates they have new functions, including swallowing and vocalization. The muscles of the third arch are innervated by cranial nerve IX and those of successive arches by cranial nerve X.

25. The cucullaris has become subdivided into trapezius, sternomastoideus, and cleidomastoideus muscles in amniotes.

26. Extrinsic integumentary muscles insert on the skin. They reach peak development as the panniculus carnosus (cutaneous maximus) of mammals and mimetics of primates. Those of the trunk are myotomal muscles. Those of the face and scalp are branchiomeric.

27. Intrinsic integumentary muscles are chiefly smooth muscles inserting on feathers and hairs.

28. Electric organs are columns of modified axial, appendicular, or branchiomeric muscle cells (electroplaxes) capable of producing, storing, and discharging electric potential. A few appear to be modified skin glands. They occur in many fishes.

LITERATURE CITED AND SELECTED READINGS

1. Adelmann, H. B.: The development of the eye muscles of the chick, Journal of Morphology **44**:29, 1927.
2. Alexander, R. M.: Functional design in fishes, ed. 2, London, 1970, Hutchinson.
3. Auffenberg, W.: A review of the trunk musculature in the limbless land vertebrates, American Zoologist **2**:183, 1962.
4. Ayers, H., and Jackson, C. M.: Morphology of the Myxinoidei. I. Skeleton and musculature, Journal of Morphology **17**:185, 1901.
5. Bock, W. J.: Experimental analysis of the avian passive perching mechanism, American Zoologist **5**:681, 1965.
6. Byerly, T. C.: The myology of *Sphenodon punctatum*, University of Iowa Studies in Natural History (First Series, no. 98) vol. 11, no. 6, 1925.
7. Ellsworth, A. H. F.: Reassessment of muscle homologies and nomenclature in conservative amniotes: the echidna, *Tachyglossus;* the opossum, *Didelphis;* and the tuatara, *Sphenodon,* Huntington, N.Y., 1975, R. E. Krieger Pub. Co., Inc.
8. Grundfest, H.: Electric fishes, Scientific American **203**:115, 1960.
9. Huber, E.: Evolution of facial musculature and cutaneous field of trigeminus, parts I and II, Quarterly Review of Biology **5**:133, 389, 1930.
10. Johnson, C. E.: The development of the prootic head somites and eye musculature in *Chelydra serpentina*, American Journal of Anatomy **14**:329, 1915.
11. Keynes, R. D.: Electric organs. In Brown, M. E., editor: The physiology of fishes, New York, 1957, Academic Press, Inc.
12. Marcus, H.: Beitrage zur Kenntnis der Gymnophionen. III. Zur Entwicklungsgeschichte des Kopfes. I Teil, Morphologische Jharbuch **40**:105, 1909.
13. Priddy, R. B., and Brodie, A. F.: Facial musculature, nerves and blood vessels of the hamster in relation to the cheek pouch, Journal of Morphology **83**:149, 1948.
14. Schmalhausen, I. I.: The origin of terrestrial vertebrates, New York, 1968, Academic Press, Inc.

11

Digestive system

In this chapter we will examine the digestive tract, the evaginations that arise from it, and the adaptive modifications associated with varying food habits, such as the gizzard of grain-eating birds, the four-chambered stomach of cud-chewers, and the capacious cecum of cellulose-eating mammals.

Mouth and oral cavity
  Teeth
  Tongue
  Oral glands
Pharynx
Esophagus
Stomach
Intestine
Ceca
Liver
Gallbladder
Exocrine pancreas
Cloaca

The digestive tract is a tube, seldom straight and often tortuously coiled, commencing at the mouth and terminating at the vent or anus (Fig. 11-1). It functions in ingestion, digestion, and absorption of foodstuffs and in elimination of undigested wastes. Major subdivisions of the tract are the oral cavity, pharynx, esophagus, stomach, and small and large intestines. Associated with the tract are accessory organs, such as the tongue, teeth, oral glands, pancreas, liver, and gallbladder. Blind pouchlike evaginations (ceca) of the tract are common. The digestive tract and accessory organs constitute the digestive system.

Differences in the anatomy of digestive tracts caudal to the pharynx are not so much correlated with whether the animal lives on land or in the water as with the nature and

abundance of the food. Is it readily absorbable when ingested, as in vampire bats, or does it require extensive enzymatic activity or mechanical maceration, as in carnivores? Is the food supply constant, so that whenever an animal opens its mouth the food is likely to be there, as in herbivores living on grassy plains, or does it have to be stalked, as a tiger stalks its prey? If so, the meal is probably bulky and must be stored until it can be gradually processed. And what is the shape of the animal's body? If it is long, like that of a cyclostome or snake, the tract will probably be relatively straight. If the trunk is short, as in turtles and frogs, the absorptive surface of the intestine must be increased by coils. Some fishes have a spiral valve (typhlosole) that increases the absorptive area without the need for coils.

The entire digestive tract is ciliated in the adult amphioxus and in many larval vertebrates. There are cilia in the stomach of many teleosts, in the oral cavity, pharynx, esophagus, and stomach of some adult amphibians, in the ceca of some birds, and in other locations in various species. They even occur transitorily in the stomach of the human fetus. However, peristalsis is chiefly responsible for passage of foodstuffs along the alimentary canal.

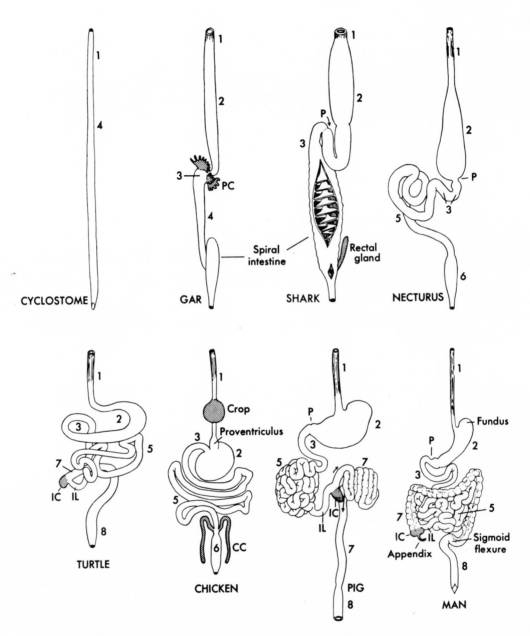

Fig. 11-1. Digestive tracts caudal to the pharynx of a few vertebrates. **1,** Esophagus; **2,** stomach; **3,** duodenum; **4,** intestine; **5,** small intestine; **6,** large intestine; **7,** colon; **8,** rectum; **CC,** paired ceca of bird; **IC,** ileocolic cecum; **IL,** ileum; **P,** pyloric sphincter; **PC,** pyloric ceca. All ceca are shown in gray.

The most anterior part of the oral cavity arises from the **stomodeum,** a midventral invagination of the ectoderm of the head (**Fig. 1-1**). A similar ectodermal invagination, the **proctodeum,** becomes the terminal segment of the digestive tract just beyond the endodermal part of the cloaca. The remainder of the digestive tract is lined by endoderm, as are any evaginations of it. Surrounding the embryonic endodermal tube is the splanchnic mesoderm, which provides the muscular and connective tissue coats of the tract and the visceral peritoneum (Fig. 4-8).

The embryonic digestive tract consists of three regions. The part containing the yolk or to which the yolk sac is attached is the **midgut.** The part anterior to the midgut is the **foregut,** and the part caudal to it is the **hindgut.** The foregut elongates to become the endodermal part of the oral cavity and the pharynx, esophagus, stomach, and most of the small intestine. The hindgut becomes large intestine and cloaca.

The embryonic gut from stomach to cloaca is attached to the dorsal body wall of the trunk by a continuous dorsal mesentery and to the ventral body wall by a ventral mesentery (Fig. 11-18). Much of the dorsal mesentery remains throughout life, but the ventral mesentery of the digestive tract disappears except at two sites. The ventral mesentery remains in all vertebrates at the level of the liver, where it is called the **falciform ligament,** and in tetrapods it remains as the **ventral mesentery of the urinary bladder.**

■ MOUTH AND ORAL CAVITY

In a restricted sense the term "mouth" refers only to the anterior opening into the digestive tract. In a broader sense the term is a synonym for oral cavity. Muscular lips are an adaptation for suckling.

The oral cavity begins at the mouth and ends at the pharynx. In fishes it is very short and almost nonexistent. It is longer in tetrapods and culminates in mammals in a sucking and masticatory organ bounded laterally by muscular cheeks. A primitive or specialized tongue lies in its floor.

The roof of the oral cavity is the **palate.** If it is a primitive (primary) palate and there are internal nares, these empty into the oral cavity anteriorly (Fig. 11-10). With formation of a secondary palate in amniotes, the nasal passageways lie above the secondary palate and open farther to the rear. In monkeys and man a **uvula** dangles from the caudal end of the soft palate into the pharynx.

In mammals a trench, the **oral vestibule,** separates the alveolar ridge of the jaw (which has the sockets, or alveoli, for the teeth) from the cheeks and lips. In many mammals, especially rodents, large membranous **cheek pouches** extend from the vestibule backward under the skin and muscle of the side of the head. The pouches are used for transporting grain and other foods.

Many birds carry seeds to the nest in a membranous **sublingual pouch** located under the tongue and emptying into the oral cavity anteriorly. When full, the pouch hangs from the rear of the oral cavity suspended in a sling formed by the mylohyoid muscle. The pouch is emptied by shaking the head.

Since the oral cavity is lined by stomodeal ectoderm anteriorly and by endoderm posteriorly, the anterior mucosa is supplied by the same nerves that innervate the ectoderm of the head, and the posterior mucosa by those nerves innervating the endoderm of the foregut.

□ Teeth

The only remnants of dermal armor that man and most other mammals have in their skin are the denticles, or teeth, on their jaws. That teeth are, indeed, derived from armor is seen in sharks, in which placoid scales show a gradual transition to teeth as they approach the edge of a jaw (Fig. 11-2). Vertebrate teeth, like placoid scales, are composed primarily of a core of **dentin** sur-

Fig. 11-2. Jaws of the Port Jackson shark, showing transition of placoid scales to teeth. (Courtesy Ward's Natural Science Establishment, Inc., Rochester, N.Y.)

rounded by a crown of **enamel** (Figs. 11-3 and 11-4). In the root there is a central canal (root canal) containing a dermal papilla. Odontoblasts in the papilla produce the dentin. They arise from ectomesenchyme. Ameloblasts in an epidermal enamel organ produce the enamel. An enamel organ is present but functionless in armadillos and a few other vertebrates, and so their teeth have no enamel. Teeth vary among vertebrates in number, distribution within the oral cavity, degree of permanence, mode of attachment, and shape.

A few species in every vertebrate class have no teeth. Agnathans, sturgeons, some toads, sirens (urodeles), turtles, and modern birds are toothless. However, one species of terns develops an embryonic set that does not erupt, and at least one genus of turtles has an enamel organ. Toothless mammals develop a set that either does not erupt or, after erupting, is soon lost.

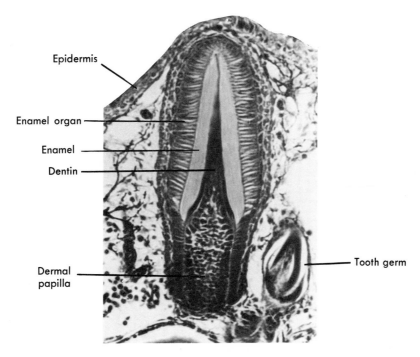

Epidermis

Enamel organ

Enamel

Dentin

Dermal papilla

Tooth germ

Fig. 11-3. Unerupted tooth of a garfish. The odontoblasts are seen lined up at the periphery of the dermal papilla. Tooth germ is a replacement for a tooth that will be shed. (Courtesy Paul F. Terranova.)

Teeth are numerous and widely distributed in the oral cavity and pharynx of fishes. They develop on jaws, palate, and even the pharyngeal skeleton. For example, the blue sucker has thirty-five to forty teeth on the last gill arch. In early tetrapods, too, teeth were widely distributed on the palate; even today most amphibians and many reptiles have teeth on the vomer, palatine, and pterygoid bones and occasionally on the parasphenoid. Teeth are confined to the jaws in crocodilians, toothed birds, and mammals, and they are least numerous among mammals. Teeth therefore have tended toward reduced numbers and a more limited distribution. Only in mammals is there a precise number of teeth in a given species.

Most vertebrates through reptiles have a succession of teeth that develop from new dermal papillae, and the number of re-placements is indefinite but numerous. It has been estimated that an elderly crocodile may have replaced its front tooth about fifty times. They and most other submammalian vertebrates replace teeth in waves that sweep from back to front and that eliminate every other tooth, as follows: tooth number 21, 19, 17, 15, and so on. When a tooth appears in position 22, a new wave eliminates teeth numbered 20, 18, 16, 14, and so on. These waves continue throughout life. In snakes the replacement wave sweeps from front to rear. There must be some survival value in this difference.

Mammals, for the most part, develop two sets of teeth, milk (deciduous) teeth and permanent teeth. Milk teeth usually erupt after birth, but in guinea pigs and some other mammals, milk teeth erupt and are shed before birth. A few mammals, such as the platypus, sirenians, and toothless

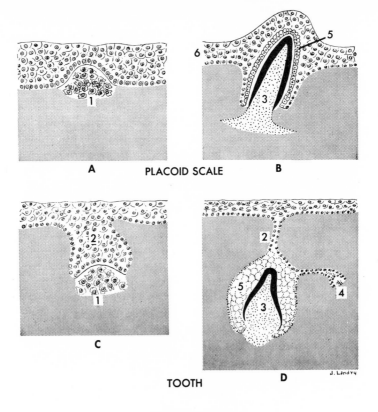

PLACOID SCALE

TOOTH

MORPHOGENESIS

Fig. 11-4. Development of placoid scale and mammalian tooth. Dermis, except dermal papilla, is gray; enamel is black. **1,** Dermal papilla; **2,** dental ridge, an ingrowth of the epidermis; **3,** dentin, produced by the dermal papilla; **4,** germ of replacement tooth; **5,** enamel organ; **6,** epidermis.

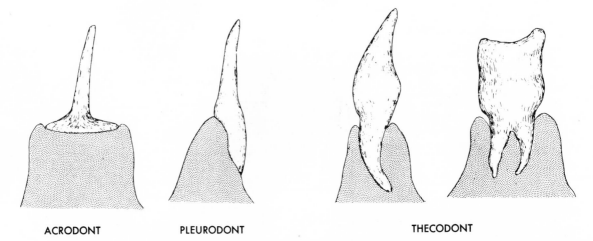

ACRODONT   PLEURODONT   THECODONT

Fig. 11-5. Variations in the relationship of teeth to jaws. Acrodont teeth are attached either at the outer surface of the jaw or, as shown, at the summit. Pleurodont teeth are attached to the inner surface of the jaw. Thecodont teeth occupy alveoli. The bone of the jaw is seen in cross section.

whales develop only a first set, and these teeth may not erupt. If they do, they are usually shed afterward. In the platypus milk teeth are replaced by horny teeth. Every species of mammal is characterized by a sequence in which teeth erupt. For example, if the permanent set on one side are numbered 1 to 8 from front to rear as in man, the sequence of eruption is 6,1,2,4,5,3,7,8. Eruption of tooth number 8, the last molar, is delayed in higher primates and this "wisdom" tooth is sometimes imperfect or absent in man.

Teeth are firmly attached to the underlying skeleton by fibrous connective tissue. In elasmobranchs they are attached to the upper and lower jaw cartilages. In bony vertebrates, jaw teeth may be attached to the outer surface or to the summit of the jawbone, as in many teleosts (**acrodont dentition**); to the inner side of the jawbone, as in frogs, necturus, and many lizards (**pleurodont dentition**); or they may occupy individual bony sockets called **alveoli** (**thecodont dentition**) (Fig. 11-5). Socketed teeth occur in crocodilians, extinct toothed birds, mammals, and many fishes, but the roots are deepest in mammals.

MORPHOLOGICAL VARIANTS

Some shark teeth bear spines for holding and tearing flesh. Others have flattened or rounded surfaces that are used in other ways (Fig. 11-2). The fangs of some poisonous snakes are specialized teeth borne on the maxillary bones. They may be permanently erect, or the jaw may move so that the teeth are brought into a striking position when the mouth is opened (Fig. 8-33). Poison from the gland pours into a groove or tube built into the tooth. For the most part, in any individual below mammals all teeth are shaped alike (**homodont dentition**). In mammals the teeth of each individual exhibit morphological varieties—*incisors, canines, premolars,* and *molars* (**heterodont dentition**). Extinct reptiles in the mamma-

lian line (synapsids) were the first to exhibit heterodont dentition (Fig. 3-18).

Incisors, located anteriorly, are usually specialized for cutting (cropping). The incisors of rodents and lagomorphs continue to grow throughout most of life, and enamel is deposited only on the anterior surface. Since enamel wears down more slowly then dentin, sharp chisellike enamel edges result. Incisors may be totally absent (sloth) or lacking on the upper jaw (Fig. 11-6, ox). Elephant tusks are modified incisor teeth (Fig. 11-6, mastodon).

Canines lie immediately behind the incisors (Fig. 11-6, dog). In generalized mammals, incisors and canines scarcely differ in appearance (Fig. 11-6, shrew). In carnivores the canines are spearlike and used for piercing prey and tearing flesh (Fig. 11-7, *B*). They are also the tusks of the walrus (Fig. 11-6). Canine teeth are always absent in rodents and lagomorphs, and so there is a toothless interval (**diastema**) between the last incisor and the first cheek tooth (Fig. 11-6, rabbit).

Cheek teeth (**premolars** and **molars**) are used for macerating food (Fig. 11-7, *A*). Molars are cheek teeth that are not replaced by a second set. They are, in reality, late arrivals of the first set. In herbivorous mammals (Fig. 11-6, ox) cheek teeth tend to be numerous, and there is little differentiation between premolars and molars. Cusps (conical elevations) in herbivores are usually not prominent, and soon wear down to broad, flat grinding surfaces. In carnivores grinding is deemphasized and the number of cheek teeth is often reduced, but cusps are prominent. At least one pair on each jaw may have very sharp cusps for cracking bones and shearing tendons (Fig. 11-6, dog, carnassial tooth).

Primitive placental mammals had 3 incisors, 1 canine, 4 premolars, and 3 molars on each side of each jaw, a total of 44 teeth. This may be expressed by the formula $\frac{3\text{-}1\text{-}4\text{-}3}{3\text{-}1\text{-}4\text{-}3}$. The same number is retained by mod-

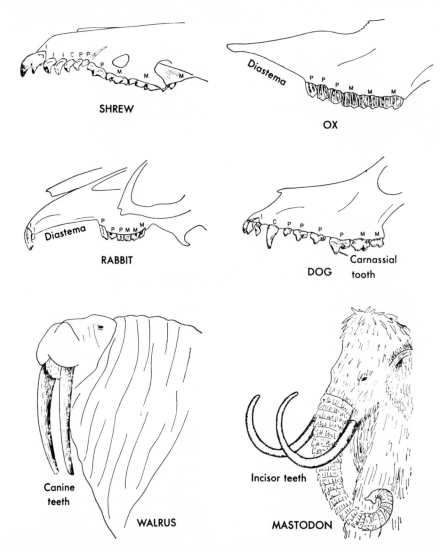

Fig. 11-6. Mammalian upper teeth, showing generalized pattern (shrew), and specializations in a herbivore (ox), in a gnawing animal (rabbit), and in a carnivore (dog). Extreme specialization of canines is seen in the walrus and of incisors in the mastodon. **I**, Incisor; **C**, canine; **P**, premolar; **M**, molar.

ern horses, although the first premolar may be missing. Dental formulas of other adult mammals are cat, $\frac{3\text{-}1\text{-}3\text{-}1}{3\text{-}1\text{-}2\text{-}1}$; man, $\frac{2\text{-}1\text{-}2\text{-}3}{2\text{-}1\text{-}2\text{-}3}$; and rabbit, $\frac{2\text{-}0\text{-}3\text{-}3}{1\text{-}0\text{-}2\text{-}3}$. The formulas for the shrew, ox, and dog can be derived from Fig. 11-6. These formulas illustrate some of the adaptive modifications displayed among mammals.

EPIDERMAL (HORNY) TEETH

Agnathans lack true teeth, but the buccal funnel and "tongue" are provided with horny teeth made of stratum corneum. Frog tadpoles develop similar teeth. In turtles, *Sphenodon*, crocodiles, birds, and monotremes, a horny "egg tooth" for cracking the egg develops on the upper jaw at the tip of

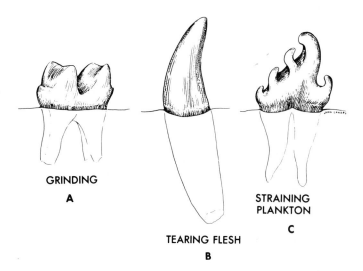

GRINDING
A

TEARING FLESH
B

STRAINING
PLANKTON
C

Fig. 11-7. Adaptive modifications of mammalian teeth, lateral views. **A,** Lower left molar from a young tapir, showing cusps for grinding. **B,** Lower left canine from a jaguar for tearing flesh. **C,** Lower right molar from a crabeater seal.

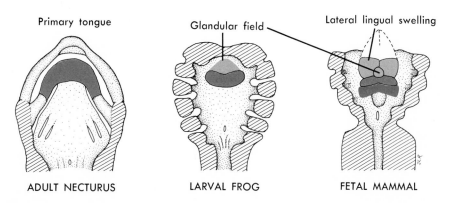

Primary tongue   Glandular field   Lateral lingual swelling

ADULT NECTURUS   LARVAL FROG   FETAL MAMMAL

Fig. 11-8. Floor of oral cavity depicting possible stages in evolution of the mammalian tongue. First arch derivatives, *light red;* second arch, *medium red;* third arch, *dark red.* Dotted outline in mammal shows final extent of lateral lingual swellings. In mammals the glandular field is also known as the tuberculum impar, and arch II endoderm is later covered by arch III endoderm.

the beak prior to hatching. (In oviparous lizard and snake embryos, the egg tooth is a transitory genuine tooth of dentin.) The adult platypus has only horny teeth. Horny beaks in some turtles and birds have serrations that simulate teeth.

□ **Tongue**

The tongue of gnathostome fishes and perennibranchiate amphibians is merely a crescentic or angular elevation in the floor of the pharynx caused by the underlying hyoid skeleton. This lean hyoid tongue is a **primary tongue** (Fig. 11-8). The tongue of agnathans (Fig. 12-5) is probably not homologous with that of jawed vertebrates.

The tongue of most amphibians consists of the primary tongue and, in addition, of a contribution, the **glandular field,** from the pharyngeal floor anterior to the hyoid arch

(Fig. 11-8, larval frog). The tongue of reptiles and mammals has a primary tongue contribution from arch II and a pair of **lateral lingual swellings** from arch I, which become stuffed with myotomal muscle. In addition, the third arch contributes by growing forward over the second arch mucosa and excluding it from the surface. In birds the lateral lingual swellings are suppressed, *intrinsic* musculature is lacking, and an entoglossal bone (Fig. 8-25, *E*) extends into the tongue. An entoglossal bone is also present in lizards (Fig. 8-25, *C*).

In turtles, crocodilians, some birds, and whales most of the tongue is immobilized in the floor of the oral cavity and cannot be extended. At the other extreme, in snakes, insectivorous lizards, and some birds the tongue is very long and darts in and out at remarkable speed. This is also true of insectivorous mammals and pollen- and nectar-eating bats, in which the tongue may be as long as the rest of the body. In most mammals the tongue is attached to the floor of the oral cavity by a midventral ligament, the **frenulum.** In man, if the frenulum extends unusually far forward, the individual is "tongue-tied," in which case the frenulum is simply cut.

The surface of the tongue of most amniotes bears papillae of one kind or another. Taste buds are associated with some of these. The spiny papillae on the tongue of a cat are used in grooming.

☐ **Oral glands**

Not many glands other than goblet cells are found in the oral cavity below mammals, and those that are present seldom produce digestive enzymes. In fact, digestive glands would probably be useless in aquatic vertebrates or birds that obtain their food from the sea, because any secretion would either be highly diluted or washed away. Terrestrial vertebrates below mammals have a few multicellular glands scattered about the roof, walls, and floor of the oral cavity, but their secretion is never copious. The largest oral glands of vertebrates are the salivary glands of mammals and the poison glands of snakes (Fig. 11-9). The latter are very much like salivary glands, histologically.

One of the rare instances of abundant oral glands in fishes is in certain species of male catfish that carry fertilized eggs in their mouth. During the breeding season their oral epithelium becomes very much folded and has many crypts supplied with large goblet cells that produce a copious secretion. These crypts serve as brood pouches

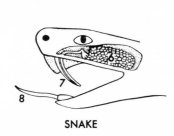

SNAKE

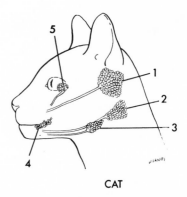

CAT

Fig. 11-9. Oral glands in a reptile and mammal. Birds have no large oral glands. **1**, Parotid; **2**, submandibular; **3**, sublingual; **4**, molar; **5**, infraorbital; **6**, poison gland of rattlesnake; **7**, maxillary tooth with groove for transfer of toxin; **8**, tongue.

that shelter the eggs. They atrophy after the eggs hatch.

Oral glands are often named according to their location. **Labial** glands open at the base of the lips; **palatal** glands open onto the palate. **Intermaxillary (internasal)** glands lie between the premaxillary bones. In frogs they consist of up to twenty-five small glands, each emptying a sticky secretion by its own duct into the oral cavity. **Sublingual** glands lie under the tongue. In *Heloderma*, the only poisonous lizard, they secrete the toxin. In mammals sublingual glands are usually classified as salivary glands, although in some species they secrete no ptyalin. The **parotid** is the largest salivary gland. Its duct crosses the masseter muscle and opens into the vestibule opposite one of the upper molars. **Submandibular** salivary glands open on papillae under the tongue. Rabbits have a fourth salivary gland, the **infraorbital,** and cats have a fifth, the **molar,** under the skin of the lower lip (Fig. 11-9). Monotremes apparently do not have salivary glands.

Saliva usually contains a watery fluid, mucin (a mixture of water-soluble glycoproteins), and ptyalin (for the digestion of starch), but not all salivary glands secrete all three. Saliva is chiefly a mammalian product, but ptyalin-secreting glands occur in lower tetrapods. Moisture is essential in order that a taste bud may function, and oral glands provide it in terrestrial animals.

■ PHARYNX

The vertebrate pharynx is that part of the digestive tract exhibiting pharyngeal pouches in the embryo. These pouches may or may not give rise to gill slits. In adult vertebrates lacking gill slits, the pharynx is that part of the foregut immediately anterior to the esophagus. The embryogenesis and basic structure of the vertebrate pharynx have been described in Chapter 1, the pharynx as a respiratory organ is discussed in Chapter 12, and taste buds and endocrine derivatives of the pharynx are discussed in Chapters 16 and 17.

In some teleosts a pair of elongated muscular tubes evaginate from the roof of the pharynx on each side near the esophagus, extend cephalad above the membranous roof of the pharynx behind the skull, and then turn caudad to terminate as blind sacs. The structures are known as **suprabranchial organs.** Elongated gill rakers from the last two gill arches form funnel-shaped baskets that extend into the entrances of the organs. Each organ is surrounded by a cartilaginous capsule to which the striated muscle in its walls is attached. The epithelium at the blind ends has many goblet cells, and the sacs contain quantities of plankton, sometimes compressed into a bolus. It may be that one function of these in gill-breathing fishes is to trap plankton from incoming water and concentrate it into mucified masses, which are then forced out and swallowed.[4] In at least one airbreathing teleost the cavity is filled with air and the epithelial lining is highly vascular and serves as an accessory respiratory membrane.[3]

The most constant features of the tetrapod pharynx are the **glottis** (a slit leading into the larynx), the openings of the auditory tubes, and the opening into the esophagus (Fig. 11-10). In mammals a cartilaginous flap, the **epiglottis,** overlies the glottis. In swallowing, the larynx is drawn forward against the epiglottis, which blocks the glottis and prevents foreign substances from entering the pathway to the lungs. In many species below mammals a flap of mucosa performs this function.

As a result of formation of a soft palate the pharynx of adult mammals consists of a **nasal pharynx** (nasopharynx) dorsal to the soft palate and an **oral pharynx** that begins at the caudal border of the soft palate and extends to the glottis and esophageal opening (Fig. 11-10, *B*). The nasal passageways (created by formation of the hard palate)

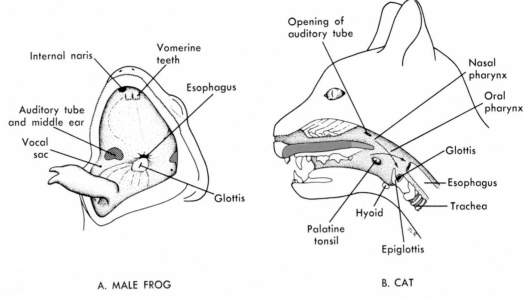

A. MALE FROG   B. CAT

Fig. 11-10. Oral cavity of **A**, amphibian, and **B**, mammal. The roof of the oral cavity in a frog is a primary palate. In cat it is a secondary palate. The latter consists of a hard (bony) palate (darker red) and a soft palate (light red). Crossed arrows indicate pharyngeal chiasma where food and water streams cross.

empty into the anterior part of the nasal pharynx, and the two auditory tubes, derived from the first pair of embryonic pharyngeal pouches, open into its lateral walls. The nasal pharynx is completely closed off from the oral pharynx during swallowing, when extrinsic muscles draw the free caudal border of the soft palate tight against the roof of the pharynx.

The oral cavity in mammals leads to the oral pharynx. The site of continuity is a narrow passageway called the **isthmus faucium.** The isthmus is bounded dorsally by the caudal border of the soft palate and laterally by the **pillars of the fauces,** which are two muscular folds on each side that arch downward from the lateral edge of the soft palate to the side of the tongue (glossopalatine arch) and pharynx (glossopharyngeal arch). Between the two pillars on each side is a **palatine tonsil,** which develops in the wall of the embryonic second pharyngeal pouch. A remnant of the pouch often re-

mains as a pocketlike crypt alongside of the palatine tonsil. The palatine tonsils are part of a ring of lymphoidal (adenoidal) masses that encircle the isthmus. Other masses are the **pharyngeal tonsils** (often called "the adenoids") in the nasal pharynx and **lingual tonsils** at the root of the tongue.

A **laryngeal pharynx** is present in some mammals, including man. It is a caudal extension of the oral pharynx dorsal to the hyoid bone and leading to the esophagus. It exists only in species in which the esophageal opening is farther caudad than the glottis. In man a fleshy **uvula** hangs from the caudal border of the soft palate into the oral pharynx.

■ ESOPHAGUS

The esophagus is a distensible muscular tube, shortest in neckless vertebrates, connecting the pharynx and stomach. In a few vertebrates it is lined with fingerlike papillae, but more often it has longitudinal folds

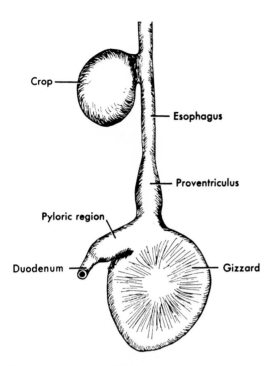

Crop —

Esophagus

Proventriculus

Pyloric region

Duodenum —

Gizzard

Fig. 11-11. Crop, esophagus, and stomach (gizzard) of a grain-eating bird.

or is smooth. The striated muscle at the cephalic end of a long esophagus is gradually replaced farther down by smooth muscle; however, striated muscle fibers may continue onto the stomach wall, especially in cud-chewing mammals that regurgitate their food for leisurely chewing.

One of the few specializations of the esophagus is the **crop** of some birds (Fig. 11-11). The crop is a paired or unpaired membranous diverticulum of the esophagus occurring primarily in grain-eaters and used for initial storage of food. An enzyme for preliminary digestion may also be secreted. In male and female pigeons and other doves a glandular part of the lining of the crop undergoes fatty degeneration under the stimulation of prolactin. The cells are then shed and regurgitated along with partially digested food as "pigeon's milk," which is fed to nestlings. A specialization of the

esophagus of sanguinivorous bats is described on p. 63.

■ STOMACH

The stomach is a muscular chamber or series of chambers at the end of the esophagus. It serves as a storage and macerating site for ingested solids and secretes digestive enzymes that partially liquify food prior to injection into the small intestine. Cyclostomes scarcely have a stomach, and a boundary between esophagus and stomach is indefinite or lacking below birds. Even in birds and mammals, the mucosa of part or all of the stomach may resemble that of the esophagus (Fig. 11-12). The stomach terminates at the pyloric sphincter.

The stomach is straight when it first develops in the embryo (Fig. 11-18) and may remain so throughout life in lower vertebrates. More often, flexures develop producing a J- or U-shaped stomach (Figs. 11-1 and 11-13). As a result, the stomach may exhibit a concave border (**lesser curvature**) and a convex border (**greater curvature**). The stomach also undergoes torsion in higher vertebrates so that it lies across the long axis of the trunk. As flexion and torsion become pronounced during development of mammalian stomachs, the dorsal mesentery of the stomach (**mesogaster**) becomes twisted and finally suspended from the greater curvature, which was originally the dorsal border of the stomach. The part of the dorsal mesentery attached to the greater curvature is then called the **greater omentum**. Because the mesentery became twisted, it encloses a **lesser peritoneal cavity** continuous with the main peritoneal cavity via an **epiploic foramen.**

The stomach of some vertebrates, especially of fishes, exhibits one or more ceca (Fig. 11-13). In birds the stomach is often divided into **proventriculus** and **gizzard** (Fig. 11-11). The proventriculus (glandular stomach) secretes a digestive enzyme, and the gizzard converts the food into a mash.

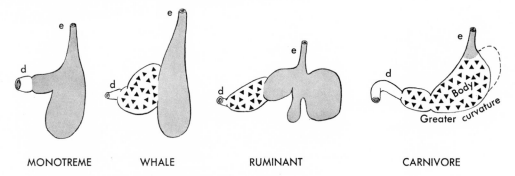

MONOTREME WHALE RUMINANT CARNIVORE

Fig. 11-12. Distribution of esophageal-like epithelium (gray), and of typical gastric glands (triangles) in the stomachs of selected mammals. **d,** Duodenum; **e,** esophagus. Broken line at far right outlines region that, in man, is called the fundus.

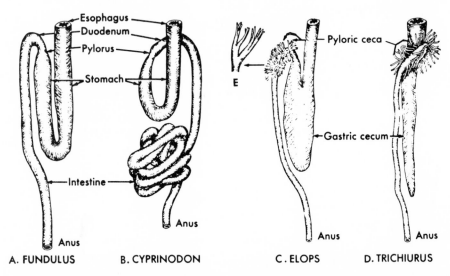

A. FUNDULUS B. CYPRINODON C. ELOPS D. TRICHIURUS

Fig. 11-13. A to **D,** Digestive tracts of four teleosts from the Gulf of Mexico. **E,** Detail of pyloric ceca from *Elops*. (Based on original sketches by Scott M. Weathersby.)

The gizzard is lined with a horny membrane, and often contains pebbles. The proventriculus and gizzard are best developed in birds that eat seeds and grain and least developed in carnivorous birds. Crocodilians also have gizzardlike stomachs (Fig. 11-14). (Birds and crocodilians had common ancestors.)

Mammalian stomachs are sometimes divided into several chambers. This is especially true in ruminants (Fig. 11-15). Grasses or grain is chewed briefly and then swallowed, after which it passes into the **rumen,** where preliminary digestion, especially by bacterial action, occurs. From the rumen, food moves into the **reticulum,** the lining of which is honeycombed (reticulated) by ridges and deep pits. In the reticulum, food is formed into a cud, which is regurgitated at will to be leisurely rechewed. After thorough chewing of the cud, the food is again swallowed. This time it passes into the **omasum,** where salivary enzymatic action continues. Finally it enters the **abomasum.**

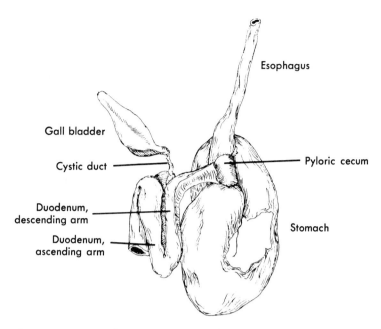

Fig. 11-14. Gizzardlike stomach and associated structures of a caiman, ventral view. The gallbladder has been displaced cephalad from its normal position between the stomach and descending arm of the duodenum. The stomach has thick muscular walls, except in the midsection (white area) where a fibrous region aids in macerating the stomach contents. Compare Fig. 11-11.

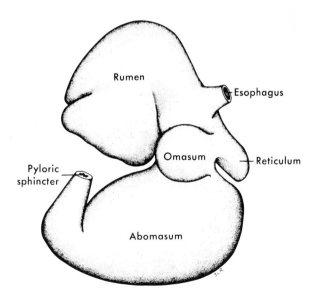

Fig. 11-15. The stomach of a calf (ruminant).

This segment exhibits the usual varieties of gastric glands, and the lining (along with that of the omasum) exhibits longitudinal ridges (**rugae**) found also in other vertebrate stomachs. The rumen, reticulum, and perhaps the omasum could equally well be considered specialized parts of the esophagus analogous to the crop of birds.

The region of the mammalian stomach adjacent to the esophagus is the **cardiac portion,** and that preceding the pyloric sphincter is the **pyloric portion.** The remainder of the stomach, when it is not multichambered, is the **body.** Gastric glands of a specific histological nature usually characterize each area—**cardiac glands** near the esophagus, **fundic glands** in the fundus, and **pyloric glands** near the pylorus. The stomach of monotremes has no typical gastric glands, and only the last segment of the stomach of whales and ruminants has them (Fig. 11-12). The human stomach is said to have about 35 million gastric glands.

■ INTESTINE

The intestine is that part of the digestive tract between the stomach and the cloaca or anus. It functions in digestion and absorption. In fishes and apodans the intestine is usually relatively straight, whereas in most tetrapods it is tortuous. The intestine of some fishes has a **spiral valve,** which compensates for a short intestine (Fig. 11-1, shark). The first part of the intestine, the **duodenum,** has characteristic glands and receives a bile duct and one or more pancreatic ducts.

The entire small intestine except the duodenum contains villi of many shapes in most vertebrates. Like the spiral valve, villi increase the absorptive area. By contracting and relaxing, they facilitate the emptying of the lacteals that are within them. The small intestine of mammals beyond the duodenum is more or less divisible into **jejunum** and **ileum** on the basis of differences in the shape of the villi and the greater number of

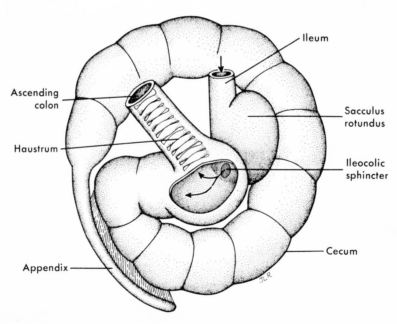

Fig. 11-16. Ileocolic junction, cecum, and vermiform appendix of rabbit. Arrows indicate available pathways.

Peyer's patches, which are lymphoid nodules, in the wall of the ileum. The small intestine is the chief site of digestion and absorption. In tetrapods an **ileocolic sphincter** (Fig. 11-16) separates small and large intestines.

It is not appropriate to identify small and large intestine in fishes, because the size often does not change and it may even become smaller near the end. In amphibians the large intestine is short and straight. In mammals, and sometimes in reptiles and birds, it is divisible into a **colon** and **rectum.** The colon begins at the ileocolic sphincter. It may have many coils. In man it ends in a **sigmoid** (S-shaped) **flexure.** The rectum is the straight terminal portion of the large intestine in the pelvic cavity.

■ CECA

Ceca of the digestive tract are blind diverticula that may occur anywhere from the esophagus to the colon. They increase the surface area of the tract. A digestive tract cecum in its simplest form is seen in the amphioxus (Fig. 2-8, *B*). The crop sac of birds is an esophageal cecum.

In modern fishes, pyloric and duodenal ceca are common (Fig. 11-13) and may be especially numerous in species lacking spiral valves. Up to 200 have been described in a mackerel. Ceca beyond the duodenum are not common in fishes or amphibians.

An ileocolic cecum is common at the ileocolic junction in amniotes—usually two in birds (Fig. 11-1, turtle, chicken, pig, and man). These ceca are sometimes long enough to equal in capacity the rest of the large intestine. They are sometimes coiled in animals that feed on cellulose, and frequently house cellulose-digesting bacteria. An ileocolic cecum may be short or absent in animals with an insectivorous or carnivorous diet. It terminates in a **vermiform appendix** in monkeys, apes, man, rodents, rabbits, and numerous other mammals (Fig. 11-16).

Ceca farther along on the colon are not rare by any means. *Hyrax*, for instance, has a large bicornuate cecum on the descending colon (Fig. 11-17).

The liver, gallbladder, and pancreas arise in the embryo as though they were going to be ceca, and the rectal gland of elasmobranchs is a cecum that secretes sodium chloride.

■ LIVER

The liver arises as a hollow diverticulum (**liver bud**) from the ventral wall of the future duodenum (Fig. 11-18, *A*). The bud, occasionally multiple, invades the ventral mesentery and then grows cephalad into the ventral mesentery of the stomach. The growing tip of the bud gives rise to numerous sprouts that become the lobes of the liver and the gallbladder. The anterior pole of the liver finally becomes anchored to the septum transversum by a coronary ligament.

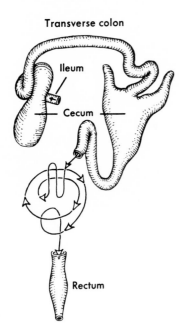

Transverse colon

Ileum

Cecum

Rectum

Fig. 11-17. Large intestine of a mammal *(Hyrax)* with several ceca. Arrows indicate direction taken by missing segment.

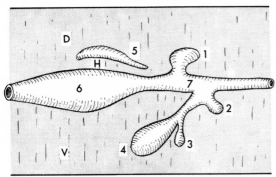

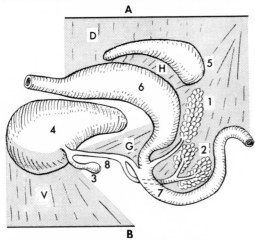

Fig. 11-18. Development of liver, pancreas, spleen, stomach, and associated mesenteries. **A.** An early stage. **B,** A later stage. **1,** Dorsal pancreatic bud from duodenum; **2,** ventral pancreatic bud from common bile duct; **3,** gallbladder; **4,** liver bud in **A** and liver in **B; 5,** spleen; **6,** stomach; **7,** duodenum; **8,** common bile duct. **D,** Dorsal mesentery; **G,** lesser omentum; **H,** gastrosplenic ligament; **V,** ventral mesentery, which remains as the falciform ligament in **B.**

Each lobe of the adult liver is drained by a **hepatic duct** that flows into a **common bile duct.** The terminal segment of the common bile duct is embedded in the wall of the duodenum for a short distance, where it is called the **ampulla of Vater.**

Although most of the embryonic ventral mesentery disappears during development, the mesentery ventral to the duodenum and

stomach that was invaded by the liver bud remains as the **hepatoduodenal ligament** (connecting the duodenum with the liver) and as the **gastrohepatic ligament** (connecting the pyloric stomach with the liver). These two ligaments constitute the **lesser omentum.** The omentum serves as a bridge transmitting the common bile duct, hepatic artery, and hepatic portal vein. The embryonic mesentery ventral to the liver remains in adults as the **falciform ligament.**

The shape of the liver conforms to the space available in the coelom. In animals with an elongated trunk the liver is elongated. In animals with short trunks it is broad and flat. It manufactures bile, which aids in fat digestion and absorption. Other roles are associated with homeostasis—maintenance of a suitable internal milieu. It helps regulate blood sugar levels by glycogenesis, glycogenolysis, and gluconeogenesis; it deaminates amino acids and hence forms ammonia, uric acid, and urea; and it manufactures several blood proteins, including certain clotting substances. The fetal liver is an important source of red blood cells, and the adult liver excretes the breakdown products of hemoglobin as bile pigments (bilirubin and biliverdin). This list of functions is not exhaustive.

■ GALLBLADDER

One sprout of the liver bud expands to become the **gallbladder** and **cystic duct.** A gallbladder develops in most vertebrates including hagfishes; however, no gallbladder develops in lampreys, many birds, rats, perissodactyls, or whales. The gallbladder serves primarily to store bile that emulsifies ingested fats, and animals lacking a gallbladder have little fat in their diet. Human beings live many years after surgical removal of the gallbladder, but they must avoid fats. Because of its embryonic origin from the liver bud, the cystic duct empties into the common bile duct.

■ EXOCRINE PANCREAS

Pancreas consists of two histologically distinct and functionally independent components—an exocrine portion secreting pancreatic juice into pancreatic ducts, and an endocrine portion (the pancreatic islands) that secretes hormones into the bloodstream.

In most teleost fishes, pancreatic tissue is distributed diffusely in the mesenteries near the duodenum. In elasmobranchs and tetrapods the pancreas consists of more or less discrete ventral and dorsal lobes.

Typically, pancreas arises as one or two ventral pancreatic buds from the liver bud and as a single dorsal bud from the foregut near the liver bud. The ventral buds invade the mesentery and form a ventral lobe or body (Fig. 11-18, *B, 2*). The dorsal bud becomes a dorsal lobe, or tail. There are variants of this pattern. In sharks the entire pancreas develops from a dorsal bud, and in mammals it develops from one ventral and one dorsal bud. It is likely that three pancreatic buds is primitive. Some teleosts have a hepatopancreas because the pancreatic bud does not separate from the liver bud.

Vertebrates may have as many pancreatic ducts as there are embryonic pancreatic buds. More often, one or more of the ducts lose their connection with the gut or bile duct, and all adult pancreatic tissue is drained by the remaining duct or ducts. Mammals illustrate the variety of conditions that exist. In sheep the duct of the dorsal pancreas loses its connection with the gut and the entire pancreas drains into the common bile duct. In some other mammals, such as pigs and oxen, the duct of the ventral pancreas loses its connection with the common bile duct, and the entire pancreas drains directly into the duodenum. In still other mammals, both ducts usually remain. When one of the two ducts is larger, as in cats and man, the other is referred to as the accessory pancreatic duct.

■ CLOACA

The cloaca is a chamber at the end of the digestive tract of most vertebrates. When present, it receives the intestine, urinary, and genital tracts, and opens to the exterior via the vent. It is shallow or nonexistent in adult lampreys, chimaeras, ray-finned fishes, and mammals above monotremes. The mammalian rectum is a derivative of the embryonic cloaca (Fig. 14-41, *C* and *E*). When there is no cloaca the intestine opens directly to the exterior via an anus. The cloaca is discussed more fully on p. 346.

☐ Chapter summary

1. Major subdivisions of the digestive tract are the oral cavity, pharynx, esophagus, stomach, and intestine. Chief accessory digestive organs are the tongue, teeth, oral glands, pancreas, liver, and gallbladder.
2. The mouth is an opening from the exterior into the oral cavity.
3. The oral cavity is shallow in fishes but becomes increasingly prominent in tetrapods. Nasal passageways open into the oral cavity in lung-breathing vertebrates lacking a secondary palate.
4. Teeth are vestiges of dermal armor. They consist of dentin formed by odontoblasts in dermal papillae and enamel formed by ameloblasts in epidermal enamel organs.
5. In lower vertebrates, teeth are more numerous, more widespread in the oral cavity, more readily replaceable, less intimately associated with the jawbones, and more alike in all parts of the mouth. A few vertebrates in every class are toothless.
6. Fishes and perennibranchiate amphibians have a primary tongue overlying the hyoid skeleton. In higher amphibians a glandular field contributes to the tongue, and in amniotes paired lateral lingual swellings contribute. Lizards and birds have an entoglossal bone.
7. Oral glands other than goblet cells are scarce below mammals. The largest are poison glands of reptiles and salivary glands of mammals.
8. The pharynx is that portion of the foregut with pharyngeal pouches in the embryo. The pouches become gill slits in gill-bearing forms; otherwise they usually disappear. In mammals the pharynx may be divided into nasal, oral, and (sometimes) laryngeal pharynxes.
9. The pharynx has openings leading to or from the spiracle, gill pouches, suprabranchial organs, oral cavity, nasal passageways, auditory tubes, larynx, esophagus, and vocal sacs, depending on the species. Lymphoid tissue is associated with it in mammals.
10. The esophagus connects the pharynx with the stomach. The crop sac is a diverticulum of the esophagus in birds.
11. The stomach is a muscular enlargement at the end of the esophagus. It is compartmentalized in birds (proventriculus and gizzard) and cud-chewing ungulates (rumen, reticulum, omasum, and abomasum). It terminates at the pyloric sphincter.
12. The intestine is digestive and absorptive. It is relatively straight in fishes, tortuous in tetrapods. Spiral valves, coils, ceca, and villi increase the absorptive area.
13. The liver, gallbladder, and pancreas arise as evaginations of the foregut. There are usually one liver bud and two or three pancreatic buds.
14. The cloaca is characteristic of most adult vertebrates. It opens to the exterior via a vent. In mammals the cloaca is divided into several passageways, one of which is the rectum. It opens to the exterior via an anus.

LITERATURE CITED AND SELECTED READINGS

1. Barrington, E. J. W.: The alimentary canal and digestion. In Brown, M. E., editor: Physiology of fishes, vol. 1, New York, 1957, Academic Press, Inc.
2. Kindahl, M. E.: Some comparative aspects of the reduction of the premolars in the Insectivora, Journal of Dental Research **46**:805, 1967.
3. Liem, K. F.: A morphological study of *Luciocephalus pulcher,* with notes on gular elements in other recent teleosts, Journal of Morphology **121**:103, 1967.
4. Miller, R.: The morphology and function of the pharyngeal organs in the clupeid, *Dorosoma petenense* (Gunther), Chesapeake Science **5**:194, 1964.
5. Oguri, M.: Rectal glands of marine and fresh-water sharks; comparative histology, Science **144**:1151, 1964.
6. Osborn, J. W.: The evolution of dentitions, American Scientist **61**:548, 1973.
7. Peyer, B.: Comparative odontology, Chicago, 1968, University of Chicago Press.

Symposium in American Zoologist

Evolution and dynamics of vertebrate feeding mechanisms, **1**:177, 1961.

12

Respiratory system

In this chapter we will look at some of the methods and mechanisms vertebrates have developed for obtaining oxygen and eliminating carbon dioxide in aquatic and terrestrial environments. We will see that many fishes breathe air and that lungs may be older than tetrapod limbs. Finally, we will examine respiratory pathways in successively higher air-breathing vertebrates and take a brief look at the mammalian diaphragm and its counterpart in reptiles and birds.

The process of obtaining oxygen from the environment and eliminating carbon dioxide is external respiration. It is accomplished via respiratory membranes that are usually part of some organ. Organs that function primarily in external respiration constitute, collectively, the respiratory system.

External respiration precedes internal respiration, which is the exchange of oxygen and carbon dioxide between the capillaries and the tissues. Because CO_2 quickly inhibits cellular activity, the continual elimination of this gas from the vicinity of the cell, and from the organism, is essential. The role of the circulatory system in the total process of respiration is therefore vital.

External respiration is carried on through respiratory membranes. Except in very early embryos, these must be highly vascular, the epithelium must be thin, the surface must be moist, and it must be in contact with the environment, or else the environment must be brought in contact with the surface.

■ ADAPTATIONS FOR EXTERNAL RESPIRATION

The chief organs of external respiration in adult vertebrates are external and internal gills, the buccopharyngeal mucosa, swim bladders or lungs, and skin. Less common adult respiratory devices include bushy or filamentous outgrowths of the pectoral fins (male *Lepidosiren*) or of the posterior trunk region and thigh (African hairy frog); the cloacal, rectal, or anal lining; and the lining of the esophagus, stomach, or even intestine. Embryos employ a variety of respiratory devices, including extraembryonic membranes (Chapter 4).

Most fishes "breathe" with internal gills, and the water usually enters through the mouth. After the mouth is closed water is forced over the gills by a pressure pump operated by the buccal and pharyngeal muscles, or it is drawn over the gills by suction created by muscular expansion of the gill chambers. The operculum also helps to create a flow. A few fishes, such as mackerel and tuna, evidently have to swim forward with the mouth open to create a flow over the gills, and they have few branchiomeric muscles. In sharks some water is admitted by the spiracle, which has a one-way valve, and in rays and skates all respiratory water enters through the spiracle, which minimizes the entrance of debris in these bottom-dwelling fishes. In a few fishes, including lampreys, water enters and also leaves by the external gill slits, and in hagfishes it enters through the naris.

Air became a source of oxygen for many fishes during the Devonian period and even before. One reason may have been that the water was warm and swampy, and it was therefore low in dissolved oxygen. The atmosphere, on the other hand, contains twenty times the oxygen that saturated water can hold. Little wonder, then, that fishes living in such an environment, whether Devonian or recent, obtain some or all of their oxygen from above the surface of the water.

Air-breathing fishes snatch bubbles of air from just above the water surface, and it comes in contact with the buccal and pharyngeal lining. If this lining is thin and highly vascular, and the blood in it is low in oxygen, oxygen is acquired at this site. A few teleosts swallow the bubble and extract oxygen in the stomach or intestine. But saccular evaginations of the phayrnx, chiefly swim bladders, provide a much larger surface for acquiring oxygen, and lungfishes and some ganoids use these. They, too, gulp air even though they have internal nares. A buccopharyngeal pump forces the air into the swim bladder. Expiration is a result of elasticity of the bladder, compression of the body wall by water pressure, contraction

of body wall muscles, and, especially, a vacuum created by lowering the buccopharyngeal floor while mouth and nares are closed. Despite acquisition of oxygen from the air most of the carbon dioxide is eliminated by the gills. Bowfins and gars "breathe" with gills in cold water, but in warm water, or in water low in oxygen for other reasons, they use their swim bladders.

Aquatic amphibians that breathe with lungs do so in the same manner as lungfishes. Air enters through the mouth even though internal nares are present, and it is pumped to the lungs. Terrestrial amphibians, on the other hand, suck in air through their nostrils by lowering the floor of the buccal cavity. Muscles open and close the nostrils during appropriate phases of the respiratory cycle.

Respiration through the skin in water and in air is employed extensively by modern amphibians. It is also used by some fishes, especially those that lack scales and therefore can have adequate capillaries close to the surface. Aquatic urodeles acquire as much as three-fourths of their oxygen from the water through the skin. Tree frogs acquire only one-fourth of their oxygen through the skin, and terrestrial species of *Rana* acquire only one-third this way. Regardless of the proportionate role of skin and lungs in oxygen uptake, most of the carbon dioxide (up to nearly 90%) is released through the skin, in amphibians. Cutaneous respiration is not important in amniotes because the thick stratum corneum insulates the capillaries from the atmosphere. Cutaneous respiration is probably more recent than branchial or pulmonary respiration; dermal scales probably had to be lost before it could be effective.

A suction pump is employed by reptiles, birds, and mammals for getting air into the lungs. The pump creates around and within the lung a gas pressure lower than atmospheric, and air rushes in. Suction is created in a variety of ways, but always with muscles. The ribs may be rotated outward and upward, the sternum may be moved, diaphragms in the coelom may be operated on, or the coelomic viscera may be displaced, as the liver is in crocodilians. In turtles the ribs are fused with the carapace and muscles of the pectoral girdle operate the pump. Exhalation may be passive as a result of elasticity of the lungs and relaxation of the pump muscles, or it may involve contraction of antagonistic muscles that force air out, although this is not the case in unlabored respiration in mammals. Respiration by rib movements seems to be very old. The earliest amphibians, ichthyostegals, believed to have been aquatic because they had lateral-line neuromast organs, had large, overlapping ribs, indicating pulmonary respiration by rib action.

■ ADULT GILLS

Although conventionally thought of as respiratory organs, gills also perform excretory functions. Chloride-secreting glands on the gills of lampreys and marine fishes that migrate between salt and fresh water excrete chloride in salt water and take in chloride in fresh water, thereby assisting in homeostasis. Gills also excrete most of the nitrogenous wastes. Fishes that acquire oxygen by aerial respiration release most of the CO_2 through their gills.

□ Cartilaginous fishes

In *Squalus acanthias* five exposed (naked) gill slits are visible on the surface of the pharynx. If a gill slit is probed, the instrument will enter a gill chamber. The anterior and posterior walls of the first four gill chambers exhibit a gill surface or **demibranch.** The last (fifth) gill chamber lacks a demibranch in the posterior wall. Hyoid cartilages support the demibranch in the anterior wall of the first gill chamber. The relationships of the remaining demibranchs to pharyngeal arches are diagrammed in Fig. 12-1. The demibranch in the anterior wall of

a gill chamber is a **pretrematic demibranch.** The one in the posterior wall of a gill chamber is a **posttrematic demibranch.** Separating the two demibranchs of a gill arch is an interbranchial septum that is strengthened by delicate cartilaginous rays. Gill rakers protrude from the gill cartilage into the pharynx and guard the entrance to the chamber. The two demibranchs of one gill arch, together with the associated interbranchial septum, cartilage, blood vessels, branchiomeric muscles, nerves, and connective tissues, constitute a **holobranch.**

Water enters the pharynx via the mouth and spiracle and passes into the gill chambers, where it comes in contact with the demibranchs. The latter consist of many gill filaments rich in capillary beds. The capillaries are supplied by afferent branchial arterioles and drained by efferent branchial arterioles. Water is forced from the gill chambers by branchiomeric muscles.

Anterior to the first gill slit of *Squalus* is a spiracle. In the embryo the spiracle is the same size as the gill slits, but it fails to keep pace in growth with the other slits. What appears to be a rudimentary or vestigial demibranch grows in its anterior wall. It is called a **pseudobranch.**

Most elasmobranchs are pentanchid, having five gill slits, but *Hexanchus,* another shark, has six and a spiracle, and *Heptanchus* has seven, the largest number of slits in any jawed vertebrate. The five gill slits in adult skates and rays are on the underside of the flattened body (Fig. 12-20), but the spiracle is located dorsally, behind the eyes, and is used for water intake. In *embryonic* rays, however, the gill slits and spiracle lie in series on the side of the pharynx. In some adult elasmobranchs the spiracle is closed by a membrane.

The holocephalan *Chimaera* has only four gill pouches, the spiracle is closed, the interbranchial septa are short and do not reach the skin, and a fleshy operculum ex-

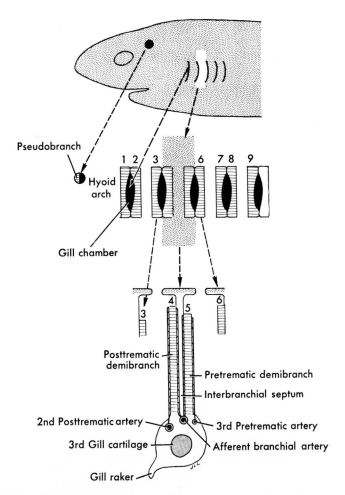

Pseudobranch

Hyoid arch

Gill chamber

1 2 3 6 7 8 9

Posttrematic demibranch

Pretrematic demibranch

Interbranchial septum

2nd Posttrematic artery

3rd Pretrematic artery

3rd Gill cartilage

Afferent branchial artery

Gill raker

Fig. 12-1. Gills in *Squalus acanthias,* a shark with a spiracle, five naked gill slits, and nine demibranchs (**1 to 9**). The fourth visceral arch, which is the second holobranch, has been excised and displayed in cross section.

tends backward from the hyoid arch and hides the gills. In several of these traits, *Chimaera* resembles teleosts.

☐ **Bony fishes**

In bony fishes five gill slits is the rule, but there are exceptions. An **operculum,** a bony flap arising from the hyoid arch, projects backward over the gill chambers (Fig. 12-2, *A*). The result is an opercular cavity that opens by a crescentic cleft (a small aperture in eels) just anterior to the pectoral girdle.

Infrequently, the left and right opercular chambers open by a common aperture in the midventral line. Opercular movements assist in the flow of water over the gills. The interbranchial septa do not reach the skin, and in teleosts they are shorter than the demibranchs (Fig. 12-2, *B*). A spiracle is present in the relatively ancient chondrosteans (Fig. 12-3, *Acipenser*), but it closes during embryonic life in modern fishes and in lungfishes (Fig. 12-3, *Gadus, Protopterus*). The demibranch on the hyoid arch is

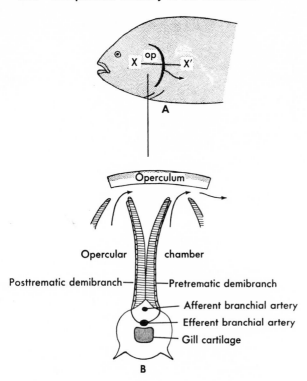

Fig. 12-2. Operculum and gill of a teleost. **A,** The operculum, **op,** extends caudad over the gills from the hyoid arch. **B,** Cross section of one holobranch in the plane **X-X'.** Arrows indicate direction of efferent water flow.

lost in most teleosts, and additional demibranchs are lost in some lungfishes.

□ Agnathans

Living agnathans have six to fifteen pairs of gill pouches. *Myxine glutinosa* usually has six pairs, but occasionally there are five or seven (Fig. 12-4). The various species of *Bdellostoma* have five to fifteen pairs. Among lampreys, *Petromyzon* has eight embryonic and seven adult pairs (Fig. 12-5). The gill pouches in agnathans are connected to the pharynx by afferent branchial ducts and to the exterior by efferent branchial ducts. *Eptatretus stouti* has a gill slit for each efferent duct, but in *Myxine* and its relatives the efferent ducts unite to open via

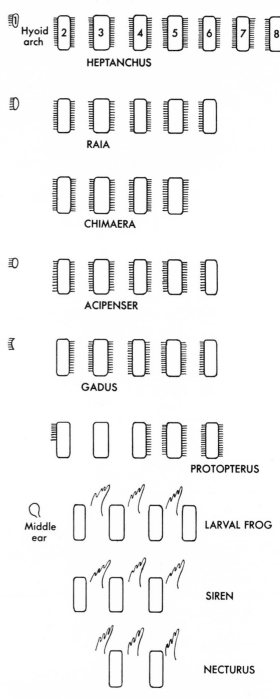

Fig. 12-3. Open pharyngeal slits in selected aquatic vertebrates and distribution of gill surfaces (horizontal lines) in fishes. *Heptanchus* is a primitive shark. *Gadus* (the cod) has a pseudobranch on the operculum, and the spiracle is closed. **1** to **8,** Pharyngeal slits. The position of external gills is indicated in the amphibians.

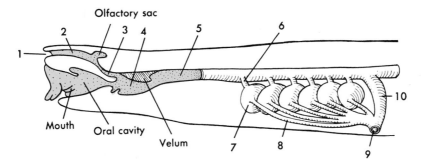

Fig. 12-4. Respiratory system in the hagfish *Myxine glutinosa*, left lateral view. **1**, Naris; **2**, nasal duct; **3**, nasopharyngeal duct; **4**, velar chamber; **5**, pharynx; **6**, afferent branchial duct; **7**, gill pouch; **8**, efferent branchial duct; **9**, common external gill aperture (present on both sides); **10**, pharyngocutaneous (pharyngeal) duct (present on left side of animal only).

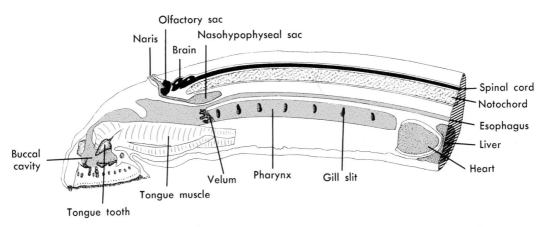

Fig. 12-5. Cephalic end of the adult lamprey *Petromyzon*, sagittal section. The duct from the naris terminates in the nasohypophyseal sac.

a common external aperture on each side (Fig. 12-4).

The pathway of flow of the respiratory water stream differs in lampreys and hagfishes. In lampreys, water enters via the external gill slits and is ejected by the same route. This is essential when the lamprey is attached by its buccal funnel to a host fish, since the nasal duct does not lead to the pharynx (Fig. 12-5). In hagfishes water enters via the naris and passes via the nasopharyngeal duct to the pharynx (Fig. 12-4).

The flow of respiratory water in hagfishes

is maintained by the pumping action of a **velar chamber** at the anterior end of the pharynx, into which the nasopharyngeal duct empties. The walls of the velar chamber are composed partly of constrictor muscles and are strengthened by posterior projections of the first branchial cartilages. The walls pulsate 50 to 100 times per minute in alert animals and 25 to 30 times per minute in sleeping animals. The action pumps water into the pharynx and creates a vacuum, which draws in additional water through the nostril.

In hagfishes, on the left side only, a **pha-**

ryngocutaneous duct connects the pharynx with the last efferent branchial duct or directly with the exterior (Fig. 12-4). Periodically, debris or particles too large to enter the afferent branchial ducts are forcefully ejected through the pharyngocutaneous duct. Embryonically, the duct arises like the gill pouches, and it is probable that the duct is a modified gill pouch.

The pharynx in lampreys becomes subdivided into an esophagus dorsally and pharynx ventrally at metamorphosis so that in adults it terminates blindly (Fig. 12-5).

■ LARVAL GILLS

Larval gills are of three kinds—external gills such as those of *Necturus*, which arise as outgrowths from the dorsal external surface of one or more gill arches; filamentous extensions of internal gills that project through gill slits to the exterior; and the internal gills of late anuran tadpoles that are hidden behind the larval operculum.

External gills usually develop before gill slits open and before any opercular fold has started to develop. They can be waved about by branchial muscles and can often be retracted. They develop in the embryonic or larval stages of most dipnoans (*Ceratodus* is an exception), in all amphibians (including apodans), and in a few ganoid fishes such as sturgeons and *Polypterus*. The latter has only one pair (Fig. 12-6, *A*). Although external gills are larval structures, there are perennibranchiate urodeles (Table 3-1) and a perennibranchiate lungfish, *Protopterus*. The latter has four pairs of larval gills and retains three pairs in a reduced state throughout life.

External gills develop in anuran tadpoles on pharyngeal arches III to V. Later, when pharyngeal pouches II to V rupture to the outside their walls become folded to form a set of internal gills. Next, from the hyoid arch a fleshy opercular fold grows backward over the gill region. The external gills then atrophy and the internal gills, hidden be-

hind the operculum, function until metamorphosis. Larval anurans are the only tetrapods that develop internal gills within an opercular chamber.

Filamentous external projections of internal gills are found in the early stages of elasmobranch development (Fig. 12-6, *C*). Their presence is correlated with development in a tightly sealed egg case or in the mother's uterus. The gills project into the egg yolk or uterine fluid, where they absorb nutrients as well as having a respiratory function. Absorption of nutrients from uterine or other maternal tissue fluids is known as histotrophic nutrition. A few viviparous chondrosteans and some teleosts develop similarly functioning filamentous gills.

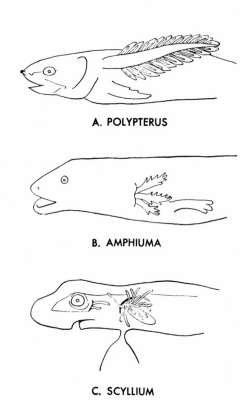

A. POLYPTERUS

B. AMPHIUMA

C. SCYLLIUM

Fig. 12-6. Larval gills of a bony fish, **A**, an amphibian, **B**, and an elasmobranch, **C**. In the *Amphiuma* the gills are resorbed before the larva escapes from the egg envelope.

■ SWIM BLADDERS AND THE ORIGIN OF LUNGS

Nearly every vertebrate from fish to man develops an unpaired evagination from the pharynx or esophagus that becomes one or a pair of sacs (swim bladders or lungs) filled with gases derived directly or indirectly from the atmosphere. The only adult vertebrates that do *not* have pneumatic sacs are cyclostomes, cartilaginous fishes, a few marine teleosts, some bottom dwellers such as flounders, and a few tailed amphibians, some of which have them as embryos. Imprints of paired sacs with ducts have even been found in one fossil placoderm. It can be said with considerable confidence that vertebrates that fail to develop pneumatic sacs have probably lost the genetic capability to induce them.

After evaginating, pneumatic sacs may retain their unpaired connection with the foregut, or the duct may close. Pneumatic sacs hereafter will be called lungs in tetrapods and swim bladders in fishes, regardless of their function.

Swim bladders may be paired or unpaired, and the **pneumatic duct** usually connects to the esophagus dorsally or ventrally, sometimes laterally (Fig. 12-7). Infrequently, it connects to the pharynx or stomach. Swim bladders lie retroperitoneally close to the kidneys and bulge more or less into the coelom (Fig. 14-14). The walls contain elastic tissue and smooth muscle, and the lining is relatively smooth. Fishes are said to be **physostomous** when the duct is open, **physoclistous** when it is closed. Most ganoids, the more primitive teleosts, and lungfishes are physostomous. Among physostomous teleosts are catfish, carp, eels, herring, pickerel, and salmon.

A principal role of swim bladders in teleosts is to serve as hydrostatic organs. The volume of gas in the sac can be reflexly

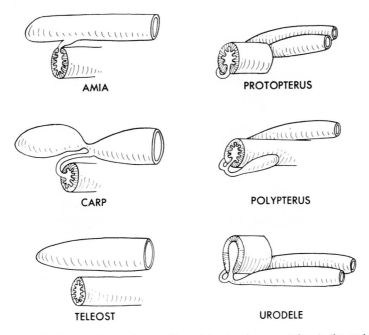

AMIA

PROTOPTERUS

CARP

POLYPTERUS

TELEOST

URODELE

Fig. 12-7. Swim bladders and lungs in aquatic vertebrates. In many teleosts the embryonic pneumatic duct later closes, and the swim bladder thereafter has no connection with the gut. *Amia* and *Polypterus* are ganoids, *Protopterus* is a lungfish, and the carp is a physostomous teleost with an anterior extension of the bladder connected with weberian ossicles.

regulated (p. 297), thereby altering the specific gravity of the fish and increasing or decreasing its buoyancy. The gas in the hydrostatic swim bladder usually comes from the blood. It is actively transported into the chamber of the bladder from a network of small arteries (rete mirabile, p. 296) in the lining of the bladder and named the **red gland.** The gland is supplied from the celiac artery, and its associated tortuous veins empty into the hepatic portal vein. The gas is reabsorbed in an area of modified epithelium near the caudal end of the bladder. In physostomes it may be absorbed by the lining of the duct or bubbled through the mouth.

The gases in swim bladders differ among fishes. Some swim bladders contain almost pure (99%) nitrogen, some up to 87% oxygen, and all contain at least traces of four atmospheric gases—nitrogen, oxygen, carbon dioxide, and argon. In deepwater fishes, nitrogen may be transported from the blood into the bladder against a nitrogen pressure of as high as 10 atmospheres.

Swim bladders perform other functions, in addition to their hydrostatic role. In one group of teleosts (Cypriniformes), a series of small bones, the weberian ossicles, connects the anterior end of the swim bladder and the sinus impar, a projection of the perilymph cavity (Fig. 16-6). Low-frequency vibrations of the gas within the swim bladder, evoked by waves of similar amplitude in the water, are transmitted by the ossicles to the membranous labyrinth. Therefore these fish can hear.

In certain herringlike teleosts (Clupeiformes) a diverticulum of the swim bladder extends into the skull, where it comes in contact with the membranous labyrinth. It has not been shown that this is a hearing device. It may be related to depth perception. In some of these species the duct of the swim bladder leads, surprisingly, to the vent.

In a few fishes, contractions of striated muscles attached to the bladder cause it to emit thumping sounds or force air back and forth between chambers separated by muscular sphincters, as in croakers and grunters.

Swim bladders function as lungs in the chondrosteans *Calamoichthys* and *Polypterus,* in holosteans, and in many dipnoans. In these air-gulping fishes, the bladder is lined by low septa and may exhibit thousands of tiny air sacs, like the lungs of lower tetrapods (Fig. 12-8). The duct opens dorsally in most ray-finned fishes, ventrally into the esophagus in dipnoans. In both, swim bladders are supplied by arteries arising from the sixth embryonic aortic arches, and in dipnoans the venous return is to the left atrium.

The striking similarities between swim bladders and lungs strongly suggest that these are the same organs adapted in some

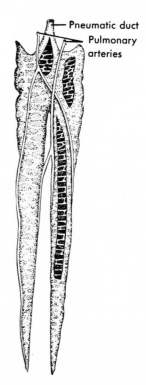

— Pneumatic duct
— Pulmonary arteries

Fig. 12-8. Swim bladders (lungs) of the African lungfish.

species for aerial respiration, in others for other functions. In the Devonian period when the fresh water was warm and periodically stagnant, and therefore low in dissolved oxygen, aerial respiration may have made the difference between extinction and survival. At that time, placoderms and most crossopterygians had pneumatic ducts. We can only speculate on which came first, respiratory or hydrostatic swim bladders, but these two conclusions seem well founded: They were functioning in aerial respiration long before vertebrates ventured onto land, and closure of the pneumatic duct in physoclistous fish is probably a mutation from a more primitive, open duct condition.

■ LUNGS AND THEIR DUCTS

Most tetrapod lungs arise as an unpaired evagination from the caudal floor of the pharynx (Fig. 1-6).* The opening in the pharyngeal floor becomes the **glottis**. The unpaired lung bud elongates only slightly before bifurcating to form the bronchi and lungs (Fig. 12-19). The primordia push caudad underneath the foregut until they bulge into the coelom lateral to the heart. As they grow into the coelom, they carry along an investment of peritoneum, which becomes the visceral pleura. The part of the lung bud between glottis and lungs develops into larynx and trachea.

□ Larynx

In a few urodeles, including *Necturus*, the larynx consists of a single pair of lateral cartilages surrounding the glottis. Most other tetrapods below mammals have two pairs of laryngeal cartilages (Fig. 12-9), **arytenoid** and **cricoid** (or, in crocodilians, arytenoid and cricothyroid). Mammals have paired arytenoid cartilages in the dorsal rim of the glottis, a ringlike cricoid, and a **thyroid** that starts out paired (Fig. 12-10). Other small

*Those of caecilians are paired from the start, but adults have only one glottis.

cartilages—cuneiforms, corniculates, procricoid, and others—develop in some species. The laryngeal cartilages of adult monotremes are all bilateral (Fig. 8-25).

Stretched across the laryngeal chamber in amphibians, some lizards, and most mammals are vocal cords, but below mammals these make few sounds discernible to man. Hippopotamuses and a few other mammals lack vocal cords. One breed of dogs, basenjis, have poorly developed vocal cords.

The larynx exhibits several interesting modifications in mammals. The thyroid and hyoid bones of the howler monkey are enormous plates that cause a goiterlike bulge in the neck (Fig. 12-11). Below the vocal cords on each side is a saclike recess, the laryngeal ventricle, or sinus of Morgagni. This is a resonating chamber that makes the weird howl of this monkey carry far into the jungle. Similar sinuses occur in some apes, and a vestigial recess is found in most mammals.

In many mammals the larynx protrudes into the nasal pharynx. In whales (Fig. 12-12) this permits rapid inhaling of air after water has been blown out of the nostrils upon surfacing, even though the oral cavity may be running over with water. In mar-

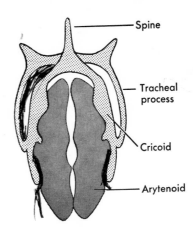

Fig. 12-9. Laryngeal skeleton of a frog. The glottis lies between the two arytenoid cartilages. The tracheal process is part of the cricoid cartilage.

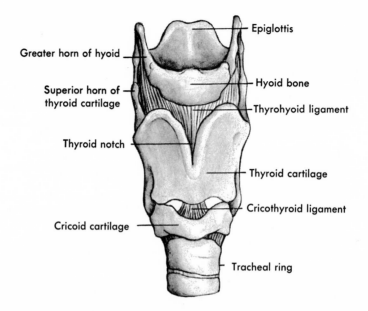

Fig. 12-10. Human larynx, frontal view. The arytenoid cartilages are located dorsally and cannot be seen from this view. (Modified from Francis, C. C., and Martin, A. H.: Introduction to human anatomy, ed. 7, St. Louis, 1975, The C. V. Mosby Co.)

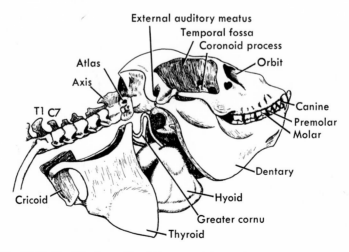

Fig. 12-11. Modification of hyoid bone and larynx in the howler monkey.

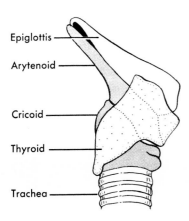

Fig. 12-12. Laryngeal skeleton of a whale. The epiglottis and arytenoid extend into the nasal pharynx.

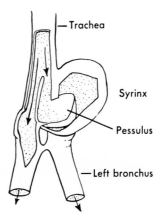

Fig. 12-13. Asymmetrical bronchotracheal syrinx of a canvasback duck. Arrows indicate path of inhaled air.

supials it prevents milk, pumped by the mother's mammary gland into the baby's esophagus, from being drawn into the lungs as the baby breathes. Of course, mammals with this arrangement cannot breathe through their mouth. Neither can air passing by this route be converted to articulate speech requiring the tongue and lips. The position of the human larynx helps to make speech possible.

A fibrocartilaginous flap, the **epiglottis,** in the floor of the oral pharynx of mammals (Figs. 12-21 and 11-10, *B*) closes off the glottis during swallowing. Crocodilians have a membranous **gular fold** in this approximate location. When elevated against the palate it closes off the oral cavity from the pharynx. This enables crocodiles to keep their mouth open under water without flooding their oral pharynx while their external nares, which protrude just above the surface of the water, are being used for breathing. When the crocodile submerges, valves in the nostrils prevent flooding of the nasal passageway.

□ **Trachea and syrinx**

The trachea is about as long as the neck. Therefore, it is short in amphibians and long

in amniotes. In birds, crocodilians, and some turtles it is longer than the neck and has to assume an S-shape. Tracheal walls are prevented from collapsing by cartilaginous or bony plates or rings. Tracheal rings are usually incomplete dorsally, and their ends are united by smooth muscle that permits changing the diameter of the tube. However, in crocodilians and birds all rings are complete. The trachea bifurcates, except in lower urodeles, to form two **bronchi** that are also stiffened by plates or rings.

Birds have a special voice box, or syrinx, at the bifurcation of the trachea. A **bronchotracheal syrinx** (Fig. 12-13) consists of a resonating chamber with walls stiffened by the last several tracheal rings and the first bronchial half rings. Folds of the membranous lining project into the chamber and a bony pessulus bearing a similunar membrane may be present. Birdsongs and birdcalls are produced when the folds and membrane are tightened by contraction of the syringeal muscles and air from the air sacs is being forced through the syrinx. Some syrinxes are simpler. Parts of the last several tracheal rings may be missing in a **tracheal syrinx,** which permits the membranous wall to vibrate. In a **bronchial syrinx**

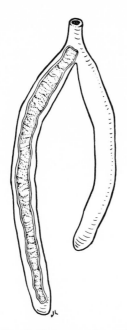

Fig. 12-14. Lungs of a necturus.

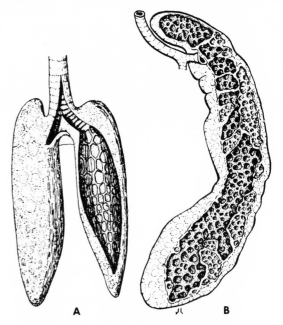

Fig. 12-15. Lungs of lizards. **A,** *Sphenodon,* a primitive lizard. **B,** *Heloderma,* a higher lizard.

the membranous wall between two bronchial cartilages folds into the chamber whenever the two cartilages are drawn together, and this fold vibrates.

□ **Amphibian lungs**

The lungs of amphibians are two simple sacs (Fig. 12-14), long in urodeles and bulbous in anurans, conforming to the shape of the pleuroperitoneal cavity. The internal lining may be smooth throughout, there may be simple sacculations in the proximal part, or the entire lining may be pocketed. The left lung of caecilians is rudimentary, and the lungs of salamanders that inhabit swift mountain streams may be only a few millimeters long. Perhaps rudimentary lungs in the latter species made it possible for them to inhabit these streams, since buoyancy would be a disadvantage in swift currents. Plethodonts do not even form a lung bud. The lungs of aquatic urodeles function mostly as hydrostatic organs, and respiration takes place through the skin. Necturus use

external gills, and only about 2% of their oxygen is obtained via lungs. Terrestrial amphibians are the chief lung-breathers among amphibians.

□ **Reptilian lungs**

In *Sphenodon* (Figs. 12-15, *A*) and snakes, lungs are simple sacs. The caudal third of the lining in snakes is septate and filled with stored (residual) air. In higher lizards (Fig. 12-15, *B*), crocodilians, and turtles, septa are so constructed that there are numerous large chambers, each with a multitude of individual subchambers. These lungs are spongy because of the numerous pockets of trapped air. The left lung in legless lizards and in snakes is rudimentary or absent except in occasional species.

An enormous diverticulum of the left lung extends into the neck of puffing adders. Inflation of this air sac causes the neck to balloon (puff up), and superinflation of the lungs causes the body to swell. Air sacs extend among the viscera in some chamel-

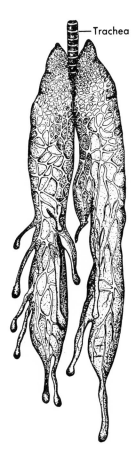

— Trachea

Fig. 12-16. Lungs of a chameleon showing saccular diverticula (air sacs).

eons, often as far caudad as the pelvis (Fig. 12-16). In some dinosaurs air sacs extended into the vertebrae.

Most reptilian lungs occupy the pleuroperiotoneal cavity, along with other viscera. In crocodilians and a few snakes and lizards they occupy separate pleural cavities set apart by a tendinous oblique septum.

□ **The lungs of birds**

The lungs and their ducts in birds are modified in contrast with those of generalized reptiles in several respects: (1) Extensive diverticula (air sacs) of the lungs invade most parts of the body. (2) The anastomosing nature of the air ducts within the lungs is such that no passage terminates blindly within the lung. (3) A syrinx supercedes the larynx as a vocal organ. (4) The lungs of birds are always isolated in pleural cavities. Air sacs and pleural cavities are not unique in birds; they occur in a few of the more specialized reptiles.

Air sacs are blind, thin-walled, distensible diverticula of the lungs that extend into all major regions of the bird's body (Fig. 12-17). They lie between layers of pectoral muscle, project among the viscera, and even penetrate the bone marrow cavity. Air sacs may be thought of as the distended blind ends of bronchi that project beyond the lung. Most birds have five or six pairs: (1) cervical sacs at the base of the neck, (2) interclavicular sacs dorsal to the furcula and sometimes united across the midline, (3) anterior thoracic sacs lateral to the heart, (4) posterior thoracic sacs within the oblique septum, (5) abdominal sacs extending caudad among the abdominal viscera, and (6) axillary sacs (less common) lying between two layers of pectoral muscle.

From the air sacs, long, slender diverticula penetrate the marrow cavity of many bones, including the centra of vertebrae, via pneumatic foramina. Pneumatic bones are generally better developed in flying birds, but this is not universally true. Most ratities lack them, as did *Archaeopteryx*.

Air sacs have a poor vascular supply and contain no respiratory epithelia; hence, they play no direct role in gaseous exchange. They do play an important accessory role in respiration. During flight they are constantly compressed by the fluctuating pressures of adjacent muscle masses and by rhythmical movements of the appendages and other body parts. They thus serve as a bellows and maintain a constant flow of air over the respiratory epithelia. At other times, movements of the ribs and oblique septum apparently suffice to operate the bellows. Air sacs are also thermoregulatory, dispelling excess heat produced by exertion during flight.

The air duct system within the avian lung is unique. A bronchus enters each lung, becomes a **mesobronchus,** and continues toward the caudal pole (Fig. 12-17). Within the lung the mesobronchus gives off ducts to the air sacs. Each mesobronchus *receives* **secondary bronchi.** The latter are interconnected by many small ducts of uniform diameter, the **parabronchi** (Fig. 12-18). From each parabronchus tiny air capillaries sprout as diverticula, which loop back into the lumen of the parabronchus. These air capil-

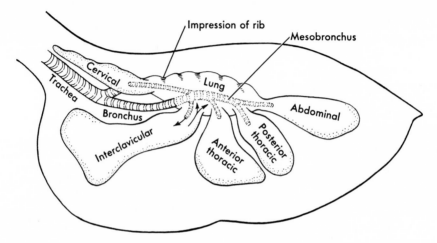

Fig. 12-17. Air sacs of a bird. Axillary sacs are not illustrated. Air leaves the sacs via recurrent bronchi (not shown), which lead to respiratory surfaces via secondary bronchi. The position of one recurrent bronchus is indicated by the double arrow.

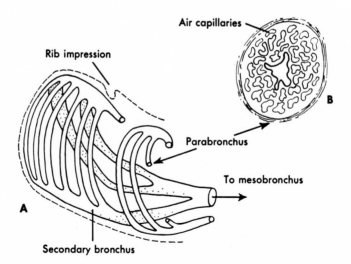

Fig. 12-18. A, Small section of bird's lung, showing parabronchi interconnecting secondary bronchi. **B,** Parabronchus transected to display the air capillaries. The parabronchus is on the airstream between an air sac and the mesobronchus. (**A** based on Locy and Larsell.[8])

laries contain the respiratory epithelium. **Recurrent bronchi** return air from the sacs to the secondary bronchi.

Inhaled air, under atmospheric pressure, passes nonstop through a mesobronchus into the air sacs, which are thereby inflated. The next compression of the sacs forces the air, via recurrent bronchi, into the secondary bronchi and parabronchi. It is prevented from reflux into the mesobronchus by valves. The air flows gently and steadily over the respiratory surfaces of the air capillaries of the parabronchi. The air then returns to the mesobronchus, bronchus, trachea, pharynx, nasal passageway, and external nares, where it escapes. Lacking blind endings as it does, the anastomosing duct system of the lungs makes possible a relatively free and steady flow of air and yet provides an ample respiratory surface. Since the air in the air capillaries is constantly and completely replaced as a result of the bellows action of the air sacs, bird lungs contain only completely fresh air. This is in contrast to the condition in other vertebrates, in which there is always residual (unexpired)

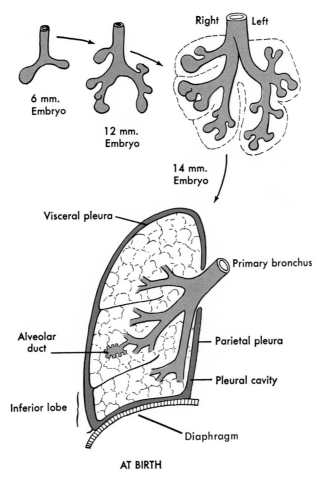

Fig. 12-19. Development of mammalian lung. Embryo lengths are applicable approximately to both the fetal pig and to man. The pleural cavity (below) is in dark red.

air partly depleted of its oxygen content. The respiratory system of birds is therefore highly efficient and well adapted to satisfy the oxygen demand resulting from sustained flight. At rest the oxygen demand is lower, and the flow is presumably less swift.

During embryonic life the bronchi, mesobronchi, secondary bronchi, parabronchi, and air sacs arise as sprouts off the unpaired lung bud. Recurrent bronchi sprout from the air sacs and invade the lung to establish connections with secondary bronchi.

□ Mammalian lungs

The lungs of mammals are multichambered and usually divided into lobes, with more lobes on the right (Fig. 12-19, 14-mm embryo). The lungs of whales, sirenians, elephants, perissodactyls, and *Hyrax* lack lobes, and in monotremes and rats, among others, only the right lung is lobed. Left and right lungs occupy separate pleural cavities.

Each bronchus penetrates a lung and divides into secondary and tertiary bronchi, which give rise to many bronchioles. The latter branch into smaller and smaller tubes. The walls of the bronchi and larger bronchioles are strengthened by irregular cartilaginous plates, which finally disappear in the smaller branches. Terminal bronchioles lead into delicate thin-walled **alveolar ducts,** the walls of which are evaginated to form clusters of **alveoli,** or respiratory pockets, estimated at 400,000,000 in man.

■ OBLIQUE SEPTUM AND DIAPHRAGM

When lungs occupy separate pleural cavities, as in some reptiles and in birds and mammals, a tendinous or tendinomuscular partition separates pleural cavities from the rest of the coelom. The partition arises as membranous folds of the dorsal and lateral parietal peritoneum that grow into the embryonic coelom until they meet. The septum transversum is also incorporated into its structure. The partition is called the oblique septum in reptiles and birds, diaphragm

in mammals. It is tendinous in reptiles, slightly muscular in birds, highly muscular in mammals. The mammalian diaphragm is dome-shaped and bulges into the thorax. Contraction of extrinsic or intrinsic muscles tightens the septum or flattens the diaphragm and, along with visceral or rib movements, increases the size of the pleural cavities. This lowers the gas pressure around, hence within, the lungs to below atmospheric pressure, and air enters the lungs.

In mammals the parietal peritoneum of each pleural cavity lines the inner surface of the thoracic wall as the **parietal pleura** (Fig. 12-19) and covers the cephalic surface of the diaphragm as the **diaphragmatic pleura.** At the root of the lung (where the bronchus and pulmonary vessels enter and leave) the parietal pleura is continuous with the **visceral pleura** that is on the surface of the lung. The space enclosed by these pleuras is the pleural cavity.

■ NARES AND NASAL CANALS

The external nares of most ray-finned fishes lead to blind olfactory sacs containing olfactory epithelium. They are often divided into an incurrent and excurrent opening by a partition. The naris in agnathans is

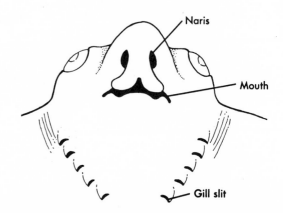

Fig. 12-20. Head of a skate, ventral view, showing oronasal groove connecting naris with mouth.

dorsomedial. In lampreys (Fig. 12-5) it leads into the nasal canal, which terminates blindly a short distance beyond the unpaired olfactory sac. In hagfishes (Fig. 12-4) a nasopharyngeal duct conveys respiratory water to the pharynx.

In lobefins the nasal canals continue beyond the olfactory sacs and open into the oral cavity as **internal nares (choanae),** but they are not used in respiration. (*Latimeria,* the living crossopterygian, lacks internal nares.)

An oronasal groove frequently connects each naris with the angle of the mouth in adult elasmobranchs (Fig. 12-20). In rays the groove becomes almost tubular as a result of folding together of the lateral walls. By the

same process, an embryonic oronasal groove is converted into nasal canals that lead to the oral cavity in lobefins, amphibians, and amniotes.

With formation of a secondary palate, the nasal canals are extended caudad. The more complete the secondary palate, the farther caudad are the internal nares. In lobefins and amphibians, which lack a secondary palate, the internal nares are forward in the oral cavity and are laterally situated. In lower reptiles and in birds they open farther caudad and nearer the midline. The internal nares open far to the rear of the oral cavity in crocodilians. In mammals they open into the nasal pharynx above the soft palate.

In mammals *olfactory* epithelium is

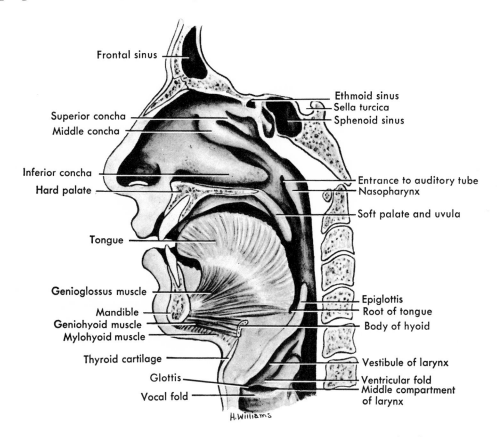

Fig. 12-21. Secondary palate and upper respiratory pathway, sagittal section, in man. (Modified from Francis, C. C., and Martin, A. H.: Introduction to human anatomy, ed. 7, St. Louis, 1975, The C. V. Mosby Co.)

found in the upper chambers of the nasal canals, and the ventral part of the canals has a ciliated, glandular *nasal* epithelium like that of the trachea. Within the nasal canals of mammals the air is warmed by venous plexuses lying under the epithelium of the turbinal elements, or conchae (Fig. 12-21). Hairs trap coarse particles and insects. Bony air sinuses open into the nasal canals. A fleshy, partially cartilaginous proboscis (nose) develops in some mammals and carries the external nares to characteristic positions (compare the location of nostrils in cats, man, and elephants). In whales, on the other hand, there is no proboscis, and the external nares are situated dorsally, although in fetal whales they are farther forward. In some whales the two nares become a median blowhole during later development.

☐ Chapter summary

1. External respiration is the exchange of respiratory gases between organism and environment. It takes place via highly vascular membranes with thin, moist epithelia.

2. Internal respiration is the exchange of oxygen between the capillaries and the tissues.

3. The chief organs of external respiration are pharyngeal gills, buccopharyngeal mucosa, skin, and swim bladders or lungs.

4. Internal gills develop in the walls of gill chambers. They usually consist of pretrematic and posttrematic demibranchs attached to a pharyngeal arch. The respiratory water stream usually enters via the mouth, but it enters via the spiracle in some elasmobranchs, via the naris in hagfishes, and via gill slits in some lampreys and some fishes.

5. Elasmobranchs have naked gill slits. An operculum covers the gill chambers in chimaeras, bony fishes, and larval anurans.

6. A spiracle is present in elasmobranchs and chondrosteans. In some species it houses a pseudobranch.

7. The hyoid demibranch tends to disappear in teleosts, and the number of demibranchs is reduced still further in lungfishes.

8. Larval gills may be external or internal. They are external in lungfishes, amphibians, and a few ganoids. Filamentous internal gills project to the exterior in larval elasmobranchs, viviparous chondrosteans, some teleosts. Anuran larvae also develop internal gills.

9. Cutaneous respiration is the chief respiratory method in aquatic amphibians and some scaleless fishes. Aquatic amphibians use lungs chiefly as hydrostatic organs.

10. Air-breathing fishes take in air by gulping it. A buccopharyngeal pump forces water over the gills or air into the lungs in many fishes and amphibians. A suction pump fills the lungs in most reptiles and in birds and mammals.

11. Pneumatic sacs arise from the foregut in nearly all vertebrates. They are called swim bladders in fishes, lungs in tetrapods.

12. Swim bladders are chiefly hydrostatic organs, but they may also serve for sound transmission, sound production, and depth perception. Their respiratory role is very ancient.

13. The larynx is supported by lateral cartilages in lower urodeles, arytenoid and cricoid (or cricothyroid) cartilages in anurans, reptiles, and birds. A thyroid cartilage is added in mammals. Other small cartilages also develop.

14. The glottis is the entrance to the larynx. The airstream may be separated from the water stream in air breathers by a gular fold in crocodilians, by an epiglottis in most mammals, and by the position of the larynx in the nasal pharynx in some mammals.

15. Vocal cords are chiefly mammalian structures but are also found in amphibians and some lizards.

16. Tracheal walls are supported by bony plates, half rings, or rings. The trachea bifurcates to form two bronchi.

17. At the base of the trachea in most birds is a syrinx.
18. Tetrapod lungs arise as a midventral evagination of the pharynx. They are present on one side only in caecilians and many legless reptiles.
19. Lungs occupy the pleuroperitoneal cavity in amphibians and lower reptiles and pleural cavities in the remaining tetrpods. An oblique septum separates pleural and peritoneal cavities in reptiles and birds. A muscular diaphragm separates them in mammals.
20. Lungs of amphibians and snakes are simple sacs. They are spongy in higher tetrapods.
21. Lungs have extensive diverticula that may even penetrate bones in birds and some reptiles.
22. In birds inhaled air flows first into an air sac, after which it enters a lung via a recurrent bronchus, passes over the respiratory epithelium in a parabronchus, and then leaves the lung via a secondary bronchus and a mesobronchus.
23. External nares lead only to olfactory sacs in cartilaginous and ray-finned fishes. In hagfishes, lobefins, and tetrapods, nasal canals connect external nares with the oral cavity or pharynx but serve as air pathways in terrestrial amphibians and amniotes only.

LITERATURE CITED AND SELECTED READINGS

1. Atz, J. W.: Narial breathing in fishes and the evolution of internal nares, Quarterly Review of Biology **27**:367, 1952.
2. Ballintijn, C. M., and Hughes, G. M.: The muscular bases of the respiratory pumps in the trout, Journal of Experimental Biology **43**:349, 1965.
3. Duncker, H.-R.: The lung air sac system of birds, New York, 1971, Springer-Verlag New York Inc.
4. Foxon, G. E. H.: Blood and respiration. In Moore, J. A., editor: Physiology of the amphibia, New York, 1964, Academic Press, Inc.
5. Gans, C.: Respiration in early tetrapods—the frog is a red herring, Evolution **24**:740, 1970.
6. Hughes, G. M., and Morgan, M.: The structure of fish gills in relation to their respiratory function, Biological Review **48**:419, 1973.
7. Johansen, K., Hanson, D., and Lenfant, C.: Respiration in a primitive air breather, *Amia calva*, Respiratory Physiology **9**:162, 1970.
8. Locy, A., and Larsell, O.: The embryology of the bird's lung, part II, American Journal of Anatomy **20**:1, 1916.
9. McMahon, B. R.: A functional analysis of the aquatic and aerial respiratory movements of an African lungfish *Protopterus aethiopicus*, with reference to the evolution of the lung-ventilation mechanism in vertebrates, Journal of Experimental Biology **51**:407, 1969.
10. Schmidt-Nielsen, K.: How birds breathe, Scientific American **225**:72, 1971.
11. Well, L. J.: Development of the human diaphragm and pleural sacs, Carnegie Institution of Washington Contributions to Embryology **35**:107, 1954.
12. White, F. N.: Respiration. In Gordon, M. S., in collaboration with Bartholomew, G. A., Grinnel, A. G., Jorgensen, C. B., and White, F. N.: Animal function: principles and adaptations, New York, 1968, Macmillan, Inc.
13. Whitford, W. G., and Hutchison, V. H.: Cutaneous and pulmonary gas exchange in ambystomatid salamanders, Copeia, no. 3, p. 573, 1966.

13

Circulatory system

In this chapter we will examine the history and ontogeny of the heart and arterial and venous channels and see how a primitive pattern has been adaptively modified to enable fishes to survive in water and tetrapods to survive in air. We will also examine circulation in fetal mammals and look briefly at lymph channels.

The circulatory system of vertebrates consists of the heart, arteries, veins or venous sinuses, capillaries or sinusoids, and blood (**blood vascular system**) and of lymph channels and lymph (**lymphatic system**). The blood carries oxygen from respiratory organs; nutrients from extraembryonic membranes, digestive tract, and storage sites; hormones and other substances associated with homeostasis and immunity to disease; and waste products of metabolism to the excretory organs. Blood also conducts heat to and from the body surface, thereby regulating and equalizing internal temperatures.

Lymph channels collect interstitial tissue fluids not taken up by the bloodstream and emulsified fats absorbed in the small intestine. Lymph vessels terminate in venous channels.

Arteries carry blood away from the heart. They have muscular and elastic walls (Fig. 13-1) capable of distention with each intrusion of blood (feel your pulse!). The smaller arteries (arterioles) dilate and constrict reflexly in response to nerve impulses and thereby assist in regulating blood pressure. They terminate in capillary beds. **Veins** commence in capillaries (other than the respiratory capillaries of the gills) and carry blood toward the heart. They have proportionately less muscle and elastic tissue and more fibrous tissue than arteries, and are therefore capable of less distention or constriction. The smaller veins are venules. They end in **capillaries.** The latter consist of endothelium only, with a lumen just large enough to accommodate red blood cells in single file. In fact, the red cells must "squeeze through" and, in so doing, become deformed. The **heart** is a pump with very muscular walls. Valves in the veins and heart prevent backflow of blood.

A **portal system** is a system of veins terminating in capillaries (Fig. 13-2). In most vertebrates, blood from the capillaries of the

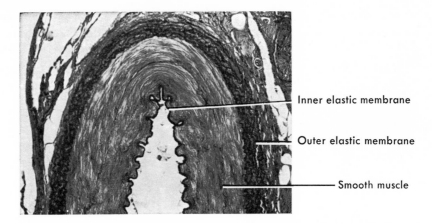

Inner elastic membrane

Outer elastic membrane

Smooth muscle

Fig. 13-1. Section of a medium-sized artery. Between the two elastic membranes lies a thick layer of smooth muscle. (From Bevelander, G., and Ramaley, J. A.: Essentials of histology, ed. 8, St. Louis, 1978, The C. V. Mosby Co.)

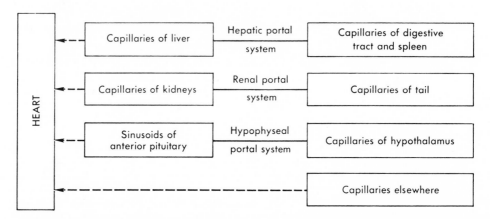

Fig. 13-2. Chief portal systems of vertebrates. The renal portal system is lacking in typical mammals, and the hypophyseal portal system has not been demonstrated in most bony fish. Channels indicated by broken lines are not part of a portal system.

tail passes via a **renal portal system** to capillaries of the kidney before returning to the heart. Blood from the digestive tract, pancreas, and spleen passes via a **hepatic portal system** to the capillaries of the liver before continuing to the heart. Except, perhaps, in bony fishes, blood from the hypothalamus passes via a **hypophyseal portal system** to the adenohypophysis before continuing to the heart.

■ HEART

The heart is a muscular pump with an inner lining (**endocardium**) of endothelium and elastic tissue, a muscular layer (**myocardium**) that is especially thick in the ventricular region, and an outer fibrous tunic (**epicardium**), which is the visceral pericardium. The tissues (as opposed to the chambers) are supplied with arterial blood by coronary arteries and drained by coronary

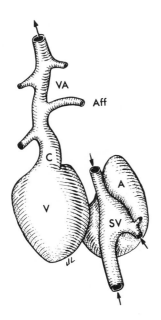

Fig. 13-3. Heart and associated vessels of the cyclostome *Myxine glutinosa,* ventral view. **A,** Atrium; **Aff,** afferent branchial artery; **C,** conus arteriosus; **SV,** sinus venosus; **V,** ventricle; **VA,** ventral aorta. Arrows indicate direction of blood flow.

veins. The heart pulsates as a result of the response of the muscle cells to electrolytes that infuse it. The rhythmicity of the pulsations is regulated by the autonomic nervous system except in hagfishes, in which no nerve fibers supply the muscle. The heart occupies the pericardial cavity, a subdivision of the coelom. In all vertebrates it is built in accordance with a basic architectural pattern.

□ Typical fishes

Hagfishes are not typical fishes, but the vertebrate heart is seen in its simplest form in these lowest vertebrates (Fig. 13-3). It exhibits a series of four chambers: **sinus venosus, atrium, ventricle,** and **conus arteriosus,** through which blood flows in that sequence.

The heart of the shark (Fig. 13-4) is typical of fishes. The sinus venosus is thin walled and has little muscle and much fibrous tissue. It receives blood from all parts of the body and is filled by suction when the ven-

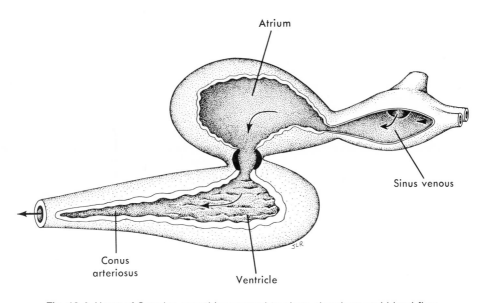

Fig. 13-4. Heart of *Squalus acanthias* opened to show chambers and blood flow.

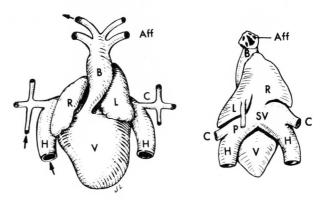

Fig. 13-5. Heart and associated vessels of necturus. Ventral view (left) and dorsal view (right). **Aff,** Common channel leading to second and third afferent branchial arteries; **B,** bulbus arteriosus; **C,** common cardinal vein; **H,** hepatic sinus; **L,** left atrium; **P,** pulmonary vein; **R,** right atrium; **SV,** sinus venosus; **V,** ventricle. A short conus arteriosus connects the ventricle with the bulbus. Arrows indicate direction of blood flow.

tricle contracts and enlarges the pericardial cavity. The sinus shows some contractility but is chiefly a collecting chamber. Its blood gushes through the sinoatrial aperture into the atrium as soon as the latter begins to relax after emptying. The caudal wall of the sinus venosus is anchored to the anterior face of the septum transversum. The atrium is a large, thin-walled muscular sac. Blood from the atrium pours into the ventricle through an atrioventricular aperture guarded by two valves. These prevent ventricular blood from being pumped back into the atrium when the ventricle contracts. The ventricle has very thick muscular walls. The anterior end of the ventricle is prolonged as a muscular tube of small diameter, the conus arteriosus, which passes to the cephalic end of the pericardial cavity, where it is continuous with the ventral aorta. Like the ventricle, the conus arteriosus is mostly cardiac muscle. A series of semilunar valves in the conus arterosus prevent backflow of blood into the ventricle. Because of its contractility and elasticity, the conus maintains a steady arterial pressure into and through the gills. In teleosts the conus is short, and its function is assumed by the **bulbus arteriosus,** a muscular expansion of the ventral aorta.

□ **Lungfishes and amphibians**

Modifications in the heart of lungfishes and amphibians are correlated with the presence of lungs and enable oxygenated blood returning from the lungs to be separated from deoxygenated blood returning from elsewhere. One modification is the establishment of a partial or complete partition within the atrium so that there is a right and left atrium (Figs. 13-5 and 13-6, Dipnoi, urodele, anuran). The partition is complete in anurans and some urodeles. The pulmonary veins empty into the left atrium so that the blood in this chamber is oxygen rich. The sinus venosus empties into the right atrium; hence, the blood in this chamber is low in oxygen. In lungless amphibians the atrium remains undivided.

A second modification is the formation of a partial interventricular septum (chiefly in lungfishes but also in *Siren,* a urodele), or of ventricular trabeculae (in amphibians). Trabeculae are shelves or ridges projecting from the ventricular wall into the chamber and running mostly cephalocaudad. Interventricular septa and ventricular trabeculae perform identical functions: they maintain separation of oxygenated and unoxygenated blood that began in the left and right atria.

A third modification is formation of a spi-

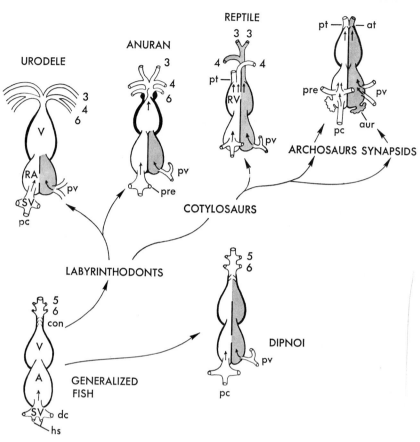

Fig. 13-6. Modifications of the atria and ventricles that result in increased separation of oxygenated and deoxygenated blood. The parts of the heart shown are **A**, atrium; **RA**, right atrium; **V**, ventricle; **RV**, right ventricle; **SV**, sinus venosus; **con**, conus arteriosus; **aur**, auricle of mammalian heart. **3 to 6**, Third to sixth aortic arches. Other vessels are **at**, aortic trunk; **dc**, common cardinal vein; **hs**, hepatic sinus; **pc**, postcava; **pre**, precava (common cardinal vein); **pv**, pulmonary veins; **pt**, pulmonary trunk. Gray chambers contain chiefly, or only, oxygenated blood.

ral valve in the conus arteriosus (also called bulbus cordis) in many dipnoans and amphibians. The valve directs oxygenated and unoxygenated blood into appropriate channels. In African lungfishes (Fig. 13-7) it shunts blood low in oxygen into the aortic arches that lead to internal gills. In anurans (Fig. 13-14) it blocks and unblocks the common entrance to the left and right pulmonary arches, shunting unoxygenated

blood to the lungs and (via cutaneous branches) to the skin.

A fourth modification shortened the ventral aorta so that it became practically nonexistent. As a result, oxygenated and unoxygenated blood that has been kept separate in the heart by septa, trabeculae, and spiral valves, moves *directly* from the heart into appropriate vessels (Fig. 13-14). Some of the changes in dipnoans and amphibians—

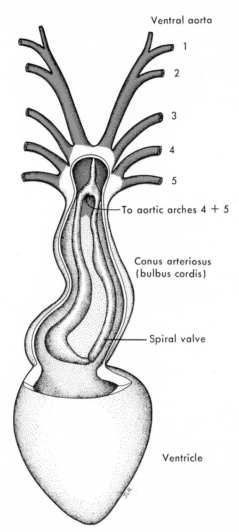

Ventral aorta

1
2
3
4
5

To aortic arches 4 + 5

Conus arteriosus
(bulbus cordis)

Spiral valve

Ventricle

Fig. 13-7. Conus arteriosus and afferent branchial arteries (1 to 5) of the lungfish *Protopterus*. The spiral valve distributes oxygen-rich blood (red) to the first three adult aortic arches and oxygen-poor blood (blue) to the last two, which support internal gills. An interventricular septum is present but not illustrated. The fourth and fifth afferent branchial arteries in this illustration are the fifth and sixth embryonic aortic arches illustrated in Fig. 13-12, *B*.

septa and trabeculae, for instance—presage similar changes in amniotes.

☐ **Amniotes**

Amniote hearts have two atria, two ventricles, and, except in adult birds and mammals, a sinus venosus (Fig. 13-8, *B*). In crocodilians the sinus venosus is partially incorporated into the wall of the right atrium. Birds and mammals exhibit a sinus venosus during early development, but it fails to keep pace with the growth of the right atrium into which it empties and finally becomes part of the wall of that chamber. Thereafter, the vessels that emptied into the sinus venosus empty directly into the right atrium. Its embryonic location is marked by the sinoatrial node of neuromuscular tissue, which plays a role in regulation of the heartbeat.

The right and left atria of adult amniotes are completely separated by an interatrial septum. Nevertheless, they are confluent during embryonic development via an **interatrial foramen (foramen ovale)**, which closes about the time of hatching or at birth. The site of the obliterated formen ovale is marked in adult hearts by a depression, the **fossa ovalis,** in the medial wall of the right atrium. The right atrium receives the sinus venosus (reptiles) or the venous blood that previously emptied into the sinus venosus (bird and mammals). It also receives blood from the coronary veins. The left atrium receives blood from the pulmonary veins.

In mammals each atrium has an earlike flap, or **auricle,** containing a blind, saclike chamber. (The term "auricle" is often applied to the atrium in lower vertebrates but should be reserved for this special appendage of the mammalian heart.) Any functional advantage of the mammalian auricle has yet to be demonstrated.

The two ventricles are completely separated only in crocodilians, birds, and mammals. In other amniotes the interventricular septum is incomplete. The internal walls of

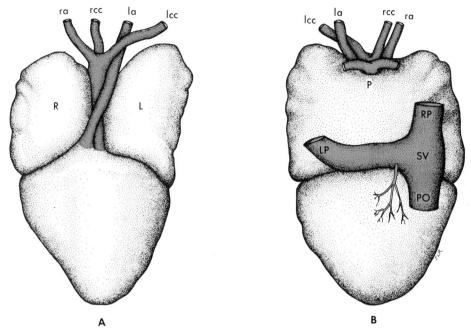

Fig. 13-8. Heart and associated vessels of *Sphenodon*. **A**, Ventral view; **B**, dorsal view. **L**, Left atrium; **la**, left aortic trunk; **lcc**, left common carotid artery; **LP**, left precava; **P**, pulmonary trunk; **PO**, postcava; **R**, right atrium; **ra**, right aortic trunk; **rcc**, right common carotid artery; **RP**, right precava; **SV**, sinus venosus. The ventricle is divided internally into two chambers. Blue vessels contain deoxygenated blood. (After O'Donoghue.[11])

the ventricles frequently exhibit interanastomosing ridges and columns of muscle (**trabeculae carneae**).

Valves guard the passage from the atria into the ventricles. The valves are fibrous flaps (muscular on the right in crocodilians and birds) connected in mammals and some lower amniotes by tendinous cords (**chordae tendineae**) to **papillary muscles** projecting from the ventricular walls (Fig. 13-9). During relaxation of the ventricle (diastole), blood from the atria falls freely past the flaps (cusps) into the ventricles. During ventricular contraction (systole), the flaps are forced upward into the atrioventricular passageway, thereby preventing reflux of blood into the atria. Both valves have one or two flaps in reptiles and birds. In most mammals the left valve has two flaps (**bi-**

cuspid, or **mitral, valve**) and the right has three (**tricuspid valve**).

Guarding the exits of the pulmonary and aortic trunks from the ventricles are **semilunar valves**, which prevent backflow into the ventricles as the latter relax. They may be vestiges of valves in the conus arteriosus of anamniotes.

□ Morphogenesis of the heart

The heart commences as two parallel tubes that are brought together and unite in the midline under the pharynx early in development. This establishes a single, pulsating tube that receives two vitelline veins at the caudal end (Fig. 15-5) and extends forward as (temporarily) two ventral aortas. At first the tubular heart is almost straight, but as development progresses it twists into

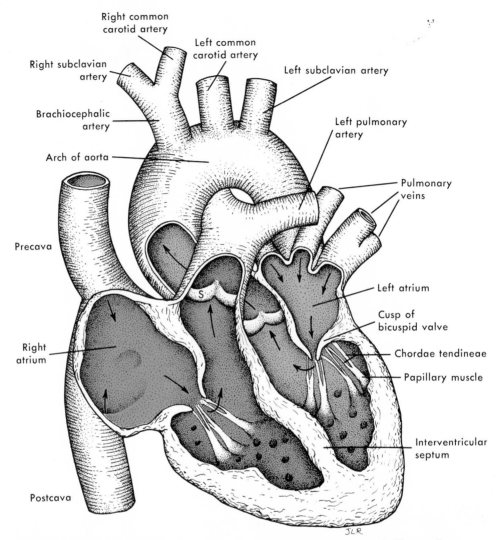

Fig. 13-9. Human heart. **S,** Semilunar valve at entrance to pulmonary trunk. The semilunar valve at entrance to ascending aorta is also shown. Blue indicates deoxygenated blood.

an S-shape so that the atrial region, previously at the caudal end, is carried dorsad and cephalad to lie where it is found in adult fishes (Fig. 13-4). In anurans and amniotes the twisting is carried further so that the atria finally lie cephalad to the ventricle(s) (Fig. 13-8). Because of the need to propel nutrients the heart is the first organ to function, and it does so even before any nerves have reached it.

■ ARTERIAL CHANNELS

Arterial channels supply most organs with oxygenated blood, although they carry deoxygenated blood to respiratory organs. In the basic pattern (Fig. 13-10) the major arterial channels consist of (1) a **ventral aorta (truncus arteriosus)** emerging from the heart and passing forward beneath the pharynx (paired early in embryogenesis); (2) a **dorsal aorta,** paired above the pharynx, and passing

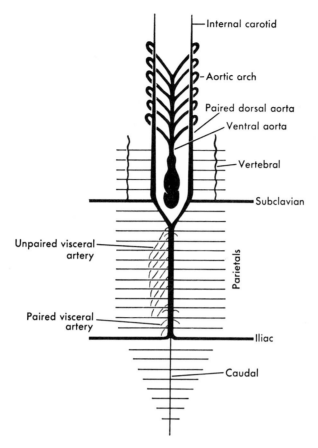

Fig. 13-10. Basic pattern of the chief arterial channels of vertebrates.

caudad above the digestive tract; and (3) six pairs of **aortic arches** connecting the ventral aorta with the dorsal aorta. Branches of these major channels supply all parts of the body. Modifications affect most prominently the aortic arches, which become adapted during embryonic development for respiration by gills or lungs.

☐ **Aortic arches of fishes**

Adaptive modifications of the embryonic aortic arches for respiration by gills may be illustrated in developing sharks (Fig. 13-11). The ventral aorta in *Squalus* extends forward under the pharynx and connects with the developing aortic arches. The aortic arches in the mandibular arch are the first to

develop. Shortly thereafter the other five pairs appear. Before the sixth pair is completed, the ventral segments of the first pair disappear, and the dorsal segments become the two efferent pseudobranchial arteries. The second pair sprout buds that become the first pretrematic arteries. Other buds sprout from the third, fourth, fifth, and sixth aortic arches and give rise to posttrematic arteries. The latter then sprout cross trunks, which grow caudad in the holobranch and by further budding establish the last four pretrematic arteries. Aortic arches II to VI soon become occluded at one site (broken lines in Fig. 13-11, *A*). The segments ventral to the occlusions become afferent branchial arteries. The dorsal segments become effer-

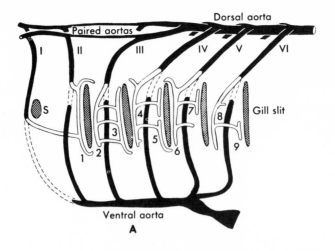

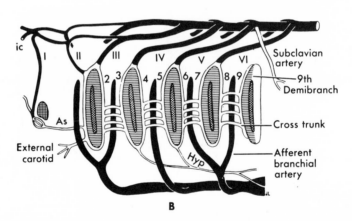

Fig. 13-11. Changes in embryonic aortic arches I to VI of *Squalus* during development, lateral view. In **A** buds (in white) off the aortic arches are establishing pretrematic and posttrematic arteries and cross trunks. Broken lines indicate sections of the aortic arches that become occluded, forcing blood into afferent branchial arterioles (not shown). In **B**, **I** has become the efferent pseudobranchial artery and **II** to **VI** have become efferent branchial arteries. **1, 3, 5, 7, 9,** Pretrematic arteries; **2, 4, 6, 8,** posttrematic arteries. **As,** Afferent pseudobranchial; **Hyp,** hypobranchial; **ic,** internal carotid; **S,** spiracle.

ent branchial arteries. In the meantime, capillary beds are developing within the nine demibranchs. Afferent branchial arterioles (not shown in Fig. 13-11) connect the afferent branchial arteries with the capillaries. Efferent branchial arterioles return oxygenated blood from the capillaries to the pretrematic and posttrematic arteries.

As a result of these modifications of the embryonic aortic arches, blood entering an aortic arch from the ventral aorta of fishes must pass through gill capillaries before proceeding to the dorsal aorta. The aortic arches have thus been modified to serve the gills.

The same developmental changes convert the six pairs of embryonic aortic arches of bony fishes into afferent and efferent bran-

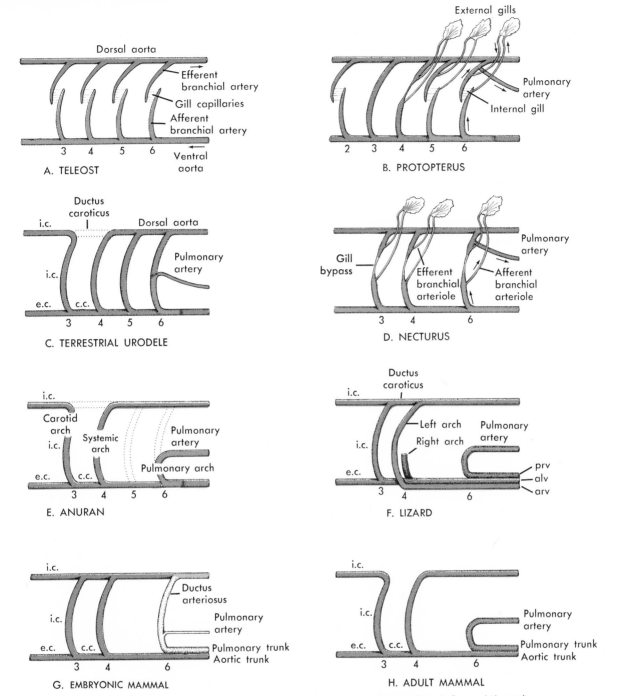

Fig. 13-12. Persistent left aortic arches in representative vertebrates. **2** to **6**, Second through sixth embryonic aortic arches. Dotted lines in **C** indicate vessel present in some species; in **E**, vessels that are functional in larvae. Arrows indicate direction of blood flow. **alv,** Aortic trunk from left ventricle; **arv,** aortic trunk from right ventricle; **cc,** common carotid artery, which is the paired segment of the embryonic ventral aorta; **ec,** external carotid; **ic,** internal carotid; **prv,** pulmonary trunk from right ventricle. In **B, C,** and **E** the ventral aorta carries venous blood (blue) during one phase of a single ventricular contraction and arterial blood (red) during the next phase. In **B** the oxygen content of each vessel depends on the extent to which oxygen is being acquired via the swim bladder. In **C** the sixth arch carries only oxygenated blood after the pulmonary artery is filled with unoxygenated blood. In **G** all blood is mixed and colors designate predominant condition.

chial arteries. The specific number converted determines the number of functional gills. In most teleosts the first and second aortic arches tend to disappear (Fig. 13-12, *A*). In *Protopterus* (Fig. 13-12, *B*) the third and fourth embryonic aortic arches do not become interrupted by gill capillaries.

In lungfishes a pulmonary artery sprouts off the left and right sixth aortic arch and vascularizes the swim bladders. This hap-

pens also in two ganoid fishes, *Amia* and *Polypterus*.* This is precisely how tetrapod lungs are vascularized!

☐ **Aortic arches of tetrapods**

Embryonic tetrapods, like fishes, construct six pairs of embryonic aortic arches

---

*In most other actinopterygians the swim bladders (lungs) are supplied from the dorsal aorta.

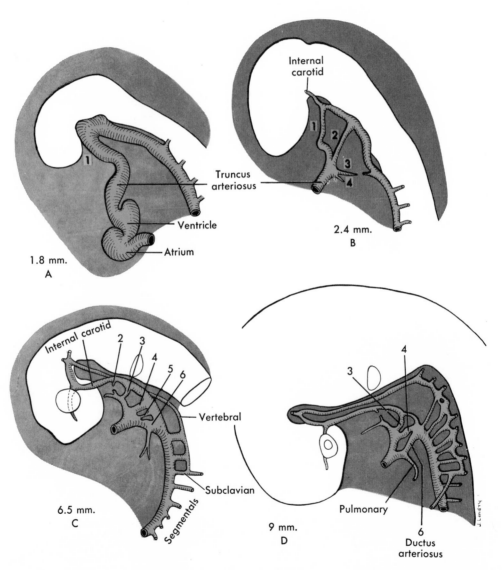

Fig. 13-13. Embryonic modifications of the aortic arches (1 to 6) of a porcupine. (Modified from Struthers.[17])

(Fig. 1-6). The first and second arches are transitory and not found in adults (Fig. 13-12, *C* to *H*). After arches I and II disappear, the third aortic arches and the paired dorsal aortas anterior to arch III are named internal carotid arteries. With the exception of a few tailed amphibians, tetrapods lose also the fifth aortic arches during embryonic life (Fig. 13-12, *E* to *H*). Pulmonary arteries sprout off the sixth arches to vascularize the lung buds (Figs. 1-6, 13-12, *C* to *H*, and 13-13, *D*). Further modifications of the fourth and sixth aortic arches and associated vessels will be discussed in amphibians, reptiles, birds, and mammals.

AMPHIBIANS

Most terrestrial urodeles retain four pairs of aortic arches (Fig. 13-12, *C*). Aquatic urodeles typically retain three because the fifth arches either drop out or unite with the fourth. In aquatic urodeles that retain gills throughout life, afferent branchial arterioles pass to the gills, efferent branchial arterioles return oxygenated blood to the arch, and a short section of each arch becomes a gill bypass (Fig. 13-12, *D*). The bypasses carry little blood as long as the animal is using its gills; however, when the water is low enough in dissolved oxygen to cause the animal to gulp air, the gills shrink, and the bypasses carry more blood. Similarly, when resorption of the gills of *Siren* is brought about by thyroid hormone injections, the animal gulps air, and the bypasses carry all the blood entering the arches. On the ventral aorta of perennibranchiates a bulbus arteriosus (Fig. 13-5) maintains a steady nonpulsating arterial pressure in the gills.

Larval anurans (tadpoles) retain four aortic arches (III through VI) for awhile. Arch VI sprouts a pulmonary artery that vascularizes the lung bud. Arches III, IV, and V supply larval gills on the third, fourth, and fifth pharyngeal arches. Gill bypasses are prominent at first, then become temporarily occluded while the external gills are functioning.

With loss of gills at metamorphosis, three changes affect the aortic arches and associated vessels of anurans (Fig. 13-12, *E*): (1) Aortic arch V disappears. (2) The dorsal aorta between aortic arches III and IV (**ductus caroticus**) disappears. As a result, blood entering aortic arch III (carotid arch) can pass only to the head. (3) The segment (**ductus arteriosus**) of aortic arch VI dorsal to the pulmonary artery disappears. (This also occurs in a few urodeles.) Blood entering arch VI can now pass only to the lungs and skin. Aortic arch IV (**systemic arch**) on each side continues to the dorsal aorta to distribute blood to the rest of the body (Fig. 13-14, *A*).

Oxygenated blood from the left atrium and deoxygenated blood from the right are kept remarkably well separated as they pass through the ventricle in amphibians. In frogs this is accomplished by ventricular trabeculae, by the movement out of the ventricle of right atrial blood first and by action of a spiral valve in the conus arteriosus (Fig. 13-14, *B*). At the start of ventricular systole the valve is flipped into a position that closes off the entrance to the systemic and carotid arches, thereby directing deoxygenated blood into the aperture leading to the pulmonary arteries. Then, as back pressure builds up within the pulmonary arteries because of filling of the lung capillaries, the spiral valve flips into an alternate position, which directs oxygenated blood into the systemic and carotid arches. Late in ventricular systole some of the left atrial blood enters the pulmonary arches. In a marine toad studied by angiocardiography, the mixing was only slightly greater.[1,12]

Adult apodans retain three complete aortic arches (III, IV, VI), although the ductus arteriosus is much reduced and carries relatively little blood. The ductus caroticus also persists in a reduced state.

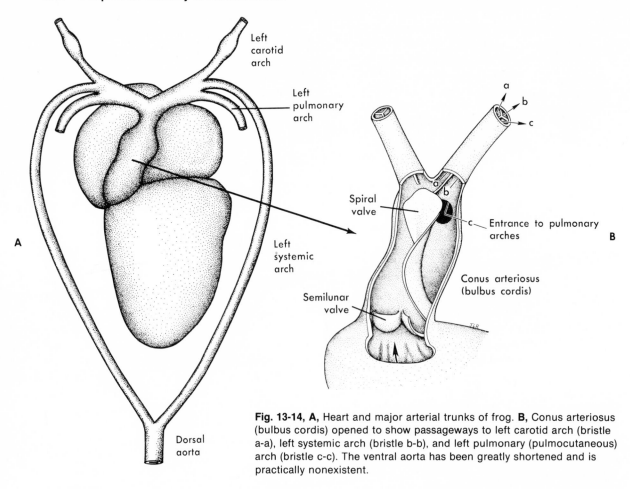

Fig. 13-14, A, Heart and major arterial trunks of frog. **B,** Conus arteriosus (bulbus cordis) opened to show passageways to left carotid arch (bristle a-a), left systemic arch (bristle b-b), and left pulmonary (pulmocutaneous) arch (bristle c-c). The ventral aorta has been greatly shortened and is practically nonexistent.

REPTILES

Modern reptiles exhibit three adult aortic arches—III, IV, and the base of VI (Fig. 13-12, *F*). Although the ductus arteriosus and ductus caroticus usually close before birth, both remain in primitive lizards; and in a few other reptiles one or the other may persist.

An innovation has been introduced in the ventral aorta of reptiles. Instead of developing a spiral valve to shunt fresh and deoxygenated blood into the proper arches, reptiles underwent a series of mutations that split the truncus arteriosus (unpaired seg-

ment of the ventral aorta) into three separate passages—two aortic trunks and a pulmonary trunk (Fig. 13-12, *F*).* The effects of these changes were as follows (Fig. 13-15,

---

*It is an attractive hypothesis that in early reptiles the truncus arteriosus was divided into two trunks that corresponded to the two ventricles. A pulmonary trunk from the right ventricle would have led to the sixth aortic arches, and an aortic trunk from the left ventricle would have led to the fourth and third aortic arches. This simple condition would then have been altered in three directions: toward the three trunks of modern reptiles, toward the condition in modern birds, and toward the condition in modern mammals.

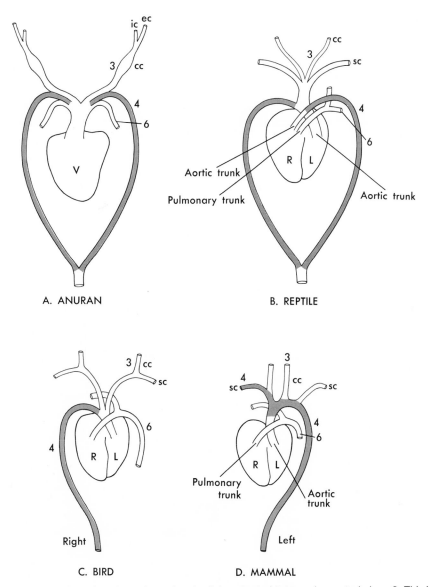

A. ANURAN

B. REPTILE

C. BIRD

D. MAMMAL

Fig. 13-15. Fate of the fourth aortic arches (red) in selected tetrapods, ventral view. **3,** Third arch (carotid); **4,** fourth arch (systemic); **6,** sixth arch (pulmonary). The relationships of the vessels have been adjusted to emphasize homologies. **cc,** Common carotid; **ec,** external carotid; **ic,** internal carotid; **sc,** subclavian artery. **L,** Left ventricle; **R,** right ventricle; **V,** ventricle. Distribution of oxygenated blood in a reptile's heart is shown in Fig. 13-16.

B): (1) The pulmonary trunk emerges from the right ventricle and connects with the left and right sixth aortic arches. Deoxygenated blood from the right atrium is therefore sent to the lungs. (2) One aortic trunk leads out of the left ventricle and carries oxygenated blood to the *right fourth* aortic arch and to the carotid arches. (3) The other aortic trunk leads out of what appears from external view to be the right ventricle and leads to the *left fourth* aortic arch. Studies of the oxygen content of this arch show that it, too, carries oxygenated blood.

Since the left systemic arch emerges from the right side of the heart, what is the mechanism whereby it carries oxygenated blood? A series of studies using cinefluoroscopy have provided the answer. In turtles, lizards, and snakes, the interventricular septum is incomplete in the vicinity where the left and right systemic arches leave the ventricle and that region is converted into a separate pocket *(cavum venosum)* by trabeculae (Fig. 13-16). Oxygenated blood from the left ventricle is directed into this pocket, which leads directly to the two systemic arches. Therefore, both left and right systemic arches receive oxygenated blood. Unoxygenated blood from the right atrium is directed by trabeculae toward the entrance to the pulmonary trunk, which is also located in a pocket *(cavum pulmonale)*. In crocodilians an opening, the *foramen of Panizza*, connects the base of the right and left systemic trunks, and part of the oxygenated blood in the right systemic arch flows through the foramen into the left.

A flow from right to left ventricle takes place in aquatic turtles and crocodilians when they remain submerged with no access to air for long periods. The advantage of this reversed shunt seems to be that energy is conserved by not sending blood to lungs that cannot oxygenate it. Under these conditions, turtles derive energy from glycolysis, an anaerobic process.[8]

Snakes and legless lizards lose their left sixth aortic arch when the left lung fails to develop, and in snakes the left third arch also disappears. The right third (carotid artery) has a bilateral distribution. All that

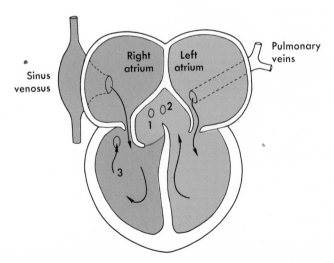

Fig. 13-16. Circulation in a turtle's heart. Red indicates oxygenated blood. **1** and **2,** Entrances to left and right systemic arches, respectively, in cavum venosum; **3,** entrance to pulmonary trunk in cavum pulmonale of right ventricle.

adult snakes have left of the original six pairs of aortic arches are the right third, left and right fourth, and the ventral part of the right sixth.

BIRDS AND MAMMALS

In birds and mammals, for the first time since the introduction of pulmonary respiration, circulatory routes have evolved in which there is no opportunity for mixing of oxygenated and unoxygenated blood. This has been achieved by closing the interventricular foramen and dividing the ventral aorta into two trunks. The pulmonary trunk (Fig. 13-15, *C* and *D*) emerges from the right ventricle and leads only to the sixth aortic arches and lungs. The aortic trunk emerges from the left ventricle and leads to the third and fourth aortic arches. *The left fourth aortic arch disappears in birds, and most of the right fourth disappears in mammals.* The part of the right fourth that remains in mammals becomes the proximal part of the right subclavian artery (Fig. 13-15, *D*).

In birds and mammals, therefore, six aortic arches develop in the embryo, and the first, second, fifth, and left fourth (in birds) or most of the right fourth (in mammals) disappear. The ductus caroticus also disappears. The ductus arteriosus functions until hatching or birth to shunt unoxygenated blood away from the lungs and into the dorsal aorta, which has branches leading to the allantois (the embryonic respiratory organ). Circulation in fetal mammals will be described later.

As a result of these modifications, all blood returning to the right side of the heart passes to the lungs. From there it returns to the left side of the heart to be recirculated (Fig. 13-17).

The left fourth aortic arch (systemic arch) of mammals is referred to simply as "the" aortic arch by mammalian anatomists. The common carotid and external carotid arteries were part of the paired embryonic ventral aortas and the internal carotids form

from the third aortic arches and paired dorsal aortas (Fig. 13-12, *H*). A few individual and species differences in the vessels arising from "the" aortic arch in mammals are illustrated in Fig. 13-18.

AORTIC ARCHES AND THE BIOGENETIC LAW

Whenever a respiratory surface develops on a pharyngeal derivative, buds off the nearest aortic arch usually vascularize that

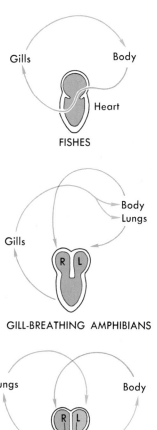

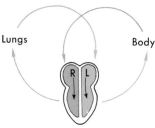

Fig. 13-17. Normal circulatory routes in fishes, gill-breathing amphibians, including necturus, and birds and mammals.

surface. When a pharyngeal arch develops an internal or external gill, the aortic arch in that pharyngeal arch vascularizes that gill. If the pharyngeal floor evaginates to form a lung bud, an aortic arch vascularizes the bud. Vascularization of pharyngeal derivatives for respiration has historical roots that extend back to the first chordates. In filter feeders, the external environment enters the pharynx, the food is separated from the oxygen-containing incurrent stream, and oxygen is acquired by the aortic arches or by buds that sprout from them. The development of six aortic arches in all vertebrate embryos and the systematic modification or elimination of first one vessel and then another in successively higher vertebrates is in accordance with the biogenetic law. This zoological principle is explained in Chapter 18.

□ Dorsal aorta

The dorsal aorta is paired in the head and pharyngeal region in early embryos and, frequently, in adults, although sometimes masquerading under other names, such as internal carotid (in which the blood flows cephalad), and ductus caroticus. The dorsal aorta of the trunk is unpaired. It continues into the tail as the caudal artery.

SOMATIC BRANCHES

A series of segmental arteries arise from the aorta along the length of the trunk. These give off vertebromuscular branches to the epaxial muscle, skin, and vertebral column and long parietal branches that encircle the body wall to the midventral line. Where there are long ribs, the parietals are called intercostal arteries. Lumbar and sacral arteries are segmentals in those regions.

The subclavian and iliac arteries are enlarged segmentals (Fig. 13-10). Subclavians arise in embryos as sprouts off the paired or unpaired dorsal aorta, or from the third (some birds) or fourth (some mammals) aortic arches close to the aorta. However, these relationships often become reoriented during later development (Fig. 13-18). Branches of the subclavian and iliac arteries pass longitudinally in the body wall to anastomose (meet and unite end-to-end). Anastomoses ensure that if one of the anastomosing vessels supplying a region becomes occluded, the vessel approaching from the opposite direction will fill the affected arterial tree beyond the occlusion. Anastomoses occur elsewhere in the body.

VISCERAL BRANCHES

A series of **unpaired visceral branches** (splanchnic vessels) pass via dorsal mesenteries to the unpaired viscera, chiefly digestive organs, suspended in the coelom. The number of such vessels is largest in generalized species such as *Necturus*. As few as three unpaired trunks—frequently celiac, superior mesenteric, and inferior mesenteric—may occur in higher vertebrates. Anastomoses between two successive visceral branches occur along the entire length of the gut. Among anastomosing visceral branches in mammals are a superior pancreaticoduodenal branch of the celiac, which anastomoses with an inferior pancreaticoduodenal branch of the superior mesenteric; a middle colic branch of the superior mesenteric, which anastomoses with a left colic branch of the inferior mesenteric; and a superior rectal branch of the inferior mesenteric, which anastomoses with a middle rectal branch of the internal iliac. Anastomoses are also common on the greater and lesser curvatures of the stomach.

Paired visceral branches of the aorta include arteries to the urinary bladder, reproductive tract, gonads, kidneys, and adrenals. A series of gonadal and renal arteries occur in lower vertebrates, several pairs in reptiles and birds, and usually a single pair in mammals.

The early embryonic dorsal aorta of amniotes ends at the level of the hind limbs by bifurcating into right and left allantoic (um-

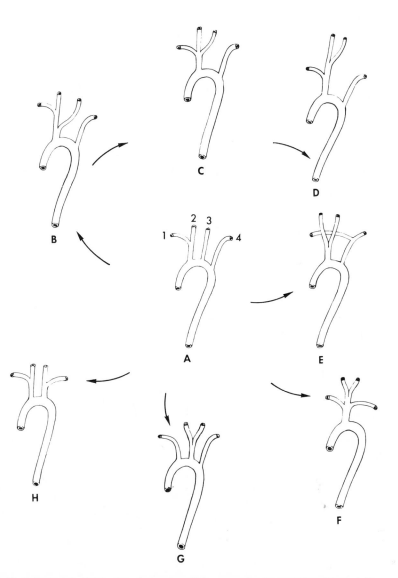

Fig. 13-18. Selected individual and species differences in the relationships of the common carotid and subclavian arteries in selected mammals. **A,** Basic pattern, seen also in man,* porcupine,* rabbit,* pig; **B,** cat,* dog,* pig,* man, and rabbit; **C** and **D,** domestic cat, **D** occurring with lower frequency; **E,** anomalous right subclavian artery in cat, man, and rat; **F,** many perissodactyls; **G,** walrus; **H,** man, porcupine, and rabbit. An asterisk indicates this to be a predominant condition in the populations examined. If there is no asterisk, the condition is common but not predominant. **1,** Right subclavian; **2,** right common carotid; **3,** left common carotid; **4,** left subclavian.

The anomalous condition shown in **E** has been induced experimentally in rats by irradiation of the fetus and also by a single injection of trypan blue into the mother immediately after mating.[5] Because all the conditions illustrated are derived ontogenetically from the same pattern, any of the variants could, and probably do, occur with measurable frequency in every mammalian species.

bilical) arteries. These carry blood to the allantois, which, with the chorion lying against the eggshell or the lining of the mother's uterus, serves as the fetal respiratory organ (Fig. 13-29). Internal iliacs sprout off the umbilical arteries as development progresses, and the umbilicals finally become branches of the iliacs.

□ Coronary arteries

The walls of all arteries and veins except the smallest are supplied with blood vessels called vasa vasorum ("vessels of the vessels"). The heart is no exception, and here the vessels are called coronary arteries and veins. In elasmobranchs the coronary arteries arise from hypobranchials that receive aerated blood from several arterial loops around the gill chambers (Fig. 13-11). In

frogs they arise from the carotid arch. In reptiles and birds they arise from the aortic trunk leading to the right fourth arch, or from the brachiocephalic. In mammals they arise from sinuslike dilations at the base of the ascending aorta just beyond the semilunar valves. In a few vertebrates, including urodeles, the coronary supply consists of many small arteries.

□ Retia mirabilia

Certain arteries along their course become highly tortuous and then straighten out again. Such structures are retia mirabilia (singular, rete mirabile), or "wonderful networks." Retia are found in the head on the carotid arteries of a variety of vertebrates. It is thought that these maintain an appropriate blood pressure within the brain or other

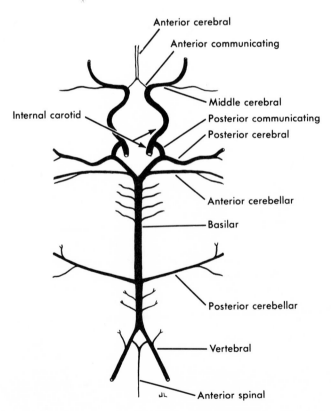

Fig. 13-19. Blood supply to the circle of Willis and brain of a mammal.

organs of the head. It is probable that the pseudobranch of *Squalus acanthias* regulates the blood pressure in the eyeball. Whales have extensive retia consisting of generous-sized arteries in the thorax in a protected position beneath the transverse processes of vertebrae and within the bony vertebral canal beside the spinal cord. These retia are all confluent and are supplied by segmental arteries and drained by vertebral arteries that are en route to the brain (Fig. 13-19). When the whale dives, the thoracic and abdominal viscera are compressed and blood is forced out of the visceral organs into the retia, which are protected from compression by their bony surroundings. These retia constitute a reservoir of blood that was oxygenated just before the whale dived. The oxygen is used by the brain during the dive, which may last for 2 hours.

Often, the tortuous artery is associated with an equally tortuous vein in such manner that artery and vein lie side by side with the blood flowing in opposite directions, or in **countercurrents.** In birds that wade in icy waters, countercurrents in retia above the thigh result in transfer of heat from the artery entering the leg to the vein emerging. This conserves body energy in the form of heat and at the same time warms returning blood to body temperature. Polar bears and arctic seals have retia that serve the same function. On the other hand, retia do not occur in arctic animals in which it would be a disadvantage—for instance, in those that are so well-insulated by feathers or hair that the metabolic heat generated by exercise needs to be eliminated rather than retained. Mammals with testes in scrotal sacs have a rete, the **pampiniform plexus,** in each inguinal canal. Heat is transferred from spermatic artery to spermatic vein, assuring that the temperature within the scrotal sac will be lower than body temperature, a necessity for the viability of sperm in those species.

In the lining of the swim bladder of some fishes, conspicuous retia, the **red glands,** maintain high gas pressures within the bladder. The pressure forces some of the gas in the bladder to enter the veins of the bladder, into which lactic acid is also being secreted by the bladder cells. Lactic acid alters the pH of the venous blood, releasing oxygen from hemoglobin and carbon dioxide from bicarbonates, and these gases are returned to the bladder by arteries that receive them by countercurrent exchange in the rete. Thus the bladder builds up pressure and volume as needed.

■ VENOUS CHANNELS

During early embryonic life the venous channels of all vertebrates conform to a single basic pattern. As development progresses, the embryonic channels are slowly modified by deletion of some vessels and addition of others. Modifications are fewer in lower vertebrates than in higher ones.

The basic pattern encompasses the following major venous streams (Fig. 13-20, A): **cardinals, renal portal, lateral abdominal, hepatic portal** (derived from **vitellines** and **subintestinals**) and **coronary** streams. Two additional major streams are characteristic of lungfish and tetrapods—a **pulmonary stream** from the lungs and a **postcaval stream** from the kidneys. The foregoing streams and their tributaries drain the entire body—head, trunk, tail, and appendages.

□ Modifications of the basic venous channels in sharks

The adult shark is an almost ideal living, swimming blueprint of the basic venous architectural pattern. A knowledge of the venous channels of sharks is therefore an excellent introduction to the venous channels of higher vertebrates and how they were modified.

CARDINAL STREAMS

The sinus venosus receives all blood returning to the heart of sharks. Most of this

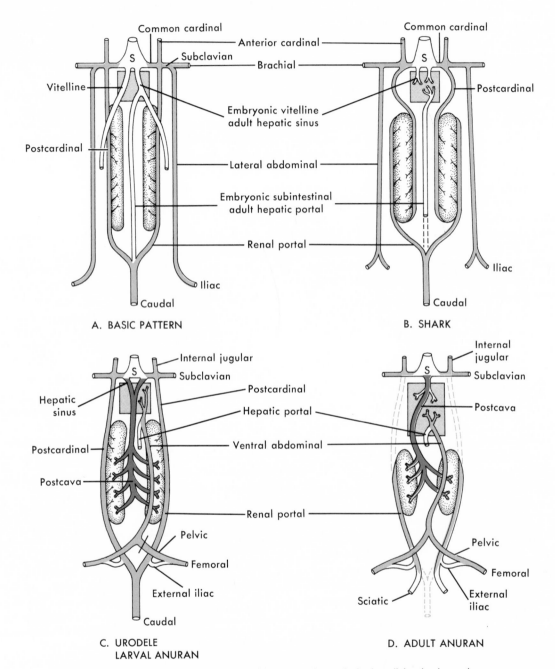

A. BASIC PATTERN

B. SHARK

C. URODELE
LARVAL ANURAN

D. ADULT ANURAN

Fig. 13-20. Modifications of the basic venous channels (colored) in sharks and amphibians. Broken lines in **B** indicate a lost segment. *Light blue,* cardinal and caudal streams; *dark blue,* postcava; *light red,* abdominal stream. The venous channels medial to the kidneys in **B, C,** and **D** arise from an embryonic subcardinal plexus.

blood, except that from the digestive organs, enters the sinus by a pair of **common cardinal veins** (Fig. 13-20, *B*) that use the transverse septum as a bridge to the heart from the lateral body walls. They appear early in development and remain essentially unchanged throughout life.

Blood from the head (except the lower jaw) is collected by a large **anterior cardinal (precardinal) vein** (or sinus) lying dorsal to the gills on each side. The anterior cardinal veins pass caudad and empty into the common cardinals (Fig. 13-20, *B*). The embryonic anterior cardinals, too, remain essentially unchanged throughout life.

The earliest embryonic **posterior cardinal (postcardinal) veins** are continuous with the caudal vein (Fig. 13-20, *A*). These embryonic postcardinals pass cephalad lateral to the developing kidneys where they receive a series of renal veins. They then empty into

the common cardinals. (Their anterior ends in sharks expand to form posterior cardinal sinuses.)

While these embryonic postcardinal veins are functioning, a network of **subcardinal** channels is forming **between** the kidneys (Fig. 13-21 shows these in a turtle). These new channels soon become confluent with the postcardinals at the anterior end of the mesonephroi, and they, too, commence to drain the kidneys. As more and more blood flows from the kidneys into the new subcardinal channels, the older postcardinals become interrupted at the anterior end of the kidney (Fig. 13-20, *B*). Thereafter, the name "posterior cardinal" is applied to the newer veins that form from the subcardinal plexus and lie medial to the kidneys. They drain chiefly the kidneys, body wall, and gonads.

RENAL PORTAL STREAM

At an early stage in development some of the blood from the caudal vein continues forward beneath the digestive tract as a subintestinal vein that drains the tract (Fig. 13-20, *A*). Later, the connection with the caudal vein is lost (Fig. 13-20, *B*). Meanwhile, afferent renal veins from the old postcardinals have invaded the mesonephric kidneys and contribute blood from the tail to the capillaries surrounding the mesonephric tubules. Finally, when the old posterior cardinals become interrupted near the cephalic ends of the kidneys (Fig. 13-20, *B*), all blood from the tail must enter the kidney capillaries.

LATERAL ABDOMINAL STREAM

Commencing at the pelvic fin from which it receives an **iliac vein,** and passing forward in the lateral body wall on each side is a **lateral abdominal vein** (Fig. 13-20, *B*). At the level of the pectoral fin it receives a **brachial vein,** after which the vessel turns abruptly toward the heart to enter the common cardinal vein. That part of the abdominal

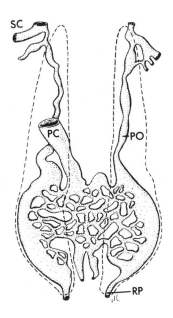

Fig. 13-21. Embryonic subcardinal venous plexus and origin of the postcava in the turtle *Chrysemys,* ventral view. Location of the mesonephros is indicated in outline. **PC,** Postcava; **PO,** left postcardinal; **RP,** renal portal vein; **SC,** subclavian vein. (Modified from De Ryke.[4])

stream between brachial and common cardinal is the **subclavian vein.** In addition to collecting blood from the paired fins, the abdominal stream also receives a **cloacal vein,** a metameric series of **parietal veins** from the lateral body wall, and minor tributaries. This basic venous channel remains unmodified during subsequent development.

HEPATIC PORTAL STREAM AND HEPATIC SINUSES

Among the first vessels to appear in vertebrate embryos are paired **vitelline,** or **omphalomesenteric, veins** (Figs. 13-20, A, and 15-5) from the yolk sac to the heart. One of the vitelline veins is soon joined by the embryonic **subintestinal vein** that drains the digestive tract. As the developing liver enlarges, the liver encompasses the vitelline veins, causing them to be broken into many sinusoidal channels. Caudal to the liver one vitelline vein disappears and the other, with the subintestinal vein, becomes the hepatic portal system. Between the liver and the sinus venosus the two vitelline veins become known as **hepatic sinuses** (Fig. 13-20, B).

☐ Venous channels in other fishes

The venous channels in other fishes are much like those of sharks. Cyclostomes have no renal portals and no left common cardinals, although both common cardinals develop in the embryo.[3] In most ray-finned fishes, abdominals are lacking and the pelvic fins are drained by the postcardinals. The venous channels of dipnoans are much like those of amphibians except that the unpaired ventral abdominal vein ends in the sinus venosus instead of the liver, and the left postcardinal is retained.

☐ Venous channels of tetrapods

The early embryonic venous channels of tetrapods are basically the same as those of embryonic sharks. We have just seen how,

by adding a vessel here and dropping one there during development, the basic pattern of a shark embryo is converted into the veins of an adult shark. We will now see how the same embryonic pattern is converted into the veins of adult tetrapods.

CARDINAL VEINS AND THE PRECAVAE

Embryonic tetrapods have postcardinals, precardinals, and common cardinals. In urodeles the postcardinals persist between caudal and common cardinals throughout life (Fig. 13-20, C). In this respect, a necturus is less modified than a shark. In anurans, most reptiles, and birds the postcardinals disappear anterior to the kidneys (Fig. 13-20, D). In mammals the right postcardinal persists under the name **azygos,**

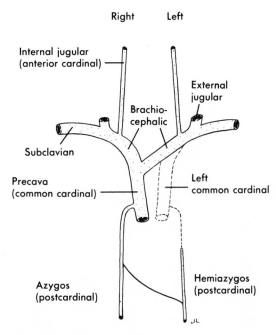

Fig. 13-22. Basic anterior venous channels of cat and man, ventral view. Broken lines indicate vessels obliterated during ontogeny. These sometimes remain as anomalies in adult mammals, including cats, pigs, and man. Internal and external jugulars are sometimes confluent. Compare channels with those of a rabbit illustrated in Fig. 13-24.

Right   Left

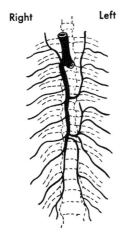

Fig. 13-23. The azygos (on the animal's right) and the hemiazygos (on the animal's left) in a rhesus monkey, ventral view. This condition is one of many variants in this species. Similar variants are seen in man. The azygos is shown flowing into the precava. (Redrawn from Seib.[16])

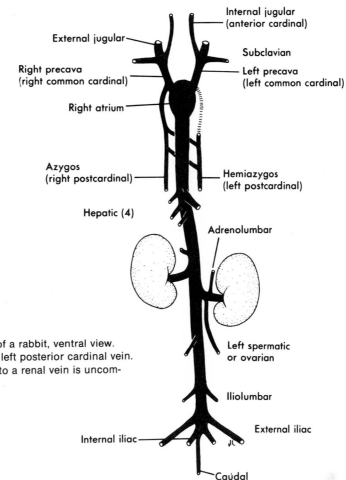

Fig. 13-24. Major systemic venous channels of a rabbit, ventral view. Broken line indicates obliterated segment of left posterior cardinal vein. Entrance of one spermatic or ovarian vein into a renal vein is uncommon in rabbits, common in cats.

and part of the left persists as the **hemi-azygos** (Figs. 13-22 to 13-24).

Common cardinal veins in tetrapods are better known as **precavae,** and anterior cardinals are called **internal jugular veins.** Although most mammals retain both the left and right precavae, some, including cats and man, lose the left precava during embryonic life (Fig. 13-22). In this case a transverse vessel, the **left brachiocephalic,** carries blood from the left side of the head and left arm to the right precava. The precava in man is better known as the **superior vena cava.** A remnant of the left precava remains as a **coronary sinus.**

THE POSTCAVA

The postcava arises during embryonic life in the subcardinal venous plexus, which receives renal veins from the kidneys (Fig. 13-21). One subcardinal channel predominates (usually the right), grows into the mesentery in which the liver is developing, and becomes confluent with the hepatic sinuses. This vessel becomes the postcava. The enlarging liver finally envelops the postcava, but does not break it into capillaries. Thus the postcava becomes an expressway from kidneys to heart via the hepatic sinuses (Fig. 13-20, *C*). In most tetrapods the two hepatic sinuses ultimately fuse to form a median

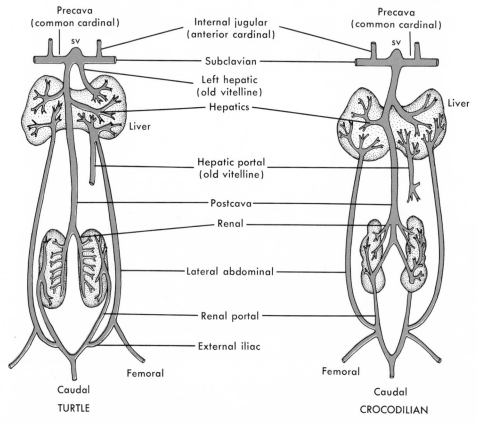

Fig. 13-25. Systemic veins of two reptiles. Only vessels of the basic pattern illustrated in Fig. 13-20, *A,* are shown. A strong branch of the renal portal vein of crocodilians continues directly to the postcava without ending in the kidney capillaries. **sv,** Sinus venosus.

vessel, which is thereafter part of the postcava (Fig. 13-20, *D*). In man the postcava is also called **inferior vena cava.**

In crocodilians, birds, and mammals, veins from the hind limbs establish direct connections with the postcava (Figs. 13-25, crocodilian, and 13-26), and the latter becomes the chief or sole drainage channel of the hind limbs.

With establishment of a postcava, blood that previously passed from kidneys to the heart via postcardinal veins now uses the postcava, and the postcardinals are reduced in size or disappear.

THE ABDOMINAL STREAM

In early tetrapod embryos paired lateral veins, presumably homologous with the ab-

dominals of sharks, commence in the caudal body wall near the hind limbs, pass cephalad in the body wall, receive veins from the forelimbs, and terminate in the cardinal veins or sinus venosus. As development progresses in tetrapods, however, this abdominal stream undergoes modifications.

Alterations in the abdominal stream in amphibians dissociate it from the anterior appendages (Fig. 13-20, *C* and *D*). The two abdominal veins fuse in the midventral line

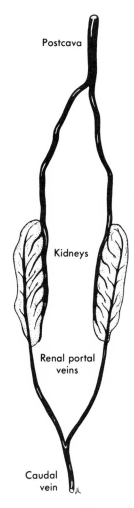

Fig. 13-27. Renal portal veins and venous drainage of the kidneys in a snake. The postcava commences at the confluence of right and left efferent renal veins.

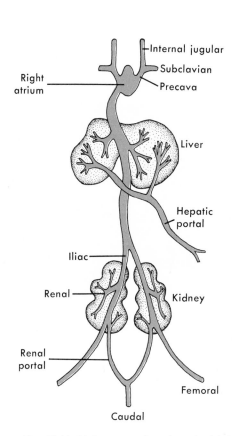

Fig. 13-26. Major systemic veins of a bird.

to form a median **ventral abdominal vein.** Blood in this median vessel finds its way into channels in the falciform ligament. Soon all the blood in the ventral abdominal is crossing the falciform ligament and emptying into the capillaries of the liver. Thereupon, the abandoned abdominal veins anterior to the liver disappear. Thus the abdominal stream no longer drains the anterior limbs.

In reptiles the two lateral abdominals do not unite to form a single vessel (Fig. 13-25). Otherwise, they undergo the same modifications as in amphibians, establishing a channel across the falciform ligament to terminate in the capillaries of the liver, and losing their connection with the anterior limbs and common cardinal veins. During embryonic life they acquire a tributary from the allantois (**allantoic vein**) as they pass along the ventrolateral body wall toward the falciform ligament. The allantoic veins disappear when the allantois is lost at hatching. Birds retain none of their embryonic abdominal stream after hatching (Fig. 13-26).

In mammals the same changes take place as in reptiles. After the allantoic veins (or **umbilical veins,** as they are now called) connect into the abdominals, the latter disappear from that point caudad. The parts that remain from navel to liver are called umbilicals. As in amphibian abdominals, the mammalian umbilicals unite to form a single umbilical vein (Fig. 13-29). It has no function but to drain the placenta; and when the umbilical cord is cut at birth, no more blood flows through it and it becomes converted into the **round ligament of the liver.** It is seen in the dissecting room as a fibrous cord in the free border of the falciform ligament extending between the umbilicus (navel) and the liver. As a result of these mutations, the very ancient abdominal stream, which at one time drained the posterior appendages, body wall, and anterior appendages, has evolved into a vessel drain-

ing only the mammalian placenta and confined to embryos.

Since the umbilical vein in embryonic mammals carries a heavy load of blood, it finally erodes a broad channel, the **ductus venosus,** directly through the liver and into the postcava (Fig. 13-29). After birth the channel becomes a ligament (**ligamentum venosum**).

RENAL PORTAL SYSTEM

The renal portal system in amphibians acquires a tributary, the **external iliac vein** (not homologous with a vessel of the same name in mammals), which carries some blood from the hind limbs to the renal portal vein (Fig. 13-20, *C* and *D*). This channel provides an alternate route from the hind limbs to the heart. The connection persists in reptiles (Fig. 13-25). (This route may be one factor that made it possible to dispense with the abdominal stream in mammals.)

In crocodilians, some blood passing from the hind limbs to the renal portal is able to bypass the kidney capillaries, going straight through the kidneys into the postcava (Fig. 13-25, crocodilian). By the time birds evolved, this had become a common pathway (Fig. 13-26).* In mammals above monotremes the renal portal system disappears as an adult structure; however, it appears transitorily in mammalian embryos.

From the foregoing it can be seen that the posterior appendages during phylogeny (fins first, limbs later) have been drained by a series of vessels (Fig. 13-28): first, the abdominal stream; then, the renal portal system; and, finally, the postcava directly. The venous realignment necessitated by the displacement of the caudal end of the mammalian nephrogenic mesoderm to form the mammalian kidney may have struck the final evolutionary blow to the renal portal system.

---

*Crocodilians and birds have a common ancestor.

HEPATIC PORTAL SYSTEM

The hepatic portal system is essentially similar in all vertebrates. It drains chiefly the stomach, pancreas, intestine, and spleen, and it terminates in the capillaries of the liver. Its origin from the embryonic vitelline and subintestinal veins has been described earlier. The abdominal stream (allantoic in birds, umbilical in mammals) becomes a tributary of the hepatic portal system, commencing with amphibians. Veins from the swim bladders are usually tributaries of the hepatic portal system in bony fishes.

Pulmonary and coronary veins of vertebrates. Pulmonary veins in lungfishes and tetrapods drain the lungs and terminate in the left atrium. In ray-finned fishes that use the air bladder as a lung, blood from the air bladder empties into the hepatic veins or common cardinal.

Coronary veins in fishes enter the sinus

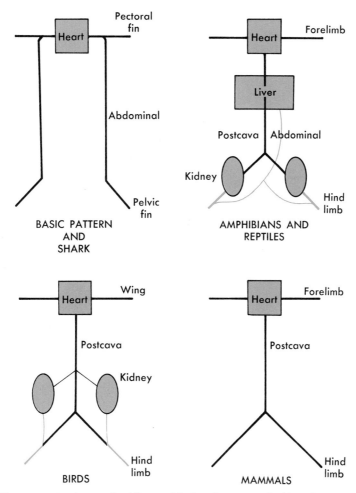

Fig. 13-28. Venous routes from paired fins and limbs, diagrammatic. Vessels represented by thin lines carry relatively less blood. Blue in a vessel indicates a portal stream. The abdominal stream is paired in reptiles.

venosus. In frogs an anterior coronary vein enters the left precava, and a posterior coronary empties into the abdominal vein near the liver. In reptiles, birds, and mammals, coronary veins empty into the right atrium (via a coronary sinus in mammals).

■ CIRCULATION IN THE MAMMALIAN FETUS AND CHANGES AT BIRTH

In the mammalian fetus, blood passes from the caudal end of the dorsal aorta into the umbilical arteries (Fig. 13-29). These ex- tend out the umbilical cord to the placenta. From the placenta oxygenated blood returns to the fetus via an umbilical vein, which tra- verses the falciform ligament to enter the liver. Some of this blood enters liver capil- laries, but most of it continues nonstop via the ductus venosus, into the postcava, and finally into the right atrium. From the right atrium most of it passes via an **interatrial foraman (foramen ovale** of the heart) into the left atrium. The rest of the aerated blood, along with blood returning to the right

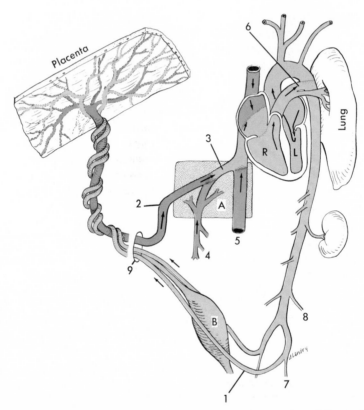

Fig. 13-29. Circulation in the mammalian fetus. **1,** Umbilical artery; **2,** umbilical vein; **3,** ductus venosus; **4,** hepatic portal vein; **5,** inferior vena cava; **6,** ductus arteriosus; **7,** internal iliac; **8,** external iliac growing into hind limb bud; **9,** umbilicus; **A,** liver; **B,** base of the allantois, which is developing into a urinary bladder; **L,** left ventricle; **R,** right ventricle. Much of the blood returning to the right atrium from the placenta passes through a foramen ovale (not illustrated) into the left atrium to be distributed via the left ventricle to the head and anterior limbs. *Darker red* indicates blood rich in oxygen; *darker blue,* low in oxygen; *light red,* mixed, considerable oxygen; *light blue,* mixed.

atrium from the head, enters the right ventricle and is pumped into the pulmonary trunk. Because the ductus arteriosus is open and functional (Fig. 13-29), most of the blood in the pulmonary trunk is shunted into the dorsal aorta. This is an advantage, since blood in the ductus arteriosus is mostly unaerated and some of it will pass down the dorsal aorta to enter the umbilical arteries leading to the fetal respiratory membranes.

Blood coming from the lungs, which is unaerated and in small quantities, enters the left atrium and, along with the blood coming into the left atrium via the interatrial foramen, passes into the left ventricle. This blood is then pumped into the ascending aorta. From this account it can be seen that the blood in the fetus is either venous (that is, lacking oxygen) or mixed except in the umbilical vein. An essentially identical allantoic circulation occurs in unhatched chicks. The embryonic bird, however, depends on the vitelline (yolk sac) circulation for nourishment.

At birth, major circulatory changes adapt the organism for pulmonary respiration:

1. The ductus arteriosus closes as a result of nerve impulses passing to its muscular wall. These impulses are initiated reflexly when the lungs are filled with air with the first gasp after delivery. In birds this is usually the day before hatching, when the imprisoned chick pecks a hole in its extraembryonic membranes and starts breathing the air entrapped between these membranes and the shell. When the chick inside the shell starts to peep, it already has air in its lungs! Shortly thereafter, all blood entering the pulmonary trunk goes to the lungs, and the ductus arteriosus soon becomes converted into an **arterial ligament (ligamentum arteriosum).**

2. A flaplike interatrial valve is pressed against the interatrial foramen by the sudden increase in pressure in the left atrium that results from the greatly increased volume of blood entering from the lungs. This valve prevents the unoxygenated blood in the right atrium from entering the left atrium, which now contains only oxygenated blood from the lungs. Within a few days the foramen ovale is permanently sealed and only a scar, the **fossa ovalis,** remains.

3. At birth, the umbilical arteries and vein are severed at the umbilicus. Thereafter, no blood passes through the umbilical arteries beyond the distal tip of the urinary bladder. From bladder to navel the umbilical arteries become converted into **lateral umbilical ligaments** in the free border of the ventral mesentery of the bladder.

4. Blood no longer flows through the umbilical vein, and this vessel becomes converted into the **round ligament of the liver.** At the same time, the ductus venosus is converted into the **ligamentum venosum.** (This occurs halfway through gestation in whales.) As a result of these changes, the fetal mammal (and bird, too) is changed from an allantoic-respiring organism to one capable of breathing air.

Failure of the interatrial foramen to close or of the ductus arteriosus to fully constrict results in cyanosis (blueness) of the skin of the newborn, since blood continues to be shunted away from the lungs and hence has the bluish color of venous blood.

■ LYMPHATIC SYSTEM

A lymphatic system is found in all vertebrates. It consists of thin-walled lymph vessels (lymphatics), fluids in transit, lymph nodes, and, in some species, lymph hearts (Fig. 13-30). In contrast to blood, lymph flows in only one direction—toward the heart.

Lymphatics penetrate nearly all the soft tissue of the body and commence as blind-end **lymph capillaries** that collect interstitial fluids. Once inside the lymph capillaries, the fluid is colorless or pale yellow, and is called **lymph.** Lymph from different areas passes into larger vessels, which finally empty into a vein. The walls of the larger lymphatics are strengthened by smooth muscle.

Lymphatics in intestinal villi collect fat absorbed from the intestine after a meal. If the meal has been particularly fatty, the lymph in these vessels is milky. For this reason, lymphatics of intestinal villi are called **lacteals,** and the lymph therein is called **chyle.** Some lymphatics in cyclostomes, cartilaginous fishes, and even man contain red blood cells. The fluid in such vessels is called **hemolymph.**

Lymph nodes are masses of hemopoietic tissue interposed along the course of lymph channels in birds and mammals only. They may be no larger than a pinhead, or they may measure several centimeters in diameter. These are the "swollen glands" that can be palpated in the neck, axilla, and groin of man when there is inflammation in the areas drained. Lymph nodes consist of a connective tissue reticulum enmeshing large numbers of lymphocytes, many of which are formed therein, and of sinusoidal passageways lined by phagocytic cells.

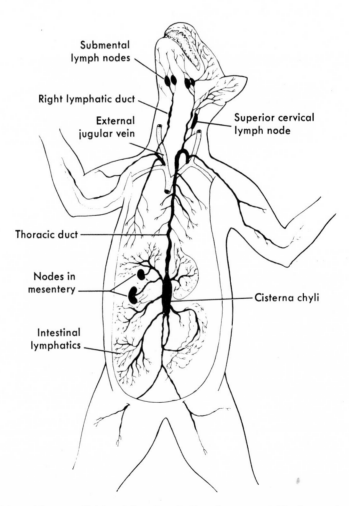

Fig. 13-30. A few of the superficial and deep lymphatics of a mammal. The large veins of the neck approaching the heart are stippled. Mammals lack lymph hearts.

Lymph usually enters a node via several afferent lymphatics, filters through the sinusoidal spaces, and leaves via a single large efferent lymphatic. Lymph nodes constitute one line of defense against disease, since they contain cells that ingest and destroy bacteria and other foreign particles. In this respect they serve as filtering beds for interstitial fluid before the latter is returned to the general circulation. In addition, nodes contain large numbers of cells that are produced in the thymus gland and that, when stimulated by antigens, produce antibodies.

Lymph channels in birds and mammals are provided with fibrous valves that assist in preventing backflow. Valves may also guard the exits from the major lymph ducts into veins.

In many fishes, and in amphibians and reptiles muscular pulsating **lymph hearts** are situated at strategic locations along the lymphatics. Frogs have four pairs, two near the thigh and two under the scapula. Urodeles have as many as sixteen pairs and caecilians as many as 100 pairs. Amphibians have more active seepage from their vascular channels than other vertebrates, and so their lymph hearts move a large volume of fluid hourly.

The flow of lymph results from numerous factors. These include, in addition to lymph hearts in ectotherms, pressure of incoming fluid, contraction of the muscular walls of the larger ducts, contraction of surrounding skeletal muscles, movements of the viscera that squeeze the lymphatics and "milk" the fluids along the vessels, and respiratory movements in reptiles, birds, and mammals, in which intrathoracic pressure rises and falls with each movement of the muscular diaphragm or oblique septum.

Sinuslike enlargements of lymph channels occur in some locations. Such are the **subcutaneous lymph sinuses** under the skin of frogs, the **sublingual lymph sac** of frogs that, when suddenly filled with lymph, causes the tongue to dart out; and the **cisterna chyli** in the abdominal cavity of many higher vertebrates, which collects chyle from lacteals.

Lymph channels in lower vertebrates empty into one or more major venous channels, including caudal, iliac, subclavian, and postcardinal veins, depending on the species. Most mammals have two major lymph channels—a **thoracic duct** commencing in the cisterna chyli and draining the abdomen, left side of the head, neck, thorax, and the left arm and shoulder and a **right thoracic duct** (sometimes several) draining the right side of the head, neck, thorax, and the right arm and shoulder (Fig. 13-30). Each empties into the subclavian vein on its own side or into an adjacent vein, such as the brachiocephalic or external jugular. Birds and a few mammals have two thoracic ducts arising from a single cisterna chyli.

Although lymph nodes have not been described in fishes, amphibians, or reptiles, other lymphoid masses of one kind or another are scattered throughout the body of all vertebrates. Lymph does not filter through these, but they contain lymph capillaries that are drained by lymph channels. They are numerous in the coelomic mesenteries. The spleen is the largest. It is absent in cyclostomes. Other lymphoid masses include mammalian tonsils, the thymus, Peyer's patches in the mucosa of the small intestine of amniotes, and the bursa of Fabricius of birds (p. 427).

☐ Chapter summary

1. The circulatory system includes the blood vascular and lymphatic systems.
2. The blood vascular system consists of the heart, arteries, capillaries, and veins.
3. The lymphatic system consists of lymphatics (including lacteals), lymph capillaries, sinuses, nodes in higher forms, and lymph hearts (absent in birds and mammals). Lymph is transported from tissue spaces, and chyle from intestinal villi to major venous channels.
4. A sinus venosus occurs in fishes, amphibians, and reptiles. It is absent in adult birds and mammals, having been incorporated in the right atrial wall.
5. A single atrium receives blood from the sinus venosus in fishes. In lungbreathing vertebrates the atrium is partitioned into two chambers by a septum, which is incomplete in lungfishes and some urodeles, complete in other tetrapods. The right atrium receives the sinus venosus or the largest systemic veins and, in amniotes, coronary veins. The left atrium receives pulmonary veins.
6. A single ventricle occurs in fishes and amphibians. Lungfishes have an incomplete ventricular septum partially dividing the ventricle in two. Amniotes have two ventricles separated by an incomplete septum in most reptiles, a complete septum in crocodilians, birds, and mammals.
7. A conus arteriosus is part of the heart in fishes. In dipnoans and amphibians it is short and is also called bulbus cordis. It is absent in adult amniotes.

8. A ventral aorta leads cephalad from the heart to all aortic arches. The unpaired segment emerging from the heart is sometimes called truncus arteriosus. In teleosts and perennibranchiate urodeles the ventral aorta exhibits a swelling, the bulbus arteriosus.
9. Oxygenated and unoxygenated blood are kept separate in the heart of dipnoans and tetrapods by partial or complete interatrial and interventricular septa, trabeculae, and spiral valves.
10. In reptiles the truncus arteriosus is split longitudinally into an aortic trunk leading from the left ventricle to the third and right fourth arches, a second aortic trunk from the right ventricle leading to the left fourth arch, and a pulmonary trunk from the right ventricle leading to arch VI.
11. In birds and mammals two trunks emerge from the heart: an aortic trunk (ascending aorta) leads to the carotid and systemic arches, and a pulmonary trunk leads to the pulmonary arteries.
12. An aortic arch is a blood vessel connecting ventral and dorsal aortas and located in a visceral arch. Typically, six pairs of aortic arches develop in each vertebrate embryo. During ontogeny the aortic arches are reduced in number, the highest vertebrates retaining the fewest. In fishes the aortic arches become interrupted by gill capillaries. In fishes and amphibians with external gills, detours from the aortic arches carry blood to the external gills. In lungfishes two gan-

oids, and tetrapods the sixth aortic arch sprouts pulmonary arteries.

13. Aortic arch I usually disappears, in part at least. Elasmobranchs retain parts of aortic arches II to VI. Modern fishes and many terrestrial urodeles retain arches III to VI. Aquatic urodeles and amniotes retain parts or all of arches III, IV, and VI.

14. The dorsal aorta between III and IV (ductus caroticus) disappears in some amphibians, most reptiles, and all birds and mammals. The dorsal segment of aortic arch VI (ductus arteriosus) disappears in anurans, most reptiles, and birds and mammals. Birds also lose the left fourth aortic arch, and mammals lose much of the right fourth. As a result of loss of the ductus caroticus, blood entering arch III (carotid) must continue to the head. As a result of loss of the ductus arteriosus, blood entering arch VI must pass to the lungs.

15. Birds and mammals lose arches I, II, V, the left or most of the right side of IV, the dorsal segment of VI on both sides, and the ductus caroticus.

16. As a result of modifications in the heart and ventral aorta and deletions in the aortic arches, a system of vessels appropriate for respiration via gills is translated phylogenetically and ontogenetically into one suitable for lung respiration.

17. There is little mixing of oxygenated and unoxygenated blood in any adult vertebrate. Mixing occurs in fetal birds and mammals.

18. Major venous channels in the basic pattern are anterior cardinals, postcardinals, common cardinals, abdominals, renal portals, hepatic portals, hepatic sinuses, coronaries, and pulmonaries. Dipnoans and tetrapods add postcavals.

19. Anterior cardinals (internal jugulars) drain the head and empty into the common cardinals.

20. Postcardinals are absent in anurans, reptiles, and birds. They persist in mammals under new names (azygos, hemiazygos).

21. Abdominal veins drain pectoral and pelvic fins in cartilaginous fishes. They lose their connection with the forelimbs in tetrapods and with the hind limbs in birds and mammals. In mammals they remain as umbilical veins. There are no abdominal veins in bony fishes.

22. The renal portal system drains only the tail in fishes. It acquires a connection from the hind limbs in amphibians. In crocodilians and birds this connection partly bypasses the kidneys and goes directly to the postcava. There is no renal portal system in cyclostomes or adult mammals above monotremes.

23. The postcava becomes increasingly prominent in higher vertebrates. Commencing in dipnoans and amphibians as an alternate route to the heart from the kidneys, it finally drains hind limbs, most of the trunk, and tail.

24. Remnants of embryonic vascular channels in adult mammals include the round ligament of the liver (remnant of left umbilical vein), ligamentum venosum (remnant of ductus venosus), ligamentum arteriosum (remnant of left ductus arteriosus), lateral umbilical ligaments (remnants of paired umbilical arteries from urinary bladder to umbilicus), and fossa ovalis (occluded foramen ovale between the atria of the fetus).

25. Gill-breathing fishes have a single circulation as follows: heart, gills, body, heart. Birds and mammals have a double circulation: left side of heart, body (except lungs), right side of heart, lungs, left side of heart. Dipnoans, amphibians, and reptiles have a functionally double circulation with essentially no mixing of blood when environmental oxygen is adequate.

LITERATURE CITED AND SELECTED READINGS

1. Angell, C. S., and Hipona, F. A.: Angiographic studies of the anuran double circulation, American Zoologist **5**:668, 1965.

2. Berg, T., and Steen, J. B.: The mechanism of oxygen concentration in the swim bladder of the eel, Journal of Physiology **195**:631, 1968.

3. Brodal, A., and Fänge, R., editors: The biology of *Myxine*, Oslo, 1963, Norway Universitetsforlaget.

4. De Ryke, W.: The development of the renal portal

system in *Chrysemys marginata Belli* (Gray), University of Iowa Studies in Natural History (New Series, no. 88) vol. 11, no. 3, 1925.

5. Fox, M. H., and Goss, C. M.: Experimentally produced malformations of the heart and great vessels in rat fetuses: transposition complexes and aortic arch anomalies, American Journal of Anatomy **102:**65, 1958.

6. Johansen, K., and Hanson, D.: Functional anatomy of the hearts of lungfishes and amphibians, American Zoologist **8:**191, 1968.

7. Kampmeier, O. F.: Evolution and comparative morphology of the lymphatic system, Springfield, Ill., 1969, Charles C Thomas, Publisher.

8. Millen, J. E., Murdaugh, H. V., Bauer, C. B., and Robin, E. D.: Circulatory adaptation to diving in the freshwater turtle, Science **145:**591, 1964.

9. Nandy, K., and Blair, C. B.: Double superior venae cavae with completely paired azygos veins, Anatomical Record **151:**1, 1965.

10. Nelsen, O. E.: Comparative embryology of the vertebrates, New York, 1953, The Blakiston Co., Inc.

11. O'Donoghue, C. H.: Blood vascular system of the tuatara, *Sphenodon punctatus*, Philosophical Transactions of the Royal Society of London **210:**175, 1920.

12. Ogren, H., and Mitchen, J.: Tracing oxygenated and unoxygenated blood through the organs of a frog by means of radioisotopes, Turtox News **45**(5): 130, 1967.

13. Ruszynák, I., Földi, M., and Szabó, G.: Lymphatics and lymph circulation: physiology and pathology, ed. 2, New York, 1967, Pergamon Press, Inc.

14. Satchell, G. H.: Circulation in fishes, Cambridge, 1971, Cambridge University Press.

15. Scholander, P. F.: The wonderful net. In Vertebrate structures and functions: readings from Scientific American with Introduction by N. K. Wessells, San Francisco, 1955-1974, W. H. Freeman and Co. Publishers.

16. Seib, G. A.: On the azygos vein in *Pithecus (Macacus) rhesus*, Anatomical Record **51:**285, 1932.

17. Struthers, P. H.: The aortic arches and their derivatives in the embryo porcupine *(Erethizon dorsatus)*, Journal of Morphology and Physiology **50:**361, 1930.

18. Yoffey, J. M., and Courtice, F. C.: Lymphatics, lymph, and lymphoid tissue, Cambridge, 1956, Harvard University Press.

19. Zweifach, B. W.: The microcirculation of the blood. In Vertebrate structures and functions: readings from Scientific American with Introduction by N. K. Wessells, San Francisco, 1955-1974, W. H. Freeman and Co. Publishers.

Symposium in American Zoologist

Functional morphology of the heart of vertebrates, **8:**177, 1968.

14

Urinogenital system

In this chapter we will see that the urinary and reproductive organs of all
vertebrates, male and female, fishes and human beings, develop in accordance with
a basic architectural pattern, and how the pattern is modified during embryogenesis.
At the end we will see how a simple chamber, the cloaca, is partitioned in male
and most female mammals to produce one passageway for urine and reproductive
products and a second for the digestive tract, and how, in female rodents and
primates, a third exit forms solely for the reproductive tract. We will also look briefly
at one way that saltwater and desert animals conserve water and how freshwater
vertebrates get rid of it.

Although the function of kidneys is different from that of gonads, their ducts are so intimately related developmentally and functionally that neither the urinary nor genital system can be discussed without reference to the other. For example, in most lower vertebrates, male kidney ducts also carry sperm; in higher vertebrates, which have a succession of embryonic kidneys, the ducts of the first kidneys carry sperm throughout life although a new kidney is formed. The cloaca, which is present in most vertebrates, is a common terminal passageway for urine and genital products, and in mammals, when the embryonic cloaca becomes partitioned to provide a separate pathway for the digestive tract, there is still a final common pathway for urine and sperm in males, and urine and the products of conception in most females. For these reasons the urinary and genital organs are discussed in a single chapter.

◼ VERTEBRATE KIDNEYS AND THEIR DUCTS

Paleontological evidence has been interpreted as indicating that the earliest vertebrates lived in fresh water and that the early stages of evolution of fishes took place in that medium. Animals that are submerged in fresh water inevitably acquire excess water either by absorbing it through the skin or by swallowing it with the food. Therefore a mechanism for elimination of excess water is necessary. On the other hand, salts are scarce in fresh water, the chief source being food. Therefore, one osmoregulatory need of freshwater organisms is to eliminate water and conserve salt.

The elimination of water and the reclamation of salt may have been the earliest function of vertebrate kidneys. Tufts of blood vessels (**glomeruli**) filtered water out of the bloodstream into the body cavity, and **convoluted tubules** with ciliated openings into the coelom collected the filtrate, retrieved any salt from it, and emptied the final filtrate into a **longitudinal duct** that passed to the cloaca.

When ancestral fishes were later adapting to salt water, they faced a different water-salt problem. Instead of accumulating too much water in their tissues, they were in danger of accumulating too much salt. The problem of survival then became one of conserving water and excreting salt. Structural modifications of the kidneys helped to solve this problem. One modification was the shortening or loss of distal segments of the kidney tubules, which eliminated one site for reabsorption of salt. Another modification was loss of the glomeruli in some marine teleosts (toadfish, sea horse, others), and in many others the glomeruli became poorly vascularized or cystic. Shortening of the tubules and loss of glomeruli resulted in increased salt excretion and water retention in marine fishes.

There are a few freshwater teleosts with aglomerular kidneys and very short tubules. Water excretion in these species is principally tubular, and they make up for loss of salt in the urine by active uptake of salt via the gills. These fishes are thought to repre-

sent former saltwater species that became adapted to a freshwater habitat.*

The concept that vertebrates arose in fresh water is not universally accepted. There are some who think that vertebrates arose in a marine environment and that the kidneys were used originally to eliminate aqueous solutions of salt, and that later, when fresh water was encountered, the same mechanism was useful in eliminating excess water.

Thus far nothing has been said about excretion of nitrogenous wastes because in no fish is the kidney of primary importance in excretion of nitrogenous compounds. These are eliminated through the gills, and, sometimes, skin, in fishes. In tetrapods, the kidneys dispose of nitrogenous wastes. Nitrogen is excreted chiefly as ammonia, which is highly soluble in water, when water is plentiful, as in freshwater teleosts and aquatic amphibians (ammonotelic animals). It is also excreted as ammonia by most marine fishes other than elasmobranchs. This is made possible by drinking sea water and rapidly eliminating the salt. It is excreted as urea in elasmobranchs and mammals (ureotelic animals), and as uric acid in a semisolid urine when water is scarce, as in terrestrial reptiles and birds (uricotelic animals).

□ **Basic plan**

From the preceding discussion it is evident that vertebrate kidneys, or nephroi, are built in accordance with a basic pattern incorporating **glomeruli, tubules surrounded by peritubular capillaries,** and a **pair of longitudinal ducts** (Fig. 14-1). Variations in kidney structure are primarily in number and arrangement of the glomeruli and tubules and in the length of the tubules.

---

*In fishes that migrate between salt water and fresh water, pituitary and other hormones "turn on" and "turn off" various segments of the kidney tubules as well as extrarenal sites of absorption and excretion of water and electrolytes.

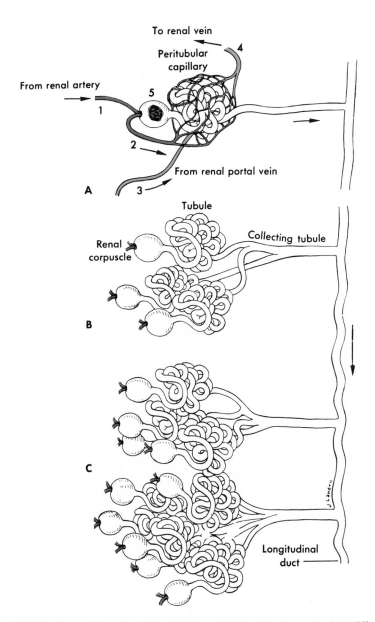

Fig. 14-1. Basic structure of vertebrate kidney. **A,** Functional renal unit. **1,** Afferent glomerular arteriole; **2,** efferent glomerular arteriole; **3,** vessel from renal portal system; **4,** tributary of renal vein; **5,** glomerulus (red) surrounded by a capsule. **B,** Primary, secondary, and tertiary tubules in a single body segment. **C,** The increased number of tubules per segment disrupts the metamerism of the kidney.

Glomeruli are tufts of capillaries, which, assisted by blood pressure, filter water and certain other substances out of the bloodstream. In some species the glomeruli are large enough to be seen with the naked eye or a hand lens. In others they are microscopic. Supplying a glomerulus (Fig. 14-3) is an afferent glomerular arteriole, and emerging from it is an efferent glomerular arteriole. The latter leads to a capillary bed that surrounds the kidney tubule. From the peritubular capillary beds venules lead to renal veins.*

The most primitive glomeruli are suspended in the coelomic cavity (Fig. 14-2, *A*). They are sometimes called "external" glomeruli to differentiate them from "internal" glomeruli, which are encapsulated by part of the kidney tubule to form a renal corpuscle (Fig. 14-2, *B*). External glomeruli are confined to cyclostomes and to the larval kidneys of lower vertebrates.

Kidney tubules are microscopic passageways that collect glomerular filtrate and conduct it to the longitudinal duct. They consist of several segments, each of which may add to, or subtract from, the glomerular filtrate that trickles through them (Fig. 14-3). By doing so they affect the amount of salt, water, and other substances that are excreted in the urine. Excepting those associated with external glomeruli, they commence as a **Bowman's capsule** that surrounds a glomerulus. The capsule plus the glomerulus constitute a **renal corpuscle.**

The more anterior tubules may exhibit a ciliated, funnel-shaped **nephrostome,** which is an opening into the coelom (Fig. 14-2, *B*). Nephrostomes are usually confined to embryos and larvae. If the embryonic tubule that exhibits a nephrostome is not lost during later development, the nephrostome may finally close. Nephrostomes are thought

<hr>

*Cyclostomes lack peritubular capillaries. The efferent glomerular vessels flow into the capillaries of the mesonephric duct or into the postcardinal vein.

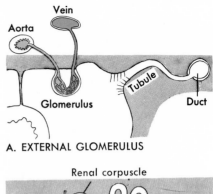

A. EXTERNAL GLOMERULUS

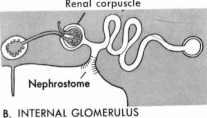

B. INTERNAL GLOMERULUS

Fig. 14-2. External and internal glomeruli.

to be vestiges of a hypothetical primitive kidney (archinephros) in which all glomeruli were external.

Kidney tubules arise from the **intermediate mesoderm.** This is a ribbon of nephrogenic tissue lying between the mesodermal somites and the lateral mesoderm just lateral to the mesodermal somites (Figs. 4-8 and 15-5) and extending uninterruptedly from the level of the heart to the cloaca. Almost the entire ribbon of intermediate mesoderm produces kidney tubules. Commencing at the anterior end, new tubules are added more and more caudad as differentiation progresses. The anteriormost tubules are always metameric, since one tubule develops at the level of each mesodermal somite. Farther back numerous tubules develop in each segment, and the metamerism is lost. The segmental tubules usually disappear later in development, with the result that the adult kidney does not extend as far forward as the embryonic kidney. Neither does it contain evidence of its earlier metamerism. Nevertheless, the

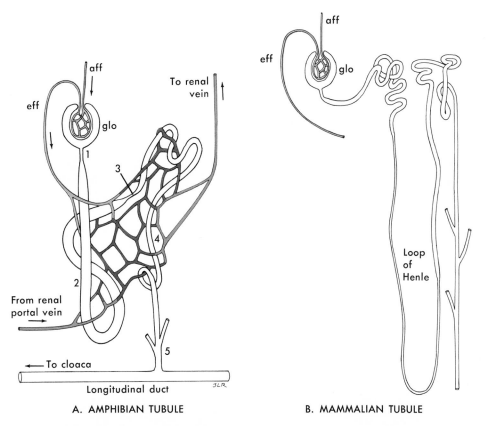

A. AMPHIBIAN TUBULE

B. MAMMALIAN TUBULE

Fig. 14-3. Kidney tubules. **A,** Mesonephric tubule of an aquatic urodele showing associated vessels. **B,** Mammalian tubule. The segments in **A** are: **1,** neck; **2,** proximal segment; **3,** intermediate segment; **4,** distal segment; **5,** collecting tubule. The mammalian tubule is characterized by the loop of Henle. **aff,** Afferent glomerular arteriole; **eff,** efferent glomerular arteriole; **glo,** renal corpuscle consisting of a glomerulus (red) and Bowman's capsule. Purple in **A** indicates mixing of arterial and venous blood in peritubular capillaries. Mammalian tubules receive no portal supply.

blood supply to lower vertebrate kidneys is from segmental arteries.

The **longitudinal ducts** commence development at the anterior end of the nephrogenic mesoderm as caudally directed extensions of the first tubules (Fig. 14-4). Each duct grows caudad until it achieves an opening into the cloaca. The duct participates in the induction of additional tubules farther and farther back. Later tubules achieve an opening into this duct unless an accessory duct is destined to drain them.

ARCHINEPHROS

Comparative studies in anatomy and embryology suggest that the earliest vertebrate kidneys extended the entire length of the coelom, that all tubules were segmental (Fig. 14-5), and that each tubule opened via a nephrostome. A single glomerulus, suspended in the coelomic cavity, is postulated for each tubule. The kidney would then have drained off the coelomic fluid, as does the excretory system of annelid worms and amphioxus. This hypothetical primitive kid-

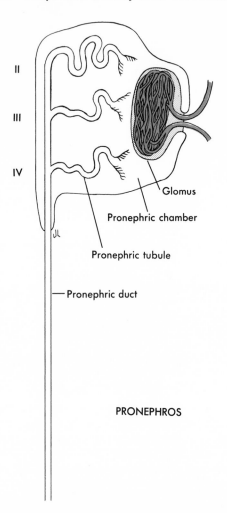

II

III

IV

Glomus

Pronephric chamber

Pronephric tubule

Pronephric duct

PRONEPHROS

Fig. 14-4. Pronephric kidney of a 15-mm larval frog. **II, III,** and **IV,** Levels of the second, third, and fourth somites. The glomus is three fused external glomeruli. The next tubule to form will be a mesonephric tubule at the level of somite VII.

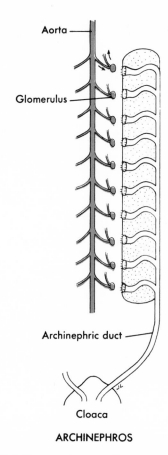

Aorta

Glomerulus

Archinephric duct

Cloaca

ARCHINEPHROS

Fig. 14-5. Hypothetical primitive kidney, also called holonephros, with external glomeruli and open nephrostomes.

ney is called an **archinephros.** The intimate association of a glomerulus and Bowman's capsule to form a renal corpuscle was probably a later development.

□ **Pronephros**

The first embryonic tubules in all vertebrates arise from the anterior end of the nephrogenic mesoderm. Because they are the first and are anteriorly located, they are

called pronephric tubules (Fig. 14-4). Pronephric tubules are segmentally arranged, one opposite each of the more anterior mesodermal somites.

Each pronephric tubule arises in the intermediate mesoderm as a solid bud of cells that later organizes a lumen and, except in birds and mammals, a nephrostome (Fig. 4-8). Associated with each pronephric tubule may be a glomerulus. The number of pronephric tubules is never large—three in frogs (Fig. 14-4), seven in human embryos opposite somites VII to XIII, and about twelve in chicks commencing at somite V. The tubules lengthen and become coiled. The region of the nephrogenic mesoderm exhibiting these segmentally arranged tubules is called the pronephros. The duct of the pronephros is the **pronephric duct.**

Pronephric tubules and pronephric glomeruli are temporary. They function only until the ones farther back are able to work. This is at the end of the larval stage in amphibians and at an equivalent stage in fishes. At that time the glomeruli lose their connection with the dorsal aorta and commence to regress. The tubules regress more slowly, and traces of them may remain in adult fishes. However, the pronephric duct does not regress. It continues to drain the tubules farther back. Although a pronephros always develops in amniotes, it is doubtful that it ever functions as an excretory organ.

□ **Mesonephros**

Under the partial stimulus of the pronephric duct acting as an inductor, additional tubules develop sequentially in the nephrogenic mesoderm behind the pronephric region. The new tubules establish connections with the existing pronephric duct. For at least several segments these tubules, too, may be segmentally disposed, exhibit the same convolutions as the ones anterior to them, and often have open nephrostomes. In fact, there is seldom justification for drawing at any specific point a boundary between the embryonic pronephros and the rest of the embryonic kidney. There is usually a gradual transition from tubules characteristic of the pronephric region to those found farther back.

In the transitional area secondary and tertiary tubules develop as buds from the initial (primary) tubule in each segment (Fig. 14-1, *B*). As these additional tubules enlarge and encroach on one another, the metamerism of the developing kidney is at first obscured and then lost altogether. Another feature of the transitional region is the development of internal glomeruli and of tubules that are longer, more convoluted, and lack nephrostomes. However, many embryonic fishes and amphibians develop nephrostomes for a long distance back, and some fishes retain a few throughout life.

With the disappearance of the pronephric region, the pronephric duct is thereafter called the **mesonephric duct,** and the newer kidney region that it serves is the mesonephros (Fig. 14-6). The mesonephros is the functional adult kidney of fishes and amphibians (Fig. 14-6 to 14-8). It is sometimes called the **opisthonephros** in these adults. The mesonephros is also a functional embryonic kidney in reptiles, birds, and mammals (Fig. 14-9).

MYXINE AND THE ARCHINEPHROS

The mesonephros of the adult hagfish *Myxine* is quite primitive.[3] The glomerular part occupies a 10-cm segment of the nephrogenic mesoderm, commencing some distance behind the regressed pronephros and terminating some distance anterior to the cloaca. This segment consists of thirty to thirty-five large renal corpuscles up to 1.5 mm in diameter, strictly segmental, and connected to the mesonephric duct by very short tubules. The tubules are so short that they probably play no role in altering the content of the glomerular filtrate by addition or reabsorption. However, there is a possibility that the mesonephric duct itself has assumed some of the functions of the missing segments of the tubules. The glomerulus associated with each corpuscle is supplied by an afferent glomerular arteriole from the dorsal aorta and drained by a vessel that flows directly into the postcardinal vein. There is no peritubular capillary network and no detectable renal portal system. Whether or not this is primitive is not known.

Between the functional kidney and regressed pronephros are a variable number of corpuscles that lack glomeruli or have lost their connection with the longitudinal duct. Caudal to the functional kidney there are additional aglomerular corpuscles. The

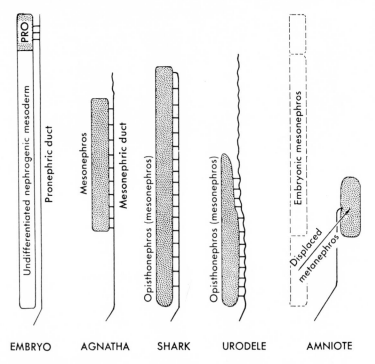

EMBRYO   AGNATHA   SHARK   URODELE   AMNIOTE

Fig. 14-6. Fate of the nephrogenic mesoderm in representative vertebrates. The pronephric duct of the embryo persists in adult anamniotes to drain the mesonephric (opisthonephric) kidney.

total number of corpuscles in *Myxine*, typical and atypical, is about seventy, and they are rigidly segmental. The presence of segmentally arranged corpuscles anterior and posterior to the main uriniferous mass supports the concept of the hypothetical archinephros.

JAWED FISHES AND AMPHIBIANS

The adult kidneys of jawed fishes and amphibians commence somewhere behind the pronephric region—just how far behind depends on how many embryonic tubules disappear in the region between the pronephros and mesonephros. In sharks and caecilians, for example, the adult kidney begins far forward and extends the length of the coelom, lying against the dorsal body wall behind the peritoneum (Figs. 14-6, shark, and 14-7). In many other fishes and in

most amphibians a longer series of transitional tubules disappears (Fig. 14-6, urodele).

In males some of the anteriormost tubules of the mesonephros have no association with glomeruli and are used to conduct sperm from testis to mesonephric duct. This part of the male mesonephros is the **epididymal kidney**, or **sexual kidney**, and the highly coiled part of the mesonephric duct that drains it is the **epididymis** (Figs. 14-7 and 14-8). The corresponding part of the mesonephros of the female may or may not be functional.

Whether extending the length of the coelom or confined to a more caudal region, and regardless of shape, the uriniferous regions of the kidneys of adult fishes and amphibians are basically alike, consisting of renal corpuscles and convoluted tubules,

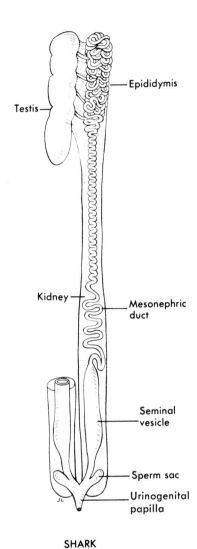

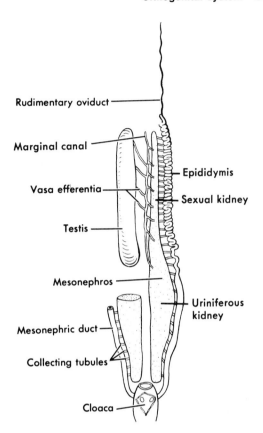

Fig. 14-7. Urinogenital system of a male shark. An accessory urinary duct, not shown, is seen in Fig. 14-21, B.

SHARK

Fig. 14-8. Urinogenital system of male *Necturus*. The narrow sexual segment of the mesonephros serves solely for sperm transport.

occasionally with nephrostomes in fishes, and emptying into a longitudinal **mesonephric (opisthonephric)** duct (Figs. 14-8, 14-15, and 14-21, *A, D, E,* and *F*). Accessory urinary ducts may also be present in fishes (Fig. 14-21, *B*, and 14-23).

The mesonephric duct may lie along the lateral edge of the kidney (Fig. 14-8), or on the ventral surface (Fig. 14-7), or embedded within it. The caudal ends of the mesonephric ducts may enlarge or evaginate to form **seminal vesicles** (Fig. 14-7) for temporary storage of sperm, or urinary bladders (Fig. 14-14). In some fishes (lamprey, shark, others) the two mesonephric ducts unite caudally to form a **urinary** or **urinogenital papilla** (Fig. 14-7). When accessory urinary ducts are present in males, the mesonephric duct may be used chiefly or entirely for sperm transport. In male sharks, for example, the accessory duct drains the caudal part of the kidney, which is where most of the urine is produced (Fig. 14-21, *B*).

AMNIOTE EMBRYOS

The mesonephros of amniote embryos has essentially the same structure as the adult kidneys of fishes and amphibians. Nephrostomes are confined to the pronephros and are rudimentary in most birds and seldom appear in mammals. In embryonic chicks the mesonephros reaches its peak of development on the eleventh day of incubation, halfway through embryonic life. In mammals it reaches its peak earlier—at 9 weeks of gestation in man.

The first mesonephric tubules in a human fetus appear after 4 weeks of embryonic life (twenty-somite stage). A wave of differentiation sweeps along the nephrogenic

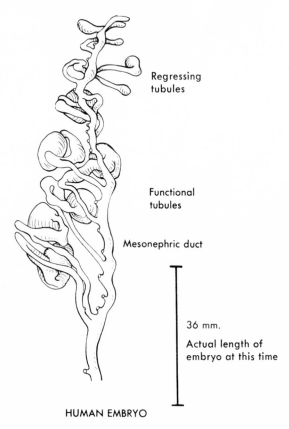

Regressing tubules

Functional tubules

Mesonephric duct

36 mm.

Actual length of embryo at this time

HUMAN EMBRYO

Fig. 14-9. Right functional mesonephros of a 36-mm human embryo. (Redrawn from Altschule.[1])

mesoderm, but even before the last mesonephric tubules at the caudal end of the series have formed, the earliest ones at the anterior end have already regressed (Fig. 14-9). The result is that at peak development of the human mesonephros there are about thirty functioning renal corpuscles, although as many as eighty have formed by that time.[1] The most caudal ones are the latest to be formed; the most anterior ones are the oldest. In man all mesonephric tubules have regressed by the time the embryo reaches 40 mm in length. The mesonephroi of various species of mammals differ in the number of mesonephric tubules formed.

Although the mesonephros is an embryonic kidney in amniotes, it functions for a short time after hatching or birth in reptiles, monotremes, and marsupials—as late as the first hibernation in some lizards and the first molt in some snakes. During the time that the mesonephros is functioning, a new kidney to be used by the amniote the rest of its life, the **metanephros,** is developing. When the metanephros begins to function, the mesonephros rapidly disappears. Soon, only remnants remain.

MESONEPHRIC REMNANTS IN ADULT AMNIOTES

Remnants of the mesonephroi are found in adult amniotes of both sexes. In mammals, remnants consist of groups of blind tubules known collectively in males as the **paradidymis** and the **appendix of the epididymis** (Fig. 14-22), both located near the epididymis; and in females as the **epoophoron** and **paroophoron** near the ovary (Fig. 14-10).

The mesonephric ducts remain as sperm ducts in male amniotes (Fig. 14-22), but they regress in females and remain only as a short, blind **Gartner's duct** (also called **ductus deferens femininus**) in the mesentery of each oviduct (Fig. 14-10).

□ Metanephros

The metanephros, or adult amniote kidney, organizes from the caudal end of the nephrogenic mesoderm, which, however, is displaced cephalad and laterad during development (Figs. 14-6 and 14-10). This is the same mesoderm that gives rise to the caudal part of the mesonephric kidney of fishes and amphibians. The number of tubules that form from this caudal section in amniotes is extremely large—up to an estimated 4.5 million. The metanephric kidney has a duct of its own, the **metanephric duct,** or **ureter.** As long as the mesonephros is functioning, the metanephric duct is merely an accessory urinary duct. When the meso-

nephros disappears, the metanephric duct becomes the sole urinary duct.

The metanephric duct arises as a **metanephric bud** off the caudal end of the embryonic mesonephric duct (Fig. 14-10). Surrounding the bud is the undifferentiated nephrogenic mesoderm, from which metanephric tubules will organize. The metanephric bud grows cephalad, carrying a metanephric blastema along with it. The caudal end of the nephrogenic mesoderm is thus displaced cephalad from its original site. The metanephric bud develops into the ureter and pelvis of the kidney. Many fingerlike outgrowths of the pelvis invade the associated nephrogenic blastema to

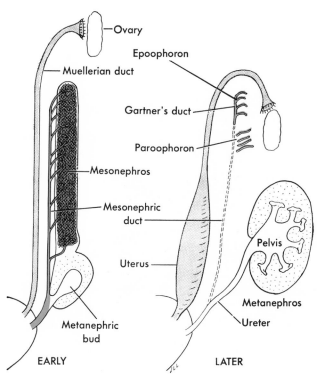

Fig. 14-10. Developmental changes in the urinogenital system of a female amniote. In the early stage (left) the mesonephric kidney and duct are present (red), and the metanephric bud has formed. The undifferentiated muellerian duct is present. In the later stage (right) the mesonephric kidney and duct have regressed, except for remnants (red), and the muellerian duct has differentiated to form a female reproductive tract.

become **collecting tubules.** Meanwhile, the nephrogenic blastema is organizing metanephric tubules that are, at first, S-shaped. One end of each tubule grows toward, and finally opens into, a collecting tubule. The other end grows toward, and surrounds, a nearby glomerulus to become Bowman's capsule.

The tubules of mammalian kidneys have a long, thin, U-shaped **loop of Henle** (Fig. 14-3, *B*) interposed between proximal and distal convolutions. As the loops of Henle elongate, they grow away from the surface of the kidney where the glomeruli are located and toward the renal pelvis. The kidney therefore has a **cortex,** containing renal corpuscles, and a **medulla,** consisting of hundreds of thousands of loops of Henle and common collecting tubules (Figs. 14-11 and 14-12). The loops of Henle secrete or reabsorb water, sodium ions, and other substances. They are the site of much water resorption when it is necessary to prevent dehydration.

The loops and collecting tubules give the renal medulla a striated appearance in frontal section. They are aggregated into one or several conical **pyramids** (Fig. 14-11), depending on the species. The pyramids taper to a blunt tip (**renal papilla**) that projects into the pelvis. Each collecting tubule drains a small number of metanephric tubules (seven to ten in man) and empties into the pelvis at the tip of a renal papilla.

The metanephric tubules of reptiles have no loop of Henle, and those of birds have only a very short equivalent segment. Their glomeruli exhibit only two or three short vascular loops, which minimizes water loss.

Many metanephric kidneys are lobulated

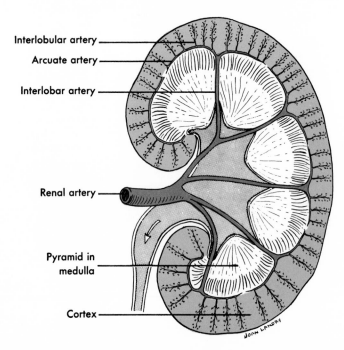

Fig. 14-11. Mammalian kidney, frontal section. The renal vein and its tributaries have been removed. Gray area is pelvis and ureter. Glomeruli are confined to the cortex (light red). Loops of Henle and common collecting tubules comprise the medulla. The kidneys of cats and rabbits have only one pyramid.

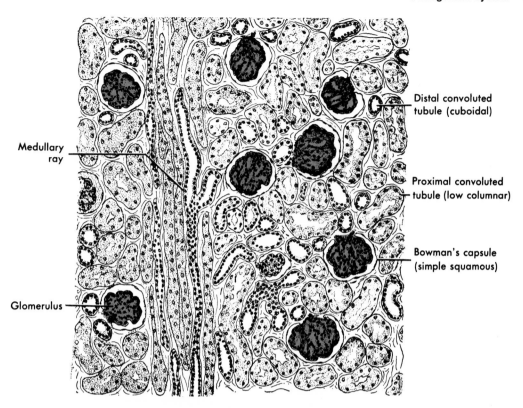

Medullary
ray

Glomerulus

Distal convoluted
tubule (cuboidal)

Proximal convoluted
tubule (low columnar)

Bowman's capsule
(simple squamous)

Fig. 14-12. Frontal section of a small area of the cortex of a mammalian kidney. Medullary rays consist of collecting tubules and loops of Henle. (From Bevelander, G., and Ramaley, J. A.: Essentials of histology, ed. 8, St. Louis, 1978, The C. V. Mosby Co.)

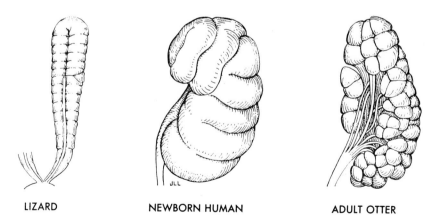

LIZARD   NEWBORN HUMAN   ADULT OTTER

Fig. 14-13. Several lobulated metanephric kidneys.

(Fig. 14-13). Lobulation occurs in human infants but later disappears. Kidneys are elongated to conform to long, slender bodies like those of snakes. The kidneys of birds are flattened against the sacrum and ilium and fit snugly against the contours of these bones. In most adult mammals the kidney is smooth, and the outline is roughly bean shaped. The renal artery, renal vein, nerves, and ureter enter or leave at a median notch, or **hilum.**

Because of their embryonic origin as buds off the mesonephric ducts, the ureters at first terminate in the latter; but as a result of further differential growth, they finally empty directly into the cloaca (reptiles, birds, and monotremes; Fig. 14-25) or into the urinary bladder (placental mammals; Fig. 14-26). In a few male reptiles, however, the ureters retain their connection with the mesonephric ducts (Fig. 14-24).

The metanephros of reptiles and birds is supplied by two or more renal arteries of segmental origin. In mammals there is usually a single renal artery. On entering the kidney of mammals, the renal artery divides into numerous branches that pass radially toward the cortex as interlobar arteries (Fig. 14-11). At the base of the cortex the interlobars give off arcuate arteries, which arch along the base of the cortex more or less parallel to the surface of the kidney. From the arcuate arteries arise tiny interlobular arteries that give off the afferent glomerular arterioles. The metanephros is drained by one or several renal veins.

The metanephric kidney of reptiles and, to a lesser degree, of birds and monotremes, receives blood from the renal portal system. The portal vessels terminate in the peritubular capillaries.

□ **Extrarenal salt excretion in vertebrates**

Vertebrates that live in an environment laden with salt, or that live in arid environments and cannot afford to use body water to carry off accumulated salts, have evolved extrarenal structures for salt excretion. Bony marine teleosts eliminate salt from chloride-secreting glands on their gills. In elasmobranchs the rectal glands perform this function. (The rectal glands of bullsharks caught in fresh water are smaller than those caught in salt water and show regressive changes.) Birds that scoop up fish from salt water as well as marine reptiles have nasal glands that excrete salt. So do terrestrial lizards and snakes that must conserve water because they live in an arid habitat. These same lizards and snakes have atrophied glomeruli, which also conserve water.

The salt-secreting nasal glands of lizards are located outside the olfactory capsule and empty into the nasal canals via small ducts. Whitish incrustations of sodium chloride and potassium can be seen in the nasal canal or at the nostrils. In sea turtles the gland secretes chiefly sodium.

The nasal gland of marine birds is a large paired gland located above the orbit. It is drained by a long duct that opens close to a nostril. A groove extends from the opening to the tip of the beak. Within 15 minutes or so after these birds have drunk water containing a load of sodium chloride and potassium, minute drops of fluid containing these salts trickle down the groove and drip or are shaken off the beak.

Sweat glands eliminate some salt in mammals, but salt loss by this route is merely incidental to secretion of water for its evaporative cooling effect. It is not a regulated route for salt excretion and, in fact, salt lost by this route usually must be replaced. Mammals eliminate excess salt via kidney tubules.

The excretion of electrolytes is regulated chiefly by hormones, especially of the pituitary and adrenal glands. Physiological adaptations of vertebrates to environments of various salinities and water content are discussed in textbooks of comparative physiology.

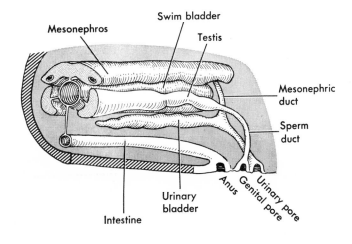

TELEOST

Fig. 14-14. Caudal end of urinogenital system of a male teleost (pike), left lateral view. The unpaired urinary bladder arises as a bud off the united bases of the two mesonephric ducts. Note absence of cloaca. (Redrawn from Goodrich.[5])

■ URINARY BLADDERS

Most vertebrates have a urinary bladder. Exceptions include chiefly the cyclostomes and elasmobranchs among fishes; snakes, crocodilians, and some lizards; and birds other than ostriches. The bladders of most fishes are terminal enlargements or evaginations of the mesonephric ducts known as tubal bladders (Fig. 14-14). The bladders of amphibians through mammals arise as evaginations of the ventral wall of the cloaca (Fig. 14-15).

In amniote embryos the evagination that gives rise to the bladder is prolonged beyond the ventral body wall as an extraembryonic membrane, the allantois (Fig. 4-10, B). Only the base of the allantois—the part proximal to the cloaca—contributes to the adult bladder (Fig. 13-29). After birth the distal part of the allantois within the body remains in mammals as a **urachus** connecting the tip of the bladder with the umbilicus. The urachus lies in the anterior border of the ventral mesentery of the bladder,

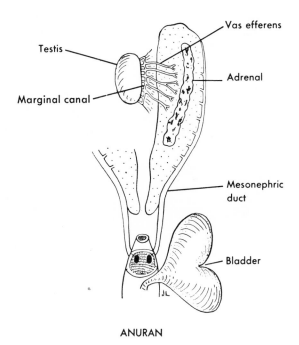

ANURAN

Fig. 14-15. Urinogenital system and adrenal of a male frog, ventral view.

along with the obliterated umbilical arteries.

Turtles and some lizards have large bladders, and some freshwater turtles have two **accessory bladders** as well (Fig. 14-28). The latter are used by females to carry water for moistening the soil when building a nest for the eggs. If they have other functions, these have not been demonstrated. In amphibians and reptiles the urine backs up into the bladder from the cloaca. In mammals the kidney ducts empty directly into the bladder, and the bladder is drained by a urethra.

The adaptive value of the tetrapod urinary bladder seems to be its capacity to store water that may be needed later. During dry weather antidiuretic hormone (ADH) from the pituitary gland of some tetrapods causes active water resorption from the bladder, and the animals cease voiding urine. ADH release is evoked by dehydration. The bladder is especially important in tetrapods living in an arid habitat, where water conservation has survival value.

■ GONADS

The embryonic gonads arise as a pair of elevated **gonadal (genital) ridges.** These are thickenings in the coelomic epithelium just medial to the mesonephroi (Fig. 14-16). The ridges are longer than the resulting mature gonad, which suggests that at one time gonads may have extended the length of the pleuroperitoneal cavity, a condition still

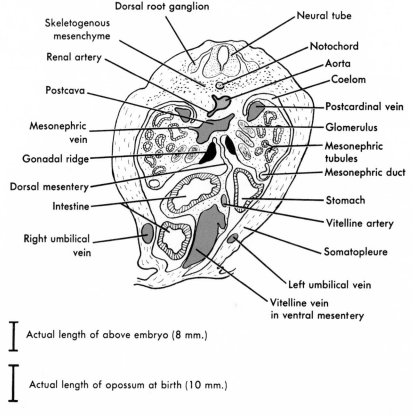

Fig. 14-16. Opossum embryo *(Didelphis)* 6½ days after cleavage, cross section. The gonadal primordia are shown in black.

present in cyclostomes. Although the gonadal ridges are paired, a few adult vertebrates have a single testis or ovary because of fusion of the two ridges across the midline (lampreys, a few teleosts), or because one of the juvenile gonads fails to differentiate (hagfishes, some viviparous elasmobranchs, some female crocodilians, some lizards, and most female birds). A few mammals, among which are the platypus and some bats, have only one ovary. As the gonads approach sexual maturity, they enlarge and usually acquire a dorsal mesentery, the **mesorchium** in males and **mesovarium** in females.

The ovary in some teleost fishes is a permanently hollow sac (Fig. 14-33). The condition results from entrapment of a small part of the coelomic cavity within the developing ovary (Fig. 14-17). The ovarian *cavity* is lined by germinal epithelium. In other teleosts the cavity within the ovary results from a secondary hollowing out of the interior of the ovary at each ovulation. The eggs or, in viviparous teleosts, the young are discharged into the cavity, which is continuous with the lumen of the oviduct (Fig. 14-33). The ovaries of most fishes other than teleosts and elasmobranchs are solid. After ovulation the ovaries shrink almost to a juvenile condition.

The amphibian ovary is also a hollow sac. The *surface* of the sac contains germinal epithelium and the eggs are shed into the coelom. After each reproductive season the ovaries regress to a resting state resembling juvenile ovaries.

The ovaries of reptiles, birds, and monotremes develop numerous irregular, fluid-filled cavities (**lacunae**) by rearrangement of the internal tissues. Such ovaries are said to be "lacunate." The mammalian ovary is compact, with no large chambers or lacunae.

In many mammals a membranous fold of peritoneum, the **ovarian bursa**, entraps part of the coelom in a small chamber along with the ovary and ostium of the oviduct. The bursa may be broadly open to the main coelom, as in cats and rabbits; it may communicate by only a slitlike passage, as in most carnivores and rats; or it may be closed completely, as in hamsters. The bursa increases the probability that all ovulated eggs will enter the oviduct.

Mature testes are usually smaller than ovaries because sperm, although more numerous, are very much smaller than eggs, especially eggs with yolk. The testes of mammals, on the contrary, are larger than ovaries because mammalian eggs (other than monotremes') lack yolk, and ova ripen a few at a time.

The embryonic testis of anurans is subdivided into an anterior portion (**Bidder's organ**), which usually disappears prior to sexual maturity, and a more caudal portion, which becomes the adult testis. Bidder's organ persists in adult male toads (Fig. 14-18) and contains large undifferentiated cells

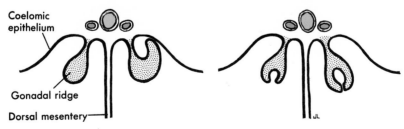

Fig. 14-17. Two methods of entrapment of coelom to form a permanently hollow ovary in teleosts. The gonadal ridges are shown in cross section.

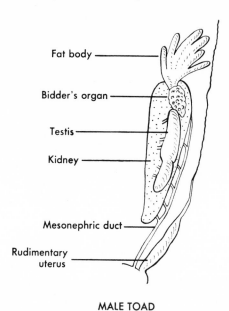

MALE TOAD

Fig. 14-18. Bidder's organ and the rudimentary female reproductive tract in a young male *Bufo,* ventral view. Only the left organs are illustrated.

resembling immature ova. If the testes are removed experimentally, Bidder's organs develop into functional ovaries, and the rudimentary female duct system enlarges under the influence of female hormones from the new ovaries.

Instances of sex reversal occur in nature in many submammalian vertebrate groups. Hens have been known to cease laying eggs, to crow, and to develop other roosterlike characteristics. This comes about when the left ovary atrophies and the right one, which is rudimentary, develops into one producing male hormones. (They apparently cannot produce sperm.)

During early development the gonads are indistinguishable as to sex, and male and female ducts appear in every embryo regardless of sex. Under the influence of sex chromosomes and hormones the early gonads develop into either testes or ovaries, and the appropriate ducts (male or female)

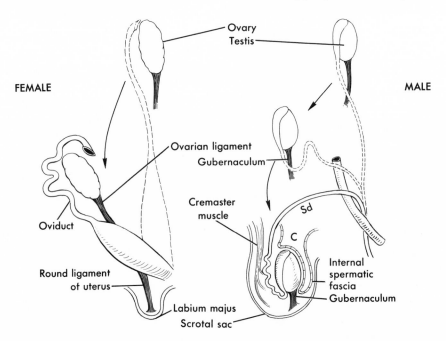

Fig. 14-19. Caudal displacement of mammalian gonads. The ovarian ligament and round ligament of the uterus collectively are homologous with the male gubernaculum. Arrows indicate the route of translocation of the right gonads, ventral view. **C,** Scrotal recess of coelom; **Sd,** spermatic duct arching over the ureter.

enlarge, whereas the other set remains rudimentary or disappears. True hermaphroditism* (production of eggs and sperm by the same individual) is common in cyclostomes and occasional in bony fishes, but it is rare among other lower vertebrates and absent among higher ones.

□ Translocation of ovaries and testes in mammals

The caudal pole of each embryonic ovary and testis is connected by a ligament to a shallow evagination of the coelom (genital swelling, Fig. 14-30), which becomes the

---

*Hermaphroditos was the son of Hermes and Aphrodite. While bathing in the mythical fountain of Salmacis, he became united in one body with the nymph living in the fountain.

scrotal sac in males, labium majus in females (Fig. 14-19). In females the cephalic part of the ligament is named ovarian ligament, and the caudal part is named round ligament of the uterus (Fig. 14-19, female). In males the ligament is the gubernaculum. Partly as a result of shortening of the ligaments, partly because elongation of the ligaments does not keep pace with elongation of the trunk, and partly for unknown reasons,* the ovaries and testes are displaced caudad toward the labia or scrotal sacs. The ovaries are not displaced as far caudad as the testes.

The testes remain retroperitoneal and de-

---

*Removal of the gubernaculum does not always prevent descent of the testes.

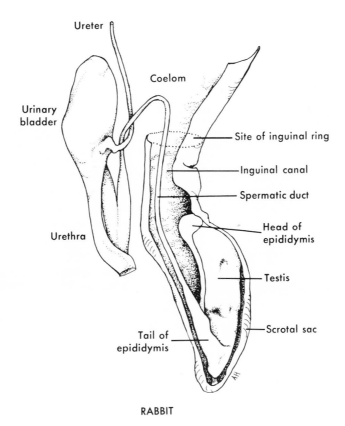

RABBIT

Fig. 14-20. Rabbit testis in scrotal cavity. The testes are retractable; therefore, the inguinal canals are broadly open to the main coelom at the inguinal ring.

scend permanently into scrotal sacs in many mammals including most marsupials, ungulates, carnivores, and primates. In others they are lowered into the sacs and retracted at will (rabbits, bats, a few rodents, some primitive primates, for example). The passage between the abdominal cavity and the scrotal cavity is the **inguinal canal** (Fig. 14-20). The opening of the canal into the abdominal cavity is surrounded by a fibrous **inguinal ring** (often the site of inguinal hernia). In species that retract their testes, the canal remains broadly open. In species in which the testes are permanently confined to the scrotum, the inguinal canal is only wide enough to accommodate the spermatic cord.

The **spermatic cord** contains the spermatic duct, arteries, veins, lymphatics, and nerves. These are all wrapped in a single sheath, the internal spermatic fascia, and all are dragged into the scrotum along with the testis. Scrotal sacs do not develop in monotremes, some insectivores, elephants, whales, and certain other mammals. In these the testes remain permanently in the abdomen.

In mammals whose testes are permanently in scrotal sacs the spermatic artery and vein lie side by side in tight coils within the inguinal canal. This rete of vessels is the **pampiniform plexus.** The arterial blood is at body temperature; the venous blood has been cooled in the scrotal sac. Heat is transferred in the plexus from arterial to venous blood so that blood reaching the testis is cool and blood returning to the internal circulation has been prewarmed. This adaptation protects the sperm of these species from temperatures that would kill them, while conserving body heat.

■ MALE GENITAL DUCTS
□ Fishes and amphibians

In the basic plan the mesonephric duct of the male transmits sperm as well as urine (Fig. 14-21, basic plan). This is the condition in some fishes and in anurans (Fig. 14-21, *D*). Connections between the mesonephroi and testes are established early in embryonic life (Fig. 14-22). Some of the anterior mesonephric tubules—a few to twenty-four or more, depending on the species—grow across the mesorchium to connect with the **rete testis,** a network of sperm passageways within the testis. These modified mesonephric tubules become **vasa efferentia,** which carry sperm from the testis to the mesonephric duct.* In modifications of the basic plan the mesonephric duct may be given over solely to the transport of either sperm or urine, depending on the species, and a new duct may develop to carry the other substance.

In many sharks the mesonephric duct is ultimately preempted by the testis and is used primarily or solely for sperm transport (Fig. 14-21, *B*). In teleosts the mesonephric duct continues to drain the kidney, and a separate sperm duct develops (Fig. 14-21, *F*). The spermatic and urinary ducts, when separate, may terminate in a urinogenital sinus or open separately to the exterior (Fig. 14-14). Cyclostomes have no spermatic ducts (see genital pores, p. 335).

Variations in the extent to which mesonephric ducts may transport sperm and urine in urodeles may be illustrated by contrasting *Necturus* and *Ambystoma.* In *Necturus* (Fig. 14-8) four vasa efferentia connect the testis to a marginal canal, and the latter sends ductules into the sexual kidney. The sexual kidney is composed of a single row of twenty-six modified mesonephric tubules, which are ciliated for sperm transport and are said to have no urinary function. These tubules drain into the mesonephric duct, which also collects urine from the kidney tubules farther back. In *Ambystoma* (Fig. 14-23) the mesonephric duct drains only the sexual kidney and collects no urine.

---

*Mesonephric tubules of female mammals form vasa efferentia ovarii that grow into a rete ovarii. These structures persist in adult human females.

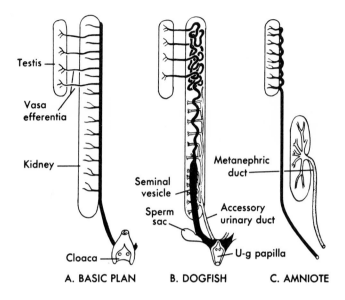

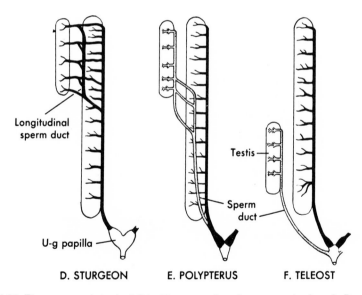

Fig. 14-21. The mesonephric duct (black) as a carrier of sperm and urine. **A,** Basic plan, carrying both sperm and urine. **B,** Carrying urine from the anterior end of the kidney only; chiefly a spermatic duct. **C,** Carrying sperm only. **D to F,** Increasing tendency toward a separate sperm duct, the mesonephric duct carrying urine. **U-g papilla,** Urinogenital papilla. For a variation in the termination of the teleost mesonephric duct see Fig. 14-14.

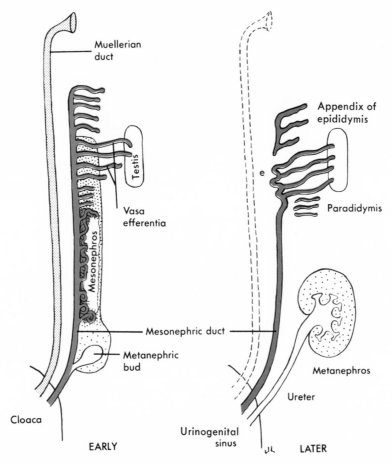

Fig. 14-22. Urinogenital system of developing male amniote. In the earlier stage (left) some of the mesonephric tubules have invaded the testis to become vasa efferentia. In the later stage (right) the mesonephros has regressed except for remnants (appendix of epididymis, paradidymis), and the muellerian duct has regressed (broken lines at right). The mesonephric duct remains to carry sperm. Mesonephric duct and tubules are red. **e,** Epididymal portion of mesonephric duct.

Instead, twelve to fourteen small urinary tubules, depending on the species, drain the uriniferous kidney and empty into a urinogenital papilla. A longitudinal duct carrying only sperm is a **vas (ductus) deferens.** The mesonephric duct has also been freed from urine transport in true salamanders and plethodonts.

☐ **Amniotes**

Although in amniotes the embryonic mesonephric kidneys disappear during em-

bryonic life, the mesonephric ducts remain in males and serve as spermatic ducts. These empty into the cloaca in reptiles and birds (Figs. 14-24 and 14-25) and into a derivative of the cloaca in mammals.

The anatomical relationships of the spermatic ducts in mammals are affected by (1) complete separation of the embryonic cloaca into a urinogenital sinus and rectum (Fig. 14-41, *E*) and (2) caudal migration of the testes. As a result of subdivision of the

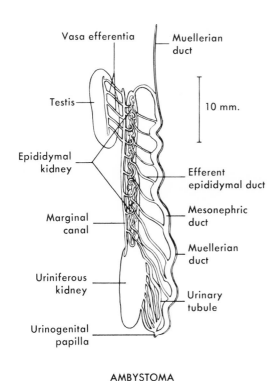

Vasa efferentia

Muellerian duct

10 mm.

Testis

Epididymal kidney

Efferent epididymal duct

Mesonephric duct

Marginal canal

Muellerian duct

Uriniferous kidney

Urinary tubule

Urinogenital papilla

AMBYSTOMA

Fig. 14-23. Urinogenital system of a male urodele, ventral view. The testis is reflected to the observer's left and the urinary tubules to the right. (Redrawn from Baker and Taylor.[2])

cloaca, the spermatic ducts finally empty into the urinogenital sinus, which is the male urethra (Fig. 14-26). As a result of caudal migration of the testes, the spermatic ducts become "caught" or "hung up" on the ureters in such a way that they must loop over the ureters en route to the urethra (Figs. 14-19 and 14-26). Near the junction of spermatic ducts and urethra in mammals are one or more accessory sex glands that produce some of the constituents of semen (Fig. 14-27).

The urethra in male mammals is frequently called **prostatic urethra** where the prostate glands empty, **membranous urethra** from prostate to penis, and **spongy urethra** within the penis.

■ GENITAL PORES

Cyclostomes lack genital ducts, and sperm and eggs are shed into the coelom and pass caudad by bodily movements and cilia. They then enter a pair of funnel-shaped **genital pores** in the caudal abdominal wall. These lead into a median papilla (genital in hagfishes, urinogenital in lampreys) that opens to the exterior just behind the anus.

Similar pores lead from the coelom to the exterior in some elasmobranchs and a number of bony fishes. They have also been described in turtles and crocodilians. In none of these, however, do these pores convey gametes, since genital ducts are present. Whether they are homologous with the genital pores of agnathans is not known, and it seems unlikely. For this reason it is preferable to call them **abdominal pores** in gnathostomes. What role abdominal pores may perform in vertebrates having genital ducts is obscure. In some marine teleosts the pores are present only in females and

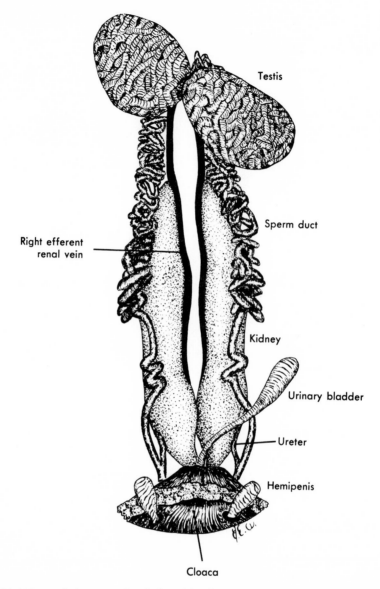

Fig. 14-24. Urinogenital organs of male lizard *Anolis carolinensis*, ventral view. The kidneys are metanephric. The sperm duct is the persistent mesonephric duct. The hemipenes are seen in an everted (erect) position.

open only during the breeding season. In these species they may play some unknown role in reproduction. On the contrary, they may be functionless vestiges that retain a hereditary responsiveness to one or more reproductive hormones.

■ INTROMITTENT ORGANS

When fertilization is internal, the male often develops intromittent (copulatory) organs for introducing sperm into the reproductive tract of the female. Intromittent organs are characteristic of reptiles and

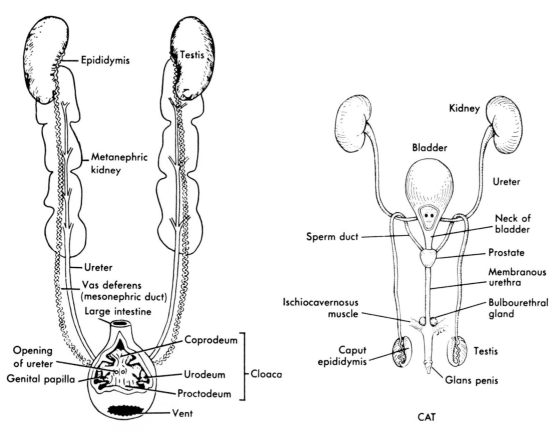

Fig. 14-25. Urinogenital system of a rooster.

Fig. 14-26. Urinogenital system of a male cat, ventral view.

CAT

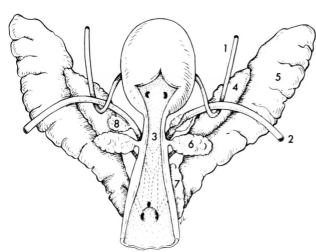

Fig. 14-27. Accessory sex organs of a male hamster, ventral view. The bladder and urethra have been opened to show entrances of ducts. **1,** Ureter; **2,** spermatic duct; **3,** urethra; **4,** coagulating gland; **5,** seminal vesicle; **6,** cranial prostate; **7,** caudal prostate; **8,** ampullary gland. A bulbourethral gland enters the urethra farther caudad.

mammals. They are also present in some fishes, in apodans, in one anuran *(Ascaphus)*, and in a few birds.

The intromittent organs of elasmobranchs are grooved, fingerlike appendages of the pelvic fins known as **claspers** (Fig. 9-7). These are inserted into the uterus of the female and direct ejaculated sperm into the uterus. In basking sharks they transfer a sperm-filled spermatophore up to 3 cm in diameter. Embedded in the fin at the base of the clasper in some sharks is a muscular siphon sac that contributes copious quantities of an energy-rich mucopolysaccharide to the seminal fluid. In many teleosts the anal fin is modified for sperm transport and is called a **gonopodium.** In *Ascaphus* the intromittent organ is a permanent tubular extension of the cloaca, resembling a tail.

Intromittent organs of amniotes are of two types—paired hemipenes and penis. Male snakes and lizards have a pair of **hemipenes,** which are pocketlike diverticula of the caudal wall of the cloaca that extend under the skin at the base of the tail. Each is held in place by a retractor muscle. During copulation the muscle relaxes and the pocket turns inside out and protrudes through the vent in an erect position (Fig. 14-24). Sperm passes along spiral grooves on its surface. Hemipenes are present, but much smaller, in females. When everted, they have been mistaken for legs in snakes.

Male turtles, crocodilians, a very few birds (swans, ducks, ostriches, a few others), and male mammals exhibit an unpaired erectile **penis.** In its simplest form (Figs. 14-28 and 14-29, reptiles) the penis is a thickening of the floor of the cloaca that consists chiefly of a mass of spongy erectile

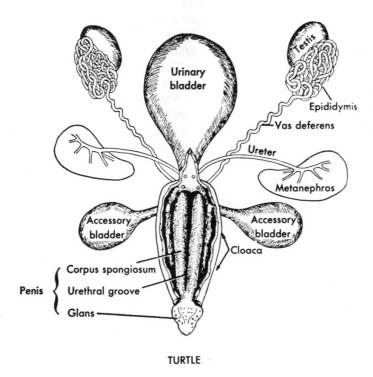

TURTLE

Fig. 14-28. Urinogenital system and cloaca of a male turtle. The dorsal wall of the cloaca has been removed, and the penis is in an extended position. The rectum, which enters the cloaca dorsal to the urinary bladder, has been removed.

tissue, the **corpus spongiosum,** containing blood sinuses and ending as a **glans penis.** When the sinuses are distended with blood, the penis is swollen and firm. The reptilian corpus spongiosum bears a urethral groove for the passage of sperm and urine (Fig. 14-28).

In mammals above monotremes the penis extends beyond the body. The grooved embryonic corpus spongiosum becomes a tube with the spongy urethra inside, and with two additional erectile masses, the **corpora cavernosa.** The glans is richly supplied with sensory endings that reflexly stimulate ejaculation, and it is sheathed by a fold of skin, the **prepuce,** except during erection. The penis of monotremes is reptilian in structure and location.

Both male and female mammalian embryos develop a **genital tubercle** between the two evaginating scrotal sacs or labia majora (**genital swellings,** Fig. 14-30). In genetic males the tubercle becomes grooved, then tubular, and elongates to form the

penis. In females no tube usually develops and the tubercle becomes the **clitoris.** The clitoris usually remains embedded in the floor of the urinogenital sinus or vagina where, like the male penis, it is erectile. In many female rodents, however, it becomes a penis-like urinary papilla (Fig. 14-42). In female hyenas the urinogenital sinus is enclosed within it and it looks exactly like a penis. Therefore, it is practically impossible to tell a female hyena from a male. Perhaps this is what hyenas are laughing about. Copulation takes place through this organ and young are delivered through it (since it contains the urinogenital sinus). The os penis and os clitoris have been mentioned in Chapter 6.

■ FEMALE GENITAL DUCTS

The typical female tract consists of a pair of muscular tubes, or **gonoducts,** extending from anterior oviducal funnels, or **ostia,** to

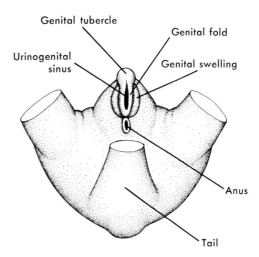

Fig. 14-30. External genitalia of sexually indifferent stage of a 29-mm human embryo (about 12 weeks of age). The cloaca has already been divided into urinogenital sinus and rectum. The genital tubercle becomes penis or clitoris, the genital swellings become scrotal sacs or labia majora, and the urinogenital sinus becomes the urethra in a male and is partitioned in a female to form the vagina and urethra.

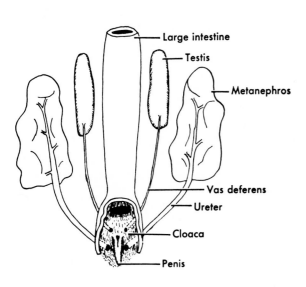

ALLIGATOR

Fig. 14-29. Urinogenital system of a sexually immature male alligator. The dorsal wall of the cloaca has been removed.

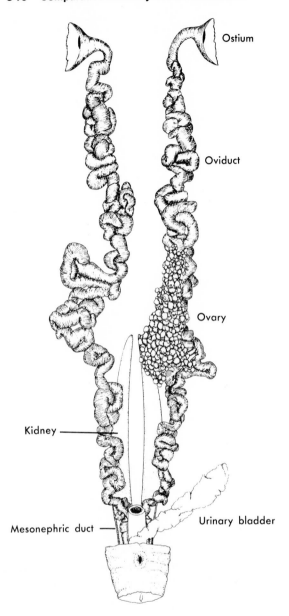

Fig. 14-31. Urinogenital system of female necturus, ventral view.

functions, such as coating the eggs with protective and nutrient substances. In animals that bear living young, parts of the ducts become **uteri** that house the young and, in euviviparous species, provide all necessary substances for the developing young prior to birth. The terminal segment of the ducts may be modified to receive the male intromittent organ and to expel the eggs or the young. When fertilization is internal, sperm penetrate the eggs in the upper reaches of the ducts. The eggs are propelled along the tract by cilia or smooth muscle.

In elasmobranchs and amphibians, ostia are derived from the union of one to three pronephric nephrostomes, and muellerian ducts arise by longitudinal splitting of the pronephric ducts. In most other vertebrates each muellerian duct arises as a longitudinal groove in the coelomic epithelium paralleling the pronephric duct.* Except at the ostium, the groove subsequently closes over to form a tube, which achieves an opening into the cloaca. The ducts eventually develop a dorsal mesentery.

Although the embryonic muellerian ducts do not fully differentiate in males, they sometimes remain as prominent structures. A complete, although rudimentary, female tract develops in some male amphibians (Fig. 14-18). After removal of the testicular hormones by orchidectomy, the rudimentary muellerian duct develops into a functional female tract consisting of oviduct and uterus. In male dogfish sharks the sperm sac at the base of each mesonephric duct is a remnant of the caudal end of the muellerian duct, and a clearly discernible vestige encircles the anterior end of the liver and ends in the falciform ligament. In male mammals, remnants include the **appendix testis,** a remnant of the anterior part of the muellerian duct, and the **prostatic sinus,** or **vagina masculina,** an unpaired sac at the junction

---

*Because of their embryonic location, muellerian ducts are sometimes called paramesonephric ducts.

the cloaca (Fig. 14-31). These tubes arise during embryonic life from a pair of longitudinal **muellerian ducts.** Muellerian ducts develop in males as well as in females but remain rudimentary or later disappear in males. Eggs are transported in the adult ducts, and certain segments perform special

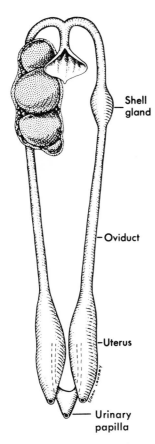

Fig. 14-32. Reproductive system of female *Squalus,* ventral view. The left ovary has been removed.

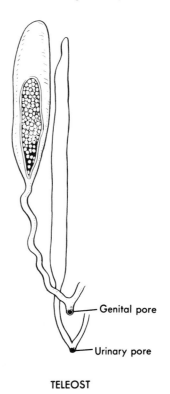

TELEOST

Fig. 14-33. Female reproductive system of a teleost. Ova are shed into the ovarian cavity.

of the spermatic ducts representing the fused caudal ends of the muellerian ducts.

☐ Fishes

In female elasmobranchs the muellerian ducts give rise to oviducts with shell (nidimental) glands and to paired uteri that open to the cloaca (Fig. 14-32). The cephalic half of the shell gland secretes albumen, and the caudal half secretes the shell. The two embryonic ostia unite to form a single adult ostium in the falciform ligament, a condition not typical of vertebrates.

The gonoducts of lungfishes resemble those of urodeles (Fig. 14-31), but the ducts of most teleosts and some ganoids are ex-

ceptional in that they are continuous with the cavity of the ovary (Fig. 14-33). Whether teleost gonoducts are modified muellerian ducts has not been settled. The two gonoducts of bony fishes frequently empty into a papilla-like **ovipositor.** A few fishes have a single gonoduct leading from a single ovary. Cyclostomes do not develop muellerian ducts. The eggs leave the coelom via genital pores.

☐ Amphibians

Muellerian ducts in amphibians give rise to oviducts that are long and convoluted (Fig. 14-31). The caudal portions may enlarge to form **ovisacs** where eggs accumulate prior to spawning. In ovoviviparous urodeles the ovisac serves as a uterus. The oviducal lining in amphibians is richly supplied with glands that secrete several jelly envelopes around each egg as it moves down the tube. The paired ostia are far forward.

Reptiles, birds, and monotremes

Crocodilians, some lizards, and nearly all birds have one gonoduct. The other is not well developed. The ducts are coiled (Figs. 14-34 to 14-36) and lined with glands that add albumen to the egg, except in snakes and lizards, whose eggs have none. In birds the albumen-secreting segment of the duct is called the **magnum** and the thick-walled shell gland is called (inappropriately) the **uterus** (Fig. 14-36). The short muscular terminal segment, called the **vagina**, secretes mucus that seals the pores of the shell to water vapor but not to oxygen (thus retarding moisture loss from the egg after it has been laid). The vagina then expels the egg. In some snakes and lizards special vaginal crypts called **spermatheca** store sperm over winter.

The female tract of egg-laying monotremes (Fig. 14-37) is reptilian. The caudal end secretes a shell onto the egg before it passes into the urinogenital sinus, which is part of the cloaca.

Placental mammals

The muellerian ducts of placental mammals give rise to **oviducts, uteri,** and **vaginas.** The embryonic muellerian ducts typically unite at their caudal ends. As a result, the adult female tract is paired anteriorly and unpaired posteriorly, terminating as an unpaired vagina. The oviducts, or **fallopian tubes** as they are more often called, are relatively short, small in diameter, convoluted, and lined with cilia. They commence at an oviducal funnel, or **ostium**, bordered by a delicate membranous fringe, the **fimbria of the oviduct.**

UTERUS

In many marsupials there is no fusion of the embryonic muellerian ducts. Therefore

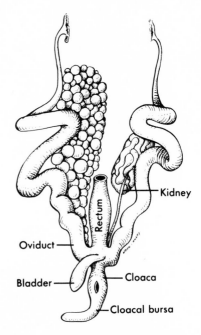

Fig. 14-34. Urinogenital system of female aquatic turtle, *Trionyx euphraticus*, ventral view. The left ovary has been removed. (Courtesy of Mohamad S. Salih, University of Baghdad.)

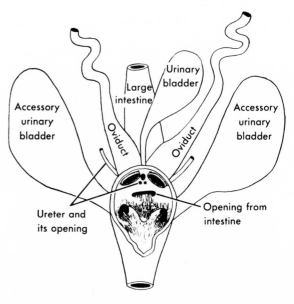

Fig. 14-35. Cloaca of a female terrestrial turtle, ventral view.

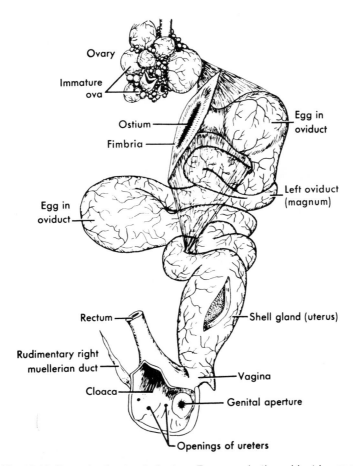

Fig. 14-36. Reproductive tract of a hen. Two eggs in the oviduct is unusual.

the female tract is double all the way to the urinogenital sinus (Fig. 14-38). Marsupials are therefore said to have a **duplex uterus,** and they have paired vaginas.

In the remaining placental mammals there are varying degrees of fusion of the caudal ends of the muellerian ducts, which may result in two **uterine horns,** a **uterine body,** and a single **vagina** (Fig. 14-39, rabbit). When there are two complete lumens within the body of the uterus, it is said to be **bipartite** (Fig. 14-40, hamster). When there is a single lumen within the body and there are two horns, the uterus is said to be **bicornuate** (Fig. 14-40, ungulates). There are species with uteri intermediate between the bipartite and bicornuate condition, and

one cannot tell, without opening the uterine body, whether it is bipartite or not. When there are uterine horns, the blastocysts implant in the horns. In some mammals one horn is much larger, and the blastocysts implant in that horn—the right in impala—even though both ovaries produce eggs.

In apes, monkeys, man, some bats, and armadillos there are no uterine horns, and the oviducts open directly into the body of the **simplex** uterus (Fig. 14-39, monkey). Except in ectopic pregnancies—pregnancies in which blastocysts implant in abnormal locations, such as the oviduct (tubal pregnancies) or coelom (abdominal pregnancies)—the usually single fetus or the twins, triplets, quadruplets, or quintuplets

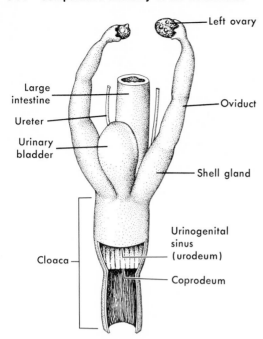

Fig. 14-37. Genital tract and cloaca of female monotreme, ventral view. The right ovary is usually smaller than the left. The cephalic half of the cloaca is divided by a partition into urinogenital sinus (receiving oviducts and ureters) and coprodeum.

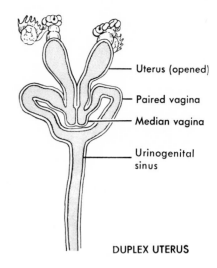

Fig. 14-38. Internal passageways of the female reproductive tract of an opossum. Compare with the external view in Fig. 14-39, marsupial.

all implant in the body of the simplex uterus.

The body of all uteri narrows to form a **cervix** (neck), the lower end of which projects into the vagina as the **lips** of the cervix. The lips surround the opening (**os uteri**) leading from uterus into vagina. The cervix must dilate under the influence of hormones for the young to be delivered.

After mating, sperm pass through the os uteri en route to the upper part of the oviducts where the sperm penetrate the egg. The uterine lining (**endometrium**) becomes highly vascular under the stimulus of hormones prior to implantation of a blastocyst. The thick, muscular layer of the uterine wall (**myometrium**) assists in ejection of the young at birth, provided it, too, has been hormonally prepared for this action.

VAGINA

Typically, the vagina is the fused terminal portion of the muellerian ducts, and it opens into the urinogenital sinus (Fig. 14-40, ungulates). In many rodents and primates, however, the vagina extends almost to the exterior (Fig. 14-39, monkey). The vaginal lining in such orders is cornified for reception of the penis.

The vagina of marsupials is unusual (Fig. 14-38). Just beyond the uteri the two muellerian ducts meet to form a **median vagina,** which may or may not be paired internally. Beyond the median vagina the two ducts continue as **paired (lateral) vaginas.** The pouchlike median vagina projects caudad and lies against the urinogenital sinus, separated by a septum. At birth the fetus is usually forced through the septum directly into the urinogenital sinus. The new passageway thus established may remain throughout life, which results in a **pseudovagina,** although it closes in opossums. As an adaptation to dual vaginas, the penis of male marsupials is forked at the tip. One tip

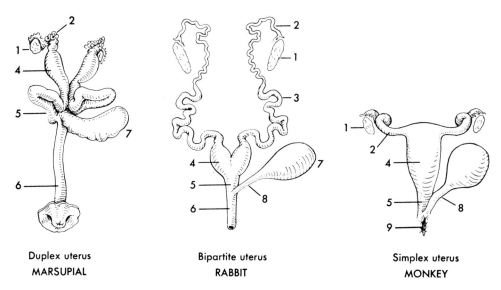

Fig. 14-39. Reproductive tracts of three female mammals. **1,** Ovary; **2,** oviduct; **3,** horn of uterus; **4,** body of uterus; **5,** vagina; **6,** urinogenital sinus; **7,** urinary bladder; **8,** urethra; **9,** vestibule of primate. In the primate (rhesus monkey) the urethra opens into the shallow vestibule just anterior to the opening of the vagina. The marsupial (redrawn from McCrady[8]) is an opossum, shown also in Fig. 14-38.

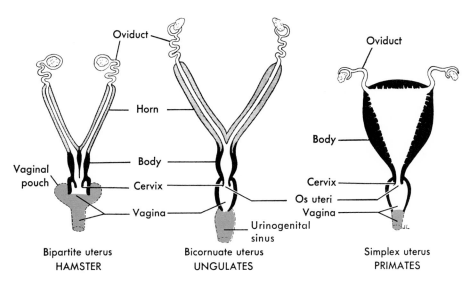

Fig. 14-40. Uterine types among mammals. Blackened regions represent fused caudal ends of the muellerian ducts; red represents the cloaca or a derivative thereof. Note the two lumens in the body of the bipartite uterus.

enters each vagina, where semen is discharged.

□ Entrance of ova into oviduct

After seeing the very large size of a shark's egg, laboratory students usually inquire how such a large egg can get into the ostium and down the relatively small oviduct. We know the answer with reference to the equally large egg of the chick or monotreme. Under the influence of hormones at the time of ovulation, the fringe (**fimbria**) of the oviducal funnel waves gently in an undulating movement. When it comes in contact with an egg, whether still in the ovary or separated from it, the fimbria clasps the egg, delicately at first and then more firmly, until the egg is engulfed by the funnel. At this time, the egg is a shapeless mass of flowing yolk (like the yolk in a fresh chicken egg) contained in a nonrigid membrane. Muscular contraction of the funnel squirts the shapeless mass into the oviduct. Thereupon, peristalsis of the wall of the oviduct moves the egg caudad. Cilia play a relatively unimportant role. In the case of the tiny eggs of mammals, however, the cilia are more important, although the fimbria also plays a role. In mammals the ovary is partially surrounded by the fimbria at all times, and this increases the probability that the egg will enter the oviduct. In mammals with an ovarian bursa the egg can go nowhere else.

■ THE CLOACA

The cloaca comes into focus whenever the digestive, urinary, or genital tracts are discussed, since this chamber is the termination of all three tracts in most vertebrates. A few adult vertebrates lack a cloaca. It becomes shallow or nonexistent in adult lampreys, chimaeras, ray-finned fishes, and mammals above monotremes. With these exceptions, a cloaca is present throughout the vertebrate series from hagfishes to monotremes and appears in its primitive condition even in the embryo of man. It opens to the exterior via the vent.

The embryonic cloaca is an enlarged terminal segment of the hindgut, which acquires an opening to the proctodeum (Fig. 1-1) when the cloacal membrane, separating the hindgut from the proctodeum, ruptures. The proctodeum usually makes some contribution to the adult cloaca, and in amphibians most of the cloaca is proctodeal in origin. The cephalic part of the adult cloaca is therefore lined with endoderm and the caudal part with ectoderm.

In many vertebrates a partition limited to the cephalic end of the cloaca, and called **urorectal fold,** separates that part into two chambers, **coprodeum** and **urodeum** (Fig. 14-37). The terminal portion of the cloaca remains undivided.* In such cases, and they are numerous among elasmobranchs, reptiles, birds, and monotremes, the large intestine opens into the coprodeum, whereas the urinary and genital ducts open into the urodeum. In placental mammals further partitioning of the cloaca takes place.

□ Vertebrates below placental mammals

The cloaca of fishes and amphibians receives the large intestine and the mesonephric ducts, and the female cloaca receives also the gonoducts. A urinary bladder opens into the ventral wall of the cloaca in amphibians (Fig. 14-31). The cloaca opens to the exterior via the vent.

The cloacas of reptiles, birds, and monotremes receive the same structures as in amphibians—large intestine, mesonephric ducts (in males only), gonoducts (in females), and urinary bladder, unless absent (Figs. 14-34 to 14-37). In addition, the ureters of reptiles, birds, and monotremes open into the cloaca, except in those few male reptiles in which the ureter retains its embryonic connection with the mesonephric duct (Fig. 14-24). The penis or clitoris, when

---

*This section is often called proctodeum (see Fig. 14-25, for example), but it is not necessarily completely homologous with the ectodermal invagination called proctodeum in the vertebrate embryo illustrated in Fig. 1-1.

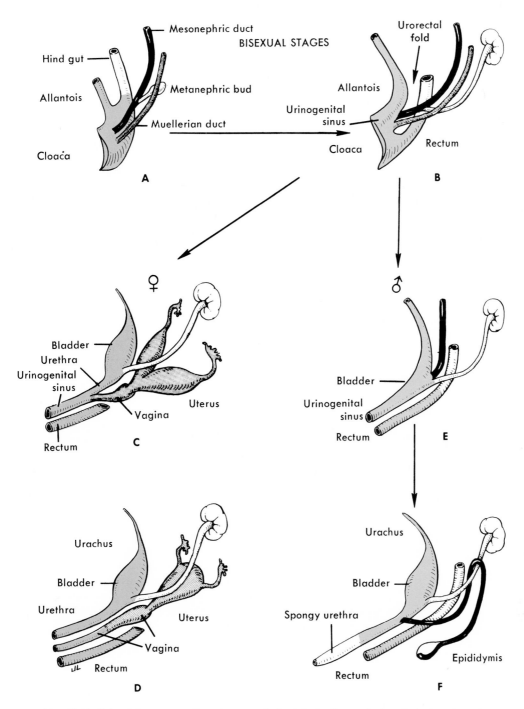

Fig. 14-41. Fate of the mammalian cloaca and allantois (red), muellerian ducts (gray), and mesonephric duct (black). **A** and **B,** Bisexual stages. Only the left muellerian and mesonephric ducts are shown. In **B** the cloaca is becoming subdivided by the urorectal fold into a urinogenital sinus ventrally and a rectum dorsally. **C,** Typical adult female mammal. **D,** Female primate, a modification of the condition shown in **C.** In **C** and **D,** the contributions of both the left and right muellerian ducts are shown. **E,** Developing male, showing reorientation of mesonephric and metanephric ducts. **F,** Adult male.

present, is embedded in the cloacal floor. A lymphoid pouch, the **bursa of Fabricius,** opens into the cloaca of young birds.

□ Fate of the cloaca in placental mammals

We have seen how, in monotremes, a urorectal fold divides the cephalic end of the cloaca into a urodeum and coprodeum (Fig. 14-37). In placental mammals the urorectal fold grows farther and farther caudad during embryonic development until it reaches the cloacal membrane separating the cloaca from the exterior. By this process the cloaca becomes completely divided into a **rectum** dorsally and a **urinogenital sinus** ventrally (Figs. 14-30 and 14-41, *B, C,* and *E*). Rupture of the cloacal membrane at two points provides an **anus** and a **urinogenital aperture** to the exterior. The early embryonic urinogenital sinus (Fig. 14-41, *B*) receives the mesonephric ducts, muellerian ducts (which are initially present in

both sexes), and the future urinary bladder (allantois), just like the urodeum of monotremes.

As development progresses in males, the muellerian ducts disappear and the urino-

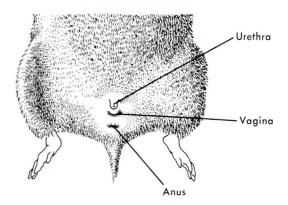

Fig. 14-42. Perineal region of a female hamster. The urethra opens at the tip of a penislike urinary papilla.

Table 14-1. Some homologous urinogenital structures in male and female mammals

Indifferent structure	Mature male	Mature female
Mesonephric duct	Ductus deferens (vas deferens) Epididymis	Ductus deferens femininus* (Gartner's duct)
Mesonephric tubules	Vasa efferentia Appendix of epididymis* Paradidymis*	Vasa efferentia ovarii* Epoophoron* Paroophoron*
Muellerian duct	Appendix testis*	Oviduct Uterus
	Vagina masculina* (prostatic sinus)	Vagina, *cephalic to urinogenital sinus*
Genital ridge	Testis Rete testis	Ovary Rete ovarii*
Gubernaculum	Gubernaculum	Ovarian ligament Round ligament of uterus
Genital swellings	Scrotal sacs	Labia majora
Genital tubercle	Penis	Clitoris
Genital folds	Contribute to penis	Labia minora
Urinogenital sinus	Urethra, *prostatic and membranous portions*	Urethra Urinogenital sinus Lower vagina in rodents and primates

*Vestigial.

genital sinus elongates (compare Fig. 14-41, *B* and *E*). The urinogenital sinus becomes continuous with the spongy urethra that has developed independently in the penis (compare Fig. 14-41, *E* and *F*). The urinogenital sinus now consists of the prostatic and membranous urethra (Fig. 14-26). The ureters become reoriented to open into the bladder, whereas the mesonephric ducts (now spermatic ducts) continue to empty into the urinogenital sinus (Figs. 14-41, *F*, and 14-26).

As development progresses in females, the mesonephric ducts disappear, and the muellerian ducts unite at their caudal ends to form the body of the uterus and the vagina (Fig. 14-41, *C*). The part of the urinogenital sinus between bladder and entrance of the vagina is the urethra (Fig. 14-41, *C*). As a result of these changes, most adult female mammals have two caudal openings to the exterior, a urinogenital aperture and an anus.

In most female primates (including man) and in some rodents, an additional partition forms in the cloaca—this one in the urinogenital sinus. It separates the urinogenital sinus into a urethra and a vagina (Fig. 14-41, *D*). As a result, the embryonic cloaca in these species becomes subdivided into three passages: urethra, vagina, and rectum. Each passageway leads to the exterior via its own aperture (Fig. 14-42). In this regard the females of these species have evolved farther than the males. The vagina in these females has a dual origin. The cephalic part is derived from the fused muellerian ducts, and the terminal part is cloacal.*

Table 14-1 summarizes the fate in adult males and females of the chief sexually indifferent urinogenital structures found in all mammalian embryos.

---

*In higher primates the urethra and vagina actually open into a shallow vestibule (Fig. 14-39) derived from the very distal end of the urinogenital sinus.

□ Chapter summary

1. Vertebrate kidneys arise from a ribbon of intermediate mesoderm extending the length of the trunk. A wave of differentiation sweeps along the ribbon and gives rise to convoluted tubules. The tubules are typically associated with glomeruli and open into a longitudinal duct that usually terminates in the cloaca.

2. Glomeruli are capillary tufts that filter water and other substances from the blood. Glomeruli sometimes dangle into the coelom (external glomeruli), but usually they are surrounded by Bowman's capsule (internal glomeruli). Their primitive function may have been water excretion.

3. Kidney tubules are tiny convoluted ductules that collect glomerular filtrate, selectively reabsorb some substances, add others, and conduct the final filtrate to the longitudinal duct.

4. The first tubules are segmental and often have open nephrostomes in lower vertebrates. These first tubules are temporary and constitute the pronephros.

5. The mesonephros organizes behind the pronephros. It is the functional kidney of adult fishes, amphibians, and embryonic amniotes. The pronephric duct usually persists to serve the mesonephros (opisthonephros), but accessory mesonephric ducts may develop.

6. The mesonephric kidney is not metameric, nephrostomes are generally lacking, and a Bowman's capsule surrounds each glomerulus. Glomerulus and capsule constitute a renal corpuscle.

7. Tubules near the anterior end of the mesonephros invade the mesorchium, grow into the testes, and serve as vasa efferentia. In male anamniotes the cephalic end of the mesonephros may be preempted for sperm transport and is then called sexual (epididymal) kidney.

8. The metanephros is the adult amniote kidney. It organizes from the caudal end of the nephrogenic mesoderm, which is displaced craniad and laterad. Its duct (ureter) arises as a bud off the mesonephric duct and usually empties into the cloaca (reptiles, birds) or urinary bladder (mammals).

9. The mammalian kidney is characterized by a loop of Henle between proximal and distal convolutions of the tubule, lack of an afferent venous supply from a renal portal system except in monotremes, and a discrete cortex.

10. The archinephros is a hypothetical, completely metameric primitive kidney.

11. Animals that must conserve water excrete certain salts extrarenally, especially via gills in fishes, rectal glands in elasmobranchs, and nasal glands in marine reptiles, shore birds, and terrestrial lizards and snakes in arid habitats.

12. Most vertebrates have urinary bladders. In fishes they are usually enlargements of the posterior ends of the mesonephric ducts (tubal bladders). In amphibians and amniotes they evaginate from the ventral cloacal wall.

13. Gonads arise from gonadal ridges. Unpaired gonads result when the paired ridges fuse

or one remains undifferentiated. Gonads tend to be displaced caudad in mammals. The ovaries of teleosts and amphibians are saccular. Those of reptiles, birds, and monotremes are lacunate. The mammalian ovary is compact. Some mammalian ovaries are enclosed within an ovarian bursa.

14. Cyclostomes lack reproductive ducts, and the eggs and sperm escape the coelom via genital pores. In other vertebrates the mesonephric duct usually conveys sperm in males, and the muellerian duct conveys eggs and the products of conception in females. The caudal ends of the muellerian ducts unit in female mammals above marsupials to form an unpaired uterine body and a vagina.

15. Uteri are duplex, bipartite, bicornuate, or simplex, depending on the extent of fusion of the muellerian ducts.

16. Most vertebrates below marsupials have a cloaca. In lampreys, chimaeras, and most ray-finned fishes, the cloaca becomes shallow or nonexistent in adults.

17. In reptiles, birds, and monotremes, the cephalic end of the cloaca is partitioned by a urorectal fold into a urodeum that receives the urinary and genital tracts and a coprodeum that receives the large intestine. The caudal end of the cloaca remains unpartitioned.

18. Above monotremes the cloaca is completely partitioned into urinogenital sinus and rectum, and the two empty independently to the exterior.

19. In a few female mammals the urinogenital sinus is subdivided into two passageways so that there are three passageways to the exterior—urethra, vagina, and rectum.

20. Male intromittent organs of fishes are chiefly modifications of the pelvic or anal fins. Turtles, crocodilians, some birds, and monotremes have an unpaired penis in the cloacal floor containing a corpus spongiosum. Mammals above monotremes have an external penis with a corpus spongiosum and two corpora cavernosa. Snakes and lizards have hemipenes.

21. Female amniotes develop a clitoris, which is a homologue of the penis.

22. Early gonads are indistinguishable as to sex, and duct systems for both sexes are present. Table 14-1 lists the chief sexually indifferent structures of mammalian embryos and the fate of these in adult males and females.

LITERATURE CITED AND SELECTED READINGS

1. Altschule, M. D.: The change(s) in the mesonephric tubules of human embryos ten to twelve weeks old, Anatomical Record **46**:81, 1930.
2. Baker, C. J., and Taylor, W. W.: The urogenital system of the male *Ambystoma*, Journal of the Tennessee Academy of Sciences **39**:1, 1964.
3. Brodal, A., and Fänge, R., editors: The biology of *Myxine*, Oslo, 1963, Norway Universitetsforlaget.
4. Fox, H.: The amphibian pronephros, Quarterly Review of Biology **38**:1, 1963.
5. Goodrich, E. S.: Studies on the structure and development of vertebrates, London, 1930, The Macmillan Co., Ltd. (Reprinted by Dover Publications, Inc., New York, 1958.)
6. Hirschfield, M. F., and Tinkle, D. W.: Natural selection and evolution of reproductive effort, Proceedings of the National Academy of Sciences U.S.A. **72**:2227, 1975.
7. Kent, G. C., Jr.: Reproductive systems of vertebrates. In Encyclopaedia Britannica, ed. 15, Chicago, 1974, Encyclopaedia Britannica, Inc., **15**:707.
8. McCrady, E., Jr.: The development and fate of the urinogenital sinus in the opossum, *Didelphys virginiana*, Journal of Morphology **66**:131, 1940.
9. Moffat, D. B.: The mammalian kidney, New York, 1975, Cambridge University Press.
10. Mossman, H. W., and Duke, K. L.: Comparative morphology of the mammalian ovary, Madison, Wis., 1973, The University of Wisconsin Press.
11. Oguri, M.: Rectal glands of marine and fresh-water sharks: comparative histology, Science **144**:1151, 1964.
12. Pang, P. K. T., Griffith, R. W., and Atz, J. W.: Osmoregulation in elasmobranchs, American Zoologist **17**:365, 1977.
13. Peaker, M., and Linzell, J. L.: Salt glands in birds and reptiles, Monographs of the Physiological Society, New York, 1975, Cambridge University Press.
14. Sadleir, R. M. F. S.: The reproduction of vertebrates, New York, 1973, Academic Press, Inc.

15

Nervous system

In this chapter we will examine the nervous system and its parts. We will see how the parts are assembled and what they do to assure survival. Much of our attention will be focused on neurons, the living cells that perform the actual conducting and secretory activities of the system.

The vertebrate nervous system plays three basic roles. It acquaints the organism with its external environment and stimulates the organism to orient itself favorably in that environment; it participates in regulation of the internal environment; and it serves as a storage site for information. These functions are accomplished by the nerves, spinal cord, and brain in association with **receptors** (sense organs) and **effectors** (chiefly muscles and glands).

The organism exists in an external environment that is sometimes friendly, sometimes inimical, and seldom neutral. An environment is friendly if it contains food for nourishment, a mate for the propagation of the species, and shelter from enemies. An environment is unfriendly if it leads to the weakening of the organism or of the species.

The organism must constantly monitor the external environment in order that it may go deeper into a friendly area or withdraw from an unfriendly one. The information is supplied by afferent (sensory) nerves commencing in sense organs. The response (body movement) is initiated by nerve impulses over efferent (motor) nerves that stimulate the skeletal muscles of the body and thus cause the fish to swim or the tetrapod to crawl, run, or fly deeper into, or out of, an area. Information from the external environment is also employed in the regulation of internal secretions, such as seasonal release of reproductive hormones (Fig. 17-4).

The organism also has an internal environment that must be continually scanned and controlled. Afferent nerves from visceral receptors carry information in the form of nerve impulses to the central nervous system; efferent nerves carry impulses from the center to visceral effectors (chiefly smooth and cardiac muscles and glands).

Memory (information storage and recall) is a function of the nervous system. Without information storage no animal could modify its behavior in accordance with experience, and every situation would be faced as if it were the first time. In other words, there could be no *conditioned* responses. As experiences multiply, information accumulates and the penalty of past errors and the rewards of successes modify behavior accordingly. The brain seems to be the chief site of information storage.

The nervous system is subdivided for convenience into central and peripheral nervous systems. The **central nervous system** consists of the brain and spinal cord. The **peripheral nervous system** consists of cranial, spinal, and autonomic nerves, their branches, and certain autonomic ganglia and plexuses. The autonomic components innervate visceral effectors.

■ THE NEURON

To understand the anatomy of the nervous system, one must be acquainted with the neuron—the living nerve cell. The neuron is to the nervous system what a muscle cell is to the muscle system: it performs the

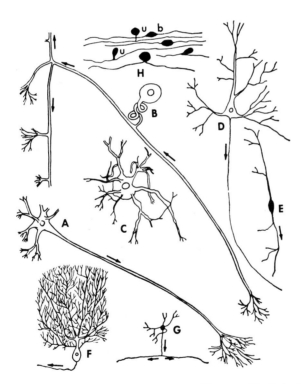

Fig. 15-1. Several morphological varieties of neurons. **A,** Motor cell body in the spinal cord; the fiber extends into the ventral root of a spinal nerve; **B,** dorsal root ganglion cell (sensory); the fiber terminates at the left in the spinal cord; **C,** sympathetic ganglion cell; **D** and **E,** pyramidal and horizontal cells from the cerebral cortex; **F** and **G,** Purkinje and granular cells from the cerebellum; **H,** a group of embryonic dorsal root ganglion cells in transition from bipolar, **b,** to unipolar, **u. A, C, D, F,** and **G,** Multipolar neurons; **E,** bipolar; **B,** unipolar. Arrows indicate direction of impulse.

specific function of the system. The neuron, rather than the nerve, transmits the nerve impulse. Neurons exhibit many shapes (Fig. 15-1), but all have a **cell body** and one or more **processes.** One process, distinguished cytologically by the absence of Nissl material (Fig. 15-2), is the **axon,** or **nerve fiber.** It transmits nerve impulses to a synapse or to an effector. Some nerve fibers extend short or long distances up or down the brain and spinal cord aggregated into functional groups called **fiber tracts** (Fig. 15-3, *A*). Nerve fibers occur also in nerves. In fact, a **nerve** is made up of bundles of nerve fibers outside the central nervous system, wrapped in a connective tissue sheath (**epineurium**) and supplied by blood vessels (**vasa nervorum**). Other processes of the neuron exhibit Nissl material and seldom extend far from the cell body. These are **dendrites.** To observe how the neuron fits into the peripheral nervous system, we will examine a sensory nerve, a motor nerve, and two mixed nerves.

A typical sensory nerve is diagrammed in Fig. 15-3, *A*. It commences in a sense organ (in this instance, the membranous labyrinth) and terminates in the brain. Like all nerves, it is made of nerve fibers. The cell bodies of sensory nerve fibers, with few exceptions, are found in a sensory ganglion on the pathway of the nerve. **A ganglion** is a group of cell bodies outside the central nervous system. A *sensory ganglion* contains sensory cell bodies. In lower vertebrates some sensory cell bodies are scattered along the nerve.

A typical motor nerve is diagrammed in Fig. 15-3, *B*. The cell bodies of most motor neurons are inside the central nervous system in a **motor nucleus.** Neurologically speaking, a **nucleus** is a group of cell bodies within the brain or cord. Motor nuclei contain the cell bodies of motor nerve fibers. The motor fibers of cranial nerve XII terminate in striated muscle. There are almost no purely motor nerves in vertebrates, since most nerves supplying striated muscles

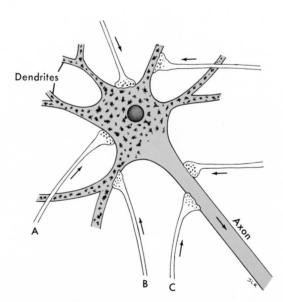

Fig. 15-2. Synaptic endings on a motor neuron. **A,** Synapse between axon terminal and a cell body; **B,** synapse between axon terminal and a dendrite; **C,** synapse between axon terminal and another axon. Nissl material is seen in cell body and dendrites.

have sensory fibers for proprioception from the muscle (Fig. 16-17).

Mixed nerves contain both sensory and motor fibers and are illustrated in Fig. 15-4. Their sensory cell bodies are in sensory ganglia, and their motor cell bodies are in motor nuclei. Most vertebrate nerves are mixed.

The site where a nerve impulse is transferred from one neuron to the next is a **synapse.** At the synapse the axon terminals of the first neuron lie in intimate contact with the cell membrane of the dendrites, cell body, or axon of the next neuron (Fig. 15-2). Nerve impulses are transmitted across the synapse by very short-lived secretions, chiefly amines, that are released from axon terminals whenever a nerve impulse arrives at the terminal. These amines are called **neurotransmitters.** Neurotransmitters are

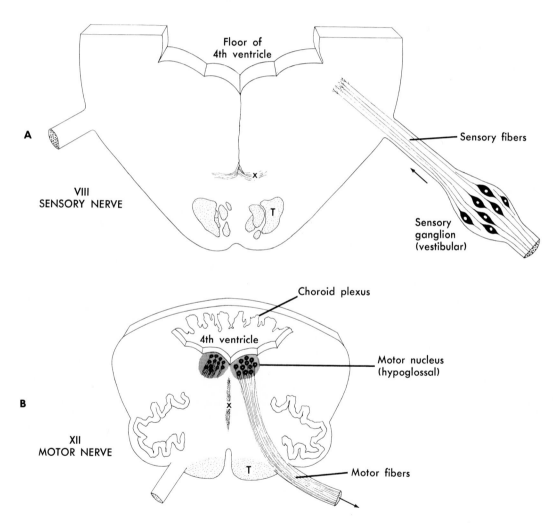

Fig. 15-3. Typical locations of cell bodies (black) of sensory and motor nerves. **A,** Sensory nerve with cell bodies in a sensory ganglion. **B,** Motor nerve (hypoglossal) with cell bodies in a motor nucleus in the brain. **T,** Descending fiber tract (corticospinal); **x,** decussating fibers. Arrows indicate direction of nerve impulses.

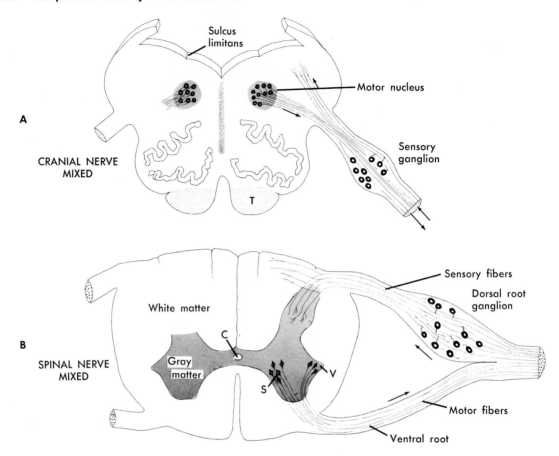

Fig. 15-4. Locations of cell bodies of mixed nerves. **A,** Mixed cranial nerve with sensory cell bodies in a sensory ganglion and motor cell bodies in a motor nucleus in the brain. Not all fiber components of this nerve are shown. **B,** Spinal nerve with sensory cell bodies in dorsal root ganglion and motor cell bodies in gray matter of cord. **C,** Central canal of cord; **S,** somatic motor nucleus in anterior horn of gray matter; **T,** descending fiber tract (corticospinal); **V,** visceral motor nucleus in lateral (visceral) horn of gray matter. Arrows indicate direction of nerve impulses.

also released from axon terminals in contact with effectors, causing the effector (muscle, gland, pigment cell) to respond.

Some neurons (**neurosecretory neurons**) with cell bodies in the central nervous system secrete small polypeptides from their axon terminals. These polypeptides, called **neurosecretions,** are hormones. Instead of terminating in synapses or at effectors, most neurosecretory fibers terminate at sinusoidal vascular channels that receive the neurosecretion (Fig. 17-1).

■ **GROWTH AND DIFFERENTIATION OF THE NERVOUS SYSTEM**

To achieve insight into the architecture of the adult nervous system, it is necessary to know how the nervous system develops.

□ **Neural tube**

An early embryonic neural tube is illustrated in Fig. 15-5. The cephalic end of the tube is the embryonic brain. The rest is future spinal cord. A typical cross section of the neural tube at this time exhibits three

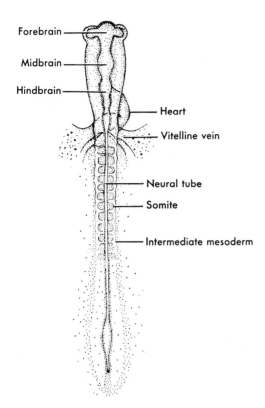

Fig. 15-5. Chick embryo of 33 hours' incubation. The optic vesicles are beginning to evaginate from the forebrain.

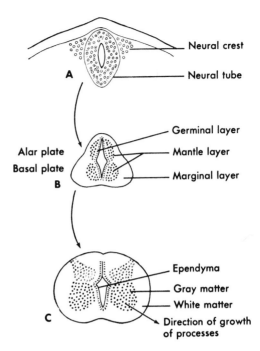

Fig. 15-6. Embryogenesis of the spinal cord. The alar plate contains association neuroblasts with which incoming (sensory) nerve fibers will synapse. The basal plate contains motor neuroblasts, some of the axons of which are growing in the direction of the arrow in **C** to become part of a nerve.

zones (Fig. 15-6, *B*): a **germinal layer** of actively mitotic cells, a **mantle layer** of cells proliferated from the germinal layer, and a **marginal layer.** Most of the mantle layer cells are **neuroblasts** (future neurons), which sprout axons and dendrites to become neurons. As the neuroblasts differentiate, their axons grow into and add to the marginal layer, which therefore consists of nerve fibers. Because many axons become surrounded by a fatty myelin sheath, the marginal layer looks white when fresh and is called **white matter.** The protoplasm of the cell bodies of the mantle layer causes this zone to look gray; hence the name **gray matter.**

Some of the nerve fibers that grow into the marginal layer turn upward or downward in the cord or brain to synapse with neurons elsewhere in the central nervous system. The earliest of these fibers extend only one segment, a primitive condition. The later and longer ones become aggregated in long ascending and descending fiber tracts, each composed of functionally related fibers.

The embryonic cord, hindbrain, and midbrain consist of an **alar** and **basal plate** located respectively above and below a sulcus limitans (Fig. 15-6, *B*, and 15-28). The alar plate receives incoming (sensory) impulses, whereas the basal plate becomes motor in function.

When all the nerve cells have been formed that will ever be formed in the cord or brain, the cells of the germinal layer cease to divide. Those that remain adjacent to the central canal become the **ependyma** (Fig. 15-6, *C*), a connective tissue lining of the central canal.

All cell bodies and processes within the central nervous system are supported by **neuroglia,** a special connective tissue chiefly of neural tube origin. The ependyma is one type of neuroglia and the only type found in cyclostomes. In higher vertebrates neuroglial cells become increasingly abundant and diversified.

☐ **Development of motor components of nerves**

Many of the axons that sprout from neuroblasts in the basal plate grow out of, and away from, the neural tube (Fig. 15-6, *C*) to make contact with striated muscles. These become motor fibers of the cranial and spinal nerves. Because these fibers sprout from neuroblasts within the central nervous system, their cell bodies are within the adult brain or cord. Notice the location of motor cell bodies in Fig. 15-3, *B*, and 15-4.

Some of the fibers that sprout from neuroblasts in the basal plate grow out of the neural tube to make contact with neuroblasts in autonomic ganglia. These are **preganglionic fibers** of the autonomic nervous system (Fig. 15-9). The neuroblasts in autonomic ganglia sprout **postganglionic fibers** that grow toward, and innervate, smooth muscles and glands. Thus all motor neurons have their cell bodies in the cord or brain with one exception—postganglionic neurons of the autonomic system have their cell bodies in autonomic ganglia.

Most neuroblasts of autonomic ganglia are migrants from neural crests (Fig. 15-6, *A*). A few, in amphibians at least, migrate outward from the basal plate of the neural tube.

☐ **Development of sensory components of nerves**

At the time the neural groove is closing to form a tube, a longitudinal ribbon of neurectoderm (ectoderm that forms nervous tissue) separates from the tube dorsolaterally on each side and quickly forms paired **neural crests** in each body segment (Figs. 4-8 and 15-6, *A*). Some of the cells of the neural crests sprout processes and become neuroblasts that give rise to sensory neurons of spinal and cranial nerves. In doing so they pass through a bipolar stage (Fig. 15-1, *H*) in which one process grows into the alar plate of the cord or brain and the other grows toward a sense organ. Thus there is established a neuronal connection between sense organ and cord or brain. Since each neural crest gives rise to a large number of sensory neurons, the result is a sensory ganglion on the nerve close to the central nervous system. Neurons whose cell bodies are in sensory ganglia are **first order sensory neurons.** They conduct an impulse from sense organ to the central nervous system and synapse inside the cord or brain with **second order sensory neurons** that conduct the impulse elsewhere (Fig. 15-9). Because most sensory cell bodies of first order sensory neurons arise from neural crests, we can make the following generalization—the cell bodies of sensory neurons of cranial and spinal nerves are typically in sensory ganglia on the pathway of the nerves.

There are three exceptions to the rule that the cell bodies of first order sensory neurons are in ganglia on the pathway of nerves—the cell bodies of olfactory nerve fibers are in the olfactory epithelium, those of optic nerve fibers are in the retina, and those of proprioceptive fibers in cranial nerves are in the brain.

The neuroblasts of olfactory nerve fibers develop from ectodermal cells in the olfactory epithelium. Their long processes grow into the nearest part of the brain, which is the olfactory bulb. Therefore, the ganglion of the olfactory nerve is in the olfactory epithelium instead of on the nerve somewhere along its pathway.

The neuroblasts that give rise to sensory fibers in the optic nerve are in the embryonic retina (optic cup, Fig. 16-4, *B*), and their long processes grow brainward along the optic stalk until they reach the optic chiasma. Therefore, the cell bodies of optic nerve fibers are in the retina and there is no ganglion on the nerve. Actually, the retina is part of the brain, since it arises as an evagination from the diencephalon and never separates from it. The term "optic nerve" is a misnomer.

The cell bodies of proprioceptive fibers in cranial nerves are not in ganglia. Neuroblasts within the embryonic midbrain sprout long processes that grow out to muscles via the cranial nerves. Therefore, the cell bodies for proprioceptive fibers in cranial nerves are in the midbrain. (The cell bodies of proprioceptive fibers of spinal nerves arise from neural crests and therefore are in dorsal root ganglia.)

Some of the sensory ganglia of cranial nerves do not arise from neural crests. Instead, a row of ectodermal thickenings, or placodes, on the side of the embryonic head above the pharynx (called **epibranchial placodes** in fishes) provide these neuroblasts. The precise contribution of these placodes to the sensory ganglia of the head varies, but they provide cell bodies in one or more of the sensory ganglia of nerves V, VII, VIII, IX, and X.

■ SPINAL CORD

The spinal cord occupies the vertebral canal and is packed in fat. It is immediately surrounded in most fishes by a connective tissue membrane, the **meninx primitiva**. In some teleosts and in amphibians, reptiles, and some birds, this primitive meninx later forms an outer fibrous **dura mater** and an inner vascular **leptomeninx**. In mammals and a few birds the leptomeninx differentiates into a weblike **arachnoid** membrane and a **pia mater**, the latter intimately applied to the cord. Thus mammalian cords and brains are surrounded by three meninges.

The spinal cord commences at the foramen magnum, but there is no landmark on the brain or cord delimiting the two. Instead, there is a gradual transition from cord to brain.

The adult cord extends to the caudal end of the vertebral column in vertebrates with abundant tail musculature. In other vertebrates the embryonic vertebral column elongates more rapidly than the spinal cord, with the result that at birth the cord is shorter than the column. In man the spinal cord terminates at the third lumbar vertebra. In frogs it ends anterior to the urostyle. In a few bony fishes the cord is actually shorter than the brain. It is only an inch or so in length in one fish that is several feet long.

When the cord is as long as the vertebral column, each spinal nerve passes directly laterad to an intervertebral foramen through which it emerges from the vertebral canal. If, however, the column subsequently elongates more than the cord, the spinal nerves must then pass caudad within the vertebral canal to reach their foramina. As a result, the more caudal spinal nerves form a bundle of parallel nerves, the **cauda equina**, within the vertebral canal (Fig. 15-7). Nonnervous elements (ependyma and meninges) of the cord continue farther caudad as a delicate strand, the **filum terminale.**

The spinal cord exhibits cervical and lumbar enlargements at the level of the anterior and posterior appendages. The enlargements result from the large number of cell bodies and fibers innervating the appendage. When one pair of appendages is particularly muscular, such as the hind limbs of massive dinosaurs, the enlargement of the cord is especially pronounced. Conversely, the spinal cord of turtles is very slender in the trunk because the thoracic and abdominal musculature is greatly reduced. In many fishes the cord exhibits a swelling, the urophysis, at its caudal end (Fig. 17-8).

The cord is flattened in cyclostomes but tends to be cylindrical or quadrilateral in higher vertebrates. In general, the neuro-

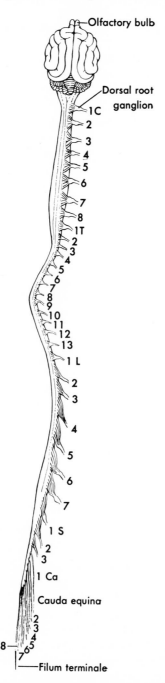

Fig. 15-7. Brain and spinal cord of a cat. The dura mater has been removed. **C,** Cervical spinal nerve; **T,** thoracic; **L,** lumbar; **S,** sacral; **Ca,** caudal spinal nerves. Note origin of spinal nerves by multiple rootlets, and enlargements of the cord in the cervical and lumbar regions.

coel, or cavity within the cord, is relatively large in lower forms and constricted in higher ones.

A cross section of a typical cord reveals the nuclei arranged in a definite pattern surrounding the central canal, where they comprise the gray matter (Fig. 15-8, high sacral). The nerve fibers occupy the periphery of the cord and constitute the white matter. The ascending and descending fibers are aggregated into fiber tracts that interconnect

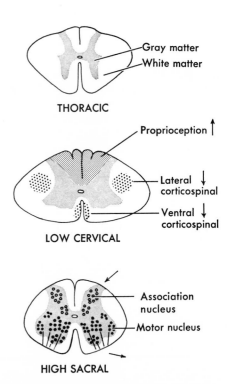

Fig. 15-8. Human spinal cord in cross section at three levels, showing a few fiber tracts (arrows) at the cervical level and a few nuclei at the sacral level. The cervical level is largest because it contains many cell bodies supplying the anterior limb, all fibers ascending to the brain from lower levels, and all fibers descending from the brain to lower levels. The corticospinal tracts carry voluntary motor impulses from the cerebral cortex. The motor horn in the thoracic region is small because there are no limb muscles to be supplied at this level. Association nuclei contain the cell bodies of second order sensory neurons.

one level of the cord with another or with the brain. Fibers for touch constitute one tract, those for voluntary motor control another, and so forth. The fiber tracts of the cord are relatively few and simple in cyclostomes. They increase in number and complexity in amphibians because of the addition of tetrapod limbs.

■ SPINAL NERVES
□ Roots and ganglia

Except in lampreys, spinal nerves arise from the cord by dorsal and ventral roots (Fig. 15-9, *B*). Each root is composed of a series of very short rootlets that unite close to the dorsal root ganglion (Fig. 15-7). The dorsal root exhibits a ganglion and is predominantly sensory. The ventral root is entirely motor. There is considerable evidence that in the earliest vertebrates (1) the dorsal and ventral roots did not unite but continued independently to their destinations, (2) the dorsal roots were mixed, (3) there were no dorsal root ganglia, and (4) primitive dorsal root ganglion cells were bipolar. These conclusions are based partly on study of the spinal nerves of lower chordates.

In an amphioxus only the dorsal root contains nerve fibers. These roots arise from the cord at the level of a myoseptum and pass into the myoseptum to be distributed to the skin (sensory) and viscera (visceral motor). The cell bodies for the visceral motor fibers are in the cord, and those for the sensory fibers are either in the cord or within the nerves scattered along much of their length. Therefore, there is no aggregation of sensory cell bodies at one location on a nerve, which means there are no sensory ganglia. Also, the sensory cell bodies remain bipolar throughout life.

The ventral roots of amphioxus do not contain nerve fibers and do not unite with the dorsal roots. They come off the cord between myosepta and are composed of bundles of very delicate, long extensions from the striated muscle fibers of the body wall.[2] These muscular filaments enter the cord through the ventral root; inside the cord they are stimulated by nerve fibers that synapse with them. Thus the somatic muscles actually "come to the cord" for their stimuli. The same condition exists in echinoderms.

In cyclostomes dorsal and ventral roots alternate and remain independent in lampreys but unite in hagfishes. Some of the cell bodies of sensory fibers are for the first time aggregated in ganglia on the dorsal root, and most of these remain bipolar. Other first order sensory cell bodies are still within the cord. Visceral motor fibers are in both roots, and the ventral root is entirely motor.

Above cyclostomes dorsal and ventral roots always unite. The dorsal root still contains numerous visceral motor fibers in many bony fishes, but in cartilaginous fishes and tetrapods most of these have been lost from that root. The cell bodies of first order sensory neurons are in dorsal root ganglia. They are bipolar in cartilaginous fishes; bipolar, intermediate, and unipolar in bony fishes; chiefly unipolar in amphibians; and almost entirely unipolar in amniotes (Fig. 15-1, *B*). The ventral root is motor (somatic and visceral), with cell bodies inside the cord.

□ Metamerism

A spinal nerve arises from each segment of the cord except near the end of the tail. These nerves are metamerically distributed to the body wall and tail. At the level where a fin or limb bud forms, they supply the appendage (Figs. 10-8 and 15-10). The segmental distribution of spinal nerves is best illustrated in fishes, since the metamerism of their body wall muscles is relatively undisturbed. Tadpoles have as many as forty pairs of spinal nerves and lose all but ten pairs when the tail is resorbed at metamorphosis.

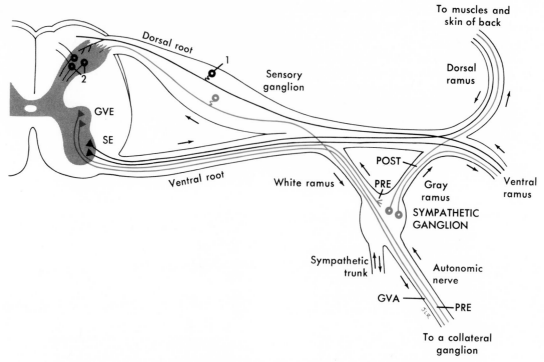

Fig. 15-9. Diagram of thoracic spinal nerve to show gray and white rami communicantes and visceral fibers (dark red). The preganglionic fiber shown in the autonomic nerve distal to the sympathetic ganglion did not synapse in the ganglion but will do so in a collateral ganglion such as the celiac (see splanchnic nerve, Fig. 15-30). **1**, Sensory cell body in dorsal root ganglion; **2**, second order sensory cell bodies in a sensory nucleus; **GVA**, general visceral afferent fiber; **GVE**, general visceral efferent nucleus in lateral horn of gray matter; **SE**, somatic efferent nucleus in ventral horn; **POST**, postganglionic fiber; **PRE**, preganglionic fiber. The dorsal ramus contains somatic sensory fibers that are not shown. Somatic fibers are black.

□ Rami and plexuses

Shortly after emerging from the vertebral canal, each typical spinal nerve divides into at least two branches (see 3 and 4 in Fig. 1-2). A **dorsal ramus** supplies the (epaxial) muscles and skin of the dorsum. A larger **ventral ramus** passes into the lateral body wall and supplies the (hypaxial) muscles and skin of the side and venter. In the thoracic and lumbar regions two additional branches, the **rami communicantes**, pass to a ganglion of the sympathetic trunk (Fig. 15-9, white ramus, gray ramus). They carry visceral fibers.

The ventral rami of successive spinal nerves often unite to form a plexus from which large nerve trunks arise. The chief plexuses are the brachial and pelvic (lumbosacral in amniotes), which supply nerves to the anterior and posterior appendages. These plexuses are relatively simple in anamniotes but become increasingly complicated in tetrapods (compare shark and mammal, Fig. 10-8). Autonomic plexuses occur on visceral pathways.

□ Occipitospinal nerves

In many fishes and amphibians one or more pairs of **occipitospinal nerves** arise between the vagal nerve and the first pair of

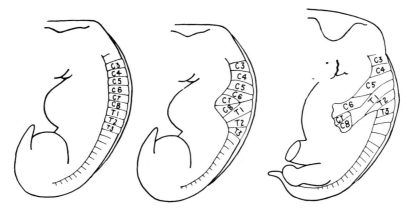

Fig. 15-10. Innervation of the skin of the forelimb by successive spinal nerves. **C**, Cervical, and **T**, thoracic somites and area of cutaneous distribution of their associated nerves.

spinal nerves. They supply the hypobranchial musculature, including the tongue when present, and usually lack sensory roots. The embryonic frog has an occipitospinal nerve immediately cephalad to the spinal nerve that supplies the tongue, but it becomes suppressed during later development. Cranial nerves XI and XII of amniotes lack sensory roots and appear to be derived in part from occipitospinal nerves.

☐ Fiber components of spinal nerves

The nerve fibers in a typical spinal nerve are of four functional varieties (Table 15-1). Three of the varieties are referred to as **general** fibers (GSA, GVA, and GVE) to differentiate them from **special** types found only in cranial nerves.

■ BRAIN

The anterior end of the embryonic neural tube in every vertebrate from fish to man exhibits three primary brain vesicles—forebrain, midbrain, and hindbrain (Fig. 15-5). The forebrain (prosencephalon) subsequently becomes subdivided into **telencephalon** and **diencephalon.** The midbrain (**mesencephalon**) develops without further

Table 15-1. Fiber components of typical spinal nerves

Components	Innervation
Sensory	
General somatic afferent fibers (GSA)	General cutaneous receptors (touch, pain, temperature, and pressure)
	Receptors on striated muscle, tendons, and bursas (proprioceptive)
General visceral afferent fibers (GVA)	Viscera, including general receptors in endoderm
Motor	
Somatic efferent fibers (SE)*	Myotomal muscle
General visceral efferent fibers (GVE)†	Smooth and cardiac muscle, and glands. Visceral fibers to skin are vasomotor, pilomotor (in mammals), or secretory (to skin glands), and they supply melanophores in lower vertebrates

*The fibers to myotomal muscle are designated simply as SE, rather than as GSE (general somatic efferent), because there are no special somatic efferent fibers.
†Autonomic fibers.

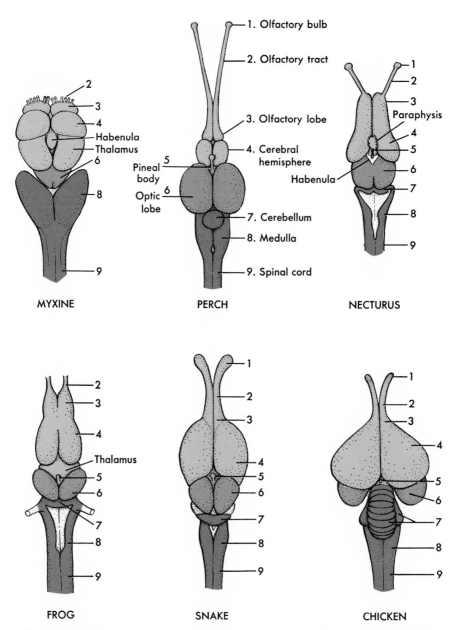

1. Olfactory bulb

2. Olfactory tract

3. Olfactory lobe

4. Cerebral hemisphere

5. Pineal body

6. Optic lobe

7. Cerebellum

8. Medulla

9. Spinal cord

Habenula
Thalamus

Paraphysis
Habenula

MYXINE   PERCH   NECTURUS

Thalamus

FROG   SNAKE   CHICKEN

Fig. 15-11. Vertebrate brains. The posterior choroid plexuses have been removed to expose the fourth ventricle in necturus and a frog. The prosencephalon, mesencephalon, and rhombencephalon are differentially colored.

subdivision. The hindbrain (rhombencephalon) subdivides into **metencephalon** and **myelencephalon.** Further differentiation of the five subdivisions involves thickening of the wall in some locations and evagination in others until the definitive brain has taken shape. The subdivisions are readily demonstrable in adults (Fig. 15-11), although to do so in mammals it is necessary to remove the overgrown cerebral hemispheres (Fig. 15-21). The enigmatic brain of the amphioxus has been discussed in Chapter 2.

☐ **Metencephalon and myelencephalon**

The myelencephalon, represented chiefly by the medulla oblongata, merges imperceptibly with the spinal cord. The area of transition is characterized internally by gradual relocation of the fiber tracts (white matter). As a result, the gray matter is dispersed into isolated masses (nuclei) in the medulla, whereas it is compact and centrally located in the cord.

The most conspicuous dorsal feature of the hindbrain is the cerebellum, a dorsal evagination of the metencephalon. The cerebellum functions in reflex control of the skeletal muscles. It receives input from the membranous labyrinth and lateral line canal system and feedback from proprioceptive receptors in muscles, joints, and tendons. It discharges motor impulses that result in muscle tonus and body posture. A dead fish falls onto its side and an unconscious bird or mammal collapses partly because the cerebellum is no longer bringing about synergistic contractions of those muscles essential to maintaining posture. The cerebellum is especially large in fishes, birds, and mammals, in which it may overlie both the medulla and the midbrain. It is relatively inconspicuous in amphibians. In cyclostomes it is not well developed and does not bulge from the brain. The cerebellum also receives regulatory impulses from the voluntary motor centers of the cerebral cortex and from other brain centers. The gray matter of the cerebellum is on its surface, a condition not found in most parts of the brain.

The other topographical features of the hindbrain are various swellings that indicate underlying nuclei, and elevated ridges or transverse bands that indicate fiber tracts. These topographical markings are most prominent in mammals and least prominent in fishes. However, one pair of these swellings becomes enormous in fishes that have taste buds scattered over the surface of the entire body. The swelling, or **vagal lobe** (Fig. 15-12), contains the visceral sensory nucleus (nucleus solitarius) that receives incoming taste fibers from cranial nerves VII, IX, and X. The exceptionally large number of second order neurons for taste causes the surface of the brain to bulge at that place.

Among mammalian markings on the hindbrain are the **pyramids** (corticospinal tracts) that carry voluntary motor impulses from higher centers, and the **pons,** which

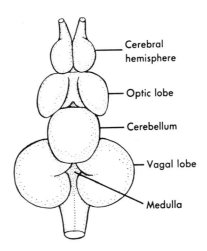

MODIFICATION FOR BOTTOM-FEEDING

Fig. 15-12. Brain of the buffalo fish *Carpiodes velifer.* Note unusual bulge (vagal lobe) on the alar plate of the medulla. Here terminate the many incoming taste fibers characteristic of bottom feeders.

contains fibers crossing (decussating) from one side to the other (Fig. 15-13).

The cavity within the hindbrain is the fourth ventricle (Figs. 15-3, *B,* and 15-17). The cerebellum is part of its roof. The rest of the roof is a membranous **tela choroidea,** a part of which hangs into the ventricle as the **choroid plexus of the fourth ventricle** (Figs. 15-3, *B,* and 15-16).

□ **Mesencephalon**

The roof, or **tectum,** of the mesencephalon has a pair of prominent **optic lobes** in all vertebrates. These bulging gray masses serve partly as optic reflex centers that receive fibers from the retina. They are especially well developed in birds, which rely on visual stimuli for much information. A pair of **auditory lobes** are found caudal to the optic lobes in the roof of the mesencephalon commencing with reptiles. These gray masses are present in fishes and amphibians but are not large enough to bulge from the surface. The auditory lobes receive impulses from the part of the membranous labyrinth sensitive to vibratory stimuli. With increased size of the cochlea, the auditory lobes enlarge. They also have other afferent connections. Other nuclei in the alar plate of the mesencephalon include the **mesencephalic nucleus of the trigeminal nerve,** which contains the cell bodies for proprioceptive fibers in all cranial nerves.

The ventral portion of the mesencephalon consists of nuclear masses and of fiber tracts connecting lower and higher levels of the brain. In mammals these tracts become massive. One pair, the **cerebral peduncles,** is just anterior to the pons.

The ventricle of the midbrain is quite large in fishes and amphibians and extends dorsally into the optic lobes as optic ventricles. In higher vertebrates the optic lobes are not hollow, and the ventricle is restricted to a narrow **cerebral aqueduct** (Fig. 15-19).

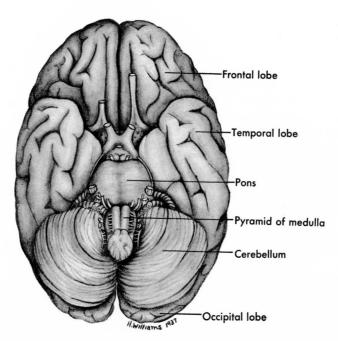

Frontal lobe

Temporal lobe

Pons

Pyramid of medulla

Cerebellum

Occipital lobe

Fig. 15-13. Brain of man, ventral view. The olfactory bulb has been cut away. (From Francis, C. C., and Martin, A. H.: Introduction to human anatomy, ed. 7, St. Louis, 1975, The C. V. Mosby Co.)

□ Diencephalon

Optic chiasma (Fig. 15-14). The site of superficial origin of the optic nerves from the brain is the optic chiasma, a ventral landmark approximating the cephalic boundary of the diencephalon.

Pituitary gland (Figs. 15-14 and 15-15). Just caudal to the optic chiasma lies the pituitary, an endocrine organ. It is attached by a stalk of brain tissue to the diencephalic floor and extends caudad beneath the mesencephalon. Its dual embryonic origin and its functions are discussed in Chapter 17.

Saccus vasculosus (Fig. 15-14). Some fishes have a thin-walled evagination of the midventral diencephalic floor just behind the pituitary. In some deep sea fishes it is larger than the pituitary and, like the latter, contains an extension of the third ventricle. However, no endocrine function can pres-

ently be assigned to the organ. It is probably a depth perceptor.

Hypothalamus. The portion of the diencephalon in the floor and ventrolateral walls of the third ventricle is the hypothalamus. The area contains a number of nuclei exerting partial control over the autonomic nervous system. In general, the anterior part of the hypothalamus is associated with parasympathetic functions, whereas the caudal part is associated with sympathetic functions. It has receptors that monitor the temperature and sodium chloride and glucose levels of blood passing through and thus plays a role in homeostasis (maintenance of a constant internal environment). It is also an important source of neurosecretions (Fig. 17-2).

Epithalamus: pineal and parapineal. The roof of the diencephalon is the epithalamus. It exhibits elevated gray masses, the **haben-**

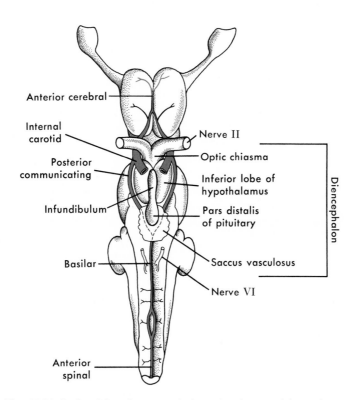

Fig. 15-14. Brain of *Squalus,* ventral view, showing arterial supply.

ulae (Fig. 15-11), which mark the location of underlying nuclei. In addition, one or two simple or elaborate evaginations arise from the diencephalic roof of nearly all vertebrates, the **pineal** and **parapineal organs** (Figs. 15-16 and 16-13), often referred to collectively as the **epiphyseal complex.**

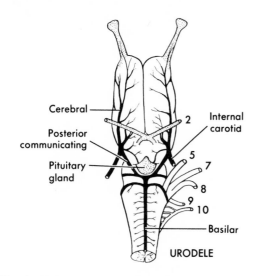

Fig. 15-15. Brain of a urodele, ventral view. Numerals identify cranial nerve roots.

The pineal is an unpaired, elongate, club-shaped or knoblike organ, sometimes threadlike or saccular, lying beneath the roof of the skull, or, in mammals, between the two cerebral hemispheres (Figs. 15-11, 15-19, and 15-21). It is connected by a stalk to the roof of the diencephalon. The stalk sometimes contains an extension of the third ventricle. In lampreys and, perhaps, larval anurans, the pineal serves as a photoreceptor. Otherwise, it functions as an endocrine gland (p. 418). It receives a motor innervation from the superior cervical sympathetic ganglion via the **nervi conarii.** The impulses are triggered, in part, by sensory input over the optic nerve. The pineal is vestigial or absent in hagfishes, one group of elasmobranchs (torpedoes), crocodilians, and some adult mammals (edentates, sirenians, porpoise, elephants, rhinoceros). It is relatively large in man.

The parapineal was widely distributed among ancient vertebrates but is now prominent only when serving as a third eye (Fig. 16-13, *C*). When present, it lies in a foramen of the skull just under the skin.

At one time the pineal and parapineal may have constituted a pair of photoreceptors.

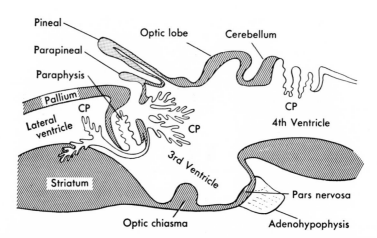

Fig. 15-16. Diencephalon and adjacent areas of a vertebrate brain, sagittal section, anterior end to the left. **CP,** Choroid plexus of lateral, third, and fourth ventricles. Based on the brain of a larval frog.

Evidence is found in their embryogenesis, in the presence of a middorsal foramen in the skulls of placoderms, and in the relationships and morphology of the two structures in lampreys (Fig. 16-13, *A*). In these cyclostomes both are present, the parapineal being connected with the left side of the brain and the pineal with the right, and both contain photosensory cells.

Thalamus (Figs. 15-11 and 15-21). The largest subdivision of the diencephalon is the thalamus, a mass of nuclei in the lateral

SHARK

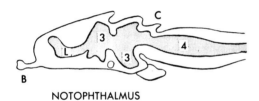

NOTOPHTHALMUS

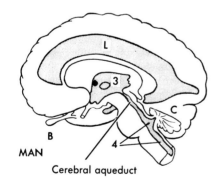

MAN

Cerebral aqueduct

Fig. 15-17. Sagittal brain sections showing the ventricles. **B,** Olfactory bulb; **C,** cerebellum; **L, 3,** and **4,** lateral, third, and fourth ventricles. The black foramen in the third ventricle in man is the interventricular foramen connecting the third and right lateral ventricles.

wall of the third ventricle. All sensory pathways ascending to the telencephalon synapse in one of the thalamic nuclei before continuing. The thalamus is small in lower vertebrates. It becomes increasingly prominent in higher forms and relays an increasing number of sensory impulses to the cerebral hemispheres. In mammals the thalamus is so enlarged that the left and right sides bulge into the ventricle and meet to form a gray (middle) commissure.

Third ventricle (Figs. 15-16 and 15-17). The cavity of the diencephalon is the third ventricle. It is continuous cephalad with the lateral ventricles. The third ventricle is laterally compressed, especially in higher vertebrates, by the thalamus. An optic recess of the ventricle extends toward the optic chiasma, and an infundibular recess extends into the pituitary stalk. The roof of the third ventricle remains thin, becomes vascularized, and hangs into the third ventricle as a choroid plexus.

☐ **Telencephalon**

The telencephalon consists of cerebral hemispheres and a rhinencephalon. In lower vertebrates the rhinencephalon is as prominent as the cerebral hemispheres. In higher ones the hemispheres increase in size, bulge forward over the rhinencephalon, and relegate this part of the brain to an inconspicuous anteroventral location. The change mirrors the increased role of the cerebral hemispheres in behavior.

RHINENCEPHALON

The rhinencephalon consists of olfactory bulbs, olfactory tracts, and olfactory lobes (Fig. 15-11). In mammals the olfactory bulbs and tracts are dwarfed and often hidden by the cerebral hemispheres (Figs. 15-18 and 15-19). The olfactory bulbs lie close to the olfactory epithelium, separated from it by the olfactory capsule. Fiber tracts connect the rhinencephalon with other parts of the forebrain.

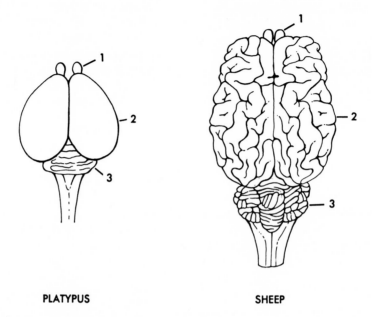

PLATYPUS   SHEEP

Fig. 15-18. Brain of a primitive mammal (platypus) lacking cortical gyri, and brain of sheep. **1,** Olfactory bulb; **2,** cerebral hemisphere; **3,** cerebellum.

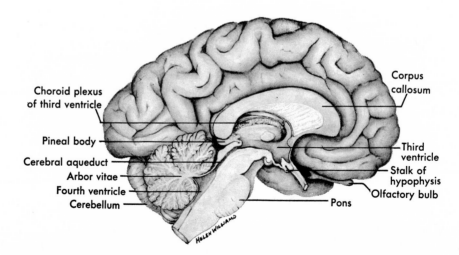

Fig. 15-19. Human brain, left half, sagittal section. (From Francis, C. C., and Martin, A. H.: Introduction to human anatomy, ed. 7, St. Louis, 1975, The C. V. Mosby Co.)

EVOLUTION OF THE CEREBRAL
HEMISPHERES

When one thinks of cerebral hemispheres, what usually comes to mind are the enlarged cerebral hemispheres of mammals, with their thick cortex of gray matter. However, the cerebral cortex is a recent acquisition. The basic pattern on which all vertebrate hemispheres are constructed and from which mammalian hemispheres have evolved is seen in fishes.

In fishes each hemisphere consists chiefly of a **paleostriatum,** so named because it is ancient. The striatum, lateral ventricle, and thin roof (**pallium**) constitute the entire cerebral hemisphere (Fig. 15-20, sturgeon, teleost). The striatum consists of motor nuclei receiving fibers chiefly from the rhinen-cephalon. Fibers from the striatum enter ancient descending tracts that terminate in the motor nuclei of cranial and spinal nerves. Thus olfactory stimuli result in reflex motor activity of many parts of the body.

The striatum of amphibians receives a larger number of sensory fibers projected forward from the thalamus than does that of fishes. It therefore receives input from a wider spectrum of receptors. Nevertheless, the number of sensory impulses relayed from the thalamus to the hemispheres is small when compared with higher vertebrates, and the cerebral hemispheres of amphibians are still preempted by olfactory stimuli.

In specialized reptiles additional nuclei, constituting a **neostriatum** (Fig. 15-20), are

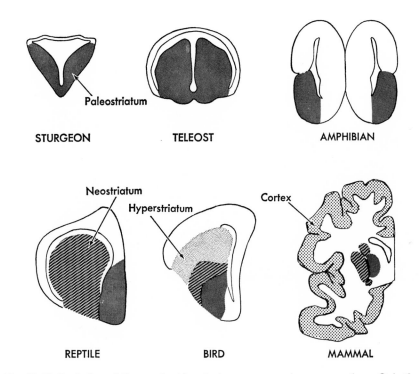

Fig. 15-20. Evolution of the cerebral hemispheres as seen in cross sections. Only the left hemisphere is shown in the lower figures. Reptiles have added a neostriatum to the old paleostriatum. Birds added a hyperstriatum. Note the striatal complex (now called basal ganglia) still present in the mammal, and the addition of a cortex on the surface of the mammalian hemisphere.

added to the striatum, and these nuclei receive many more sensory fibers from the thalamus. A trace of cerebral cortex appears on the surface of the pallium. Because of the added cell bodies and synapses, the hemispheres of reptiles are larger. They bulge laterally, dorsally, and backward over the diencephalon. Because of the increased flow of sensory information into the hemispheres, they now assume increased control over motor activity.

Bird hemispheres are essentially reptilian. The striatum reaches peak development, and additional strata of nuclei, the **hyperstriatum,** are superimposed on the neostriatum (Fig. 15-20). To the hyperstriatum come many sensory impulses, which, after being relayed to the older striatum, result in stereotyped behavior such as nest-building, incubation of eggs,

and care of the young. The cerebral cortex is better developed than that in reptiles, but almost complete ablation of the cortex has little observable effect. The olfactory lobes are very small, and smell has less influence on behavior.

In mammals the striatum and other nuclear masses, now called collectively the basal ganglia, continue to play an important, although not fully understood, role in the nervous system. But the cerebral cortex on the greatly expanded roof (pallium) has become the most conspicuous part of the mammalian brain. As a result of the enormous upward, over, and backward growth of the pallium, the striatum, diencephalon, and midbrain are all hidden from dorsal view (Fig. 15-19). Removal of the overgrown cerebral hemispheres will reveal the primitive relationships of the rhinencephalon,

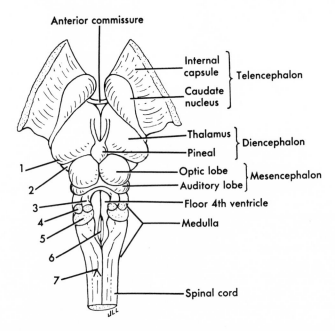

Fig. 15-21. Brain stem of a sheep. The cerebral hemispheres and cerebellum have been cut away to reveal the primitive vertebrate structures. **1,** Location of lateral geniculate body of thalamus; **2,** medial geniculate body; **3 to 5,** anterior, middle, and posterior cerebellar peduncles, which carry fibers to and from the cerebellum (the peduncles had to be cut when the cerebellum was removed); **6,** hypoglossal trigone in the floor of the fourth ventricle marking the location of the hypoglossal nuclei; **7,** posterior funiculus containing ascending fibers for proprioception. The caudate nucleus is part of the striatum of the cerebral hemisphere.

striatum, thalamus, midbrain, metencephalon, and medulla (Fig. 15-21).

The mammalian cortex has at least four
roles. (1) It is the highest center to which
sensory impulses may pass. These impulses
give rise to sensations of a discriminative
(epicritic) nature and contribute to the esthetic enjoyment of sensory stimuli. (2) It
appears to be one of the locations where
past experiences are stored as memory. (3)
It is a center where data, incoming or recalled, may be correlated, analyzed, and
employed in making choices. (4) It is the
highest center from which motor activity
may be initiated. The cortex is therefore
the "thinking" part of the brain. The manner
in which the cerebral cortex is employed
in the solution of human problems will determine the future fate of civilization insofar
as it is under the control of man.

In many mammals, but not all, the cerebral cortex becomes so voluminous that it is
folded into numerous ridges (**gyri**) and
grooves (**sulci**) (Fig. 15-19). Under the cortex in the roof of the ventricles except in
monotremes and marsupials lies a broad
transverse sheet of commissural nerve
fibers, the **corpus callosum.** It connects the
cortices of the two hemispheres (Fig. 15-
19).

LATERAL VENTRICLES

The cavities of the cerebral hemispheres
are the lateral ventricles. They are continuous with the third ventricle via an interventricular foramen (Fig. 15-17). The location
of the foramen marks the original cephalic
end of the embryonic neurocoel before
evagination of the cerebral hemispheres
took place. The roof of the ventricles in
mammals includes the corpus callosum and
the pallium with its cortex. The floor consists, in part, of the striatum and associated
nuclei. Separating the left and right lateral
ventricles is a thin, double-walled vertical
partition, the **septum pellucidum.** The lateral ventricles, like the rest of the neurocoel, are lined by an ependyma that is

extensively ciliated throughout life in many
species, including man.

PARAPHYSIS

At the caudal end of the telencephalon of
many vertebrates, the thin, nonnervous roof
forms a saclike evagination, sometimes very
large, called the paraphysis (Fig. 15-16). It
often becomes smaller or disappears during
ontogeny and, when present, is easily over-
looked. The vessels of the paraphysis receive a rich autonomic nerve supply, and
a neurosecretory tract leads from the para-
ventricular nucleus of the hypothalamus to
the base of the paraphysis. The function of
the paraphysis is not yet clear.

CHOROID PLEXUSES AND
CEREBROSPINAL FLUID

The cavities of the brain and spinal cord
are filled with lymph-like cerebrospinal
fluid secreted partly by choroid plexuses. A
typical choroid plexus consists of the thin
ependymal roof of the ventricle and of the
pia mater (or leptomeninx in fish), invaded
by a rich vascular plexus. A choroid plexus
hangs into the third and fourth ventricles
(Fig. 15-3, *B*). From the third ventricle it
extends forward into the lateral ventricles
(Fig. 15-16).

Cerebrospinal fluid secreted into the ventricles moves sluggishly caudad into the
neurocoel of the spinal cord partly by ciliary
action. From the fourth ventricle the fluid
also passes into the submeningeal spaces
via a pair of **lateral apertures of the fourth
ventricle** under cover of the cerebellum. A
median aperture is also present in many
vertebrates. From the submeningeal spaces
the cerebrospinal fluid passes outward
along the roots of the cranial and spinal
nerves for short distances. It also passes
centrally along the nerve rootlets into the
cord and brain and bathes each motor
neuron. It seeps to the inner ear, where it
contributes to the perilymph.

Cerebrospinal fluid is removed by lymph
channels, especially along the roots of the

spinal nerves. In higher vertebrates it is also removed by clusters of macroscopic **arachnoid villi** that penetrate the dura mater and hang into the large venous sinuses of the brain.

Cerebrospinal fluid assists in protecting the central nervous system from concussion. It also exchanges metabolites with the tissues it bathes. In the saccus vasculosus of fishes the cerebrospinal fluid probably stimulates sensory neurons, thereby initiating a train of impulses to the cerebellum.

■ CRANIAL NERVES

The first ten cranial nerves of all vertebrates are distributed in accordance with a basic pattern. For convenience, these nerves may be grouped as follows: predominantly sensory nerves (I, II, and VIII), eyeball muscle nerves (III, IV, and VI), and branchiomeric nerves (V, VII, IX, and X). Amniotes have two additional cranial nerves (XI, XII), which are purely or predominantly motor. The basic pattern of cranial nerve distribution is seen in fishes. Variations in the distribution of these nerves in tetrapods result from adaptation to life on land.

Cranial nerves do not have dorsal and ventral roots. Nerves supplying myotomal muscles, except nerve IV, have ventral roots only, and branchiomeric nerves and nerve XI have **lateral roots.** The remaining cranial nerves have varied origins.

□ Predominantly sensory cranial nerves*
NERVE I (OLFACTORY)

In all vertebrates the cell bodies of the olfactory nerve fibers are located in the olfactory epithelium. The fibers terminate in the olfactory bulb (Fig. 15-22). In *Squalus*

---

*Nerves II and VIII, although functionally sensory, contain a number of efferent fibers from the brain to the deep layer of the retina or to the vestibular hair cells. These fibers apparently influence the discharge of sensory impulses from the receptor.

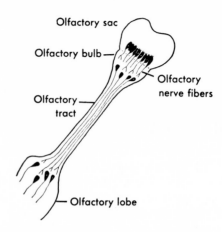

Fig. 15-22. Olfactory sac (containing olfactory epithelium) and rhinencephalon of a dogfish shark, showing location of cell bodies of sensory fibers. The olfactory bulb contains second order sensory neurons.

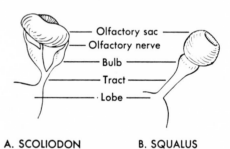

A. SCOLIODON   **B. SQUALUS**

Fig. 15-23. Rhinencephalon (olfactory bulb, tract, lobe) of two sharks. The olfactory sacs are not part of the brain. Olfactory nerve fibers connect the sac and bulb and form a discrete nerve in *Scoliodon* but not in *Squalus.*

the olfactory epithelium lies so close to the bulb that an olfactory nerve cannot be distinguished as an anatomical entity. In *Scoliodon* (Fig. 15-23, A), another shark, the sac containing the olfactory epithelium is sufficiently removed from the olfactory bulb that an olfactory nerve is demonstrable. In most vertebrates, one or more short bundles of olfactory fibers (**filia olfactoria**) extend between the olfactory epithelium and the bulb, and these constitute collectively the olfactory nerve.

In mammals the olfactory epithelium is in the upper part of the nasal passage, separated from the olfactory bulb by the cribriform plate of the ethmoid bone (derived from olfactory capsule). The foramina in the ethmoid bone (Fig. 8-6, *B*) transmit the filia olfactoria. When the brain of a vertebrate is lifted from the cranial cavity, the olfactory nerve bundles are torn, and only stumps remain attached to the brain.

The olfactory nerve frequently has a separate division that supplies the vomeronasal organ. When present, it is called the **vomeronasal nerve.**

A terminal nerve lies close to the olfactory bulb and tract in all vertebrates including man. It arises from the ventral surface of the forebrain and supplies general sensory and vasomotor fibers to a rather small area of the nasal epithelium and mucosa, but it has no olfactory function. The terminal nerve appears to be a vestige of a branchiomeric nerve that was present at one time anterior to the trigeminal.[7]

NERVE II (OPTIC)

The cell bodies are found in the retina. The optic nerve emerges from the rear of the eyeball and extends to the optic chiasma (Fig. 15-14), where the **optic tracts** commence. Except in mammals, all the optic nerve fibers decussate in the chiasma to enter the optic tract on the opposite side of the brain. In mammals, only those fibers from the nasal side of the retina cross. As a result, there is overlap of the visual fields for binocular vision.

NERVE VIII (VESTIBULOCOCHLEAR)

The eighth nerve in lower vertebrates has an anterior and a posterior branch. The anterior branch innervates the ampullae on the anterior vertical and horizontal semicircular ducts, and the utriculus. The posterior branch innervates the ampulla on the posterior vertical semicircular duct, and the sacculus and lagena.

Commencing with amphibians, the la-

gena enlarges and differentiates to become the cochlea for hearing. As a result, the posterior branch of cranial nerve VIII becomes greatly enlarged and is called the **cochlear** nerve, but it continues to carry fibers from other parts of the membranous labyrinth. The anterior branch is then called the **vestibular** nerve. Both branches enter the medulla and have a ganglion (**cochlear** or **vestibular**) on their pathway. The cochlear ganglion lies on the spiral membrane within the cochlea and is also known as the **spiral** ganglion.

In some mammals (rat and mouse, but not bat or cat) large cell bodies are also distributed along the entire length of the eighth nerves, and their fibers extend into the medulla. These are cell bodies of second order sensory neurons. They are stimulated by collateral branches of incoming fibers. The mechanism reinforces stimuli from the labyrinth.

☐ **Eyeball muscle nerves**
NERVES III, IV, AND VI (OCULOMOTOR, TROCHLEAR, AND ABDUCENS)

The third, fourth, and sixth nerves supply the superior oblique (IV), external rectus (VI), the four remaining extrinsic eyeball muscles (III), and certain other myotomal muscles of the eyes (Table 10-2, p. 221). These eyeball muscle nerves resemble spinal nerves that have lost their dorsal roots. In addition to somatic motor fibers, the nerves contain sensory fibers for proprioception from the muscles innervated.

Nerve III arises ventrally from the mesencephalon. Nerve IV is the only nerve arising dorsally from the brain (posterior end of the mesencephalon or anterior roof of the fourth ventricle) and one of the few nerves with motor fibers that decussate (cross) before emerging. Nerve VI emerges ventrally at the anterior end of the hindbrain. Nerves IV and VI are the smallest of the cranial nerves, having the fewest fibers. The cell bodies of all the fibers in these nerves, both motor and proprioceptive, are in nuclei within the central nervous system. Therefore, the nerves have no sensory ganglia.

AUTONOMIC FIBERS IN NERVE III

Nerve III contains visceral motor fibers that end in the ciliary ganglion of the autonomic nervous system (Fig. 15-30). From the ganglion, postganglionic fibers pass to the sphincter muscles of the iris diaphragm and to the ciliary body of the eye.

☐ **Branchiomeric nerves**

One of the characteristics of vertebrates is the embryonic development of a series of visceral arches separated by pharyngeal pouches. The fate of the skeleton and muscles of these arches has already been discussed. Whether the animal is to become fish or tetrapod, the muscles derived from the first arch are supplied by cranial nerve V; those from the second arch, by nerve VII; from the third arch, by nerve IX; and from succeeding arches, by nerve X. Since V, VII, IX, and X innervate branchiomeric muscles, they are branchiomeric nerves.

The branchiomeric nerves are mixed nerves. In addition to innervating branchiomeric muscles, branchiomeric nerves have other important motor and sensory functions. Some of their motor fibers are com-

ponents of the autonomic nervous system. Their sensory fibers supply several groups of sense organs, and each nerve includes proprioceptive fibers. (The route taken by proprioceptive fibers from some of the branchiomeric muscles has not been ascertained.) The distribution of branchiomeric nerves in *Squalus* illustrates the basic pattern (Fig. 15-25). Alterations in tetrapods are chiefly the result of elimination of gills and other adaptations to terrestrial life.

NERVE V (TRIGEMINAL)

The fifth nerve arises from the anterior end of the hindbrain and typically exhibits three divisions: **ophthalmic, maxillary,** and **mandibular.** In fishes the ophthalmic may be subdivided into superficial and deep ophthalmic nerves and the maxillary may be called infraorbital. All branches contain sensory fibers. Only the mandibular branch contains motor fibers.

Cranial nerve V is sensory to the ectoderm of the head, including the teeth, anterior part of the tongue, and nasal epithelium, for general cutaneous sensation (Fig. 15-24). The mandibular branch also contains proprioceptive fibers. The cell bodies of all sensory fibers except proprioceptive are found in the **trigeminal ganglion** unless, as in some lower vertebrates, the ophthalmic division has its own ganglion.

The mandibular nerve is motor to all muscles derived from the first pharyngeal arch. The predominant distribution is therefore to

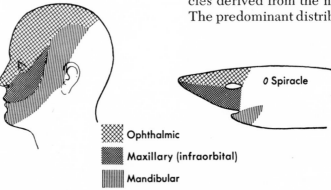

Ophthalmic

Maxillary (infraorbital)

Mandibular

Fig. 15-24. Cutaneous distribution of the trigeminal nerve of vertebrates.

muscles of the jaws. However, mammals have a tensor tympani muscle attaching to the malleus (derived from the embryonic Meckel's cartilage), which is therefore innervated by the fifth nerve. Table 10-5 (p. 230) gives the motor distribution of cranial nerve V.

NERVE VII (FACIAL)

The seventh nerve arises from the anterior end of the hindbrain in close association with the fifth. In sharks the fifth and seventh nerves have common (trigeminofacial) roots.

Nerve VII is sensory to the neuromast organs on the head of fishes and aquatic amphibians. With adaptation to land, these sensory fibers were lost. Branches also supply taste buds in the pharynx at the level of the first and second arches, any taste buds on the external surface of fishes, and taste buds on the anterior part of the tongue in tetrapods. Nerve VII also contains general sensory fibers from the endoderm of the second arch and proprioceptive fibers from the muscles of the arch. The cell bodies of all sensory fibers, except proprioceptive, are in the **facial ganglion.**

The facial nerve is motor to muscles of the second arch (Table 10-5). These include the mimetic muscles of mammals. Since the stapes is a derivative of the hyoid arch, the stapedial muscle of mammals is also innervated by the seventh nerve. The seventh nerve in mammals also contains visceral motor fibers to the submandibular and sphenopalatine ganglia (Figs. 15-27 and 15-30). The submandibular ganglion innervates the submandibular and sublingual salivary glands. The sphenopalatine innervates the lacrimal gland and mucous membranes of the nose.

NERVE IX (GLOSSOPHARYNGEAL)

The ninth nerve arises from the medulla. In sharks it has three major branches and is a typical branchiomeric nerve. These branches are **pretrematic** (sensory), **pharyngeal** (sensory), and **posttrematic** (mixed).*
The pretrematic branch supplies the anterior demibranch of the first gill chamber for general sensation (Fig. 15-25). The pharyngeal branch supplies taste buds and general visceral receptors in the pharyngeal mucosa at the level of the third visceral arch. The posttrematic branch is sensory to the demibranch in the posterior wall of the first gill chamber, motor to the muscles of the third visceral arch, and proprioceptive from those muscles. A small **lateral-line branch** of IX supplies a short segment of the lateral-line canal at the junction of head and trunk.

Preganglionic fibers of the autonomic nervous system are present in IX. They have not been fully explored in all vertebrates, but in mammals they innervate the otic ganglion, from which postganglionic fibers pass to the parotid salivary glands.

With loss of gills and neuromast organs during adaptation to life on land, cranial nerve IX lost many fibers. It continues to supply surviving taste buds in the third arch mucosa (on the posterior part of the tongue of mammals) and general receptors on the posterior part of the tongue and in the upper pharynx. Of the branchiomeric muscles of the third arch, only a stylopharyngeus remains in mammals.

The sensory cell bodies of nerve IX, except those for proprioception, are found in the **petrosal ganglion** of lower vertebrates, and in the **superior** and **inferior glossopharyngeal ganglia** of mammals. The superior ganglion is derived from a neural crest. The inferior ganglion is derived chiefly from an ectodermal placode. In fishes the lateral line branch has its own ganglion.

---

*The maxillary and mandibular branches of nerve V in fishes probably represent pretrematic and posttrematic branches of that nerve. The pretrematic branch of VII in fishes has joined the maxillary of V to form an infraorbital trunk and the posttrematic branch of VII (behind the spiracle) is the hyomandibular nerve. The pharyngeal branch of VII in sharks is the palatine.

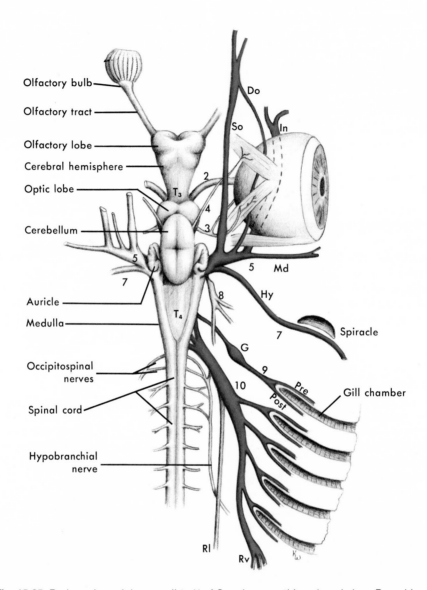

Olfactory bulb

Olfactory tract

Olfactory lobe

Cerebral hemisphere

Optic lobe

Cerebellum

Auricle

Medulla

Occipitospinal nerves

Spinal cord

Hypobranchial nerve

Do

So

In

T_3

2

4

3

5

7

T_4

5   Md

8   Hy

7   Spiracle

G

9   Pre   Gill chamber

10   Post

Rl

Rv

Fig. 15-25. Brain and cranial nerves II to X of *Squalus acanthias,* dorsal view. Branchiomeric nerves are shown in red. **Do,** Deep ophthalmic; **G,** petrosal ganglion; **Hy,** hyomandibular; **In,** infraorbital; **Md,** mandibular; **Pre,** pretrematic; **Post,** posttrematic; **So,** superficial ophthalmic; **Rl,** ramus lateralis (lateral line nerve) of vagus; **Rv,** ramus visceralis of vagus; T_3 and T_4, tela choroidea of third and fourth ventricles. The pineal body has been removed. The three eyeball muscles shown are, commencing anteriorly, superior oblique, superior rectus, and lateral rectus.

NERVE X (VAGUS)

The vagus arises from a series of rootlets along the lateral aspect of the medulla. The branchiomeric portion of the vagus in *Squalus* consists of a series of four trunks (Fig. 15-25) (more in elasmobranchs with more gill chambers), each of which exhibits a pretrematic, posttrematic, and pharyngeal branch distributed as in nerve IX. The pretrematic branches supply the anterior walls of the last four gill chambers. Posttrematic branches supply the posterior walls of these chambers and are motor to arches IV to VII. It is therefore the chief respiratory nerve of fishes. The pharyngeal branches supply the pharyngeal epithelium for taste and general sensation.

In addition to branchiomeric components, the vagus in *Squalus* has two other major trunks. The **ramus lateralis** is sensory to the lateral-line canal all the way to the tip of the tail, and the **ramus visceralis** supplies afferent and efferent visceral fibers to some of the viscera.

During the process of adapting to land the vagus lost those functions associated solely with life in the water but retained other functions. The prominent lateral-line branch disappeared. The sensory branches to the gill chambers were lost. However, general receptors in the mucosa of the pharynx, as well as surviving taste buds in the vicinity of the glottis, continue to be supplied by the vagus. Surviving also are the motor branches to those branchiomeric muscles of the fourth and successive arches that assumed new functions on land. These are chiefly the cricothyroid, cricoarytenoid, and thyroarytenoid muscles. Since much of the distribution of the vagus has been lost in tetrapods, the ramus visceralis has become the major component of the nerve. It continues to supply afferent and efferent fibers to the heart and some other viscera.

In fishes as well as in tetrapods the vagus supplies preganglionic fibers to certain terminal ganglia of the autonomic system in the trunk (Fig. 15-30). Included in amniotes are autonomic fibers contributed by the internal ramus of the accessory nerve (Fig. 15-26). The vagal nerve also contains proprioceptive fibers.

The cell bodies of all sensory fibers of the vagal nerve, except those for proprioception, are found in one or more ganglia. In some elasmobranchs each of the four or more branches to the gills has its own **epibranchial ganglion,** and it is likely that these branches were at one time four separate cranial nerves. In birds and mammals nerves IX and X have two sensory ganglia, a **superior** and **inferior.** The superior ganglia are derived from neural crests. The inferior ganglia are derived from epibranchial placodes.

The innervation of the mucosa of the oral cavity and pharynx by nerves V, VII, IX, and X, *in that sequence* in all vertebrates from fish to man, demonstrates the negligible effects of life on land on the sensory innervation of the pharyngeal endoderm.

☐ **Accessory and hypoglossal nerves**
NERVE XI (ACCESSORY)

An accessory nerve constitutes an eleventh cranial nerve in amniotes. It is purely motor. In mammals it has a series of **cranial roots,** which arise from the medulla (Fig. 15-26), and a **spinal root** from the spinal cord. It appears to be derived from a primitive posterior-most branchial nerve combined with one or more occipitospinal nerves, which are otherwise missing in amniotes.

The nucleus of origin of the spinal root of the accessory nerve occupies several segments of the cord—typically five or six in man, seven in horses, and fewer in many mammals. The spinal rootlets unite to form a common trunk that passes cephalad close to the cord and enters the cranial cavity via the foramen magnum. Within the cranial cavity the spinal root joins the cranial roots to form the eleventh nerve (Fig. 15-26).

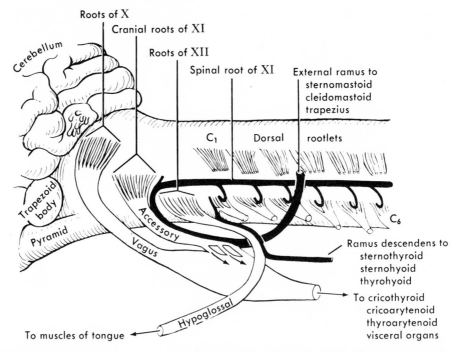

Fig. 15-26. Vagus, spinal accessory, and hypoglossal nerves of a cat. The hypoglossal rootlets are in series with the ventral roots of the spinal nerves. Components in black are spinal nerve contributions. **C₁**, Dorsal rootlets of the first cervical spinal nerve; **C₆**, ventral rootlets of the sixth cervical spinal nerve. Nerve XI contributes an internal ramus (arrows) to the vagus.

Near the jugular foramen the cranial root fibers join the vagus to be distributed with the latter as the **internal ramus** of the accessory nerve. The internal ramus is composed of preganglionic fibers of the autonomic system. The fibers of spinal origin form the **external ramus,** which supplies the trapezius, sternomastoid, and cleidomastoid muscles.

NERVE XII (HYPOGLOSSAL)

The twelfth nerve of amniotes is motor except for proprioceptive fibers. It arises from the caudal end of the brain by a series of motor rootlets. It innervates myotomal muscle of the tongue (genioglossus, styloglossus, hyoglossus, and lingualis). On emerging from the hypoglossal foramen, the nerve in mammals is joined by fibers from the first one or two cervical spinal nerves (Fig. 15-26). These fibers are distrib-

uted to the geniohyoid and other hypobranchial muscles.

That the hypoglossal is a cranial rather than a spinal nerve is dictated solely by the location of the foramen magnum. *Actually,* it is a spinal nerve that became "locked up" in the braincase. Like spinal nerves, the hypoglossal develops an embryonic dorsal root and ganglion (**Froriep's ganglion**). However, root and ganglion later disappear.

The twelfth nerve is a derivative of the occipitospinal series of fishes and amphibians, as may be deduced from the following facts: (1) Occipitospinal nerves have been reduced in number above fishes, whereas the number of cranial nerves has increased. (2) The twelfth nerve, like many occipitospinal nerves, lacks a dorsal root. (3) The occipitospinal nerves of lower verte-

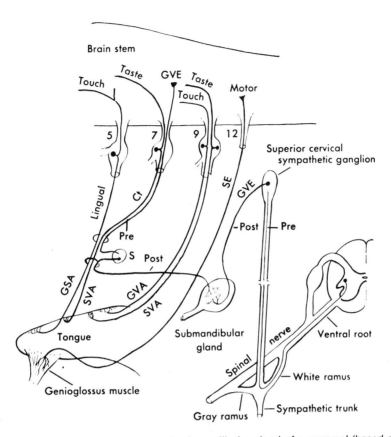

Fig. 15-27. Innervation of the tongue and submandibular gland of a mammal (based on cat and man). **5, 7, 9,** and **12,** Cranial nerves; **Ct,** chorda tympani; **Pre** and **Post,** preganglionic and postganglionic fibers of the autonomic nervous system; **S,** submandibular ganglion of the autonomic system. A key to the fiber components (**GSA, SE,** and so forth) is given in Tables 15-1 and 15-2.

brates supply hypobranchial muscles, and the tongue is hypobranchial muscle. (4) Whereas the tongue muscle is supplied by the last cranial nerve in amniotes, it is supplied by the first spinal nerve in anamniotes.

□ **Innervation of the mammalian tongue**
(Fig. 15-27)

The innervation of the mammalian tongue illustrates how a single organ may be served by numerous nerves, depending on the ontogenetic and phylogenetic history of its parts. The mucosa of the anterior part of the tongue is first arch endoderm and is therefore innervated by nerve V for general

sensations. The taste buds on this part of the tongue are innervated by nerve VII, which supplies taste buds on the first and second arches in fishes. The mucosa on the posterior part of the tongue is innervated by nerve IX for both general sensation and taste because of the origin of this mucosa from the third visceral arch. The muscles of the tongue are myotomal and are innervated by nerve XII.

Although four cranial nerves innervate the tongue, only three nerves may be traced into it, since the taste fibers from nerve VII (in the chorda tympani) enter the tongue as part of the lingual branch of nerve V.

Table 15-2. Special fiber components of cranial nerves*

Components	Innervation and nerve
Special somatic afferent fibers (SSA)	Special somatic receptors Retina (II) Membranous labyrinth (VIII) Neuromast organs (VII, IX, and X)
Special visceral afferent fibers (SVA)	Special visceral receptors Olfactory epithelium (I) Taste buds (VII, IX, and X)
Special visceral efferent fibers (SVE)	Branchiomeric muscle (V, VII, IX, X, and XI)

*In addition to these special components, cranial nerves other than I, II, and VIII have one or more of the general components listed in Table 15-1.

□ **Fiber components of cranial nerves**

It has been mentioned earlier (p. 363, and Table 15-1) that the fibers in spinal nerves may be classified in four functional categories (GSA, GVA, SE, and GVE), each supplying specific types of general receptors or effectors. One or more of these components may be found also in most cranial nerves: SE in III, IV, VI, and XII; GVE (autonomic fibers) in III, VII, IX, X, and XI; GSA chiefly in V; and GVA in VII, IX, and X. In addition to the foregoing types of *general* fiber components, certain cranial nerves contain *special* components of which there are three types—SSA, SVA, and SVE (Table 15-2). Each of the foregoing nerve fibers commences or terminates in the brain or cord in a column of gray matter preempted by that specific component (Fig. 15-28).

■ **AUTONOMIC NERVOUS SYSTEM**

The autonomic nervous system is that part of the nervous system that innervates glands and smooth and cardiac muscle. It consists chiefly of autonomic nerves, plexuses, and

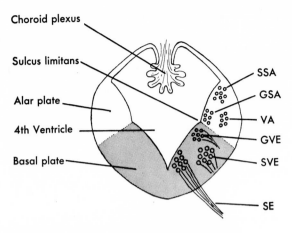

Fig. 15-28. Cross section of medulla showing location of certain nuclei. Sensory nuclei are in the alar plate; motor nuclei, in the basal plate. The two plates are delimited by the sulcus limitans. **Se,** Somatic motor fibers from cell bodies in somatic motor column. **VA,** Sensory nucleus for visceral input, except smell. The remaining nuclei are identified in Tables 15-1 and 15-2.

ganglia (Figs. 15-29 and 15-30). It does not, however, constitute an anatomical entity; that is, it cannot be completely dissected away from the rest of the nervous system, since its components commence inside the central nervous system and emerge via cranial or spinal nerves. It is entirely a visceral motor system. However, sensory fibers from the viscera use autonomic pathways to reach the spinal cord and brain.

Two motor neurons in series conduct the impulse from the brain or cord to a typical visceral effector. The first neuron (**preganglionic neuron**) of the chain has its cell body in a visceral efferent nucleus in the central nervous system, and the preganglionic fiber terminates in an autonomic ganglion. The second neuron (**postganglionic neuron**) has its cell body in an autonomic ganglion. Its fiber extends to the effector (see Fig. 15-9 and nerve III in Fig. 15-30).

The autonomic system is composed of (1) the **thoracolumbar (sympathetic) system** emerging from the cord via thoracic and

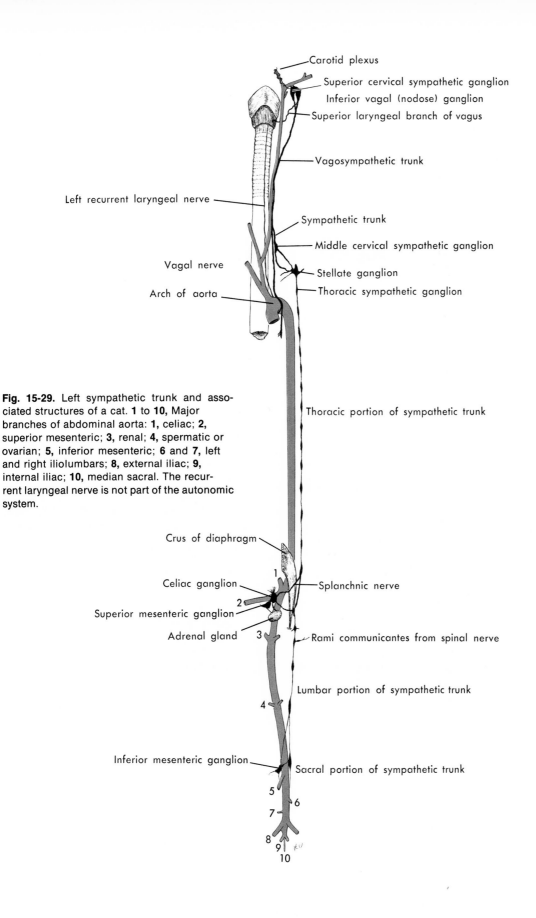

Carotid plexus

Superior cervical sympathetic ganglion

Inferior vagal (nodose) ganglion

Superior laryngeal branch of vagus

Vagosympathetic trunk

Left recurrent laryngeal nerve

Sympathetic trunk

Middle cervical sympathetic ganglion

Vagal nerve

Stellate ganglion

Arch of aorta

Thoracic sympathetic ganglion

Thoracic portion of sympathetic trunk

Fig. 15-29. Left sympathetic trunk and associated structures of a cat. **1 to 10,** Major branches of abdominal aorta: **1,** celiac; **2,** superior mesenteric; **3,** renal; **4,** spermatic or ovarian; **5,** inferior mesenteric; **6 and 7,** left and right iliolumbars; **8,** external iliac; **9,** internal iliac; **10,** median sacral. The recurrent laryngeal nerve is not part of the autonomic system.

Crus of diaphragm

Celiac ganglion

Splanchnic nerve

Superior mesenteric ganglion

Adrenal gland

Rami communicantes from spinal nerve

Lumbar portion of sympathetic trunk

Inferior mesenteric ganglion

Sacral portion of sympathetic trunk

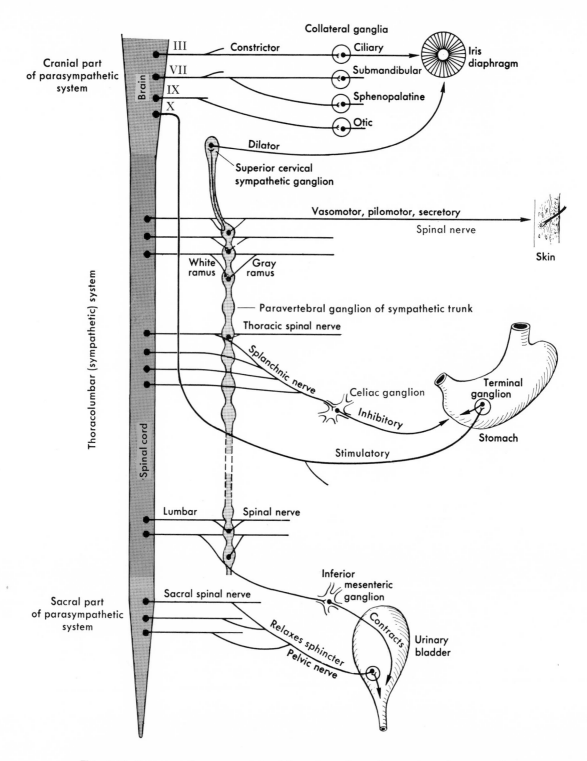

Fig. 15-30. Representative components of the autonomic nervous system of a mammal. Innervation of iris diaphragm, skin, stomach, and urinary bladder. Arrows emphasize dual control exerted elsewhere than in the skin by craniosacral and thoracolumbar systems. Preganglionic fibers are those with a cell body (black dot) in the central nervous system. Postganglionic fibers are those with a cell body in a ganglion. Other spinal nerves in addition to those shown contain autonomic fibers.

lumbar nerves and (2) the **craniosacral (parasympathetic) system** emerging from the brain via cranial nerves III, VII, IX, X, and XI and from the cord via several sacral spinal nerves. Most visceral structures, except those in the skin, are supplied by postganglionic fibers from both systems (see Fig. 15-30, iris diaphragm, stomach, urinary bladder). The stimulatory effects of one system modulate the inhibitory effects of the other to bring about an appropriate response.

Autonomic ganglia may be classified in three categories: paravertebral, collateral, and terminal. **Paravertebral (chain) ganglia** lie close to the vertebral column and, except in some fishes, are connected to form a longitudinal **sympathetic trunk** (Fig. 15-30). There is approximately one paravertebral ganglion for each thoracic and lumbar nerve and several in the neck. Thoracic and lumbar paravertebral ganglia are connected to the nearest spinal nerve by a **white ramus communicans,** which conducts preganglionic fibers from the spinal nerve to the ganglion (Fig. 15-9).* It also conducts visceral afferent fibers *to* the spinal nerve. A **gray ramus communicans,** absent in elasmobranchs, returns some postganglionic fibers to the spinal nerve to be distributed to the body wall and limbs as vasomotor, pilomotor, and secretory fibers (Figs. 15-9 and 15-30). Other postganglionic fibers supply viscera in the head, neck, and thorax.

Collateral ganglia are in the head outside the skull and in the abdominal cavity close to a branch of the aorta. They are not part of a chain. Collateral ganglia in the head and their innervation are listed in Table 15-3. Collateral ganglia in the trunk (**celiac, superior mesenteric, inferior mesenteric,** others) supply postganglionic fibers to abdominal and pelvic viscera. They receive preganglionic fibers from spinal nerves via

---

*There are no white rami in the neck or sacral region.

Table 15-3. Innervation and peripheral distribution of autonomic ganglia of the head of mammals

Ganglion	Receives fibers from	Projects fibers to
Ciliary	Oculomotor nerve	Ciliary body of eye Sphincter muscles of iris
Submandibular	Facial nerve	Submandibular gland Sublingual gland
Sphenopalatine	Facial nerve	Lacrimal gland Glands of nose and pharyngeal mucosa
Otic	Glossopharyngeal nerve	Parotid gland

autonomic nerves such as the splanchnic and pelvic (Fig. 15-30).

Terminal ganglia are embedded in the walls of the organ innervated. They occur only in the trunk at the endings of preganglionic fibers of the parasympathetic system. Cell bodies in these ganglia send very short postganglionic fibers to the innervated tissue.

On emerging from autonomic ganglia, postganglionic fibers (other than those returning to spinal nerves via a gray ramus) form plexuses on the surface of nearby blood vessels and accompany these vessels to the organs. For example, the carotid and celiac plexuses consist of fibers emerging from the superior cervical and celiac ganglia, respectively.

In all vertebrates the autonomic nervous system is primarily involuntary; that is, motor impulses are initiated reflexly by visceral afferent impulses. We are seldom aware of the stimulus or the response. Some voluntary control is possible by the use of biofeedback.

☐ Chapter summary

1. The neuron is the functional unit of the nervous system. It consists of a cell body and one or more processes. The long process is an axon, or nerve fiber.

2. A nerve is a bundle of nerve fibers outside the central nervous system. A tract is a bundle of nerve fibers within the central nervous system.

3. A nucleus is a group of similarly functioning cell bodies inside the central nervous system. Nuclei constitute the gray matter of the brain and cord.

4. A ganglion is a group of cell bodies outside the central nervous system. Ganglia are sensory or autonomic (motor).

5. Neurotransmitters are amines that transmit nerve impulses across synapses. Neurosecretions are polypeptide hormones secreted by neurosecretory neurons.

6. The cell bodies of motor neurons are inside the brain or cord with one exception: the cell bodies of postganglionic fibers of the autonomic nervous system are in autonomic ganglia.

7. Cell bodies of first order sensory neurons in higher vertebrates are found in sensory ganglia on the nerves with three exceptions: the cell bodies of olfactory nerve fibers are in the olfactory epithelium, the cell bodies of optic nerve fibers are in the retina, and those of proprioceptive fibers of cranial nerves are in the mesencephalon.

8. Most spinal nerves exhibit sensory ganglia on their dorsal roots. The following cranial nerves have sensory ganglia: V (trigeminal), VII (facial), VIII (cochlear and vestibular), IX (petrosal; in mammals, inferior and superior glossopharyngeal), and X (in mammals, inferior and superior vagal).

9. Spinal nerves are metameric in origin and distribution. Most spinal nerves exhibit dorsal and ventral roots and dorsal, ventral, and communicating rami. Ventral rami often unite to form simple or complicated plexuses.

10. There is evidence that in the earliest vertebrates dorsal and ventral roots did not unite, dorsal roots were mixed and ventral roots were motor, and the cell bodies of the sensory fibers were not aggregated in ganglia. These conditions are found in lampreys, except that some sensory cell bodies aggregate on the dorsal root. Above cyclostomes, dorsal and ventral roots unite, ventral roots are motor, and dorsal roots are chiefly or wholly sensory.

11. Spinal nerves contain the following fiber components: GSA, SE, GVA, and GVE. Cranial nerves may contain one or more of the preceding and also one or more of the following: SSA, SVA, and SVE.

12. Occipitospinal nerves lacking sensory roots and supplying hypobranchial musculature arise between the vagus and first typical spinal nerves. They are more numerous in lower vertebrates and are represented in amniotes by part of nerve XI and by nerve XII.

13. Anamniotes have ten pairs of cranial nerves, amniotes have twelve. Nerves I, II, and VIII are sensory and supply special re-

386

ceptors of the head. Nerves III, IV, and VI are mixed and supply the myotomal muscles of the eyeball. Nerves V, VII, IX, and X supply the jaws and gill arches in fishes and visceral arch derivatives in tetrapods.

14. Cranial nerve V is the chief nerve for cutaneous sensation on the surface of the head and the ectodermal part of the oral cavity. Nerves VII, IX, and X supply neuromast organs and taste buds.

15. Nerve XI is derived partly from the caudal end of the branchiomeric series and partly from occipitospinal nerves; hence it has cranial and spinal roots. The internal ramus contains GVE fibers that are distributed with the vagus. The external ramus supplies the trapezius and sternocleidomastoid muscles.

16. Nerve XII represents one or more occipitospinal nerves and supplies the muscles of the tongue.

17. The terminal nerve is a vestige of an anterior branchiomeric nerve. It supplies a restricted region of the nasal mucosa with general sensory and vasomotor fibers.

18. The autonomic nervous system innervates smooth and cardiac muscles and glands. Preganlionic fibers of the craniosacral (parasympathetic) division emerge from the brain via cranial nerves III, VII, IX, X, and XI and from the sacral region of the cord via sacral spinal nerves. Preganglionic fibers of the thoracolumbar (sympathetic) division emerge from the cord via thoracic and lumbar spinal nerves.

19. The autonomic ganglia of the head and their associated nerves are ciliary (III), sphenopalatine (VII), submandibular (VII), and otic (IX). These are parasympathetic ganglia.

20. Autonomic ganglia are paravertebral (sympathetic chain), collateral (in head or near abdominal aorta), and terminal (in trunk close to or within the organ innervated). They contain cell bodies of postganglionic neurons.

21. Most viscera are supplied by sympathetic and parasympathetic fibers, but skin receives only sympathetic fibers (vasomotor, pilomotor, secretory).

22. A meninx primitiva surrounds the brain and cord in some fishes. A dura mater and leptomeninx develop in most vertebrates. In a few birds and in mammals the leptomeninx differentiates into pia mater and arachnoid membranes.

23. The spinal cord often exhibits cervical and lumbar enlargements and, in fishes, a urophysis. When the cord is shorter than the vertebral column, the cord terminates in a filum terminale surrounded by a cauda equina.

24. Cerebrospinal fluid is secreted by choroid plexuses in the lateral, third, and fourth ventricles. The fluid fills the brain ventricles, central canal of the spinal cord, and escapes to the meningeal spaces via foramina in the roof of the fourth ventricle.

25. The brain has three major subdivisions: prosencephalon (forebrain), mesencephalon (midbrain), and rhombencephalon (hindbrain). The more prominent brain stuctures are outlined in the following chart.

PROSENCEPHALON
FOREBRAIN

Telencephalon	**Diencephalon**
Rhinencephalon	Epithalamus
Olfactory bulbs	Habenulae
Olfactory tracts	Pineal
Olfactory lobes	Parapineal
Cerebral hemispheres	Thalamus
Corpora striata	Hypothalamus
(basal ganglia)	Optic chiasma
Corpus callosum	Infundibular stalk
Neocortex on pallium	and pituitary
Paraphysis	Saccus vasculosus
Lateral ventricles	Third ventricle

MESENCEPHALON
MIDBRAIN

Optic lobes ⎫
Auditory lobes ⎬ Tectum (roof)
Cerebral peduncles ⎭
Cerebral aqueduct

RHOMBENCEPHALON
HINDBRAIN

Metencephalon	**Myelencephalon**
Cerebellum	Medulla oblongata
Pons	Vagal lobes
	Pyramids
	Fourth ventricle

LITERATURE CITED AND SELECTED READINGS

1. Ariëns Kappers, C. U., Huber, G. C., and Crosby, E. C.: The comparative anatomy of the nervous system of vertebrates, including man, New York, 1936, The Macmillan Co. (Republished by Hafner Publishing Co., Inc., 1960.)
2. Flood, P. R.: A peculiar mode of muscular innervation in amphioxus. Light and electron microscopic studies of the so-called ventral roots, Journal of Comparative Neurology **126:**181, 1966.
3. Kuhlenbeck, H.: The central nervous system of vertebrates, vol. 3, part I, Basel, 1970, S. Karger AG.
4. Noback, C. R.: The human nervous system, New York, 1975, McGraw-Hill Book Co.
5. Norris, H. W., and Hughes, S. P.: The cranial, occipital, and anterior spinal nerves of the dogfish, *Squalus acanthias,* Journal of Comparative Neurology **31:**293, 1920.
6. Pearson, R., and Pearson, L.: The vertebrate brain, New York, 1976, Academic Press, Inc.
7. Sarnat, H. B., and Netsky, M. G.: Evolution of the nervous system, New York, 1974, Oxford University Press.

Symposium in American Zoologist

Recent advances in the biology of sharks, Section III, Central nervous system and sense organs, **17:**411, 1977.

16

Sense organs

In this chapter we will study the special sense organs that vertebrates use to monitor the external environment. We will learn the more gross aspects of their structure and consider their embryogenesis, evolution, and adaptive value for life in water or on land. We will also briefly examine receptors that monitor the activity of striated muscles and a few receptors that monitor the internal environment.

Special somatic receptors
  Neuromast organs
  Membranous labyrinth
  Pit receptors of reptiles
  Saccus vasculosus
  Light receptors
Special visceral receptors
  Olfactory organs
  Organs of taste
General somatic receptors
  General cutaneous receptors
  Proprioceptors
General visceral receptors

Natural selection has resulted in the evolution of a large variety of simple and complex sense organs, or receptors, for monitoring the external and internal environment. Sense organs are transducers of energy. They transduce, or change, mechanical, electrical, thermal, chemical, or radiant energy into nerve impulses in sensory nerve fibers. The energy constitutes the stimulus.

Receptors, other than chemoreceptors, that provide information about the external environment and the individual's orientation in it are **somatic receptors.** As a result of the information, the animal makes appropriate striated muscle responses that tend to assure survival. The sound of an approach-

ing enemy, for example, initiates appropriate striated muscle activity. Receptors that provide information about the internal milieu—the environment within the animal—are **visceral receptors.** As a result of the information the animal adjusts its internal environment. An abbreviated list of somatic and visceral receptors is provided in the outline at the beginning of the chapter.

Special receptors have a limited distribution in the body and, above aquatic amphibians, are confined to the head. General receptors are widely distributed over the surface and in the interior. In this chapter we will first discuss special somatic and visceral receptors, then a few general receptors.

■ SPECIAL SOMATIC RECEPTORS
□ Neuromast organs

Neuromasts are receptors in the skin of fishes and aquatic amphibians that monitor mechanical, electrical, thermal, and, perhaps, chemical, components of the surrounding water (Fig. 16-1). They are found singly or in groups or linear series. The cephalic and lateral-line canals of fishes and aquatic amphibians, the ampullae of Lorenzini of sharks, the closed vesicles of Savi of electric rays, and the pit organs of fishes are

389

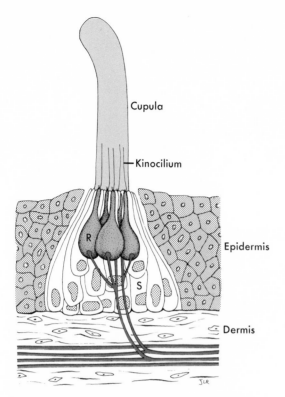

Fig. 16-1. A neuromast organ in the epidermis of a necturus. **R,** Hair cell (receptor cell) innervated by a heavily myelinated sensory nerve fiber (dark red) and having a single kinocilium and three or four short delicate stereocilia extending into the cupula; **S,** supporting cell.

variants or aggregates of neuromasts (Fig. 16-2, *A*). Although neuromasts in different locations show minor structural variety, all are fluid-filled pits. The least differentiated neuromasts (external, or naked, neuromasts) lie in the epidermis and open onto the surface (Fig. 16-1). Others lie under the skin in sunken canals that are usually embedded in dermal bone (Fig. 16-2, *B*). The canals have pores leading to the surface at intervals. The fish *Amia* has as many as 3,700 neuromast pores on the head alone.

A neuromast exhibits two cell types, **hair cells (receptor cells)**, and **supporting cells** (Fig. 16-1). Sensory nerve fibers terminate on the hair cells. Each hair cell has a num-

ber of short stereocilia and one long kinocilium that projects into fluid or into a **cupula** that projects into fluid. Mechanical changes in the fluid displace the cupula and cilia, and the shearing effect depolarizes the hair cells and stimulates the sensory nerve fiber.

The most widely studied neuromast organs are the cephalic and lateral-line canal system. In cyclostomes and amphibians the canals, consisting of linear series of neuromasts, are shallow open grooves on the surface of the head and trunk. In sharks the canals are closed on the head and trunk, but they are open grooves on the tail. In chimaeras the canals form elevated ridges on the surface. In bony fishes the canals lie on the surface of, or are even embedded in, scales or dermal bone. Grooves in the dermal armor of the earliest fossil fishes indicate that at one time the system was a network of superficial canals that covered the entire body. The network has been reduced in modern fishes by loss of many branches and by interruption of the canals at one or more locations. In some recent fishes the canals are restricted to the head.

In the salamander *Notophthalmus*, when the larva is metamorphosing into a red eft and migrating to land, the neuromast organs become buried under the proliferating stratum corneum. Later—several years later in some localities—when the eft returns to the water as a sexually mature newt, the stratum corneum is shed, and the system is again exposed.

The function of the neuromast system has been the object of much early speculation and, subsequently, of extensive sophisticated research. Two basic questions have been posed: What is the nature of the stimuli that evoke a response in the sensory cell? What role does the information play in the behavior of the organism?

There is considerable agreement that at least part of the neuromast system responds to water currents, to compression waves in the water, to hydrostatic pressure, and to

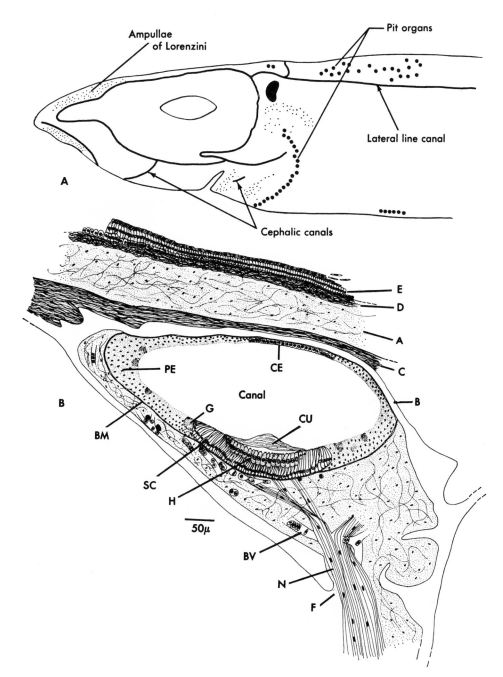

Fig. 16-2. A, Distribution of neuromast organs in skin of a shark. The precise distribution of the several types varies among species. **B,** Lateral-line canal of bony fish in cross section at the level of a neuromast organ. The canal runs longitudinally under the skin embedded in dermal bone *(B)*. *E,* Epidermis; *A, C,* and *D,* dermis. The epithelial lining of the canal rests on a basement membrane *(BM)* and exhibits a cuboidal epithelium *(CE)*, a pseudostratified epithelium *(PE)*, goblet cells *(G)*, and the neuromast organ. The latter is composed of sustentacular cells *(SC)*, sensory cells *(H)*, and a cupula *(CU)*. The sensory nerve *(N)* innervating the neuromast organ penetrates the bone via a foramen *(F)*. A blood vessel *(BV)* is shown approaching the receptor. (From Branson and Moore.[4])

weak electric potentials. Water currents include flow past the organism and currents caused by passing prey, by swimming movements, or by reflection of waves from nearby objects. The information would enable an organism to orient itself appropriately in flowing water (rheotaxis), to avoid enemies, and to participate in schooling.

With reference to electroreception, most animals produce enough electrical potential when their muscles are contracting to make their presence detectable by electroreceptive neuromasts at short range. Electroreception has been demonstrated in the ampullae of Lorenzini. Salinity changes might also be a source of electric potential stimuli.

Investigators are hesitant to ascribe sound detection to the system because of the semantic problem of a definition of "sound." Also, there is a possibility that waves of a length usually associated with sound may not be those that activate the system.

The function of the system is probably broad, and its predominant biological role probably varies with habitat—swiftly flowing streams as opposed to the depths of the ocean, for example. Although the precise role of the system remains to be demonstrated, there is no doubt that the neuromast system, along with chemoreceptors and visual organs, constitutes a very important site of input of information into the central nervous systems of aquatic vertebrates.

Neuromast organs arise from a linear series of embryonic dorsolateral **ectodermal placodes** of the head, trunk, and tail, which subsequently sink into or under the skin. The membranous labyrinth is a highly specialized neuromast organ. All neuromast organs other than the membranous labyrinth are innervated by cranial nerves VII, IX, or X, regardless of their location. The neuromast organs and the membranous labyrinth (innervated by nerve VIII) are often referred to collectively as the **acousticolateralis system.**

□ Membranous labyrinth

All vertebrates exhibit a pair of fluid-filled membranous labyrinths, or inner ears, embedded in the skull lateral to the hindbrain. The fluid is **endolymph.** In fishes other than cyclostomes, each labyrinth consists of three **semicircular ducts, a utriculus,** and a **sacculus** (Fig. 16-3). Tetrapods have, in addition, a **cochlear duct (cochlea).** The anterior and posterior semicircular ducts are vertical and lie at right angles to one another. The third semicircular duct (lateral duct) lies in a horizontal plane. Each duct exhibits a small ampulla. Emerging dorsally from the sacculus or utriculus is an **endolymphatic duct** that usually terminates blindly in a small or large **endolymphatic sac** (Fig. 16-3, man). However, in elasmobranchs the endolymphatic ducts open onto the surface of the head via endolymphatic pores. Lampreys have only the anterior and posterior semicircular ducts, and hagfish have only the posterior, although there are two ampullae.

Elevated patches of receptor sites called cristae are found in the ampullae. Others, called maculae, are found in the sacculus and utriculus. An outpocketing of the posterior wall of the sacculus, the **lagena,** contains another macula, which, in tetrapods, expands greatly to become the **cochlea,** an organ for detecting sound. These receptor sites contain hair cells and supporting cells with cilia that extend into the endolymph. Branching among the bases of the hair cells or ending on them are the sensory endings of cranial nerve VIII. Within the endolymph are calcareous concretions, the **otoliths** (calcium carbonate in association with a protein, in man). The number, size, and shape of the otoliths (also called otoconia) vary with the species. They are displaced by movements of the head in certain planes and provide some of the stimuli received by the hair cells.

The membranous labyrinth arises during embryonic life as an ectodermal placode in

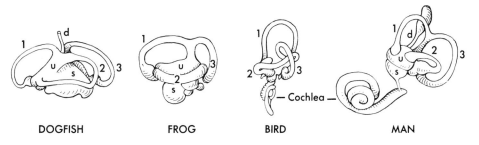

Fig. 16-3. Left inner ears of representative vertebrates. **1,** Anterior semicircular duct; **2,** lateral semicircular duct; **3,** posterior semicircular duct; **d,** endolymphatic duct and sac; **s,** sacculus; **u,** utriculus.

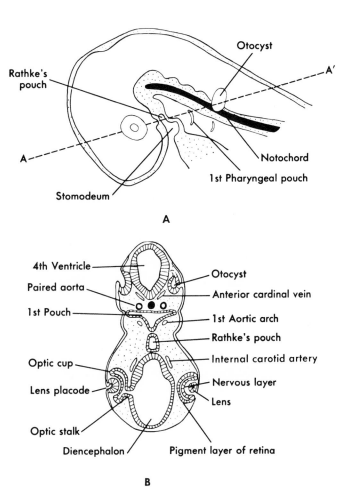

Fig. 16-4. Origin of inner ear (from otocyst), retina (from optic cup), and lens (from lens placode). **A,** Head region of embryo. **B,** Cross section of head at level of A—A'. Left side of **B** is slightly earlier than right side.

line with the placodes of the lateral-line canal system when the latter are present. The placode, like deep neuromasts, sinks under the skin to become a fluid-filled vesicle, or **otocyst** (Fig. 16-4). However, the otocyst develops into an organ of much greater complexity than other neuromast organs. The membranous labyrinth is surrounded by **perilymph.** A short, blind perilymphatic duct is present in some vertebrates.

EQUILIBRATORY FUNCTION OF THE LABYRINTH

The membranous labyrinth receives information concerning acceleration and deceleration of the head in motion (as in tilting or turning it) and the direction of rotation. The endolymph acts as a plastic body and presses on specific hair cells. The information is transmitted to the brainstem and cerebellum, where reflex movements of the eyeballs are initiated so that the eyes are always looking in the same direction in which the head is directed. (It takes conscious effort to turn your head swiftly to the left while continuing to look straight ahead. Try it!) That the labyrinth controls eyeball movements can be demonstrated by spinning someone on a revolving chair rapidly a number of times and then stopping the chair

abruptly. An observer will note that the eyeballs continue to exhibit rapid jerky side-to-front-to-side movements (nystagmus), which finally cease.

Information from the labyrinth also results reflexly in compensatory changes in position of the neck, trunk, and appendages that restore the body to proper orientation, as when a cat is turned upside down and dropped, and lands on its feet. The labyrinth therefore plays a role with the eyes and proprioceptive receptors in reflexly maintaining body posture and therefore balance or equilibrium.

AUDITORY FUNCTION OF THE LABYRINTH

The labyrinth has an auditory function as far down the phylogenetic scale as fishes. This function in fish is subserved by the **lagena.** The lagena has its own macula, which, at least in some fishes, responds to longitudinal sinusoidal waves of the frequency of sound waves.

The lagena evolved in tetrapods into the cochlea, the number of hair cells increased, and the sensory epithelium became the **organ of Corti** (Fig. 16-5). The branch of nerve VIII that supplied the lagena became the cochlear nerve.

In Cypriniformes, an order of mostly freshwater teleosts that includes catfishes and carp, sound waves in the water evoke waves of similar frequency in the gas in the swim bladder, and these are transmitted by a series of **weberian ossicles** to the inner ear. Weberian ossicles (Fig. 16-6) are modified transverse processes of the first three (occasionally four or five) trunk vertebrae. The ossicles commence at the swim bladder and end against the **sinus impar,** an extension of the perilymphatic space. In herringlike fishes the swim bladder has tubular extensions in direct contact with the sinus impar. Weberian ossicles perform the same function as the malleus, incus, and stapes of mammals. These fish can hear.

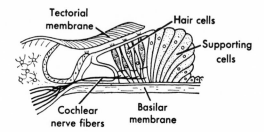

Fig. 16-5. Organ of Corti from cochlear duct. (From Schottelius, B.A., and Schottelius, D.D.: Textbook of physiology, ed. 17, St. Louis, 1973, The C. V. Mosby Co.)

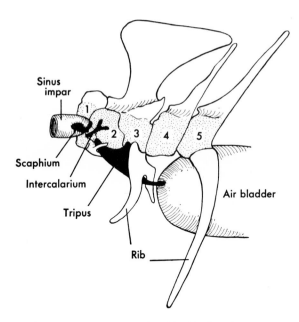

Fig. 16-6. Weberian ossicles (black) of a teleost. **1** to **5,** Centra of first five vertebrae. The sinus impar is an extension of the perilymphatic space. The ossicles are connected by modified intervertebral ligaments in life.

MIDDLE EAR OF TETRAPODS

The most common route of conduction of sound waves to the labyrinth in tetrapods is from the eardrum (**tympanic membrane**) via the hyomandibular segment of the hyoid skeleton, known in tetrapods as the **columella,** or **stapes.*** This cartilage or bone is located in the middle ear cavity, or **cavum tympanum** (Fig. 16-7).

The middle ear cavity arises as an evagination of the first pharyngeal pouch, which grows toward the hyomandibula and partially surrounds it. Erosion of the mesenchyme surrounding the hyomandibula also occurs. As a result, the hyomandibula (columella, or stapes) becomes isolated in the middle ear cavity. In mammals the posterior

---

*Although the terms are used here as synonyms, the columella and stapes or stapes complex are derived in part from elements in addition to the hyomandibula.

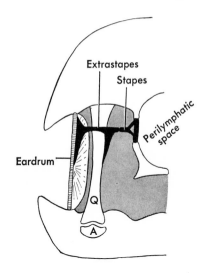

Fig. 16-7. Stapes complex (black) and middle ear cavity (gray) of a lizard. **A,** Articular bone; **Q,** quadrate bone. These two bones become ear ossicles in mammals.

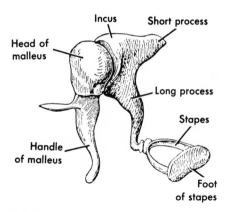

Fig. 16-8. Middle ear ossicles of a mammal. (From Schottelius, B.A., and Schottelius, D.D.: Textbook of physiology, ed. 17, St. Louis, 1973, The C. V. Mosby Co.)

tips of the embryonic upper and lower jaw cartilages are also encompassed by the expanding middle ear (Fig. 8-24). Mammals therefore have three bones in the middle ear cavity—the **malleus, incus,** and **stapes** (Fig. 16-8). The foot of the stapes is attached to a **secondary tympanic membrane** stretched across an **oval window** in the wall of the bony otic capsule. Vibration of this membrane is transmitted to the perilymph and from there to the endolymph of the inner ear. The middle ear cavity remains in communication with the pharynx throughout life via the **auditory (eustachian) tube.**

Although conduction of sound via a drum is the predominant route in tetrapods, a drum is not always employed. Urodeles, apodans, a few anurans, and limbless lizards and snakes have no eardrum, no middle ear cavity, and even the columella is often vestigial. (Of course, aquatic urodeles have a lateral line system, which compensates in part for lack of auditory structures.) Since in the absence of a drum the columella may end at the squamosal or quadrate bone, sound conduction via bone in these organisms is enhanced if the floor of the buccal cavity, the lower jaw, or the anterior limbs are in contact with a dense substrate—water or earth. The hyomandibular route to the membranous labyrinth was available long before fishes emerged onto land, since the hyomandibula articulates with the otic capsule as a primitive condition (Fig. 8-1). An intermediate stage between fishes and tetrapods in the evolution of a sound-conducting hyomandibula is seen in primitive urodeles (Fig. 8-28, *B*).

Some larval amphibians have a long slender rod extending between their lung and otic capsule and performing the same function as weberian ossicles. Amphisbaenians have a similar rod just under the skin of the lower jaw. These rods are not homologous.

OUTER EAR OF TETRAPODS

The eardrum is on the surface of the head in amphibians and most reptiles (Fig. 16-9). In crocodilians, birds, and mammals, it is deeper in the head, at the end of an air filled passageway, the **outer ear canal (external auditory meatus).** In mammals an appendage, the **pinna,** collects sound waves and directs them into the canal.

□ **Pit receptors of reptiles**

Snakes and lizards have receptors in the form of pits that open to the surface between epidermal scales. These organs are of several varieties, the most common being **apical pits.** They are scattered over the surface

Fig. 16-9. Head of an iguanid lizard showing eardrum just behind angle of jaws.

of the body, especially on the trunk. As their name denotes, they lie at the apex (posterior free border) of the scales. There are usually one or two apical pits associated with each scale, but there may be as many as seven. Frequently, a filamentous hairlike bristle projects from the pit. Since naked and encapsulated nerve endings in reptiles are buried under dense cornified scales, apical pits probably provide better sites for the input of stimuli, probably tactile.

A single pair of more specialized pit receptors is found on the head of snakes in the family Crotalidae and on boas and their relatives. In crotalid snakes the pits are directed forward and may be several millimeters wide and twice as deep. Hence they are easily seen. For this reason, crotalids are called pit vipers. The pits of pit vipers are **loreal pits,** since they are located at the posterior end of the loreal scale. (The loreal scale is found in the lore of reptiles, the region between the external naris and the eye, as shown in Fig. 5-16.) The pits in boas are slitlike and less obvious. Since they are associated with a labial scale, they are called **labial pits.**

Loreal and labial pits are thermal receptors that respond to radiant heat. Those in pit vipers are considerably more sensitive. Physiological and behavioral studies have shown loreal pits detecting temperature changes of 0.001° C at a distance of several feet. Therefore they can detect the presence of warm-blooded animals on which they prey, if the animal is only slightly warmer than the environment. Because of this sensitivity, they can locate prey and strike accurately in the dark.

The pit receptors of reptiles bear a striking morphological resemblance to the external neuromasts of fishes. It is therefore an attractive hypothesis that pit receptors are evolutionary derivatives of neuromasts. On the other hand, their innervation by spinal nerves or by cranial nerve V is not consistent with this hypothesis.

□ **Saccus vasculosus**

Elasmobranch fishes, ganoids, and teleosts have a highly vascular, thin-walled saccular evagination of the floor of the diencephalon immediately behind the pituitary gland. This is the saccus vasculosus, or infundibular organ (Fig. 15-14). It is best developed in deep sea fishes, least developed in shallow freshwater forms.

The saccus vasculosus contains an extension of the third ventricle. Its epithelial lining contains hair cells and supporting cells. The cilia of the hair cells project into the cerebrospinal fluid, and nerve fibers pass from the hair cells to the hypothalamus and other brain centers. There is strong evidence that the saccus vasculosus monitors the pressure of the cerebrospinal fluid (which varies with depth) and uses the information to regulate the volume of gas in the swim bladder via sympathetic and vagal nerve fibers. Cyclostomes and lungfishes lack a saccus vasculosus.

□ **Light receptors**

Light receptors, or **photoreceptors,** as they are also called, are sensitive to radiation in a narrow spectrum of wavelengths. The limits of the spectrum vary among species. Vertebrates as a group have two sets of photoreceptors—lateral (paired) eyes and median (unpaired) eyes. In addition to providing visual stimuli, light also regulates daily and seasonal rhythms in the endocrine system via the hypothalamus (Fig. 17-4).

LATERAL EYES

The receptor site of the lateral eye is the **retina,** a membrane rich in nervous tissue and synapses at the rear of a fluid-filled vitreous chamber of the eyeball (Fig. 16-10). The retina arises from the embryonic forebrain as an evagination that soon invaginates to become a double-walled **optic cup** (Fig. 16-4, *B*). The optic cup retains an attachment to the brain via the **optic stalk.** The layer of the optic cup that light first

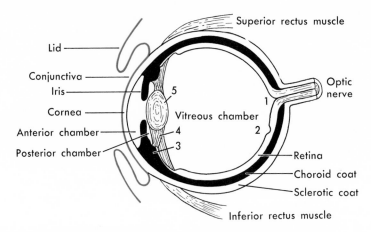

Fig. 16-10. Vertebrate eyeball in sagittal section. **1,** Blind spot; **2,** fovea; **3,** muscular ciliary body; **4,** suspensory ligament; **5,** lens. Red represents the conjunctiva.

strikes becomes the nervous layer of the retina. Some of the retinal cells differentiate to become rods and cones, which are the actual photoreceptors. Others give rise to bipolar neurons, and still others to cell bodies of optic nerve fibers (Fig. 16-11). Optic nerve cell bodies (called ganglion cells by the histologist) sprout long processes that grow along the optic stalk and into the brain. These processes constitute the optic nerve. From this description it can be seen that the retina arises from the brain and never becomes completely detached from it.

The embryonic mesenchyme surrounding the optic cup forms a pigmented **choroid**

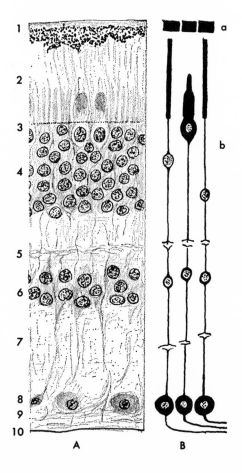

Fig. 16-11. Nervous layer of retina. **A,** Histological appearance; **B,** areas of synapse, diagrammatical. The light enters the retina at layer **10** (base of picture) and passes through the other layers to the rods and cones. **1** and **a,** Pigmented epithelium; **2,** layer of rods and cones; **3,** external limiting membrane; **4,** nuclei of the rods and cones; **5,** outer molecular layer; **6,** layer of bipolar cell bodies; **7,** inner molecular layer; **8,** layer of ganglion cells (cell bodies of optic nerve fibers); **9,** layer of optic nerve fibers; **10,** internal limiting membrane; **b,** rods and cones. (From Bevelander, G., and Ramaley, J. A.: Essentials of histology, ed. 8, St. Louis, 1978, The C. V. Mosby Co.)

coat and a fibrous **sclerotic coat,** or **sclera** (Fig. 16-10). The choroid coat is perforated anteriorly by a circular or slitlike aperature, the **pupil.** The part of the choroid coat surrounding the pupil is the **iris diaphragm.** Embedded in the diaphragm of most vertebrates are two sets of intrinsic eyeball muscles (smooth muscles, except in reptiles and birds). **Dilator muscles** are radially arranged and increase the diameter of the pupil. **Constrictors** are arranged in circular fashion and by contracting reduce the diameter of the pupil. (In most fishes the diameter of the pupil does not change.) Around the periphery of the iris diaphragm is a muscular ring, the **ciliary body.** Suspensory ligaments connect the ciliary body with the lens. Cyclostomes lack a ciliary body.

The sclerotic coat of the eyeball can be seen as the white of the eye. In front of the pupil the sclerotic coat is transparent, and this part is the **cornea.** Looking through the cornea, one can see the pigmented iris diaphragm around the pupil. The pupil appears black because there is no light emerging from the eyeball. Inserting on the sclerotic coat are the extrinsic muscles (rectus and oblique) that rotate the eyeball in its socket. The muscles are essentially identical from fish to man except that in birds they are so poorly developed that they are practically useless. As a result, a bird must move the entire head to change the direction of its gaze. The sclerotic coat may be fibrous, but more frequently cartilaginous or bony plates or ossicles develop within it, reinforcing the wall of the eyeball (Fig. 8-3).

The lens arises as a thickened placode of surface ectoderm that sinks into position in front of the optic cup. The lens placode is induced to form by organizers, or inductor substances, released by the embryonic retina. That the retina serves as the inductor may be demonstrated in an amphibian embryo by exchanging undifferentiated ectoderm from the thigh with that at the site where a lens will form. A lens will then be induced in the tissue transplanted to the head but not in the potential lens ectoderm that was removed from the influence of the retina. Removal of the retina will result in no lens placode.

The chamber behind the lens (**vitreous chamber**) becomes filled with a jellylike viscous refracting substance, the **vitreous humor.** The chamber between the lens and iris diaphragm (posterior chamber) and that between the iris diaphragm and cornea (anterior chamber) are filled with **aqueous humor.** In teleosts, snakes, lizards, and birds, a pigmented and highly vascular conical or fan-shaped projection of the choroid coat extends into the vitreous chamber nearly to the lens from a position near the entrance of the optic nerve. It is known as the **falciform process** in fishes, **conus papillaris** in reptiles, and **pecten** in birds. It is thought that these may alert the animal to nearby moving objects by casting a shadow on the retina. Their intense vascularity has also led to the hypothesis that they participate in metabolic exchanges between the blood and the interior of the eyeball.

Accommodation of the eye for near or far vision is effected differently in different vertebrates. In lampreys a **cornealis muscle** pulls on the cornea from one side, altering its curvature. In teleosts a muscular knob, the **campanula,** extends from the falciform process (described above) to the lens and draws the lens backward. In sharks, amphibians, and reptiles other than snakes, **protractor muscles** draw the lens forward. In most snakes increased pressure in the vitreous humor generated by muscles at the root of the iris forces the lens forward. In reptiles, birds, and mammals, the lens is elastic and resilient and its shape is altered by **ciliary muscles** located at the base of suspensory ligaments (Fig. 16-10). The ciliary muscles are striated in reptiles and birds, smooth and, therefore, slower to respond in mammals.

The surface of the eyeball that you can

touch with your finger is covered with transparent skin, the **bulbar conjunctiva.** This is directly continous with the **palpebral conjunctiva** on the inner surface of the lids. The palpebral conjunctiva is continous with the typical skin that covers the lids. In snakes and many lizards the upper and lower lids are permanently united in a "closed" position and are transparent, forming a **spectacle.** Each time a snake molts, the stratum corneum on the surface of the spectacle is shed, and this carries away any scratches. Lizards sometimes clean the spectacle with their tongue.

Epidermal glands in the orbit of terrestrial vertebrates keep the conjunctiva moist and clean. The **lacrimal gland** secretes a watery fluid ("tears"). It usually lies against the lateral wall of the orbit. It is poorly developed in some reptiles and birds, but it is enormous in marine turtles, in which it serves for salt excretion. A **harderian gland** secretes a more viscous fluid and is usually located against the medial wall of the orbit. It is absent in some mammals, especially those that live permanently in water. Some mammals, cats for instance, also have an **infraorbital gland.** The fluids secreted by orbital glands usually drain into a nasolacrimal canal that leads to the nasal cavity or to the vomeronasal organ.

Vertebrates that live in caves or other dark recesses (some fishes, cave salamanders, caecilians, and moles, for example) are frequently blind, or the eyes may even be vestigial. Frequently the lids fail to open. In hagfishes no eyeball whatsoever differentiates.

MEDIAN EYES

Many vertebrates below birds have a functional third eye on the top of the head (Fig. 16-12). These include lampreys, ganoid fishes, a few teleosts (especially larvae), anuran larvae, some adult anurans, and many lizards. The median eye is part of a **pineal (epiphyseal) complex** that consists of

Fig. 16-12. Third eye (parapineal eye) in center of head of an iguana. The cornea (semitransparent interparietal scale) is seen in the middorsal line between the caudal ends of the lateral eyes.

two evaginations from the roof of the diencephalon. The more anterior one becomes a parapineal and the posterior one becomes a pineal body (Fig. 16-13, *B*). It is usually the parapineal that is photosensitive in vertebrates with a third eye, but in lampreys the pineal and parapineal are both light sensitive.

In lampreys (Fig. 16-13, *A*) the pineal ends as a hollow knob beneath an area of skin devoid of pigment (the cornea) between the lateral eyes. The upper wall of the knob consists of several layers of cells that form a lens. The lower wall (retina) contains photosensitive cells and, beneath these, ganglion cells with long processes that pass down the stalk to sensory nuclei in the right side of the diencephalon. The parapineal is essentially similiar, and its descending fibers terminate on the left side. The pineal is dominant.

The parapineal eye of lizards, or **parietal**

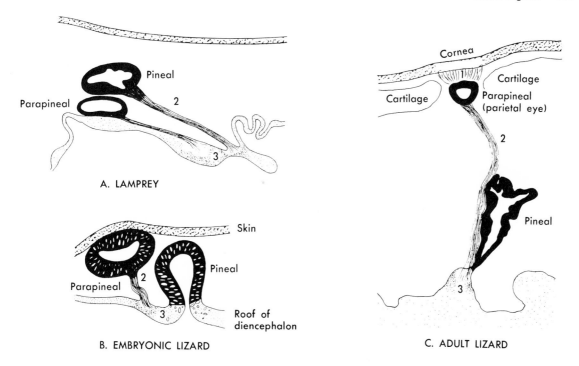

Fig. 16-13. Pineal and parapineal organs of lamprey and embryonic and adult lizard. **1,** Cornea; **2,** fiber tract; **3,** location of habenular nuclei. The parapineal of the lizard lies within the parietal foramen. (**B** and **C** modified from Nowikoff.[11])

eye, as it is sometimes called, lies in the parietal foramen immediately under a single, translucent epidermal scale in the midline of the head (Fig. 16-13, *C*). It consists of a cornea, lens, retina with photoreceptive cells resembling those of the vertebrate retina, ganglion cells, and a sensory fiber tract that extends down the epiphyseal stalk and enters the roof of the diencephalon.

In larval frogs the third eye is called **frontal organ** or **stirnorgan.** At metamorphosis the photoreceptive part of the pineal complex regresses, leaving only a glandular component that produced a hormone (melatonin) in the larva. In at least one tree frog, however, the third eye persists throughout life. Some authorities are uncertain whether the tadpole's third eye is the pineal or parapineal part of the pineal complex.

Median eyes, unlike lateral eyes, do not form retinal images. Instead, they monitor the duration of photoperiods and, probably, the intensity of solar radiation. Lizards use the input to regulate the duration of their exposure to the sun. Metabolic processes in ectotherms are accelerated by elevated body temperatures, and their temperature depends on how much solar radiation they receive. Input via median eyes is also partly responsible for daily and annual body rhythms such as daily spontaneous motor activity and seasonal recrudescence of gonads. Third eyes were present in all major groups of Devonian fishes and in many early amphibians and reptiles before being lost in more recent vertebrates.

■ SPECIAL VISCERAL RECEPTORS

There are two functional varieties of special visceral receptors, those for smell (olfactory) and those for taste (gustatory). Both are chemoreceptors that are sensitive to amino acids. The olfactory system seems to be less specific. The chief difference in the responses involves the side chains of the acids.

□ **Olfactory organs**

The epithelium of the olfactory organ in gnathostomes arises as a pair of ectodermal placodes that form just cephalad to the stomodeum. The placodes sink into the head to form a pair of olfactory, or nasal, pits (Fig. 1-11). The ectodermal lining of the pits differentiates into olfactory, supporting, and mucous cells. The olfactory cells sprout processes that grow toward and penetrate the olfactory bulb of the forebrain. There are an estimated 50 million olfactory cells in man. Their processes are the afferent fibers of the olfactory nerve. A special histological variety of branched mucous glands is characteristic of the olfactory epithelium of all vertebrates except fishes. They are called **Bowman's glands,** and they keep the olfactory epithelium moist. One theory of olfaction is that for a substance to act as an odorant, it must have a stereochemical configuration conforming to that of a binding substance on the surface of an olfactory cell.[2]

In fishes the olfactory epithelium becomes surrounded by connective tissue that forms an olfactory sac. The sacs are blind except in lobe-finned fishes, which have internal nares. A current of water into and out of the blind sacs is assured because each external naris is partitioned into incurrent and excurrent apertures so situated that the forward motion of the fish propels a stream of water into one aperture and out the other. The mucosa containing the olfactory epithelium may exhibit folds that increase the surface area. The olfactory cells monitor the water stream and are stimulated by odorants that may have their source in potential food, mates, or enemies. Probably the most primitive response to olfactory stimuli is reflex contraction of locomotor muscles, which propel the fish closer to, or farther from, the source of the odorant.

In lungfishes and tetrapods the olfactory pits push deep into the head to acquire an opening into the oral cavity or pharynx. The openings are internal nares. When this occurs, the olfactory epithelium is confined to a portion of the lining of this newly established nasal canal, so that it is appropriate in tetrapods to distinguish an olfactory epithelium and a respiratory epithelium. The olfactory epithelium contains olfactory cell bodies just as in fishes but in tetrapods it monitors an airstream instead of a water stream. Odorants in the airstream dissolve on the moist olfactory epithelium and stimulate the olfactory cells.

Olfactory mechanisms are well developed in fishes and mammals, least developed in birds, which therefore have a poor sense of smell. In some whales the olfactory nerves disappear during embryonic life. It seems logical that the ancestors of whales had a functional olfactory apparatus, but a mammal trying to smell something under water would drown! Loss of the olfactory nerve—a result of mutations—was therefore no disadvantage to the whale.

VOMERONASAL (JACOBSON'S) ORGANS

There has been a tendency among tetrapods, and even among fishes, for a ventral segment of the olfactory epithelium to become more or less isolated from the nasal passageway, to the extent of becoming an independent olfactory organ in some species. The isolated olfactory area is called a vomeronasal organ because of its frequent location above the vomer bone. It is also known as Jacobson's organ. The epithelium arises from the nasal pits.

In urodeles the vomeronasal organs are a pair of deep grooves in the ventromedial floor of the nasal canal. In anurans they are blind sacs that open to the canal. In lizards

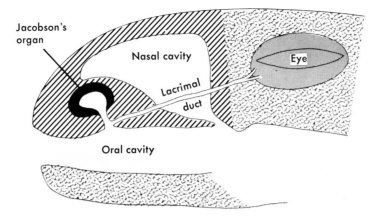

Fig. 16-14. Anatomical relationships of the vomeronasal (Jacobson's) organ in a generalized lizard. Only the left organ is shown. Diagonal lines designate a parasagittal section. The lacrimal duct is also called nasolacrimal canal.

and snakes the organs reach an evolutionary peak. They lose their connection with the nasal canal, and achieve an opening into the anterior roof of the oral cavity (Fig. 16-14), becoming two moist pockets into which the two tips of the forked tongue are thrust each time it darts out of the mouth and back. The organs monitor chemicals that accumulate on the tongue.

Vomeronasal organs are vestigial in some turtles and absent in crocodilians and birds. They are well developed in monotremes, marsupials, and generalized insectivores, in which they are tubular structures above the false palate, opening into the nasal cavity, or into the oral cavity via incisive foramina. Cats and some rodents also retain vomeronasal organs, but in most mammals they are vestigial. In man the vomeronasal organs and nerves reach maximal size about the fifth month of gestation and then regress. The epithelium of the vomeronasal organ is supplied by the vomeronasal nerve, a separate division of the olfactory nerve.

☐ Organs of taste

Taste buds (gustatory organs) have the same basic morphology as neuromasts (Fig. 16-15). Each bud is a barrel-shaped fascicle

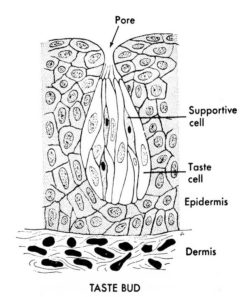

Fig. 16-15. Taste bud on the tongue of a monkey. The innervation is not shown.

of elongated taste cells and reserve taste cells (supporting cells) arranged in a pit around a central canal that opens to the surface via a taste pore. Extending from each taste cell into the canal is a cilium. Surrounding the base of the taste cells and in contact with them are sensory nerve endings. Chemicals with an appropriate configuration in solution in the bud alter the chemistry of the taste cells, which induces nerve impulses in the sensory fibers. The impulses pass to the brain. The functional life of a taste cell is only about 10 days, at which time they die from "wear and tear." They are replaced by reserve taste cells, which meanwhile have been serving as supporting cells. New reserve cells are constantly proliferated from the epithelium.

In fishes, taste buds are widely distributed in the roof, side walls, and floor of the oral cavity and pharynx, where they monitor the water stream to the gills. In bottom feeders or scavengers such as catfish, carp, and suckers, taste buds are distributed over the surface of the body to the tip of the tail. They are especially abundant on the "whiskers" of catfish.

In most tetrapods taste buds are restricted to the tongue, posterior palate, and pharynx. They are most abundant in mammals, least abundant in birds. There are more taste buds in a human embryo than at 7 years of age, as a result of failure to replace some that die. There is evidence that the mammalian fetus makes some taste discriminations before birth.

Taste buds from fishes to man are supplied by branches of cranial nerves VII, IX, and X, in that sequence, from mouth to pharynx. Hence, in man, taste buds on the anterior surface of the tongue are supplied by nerve VII, those on the posterior surface of the tongue by nerve IX, and those in the vicinity of the glottis by nerve X. All taste buds on the external surface of the body of fishes are also supplied by nerve VII. The exaggerated size of the sensory nucleus receiving the many taste fibers in such fishes is illustrated in Fig. 15-12.

■ GENERAL SOMATIC RECEPTORS

There are two categories of general somatic receptors—general cutaneous receptors and proprioceptors. The latter supply striated muscles, joints, and tendons.

□ General cutaneous receptors

The skin of all vertebrates contains free, (that is, naked) sensory endings that ramify among the epidermal cells everywhere on the surface and are stimulated by contact (Fig. 16-16). These are probably the oldest cutaneous endings in vertebrates, and in cyclostomes they are the only ones. They give rise to what has been called protopathic sensation. This is a crude, poorly localized, phylogenetically ancient sensation. It is purely protective. It is not necessary that a fish know the texture of whatever touches it or whether the object is warm or cool. The mere fact of contact indicates possible danger. The impulses ascend to the thalamus, where reflex motor activity is initiated to reorient the animal so that it avoids the stimulus. In mammals these endings give rise to vague and sometimes unpleasant sensations including pain, as in toothache. Free nerve endings for touch entwine around the base of each hair. Touch a single hair on your arm and note the sensation. Localization of the stimulus is a function of the cerebral cortex.

In addition to free endings ramifying in the epidermis, tetrapods have acquired encapsulated bulblike endings in the dermis (Fig. 16-16). They consist of endings associated with epithelial-like cells surrounded by a thin or thick connective tissue capsule. They may have evolved from free endings by the addition of capsular cells and by withdrawal into the dermal papillae or even deeper. Encapsulated endings seem to have evolved along with the cerebral hemispheres and are associated with fine touch and temperature discrimination (epicritic sensation). Birds and mammals have the largest number and variety of such corpuscles, including those on the beak of birds, at the end of the snout in mammals that rout in the soil, and on external genitalia and

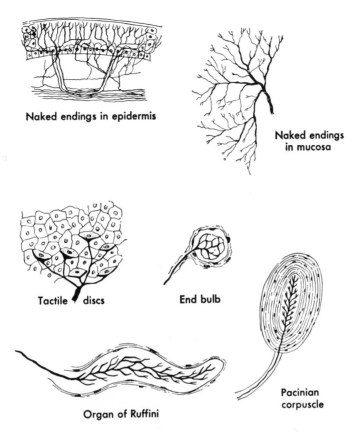

Naked endings in epidermis

Naked endings in mucosa

Tactile discs

End bulb

Organ of Ruffini

Pacinian corpuscle

Fig. 16-16. General somatic and visceral receptors. Encapsulated endings include pacinian corpuscles for touch, end bulbs (of Krause) for cold, and organs of Ruffini for heat.

other erogenous areas. Among invertebrates, the amphioxus has encapsulated endings. This is probably an instance of evolutionary convergence.

☐ Proprioceptors

Skeletal muscles, tendons, and the bursas of joints are supplied with sensory endings that are stimulated when striated muscles contract. The sensation evoked is known as proprioception, kinesthesia, or deep muscle sensibility. The sensory endings in most skeletal muscles of higher vertebrates are found in **muscle spindles.** These are tiny fusiform bundles of fewer than a dozen striated muscle fibers that are wrapped together and located among and parallel to

typical striated muscle fibers near the end of a muscle. Each muscle fiber in the spindle (intrafusal fiber) is supplied with at least one sensory (proprioceptive) nerve fiber and a gamma motor fiber (Fig. 16-17). The proprioceptive fiber monitors the contractions induced by the gamma motor fiber. Pacinian corpuscles serve as proprioceptive receptors in the bursas of joints, and unencapsulated proprioceptive endings are found in tendons.

Although some of the proprioceptive impulses reach centers of consciousness—the cerebral cortex in mammals—most of them are shunted into more ancient centers, particularly the cerebellum. Here they provide feedback for the reflex regulation of motor

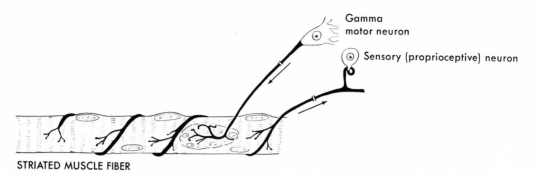

Fig. 16-17. Innervation of an intrafusal muscle fiber of a muscle spindle. The gamma motor neuron stimulates the intrafusal fiber to contract. The contraction is monitored by a sensory (proprioceptive) neuron that has an annulospiral ending wrapped around the intrafusal fiber. The proprioceptive fiber synapses in the central nervous system with motor cell bodies of regular (alpha) motor nerve fibers (not shown) that reflexly stimulate regular (nonintrafusal) muscle fibers in the muscle. The mechanism is essential for muscle tonus.

impulses that are operating the muscles. The result of the reflex arc is coordinated (synergistic) muscular activity associated with maintenance of posture, locomotion, and, in man, performance of skilled activities such as using tools and playing a piano. To demonstrate proprioception, close your eyes, then extend your arm or leg. Your awareness of the change in position is an example of conscious proprioception.

Proprioceptive fibers from muscles in the neck, trunk, and appendages enter the spinal cord via the dorsal roots of spinal nerves. Large, ascending fiber tracts in the cord (Fig. 15-8) conduct proprioceptive impulses to the brain. Fibers from striated muscles in the head enter the brain via cranial nerves. Amphioxus and lower vertebrates have simpler proprioceptive endings than those described above.

■ GENERAL VISCERAL RECEPTORS

General visceral receptors are mostly naked (that is, unencapsulated) sensory endings (Fig. 16-16) in the mucosa of the internal tubes and organs of the body, in cardiac muscle, in smooth muscle including that of blood vessels, and in the capsules,

mesenteries, and meninges of the viscera. Pacinian corpuscles are also found in mesenteries and some coelomic viscera. General visceral receptors are stimulated mechanically by stretching or chemically by the presence of certain substances, such as acid in the pyloric stomach. They are also stimulated by tactile and thermal stimuli in the pharyngeal and esophageal region, at least.

Most of the sensory input from general visceral receptors gives rise to no conscious sensation, but to reflex control of smooth muscles and glands. Because essentially the same kind of monitoring of the internal environment must take place in all vertebrates whether in water or on land, general visceral receptors are not subject to so many selective pressures as are somatic receptors. Therefore, visceral receptors vary little from fish to man.

Tetrapods have **vascular monitors** associated with the carotid arteries and systemic arch. These chemoreceptors and baroreceptors monitor the oxygen content and pressure of blood coming from the heart. Three such receptors have been described in the cervical region of tetrapods. These are the carotid body, the carotid sinus, and the aortic body. Similar structures have not been described in fishes.

The **carotid body,** located close to the common carotid or internal carotid artery, receives an arterial supply from one of these vessels. It is richly supplied with sensory endings of cranial nerve IX and monitors the oxygen and, perhaps, carbon dioxide in blood passing through the organ. Anoxia evokes a strong discharge of sensory impulses, which reflexly increase the respiratory rate. It may be responsible for the feeling of being "out of breath." In turtles, birds, and mammals, the carotid body is a discrete organ typically flattened against the wall of the carotid vessel. In lizards it is in the adventitia of the carotid artery. None has been described in snakes or crocodilians, but a diffuse receptor undoubtedly exists, since even in mammals the carotid body is sometimes broken into tiny nodules microscopic in size. Amphibians exhibit a structure called the carotid labyrinth, but its homologies are not known.

The **carotid sinus** is a bulbous swelling of the common carotid or internal carotid artery close to the origin of the latter. The walls of the sinus exhibit localized thinning and are innervated by sensory endings of cranial nerve IX. The sinus serves as a baroreceptor that monitors arterial blood pressure. Lowered pressure evokes a sensory discharge that provides input to the cardiovascular regulatory center of the medulla. The carotid sinus nerve innervates both the sinus and the carotid body. It contains sensory fibers of nerve IX.

The **aortic body** has been described only in mammals. It lies close to the arch of the aorta. It is histologically identical with the carotid body and has the same function, although it is innervated by sensory fibers in nerve X. Other vascular monitors include osmoreceptors in the hypothalamus and elsewhere.

☐ Chapter summary

1. Receptors are transducers of mechanical, electrical, thermal, chemical, or radiant energy.
2. Somatic receptors provide information about the external environment other than via taste and smell, and about skeletal muscle action (proprioception).
3. Visceral receptors are for taste, smell, and monitoring the internal milieu.
4. Special receptors have a limited distribution in the body. General receptors are widely distributed.
5. Neuromast organs are fluid-filled pits, ampullae, or canals lined with ciliated hair cells and supporting cells. They are found in fishes and aquatic amphibians and function in electroreception and mechanoreception, at least. They are innervated by cranial nerves VII, IX, and X.
6. The membranous labyrinth, or inner ear, is a neuromast-like complex consisting of semicircular ducts, sacculus, utriculus, and lagena, or cochlea. It always has an equilibratory role and often has an auditory role. Endolymphatic and perilymphatic ducts are often associated with it.
7. Stimulation of the semicircular ducts results in reflex movements of the eyeballs and muscles that maintain posture.
8. Stimulation of the sacculus and lagena of many fishes and of the cochlea of tetrapods results in hearing. Auditory stimuli most often reach the endolymph by bone conduction, chiefly via weberian ossicles in cypriniform fishes and middle ear ossicle(s) in tetrapods.
9. Inner ears are found in all vertebrates; middle ears are characteristic of tetrapods; outer ears are found only in amniotes; and pinnas are found only in mammals.
10. Snakes and lizards have pit receptors of uncertain homology. Apical pits are found at the apex of epidermal scales and are probably mechanoreceptors. Loreal and labial "pit organs" are found on the head of crotalid snakes and boas and are thermoreceptors for radiant heat.
11. The saccus vasculosus is a midventral evagination of the diencephalon caudal to the pituitary in fishes. The epithelium has hair cells and supporting cells of unknown function.
12. Vertebrates have two varieties of light receptors, paired (lateral) eyes and unpaired (median) eyes. The latter are part of the pineal (epiphyseal) complex.
13. Median eyes usually have a cornea, lens, and retina but do not form an image. They are found in lampreys, some jawed fishes, larval anurans, snakes, and some lizards. The parapineal serves as a photoreceptor more frequently than the pineal.
14. Organs for smell are special visceral chemoreceptors. They are the olfactory sacs of fishes, the olfactory epithelium of the nasal canal of tetrapods, and the vomeronasal organs of many tetrapods. All are innervated by fibers of the olfactory nerve.
15. Taste buds (gustatory organs) are special visceral chemoreceptors. They are distributed over the surface of the head, trunk, and tail of some fishes, confined to the head in others, and restricted to the buccal cavity and pharynx in tetrapods. They are least

abundant in birds. They are supplied by nerves VII, IX, and X in all vertebrates.

16. General cutaneous receptors include naked endings that ramify among epidermal cells, and encapsulated endings in or beneath the dermis. They modulate chiefly touch, temperature, pain, and deep pressure sensibility from the skin.

17. Proprioceptors are found in skeletal muscles, tendons, and bursas. In higher vertebrates those in muscles are in muscle spindles. Proprioceptors provide feedback from skeletal muscles, which facilitates muscle coordination.

18. General visceral receptors are mostly naked endings in visceral sites. They are chiefly stretch receptors and chemoreceptors.

19. Carotid bodies are chemoreceptors that monitor the oxygen content and, perhaps, carbon dioxide, in the blood. The carotid sinus is a baroreceptor that monitors blood pressure. These two receptors are widely distributed among amniotes and are innervated by nerve IX.

20. The unpaired aortic body lies on the arch of the aorta of mammals. It has a function identical with that of the carotid bodies and is innervated by nerve X.

21. Osmoreceptors are found in the hypothalamus and elsewhere.

LITERATURE CITED AND SELECTED READINGS

1. Adams, W. E.: The comparative morphology of the carotid body and carotid sinus, Springfield, Ill., 1958, Charles C Thomas, Publisher.
2. Amoore, J. E., Johnston, J. W., Jr., and Rubin, M.: The stereochemical theory of odor, Scientific American 210:42, February, 1964.
3. Bellairs, A.: Observations on the Jacobson organ, Journal of Anatomy 76:168, 1942.
4. Branson, B. A., and Moore, G. A.: The lateralis components of the acoustico-lateralis system in the sunfish family Centrarchidae, Copeia, no. 1, p. 1, 1962.
5. Bullock, T. H.: Seeing the world through a new sense: electroreception in fish, American Scientist 61:316, 1973.
6. Cahn, P. H., editor: Lateral-line detectors, Bloomington, 1967, Indiana University Press.
7. Denison, R. H.: The origin of the lateral-line sensory system, American Zoologist 6:369, 1966.
8. Dodt, E.: The parietal eye (pineal and parietal organs) of lower vertebrates. In Jung, R., editor: Handbook, of sensory physiology, part B, vol. 7/3B, Berlin, 1973, Springer-Verlag.
9. Eakin, R. M.: The third eye, Berkeley, 1973, University of California Press.
10. Hopkins, C. D.: Electric communication in fish, American Scientist 62:426, 1974.
11. Nowikoff, M.: Untersuchungen über den Bau, die Entwicklung und die Bedeutung des Parietalauges von aurien, Zeitschrift für wissenschaftliche Zoologie 96:118, 1910.
12. Prince, J. H.: Comparative anatomy of the eye, Springfield, Ill., 1956, Charles C Thomas, Publisher.
13. Prosser, C. L., editor: Comparative animal physiology, ed. 3, vol. 2, Sensory, effector, and neuroendocrine physiology, Philadelphia, 1973, W. B. Saunders Co.
14. Szabo, T.: Sense organs of the lateral line system in some electric fishes of the Gymnotidae, Mormyridae and Gymnarchidae, Journal of Morphology 117:229, 1965.
15. Wurtman, R. J., Axelrod, J., and Kelly, D. E.: The pineal, New York, 1968, Academic Press, Inc.

Symposia in American Zoologist

The vertebrate ear, 6:368, 1966.
Vertebrate olfaction, 7:385, 1967.
Vertebrate sound production, 13:1137, 1973.
Recent advances in the biology of sharks, Section III, Cranial nerves and sense organs, 17:411, 1977.

17

Endocrine organs

In this chapter we will see that vertebrates from fishes to man have the same endocrine array, that the endocrine organs arise from the same embryonic precursors in all, that they are similarly regulated, and that they synthesize the same basic molecules. We will meet a few examples of anatomical and molecular mutations that have occurred and note the survival value of a few of these in the environments in which vertebrates live and reproduce.

An endocrine organ produces one or more hormones. **Hormones** are products of specific groups of cells that have a regulatory effect on other specific cells, called "target cells." Target cells may be adjacent to the cells producing the hormone, or remote. More often they are remote.

Some endocrine organs apparently evolved from clusters of hormone-producing cells located at one time in the lining of the digestive tract. This seems to be true of the thyroid, for example, as we will see

shortly. And, in fact, the thyroid still has a duct leading to the pharynx in some adult vertebrates. Cells that secrete insulin are found usually, but not always, in the pancreas of vertebrates; however, in many invertebrates they are in the gastrointestinal lining. They are close to that location in some cyclostomes. Also, some of the hormones that regulate digestive organs—gastrin (which causes the stomach lining to release gastric juice), cholecystokinin (which causes the gallbladder to empty), and pancreozymin (which causes the pancreas to release certain digestive enzymes)—are still produced by cells in the lining of the digestive tract. Since these "gastrointestinal" hormones are not produced in conventional endocrine organs we will not discuss them further.

Some vertebrate endocrine organs, the pituitary and adrenal, for example, evolved phylogenetically from two entirely separate endocrine tissues that became anatomically associated during subsequent evolution. Evidence will be presented to support this idea. Finally, the brain and spinal cord produce hormones. We will begin this chapter with a discussion of the endocrine roles of the nervous system. Then we will discuss endocrine tissues derived from ectoderm, mesoderm, and endoderm, in that order.

■ ENDOCRINE ROLES OF THE NERVOUS SYSTEM

The hormones of the nervous system are **neurosecretions.** They are produced by a special kind of neuron called a **neurosecretory neuron.** Neurosecretions should not be confused with neurotransmitters (p. 355); neurosecretions are small polypeptides, neurotransmitters are amines.*

---

*In some instances neurosecretions serve also as neurotransmitters, and neurotransmitters—epinephrine, for example—serve in certain locations as hormones.

Neurosecretions are synthesized in the cell bodies of neurosecretory neurons. The cell bodies are in neurosecretory nuclei (Fig. 17-1). The neurosecretions move along the axons, or **neurosecretory fibers,** bound to stainable proteins (neurophysins) and accumulate at axon terminals until released. The terminals lie in intimate association with sinusoidal vascular channels, and the two constitute a **neurohemal organ.** The major neurohemal organs of vertebrates are the posterior lobe of the pituitary, median eminence, and the urophysis at the end of the spinal cord of fishes. The release of neurosecretions into the vascular channels is regulated reflexly.

Most vertebrate neurosecretions are produced in the hypothalamus (Fig. 17-2). Some hypothalamic neurosecretions are re-

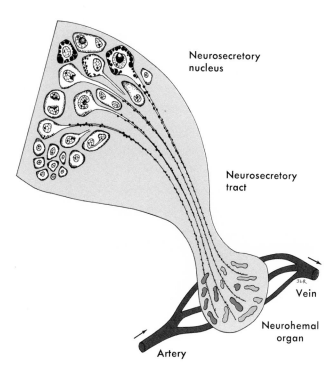

Neurosecretory nucleus

Neurosecretory tract

Vein

Neurohemal organ

Artery

Fig. 17-1. Neurosecretory neurons in a functional neurosecretory unit. The neurohemal organ contains sinusoidal vascular channels. The chief neurohemal organs in vertebrates are urophysis, posterior lobe of the pituitary, and median eminence of the diencephalic floor.

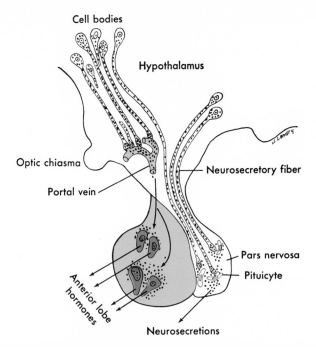

Fig. 17-2. Hypothalamic neurosecretory neurons. Cell bodies in the hypothalamus manufacture neurosecretions (black granules) that flow along the axons (neurosecretory fibers) and are discharged into the hypophyseal portal vein or into vascular channels in the pars nervosa (posterior lobe) of the pituitary. Neurosecretions released into the portal vein help regulate the hormone-producing cells (dark red) of the anterior lobe. Those released in the pars nervosa affect tissue remote from the pituitary.

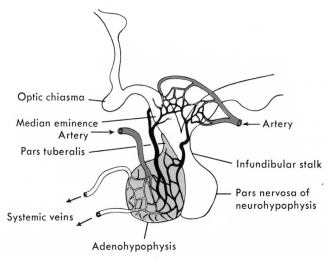

Fig. 17-3. The hypophyseal portal system (black) of mammals, schematic. Arrows indicate direction of blood flow.

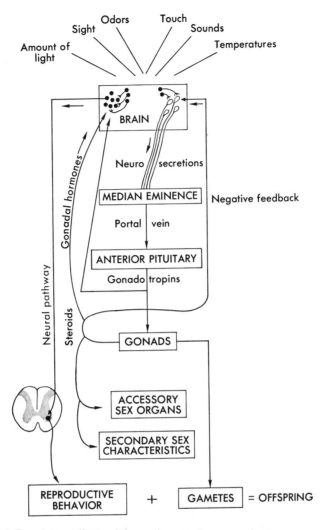

Fig. 17-4. Regulatory effects of the environment on reproduction.

leased in the median eminence, where they enter the hypophyseal portal vein (Figs. 17-2 and 17-3). This vein carries them to the adenohypophysis, which is their target. Others are released in the posterior lobe of the pituitary (Fig. 17-2) where they enter the general circulation.

The neurosecretions of the hypothalamus are regulated partly by the cyclical external environment. The two most obvious cycles are the daily (circadian) rhythm of light (photoperiod) and darkness, and the 12 month (annual) cycle of seasonal changes in temperature, day length, rainfall, salinity, and other variables. Light, temperature, and other components of the external environment are monitored by sense organs, and the information is transmitted to the hypothalamus over afferent nerves. Some of the secretions of the hypothalamus regulate the anterior lobe, and the latter secretes hormones that promote body activities favorable for the species under existing conditions. One effect (Fig. 17-4) is promotion of gametogenesis and appropriate reproductive behavior (migration, territory defense, mating behavior, nest building, care of the eggs and young) at the precise time of year when environmental conditions are most suitable for the survival of offspring at birth. Such adaptive neuroendocrine reflex arcs (receptor–afferent nerve–brain–neurosecretions–median eminence–hypophyseal portal system–effector) are evidently the result of natural selection. Neurosecretions are found as far down in the animal kingdom as coelenterates.

■ ENDOCRINE ORGANS DERIVED FROM ECTODERM
□ Pituitary gland

The pituitary gland, or hypophysis, lies underneath the diencephalon cradled, except in cyclostomes, in a depression in the sphenoid area of the skull. Because of its shape, the depression is called the sella turcica. The gland consists of two major subdivisions with different embryonic origins: a **neurohypophysis** derived from the floor of the diencephalon and an **adenohypophysis** derived from the roof of the stomodeum (Fig. 17-5). Their major subdivisions, with some common synonyms in parentheses, are as follows:

Neurohypophysis (pars neuralis)
    Median eminence
    Infundibular stalk
    Pars nervosa (posterior lobe; neural lobe)
Adenohypophysis (pars buccalis)
    Pars intermedia
    Pars distalis (anterior lobe)
    Pars tuberalis
    Ventral (inferior) lobes of elasmobranchs

NEUROHYPOPHYSIS

The neurohypophysis is that part of the pituitary gland that arises from the floor of the diencephalon (Fig. 17-5). Since during development the floor tends to evaginate ventrally, it contains a shallow or deep recess of the third ventricle. The recess is most prominent in amniotes, in which the diencephalic floor is drawn out into an elongated infundibular stalk at the end of which is the **pars nervosa,** or **posterior lobe** (Fig. 17-5, *C*). The posterior lobe is a neurohemal organ and produces no hormones. The neurosecretions released there are octapep-

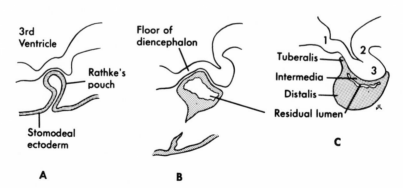

Fig. 17-5. Embryogenesis of amniote pituitary. **A,** Rathke's pouch stage. **B,** Isolation of adenohypophyseal anlage (gray) in contact with the floor of the diencephalon. **C,** Young pituitary consisting of adenohypophysis (gray) and neurohypophysis (white). **1,** Median eminence; **2,** infundibular stalk; **3,** pars nervosa (posterior lobe). The subdivisions of the adenohypophysis are labeled.

tides. They are transported to the heart for distribution throughout the body.

Arginine vasotocin is the oldest "posterior lobe hormone." It is the only one in cyclostomes and is found in all higher vertebrate classes, although in mammals it appears to be confined to fetuses. In terrestrial vertebrates it prevents dehydration. It does this by acting on kidney tubules, causing them to reclaim water from glomerular filtrate, and by bringing about resorption of water stored in urinary bladders. (The bladder is not an excretory organ, in the functional sense. It is a reservoir for water that, under arid conditions, becomes precious.) Arginine vasotocin also causes absorption of water from soil through the skin in amphibians and lungfishes that burrow during arid conditions. **Arginine vasopressin,** a mutant molecule, regulates water and salt excretion in mammals and, for this reason, is often called "antidiuretic" hormone (Fig 17-6). Another mutant, **oxytocin,** induces uterine contractions during birth of mammalian young and causes the "letdown" of milk into the nipples during nursing. What these and other mutants are able to do for an animal depends on the ability of a specific group of cells, whether contractile, secretory, or otherwise, to respond to (that is, to have molecular receptor sites for) the circulating mutant. A mutant molecule could have no effect on a species that has no cells competent to respond.

Just behind the optic chiasma the diencephalic floor exhibits a swollen area, the **median eminence** (Fig. 17-3). This, too, is a neurohemal organ. The neurosecretions released here pass via the hypophyseal portal vein to the adenohypophysis (Figs. 17-2 and 17-3). Hypophyseal portal systems are rarely found in teleosts. The anatomical relationship of neurohypophysis to adenohypophysis has become so intimate that no portal veins are needed (Fig. 17-7, trout).

ADENOHYPOPHYSIS

The adenohypophysis arises as a bud of ectodermal cells from the roof of the stomodeum. In amniotes and some lower fishes and selachians the bud is hollow and is known as Rathke's pouch. In other fishes and in amphibians the bud is solid. When the anlage of the adenohypophysis has made intimate contact over a broad area with the floor of the brain, the connection between the stomodeum and adenohypophysis usually disappears. However, it remains as an open ciliated duct leading to the buccal cavity in *Calamoichthys, Polypterus,* and some primitive teleosts. Rathke's pouch may remain in the adult gland as a residual lumen (Fig. 17-5, *C*).

The origin of the adenohypophysis from

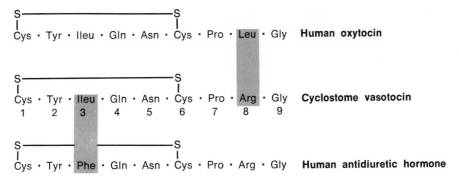

Fig 17-6. Two evolutionary mutants of the posterior lobe hormone of cyclostomes. Gray bars designate the only differences between cyclostome and human posterior lobe hormones.

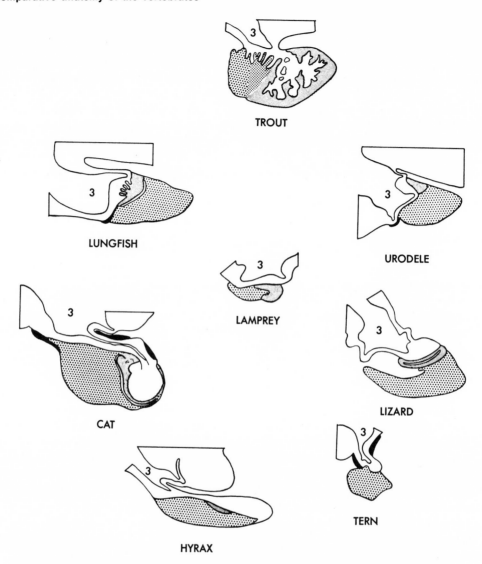

Fig. 17-7. Pituitaries of representative vertebrates, sagittal views. Dots represent the pars distalis, gray denotes the pars intermedia, and black denotes the pars tuberalis. The pars distalis of teleosts (trout) exhibits two cytological regions: a rostral part (large dots) and a proximal part (small dots). The neurohypophysis and associated parts of the brain are white. The infundibular stalk contains a recess, **3,** of the third ventricle.

the stomodeum suggests that primitively the adenohypophysis may have secreted into the buccal cavity. It was believed at one time that the pituitary (*pitua* means phlegm) was the source of phlegm that falls into the throat. Although they were wrong about the source of phlegm, they may have been naively correct about the primitive site of secretion of the pituitary!

The adenohypophysis typically exhibits three regions—pars intermedia, pars distalis, and pars tuberalis (Fig. 17-5). The pars tuberalis is a paired cephalic extension of the distalis along the floor of the diencephalon or up the infundibular stalk. No pars tuberalis forms in most fishes, snakes, or lizards, but the rostral part of the distalis in teleosts (Fig. 17-7, trout) may be a homologue. In elasmobranchs the adenohypophysis has a fourth subdivision, the ventral (inferior) lobes.

The pars intermedia lies in intimate contact with the neurohypophysis. However, in some birds and in cetaceans, manatees, and a few other mammals, no recognizable pars intermedia develops. In apes and man the intermedia grows smaller in size with age, after being relatively large in the embryo. Size differences are probably related to the function of intermedin, also known as **chromatophorotropic hormone** and **melanophore-stimulating hormone,** a product of the intermedia. The hormone causes pigment granules in some chromatophores to disperse, thus darkening the skin. Physiological color changes (p. 115) are characteristic only of cold-blooded animals. Intermedin also initiates melanogenesis in mammalian hair follicles to produce morphological color changes.

The pars distalis secretes the following hormones:

Somatotropin (growth hormone)
Thyrotropin (thyroid-stimulating hormone)
Adrenocorticotropin
Gonadotropins
  Follicle-stimulating hormone
  Luteinizing hormone—in males known as
    interstitial cell–stimulating hormone
Prolactin

Somatotropin stimulates the synthesis of proteins from amino acids; hence it is a general growth-promoting hormone. Thyroid-stimulating hormone stimulates the thyroid gland to accumulate iodine, to synthesize thyroid hormone, and to release thyroid hormone into the circulatory channels. Adrenocorticotropin affects certain zones of the adrenal cortex. Follicle-stimu-

lating hormone acts on ovarian follicles. Luteinizing hormone induces the ovulated ovarian follicle to develop into a new endocrine body, the corpus luteum. In males luteinizing hormone is better known as interstitial cell–stimulating hormone. It induces the interstitial cells of the testes to produce testosterone.

Prolactin is an ancient hormone found in all vertebrates either as a separate molecule or as part of a hormonal complex. Along with other hormones, it initiates the secretion of milk by the mammary gland, and this accounts for the name. Prolactin has the same effect on the gland of the crop sac of pigeons. The crop sac secretes a nutritive "pigeon milk," which is regurgitated and fed to nestlings by the parent. In some teleost fishes, prolactin stimulates the secretion of parental mucus, which is eaten by the young for nourishment. It has also been implicated in the osmoregulatory adjustments necessary for **anadromous** (migratory) saltwater fishes to survive in fresh water. This same hormone causes red efts to migrate to the ponds where they will mate when their gonads have reached maturity. In all vertebrates prolactin has a role in inducing certain parental behavior patterns, including the building of nests (simple in most cold-blooded vertebrates, sometimes complex in warm-blooded ones); cleaning the young; protection, turning, and incubation of eggs; and protection of the young, including transporting them about in some species. The diverse effects of prolactin, only a few of which are cited, are probably attributable to a single metabolic role of the molecule.

☐ **Urophysis**

Fishes other than cyclostomes have a neurohemal organ at the base of the tail (Fig. 17-8). It is called the urophysis, or caudal neurosecretory organ. It receives the axon terminals of neurosecretory fibers whose cell bodies, sometimes called Dahlgren

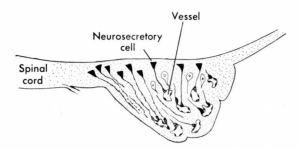

Fig. 17-8. Urophysis of a carp.

cells, are in the gray matter of the cord. The neurosecretions affect a number of functions, including thermoregulation, but its precise role is not yet clear.

□ Pineal body

The pineal organ, an evagination of the roof of the diencephalon, produces the amine **melatonin.** Mammalian melatonin causes melanin granules in dermal, but not epidermal, melanophores of larval amphibian skin to aggregate, thereby blanching the skin. The effect is opposite that of intermedin. Melatonin also has a gonadal-regulating activity (inhibitory) in some species.

Experimental evidence indicates that light affects the synthesis of melatonin, impeding it in some species, though apparently not in all. The effect may be direct, when the pineal is located under translucent skin; otherwise, the effect is via the optic nerves, sympathetic trunk, and nervi conarii (p. 368). Amphibian larvae with pineal organs intact become pale if placed in the dark. Pinealectomy abolishes this response. Additional discussions of the pineal will be found in Chapters 15 and 16.

□ Adrenal medulla and aminogenic tissue

The adrenal in most vertebrates is a gland on the ventral surface of the kidney or near its cephalic pole. In mammals it consists of two components, a peripheral cortex and a central medulla (Fig. 17-9). However, cortex

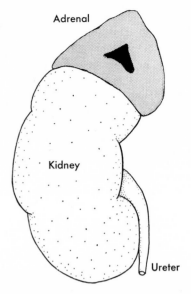

Fig. 17-9. Adrenal of man. The cortex (steroidogenic tissue, gray) surrounds the medulla (aminogenic tissue, black).

and medulla are two entirely different glands separated from each other in many fishes and more or less interspersed among one another in other vertebrates below mammals (Fig. 17-10). Therefore, we must discuss the adrenal as two separate glands, which it really is.

One component of the adrenal complex is ectodermal. This tissue, homologous with the adrenal medulla of mammals, synthesizes catecholamines, chiefly epinephrine (adrenaline). Because of its staining reaction, this aminogenic component is called chromaffin tissue.* It arises from neural crests.

In lampreys and some teleosts, the aminogenic cells lie in clusters scattered along the postcardinal vein. In sharks and rays they usually form one or several masses be-

---

*Chromaffin tissue is widely distributed in the trunk. Masses occur in close association with sympathetic ganglia where they are called paraganglia, and in the gonads, kidneys, heart, and other viscera. However, not all chromaffin tissue is aminogenic.

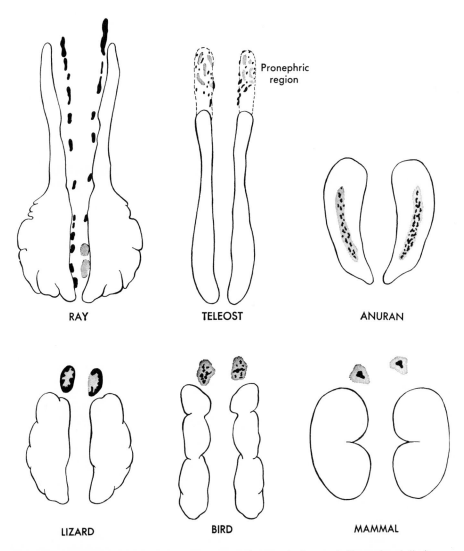

Fig. 17-10. Adrenal components in selected vertebrates. Aminogenic tissue (medulla in mammals) is shown in black, steroidogenic tissue (cortex in mammals) in gray. The kidneys are shown in outline.

tween the caudal ends of the kidneys (Fig. 17-10, ray). In most teleosts they are near the cranial end of the kidneys, frequently in association with vestiges of the pronephroi (Fig. 17-10, teleost). In this location they are more or less interspersed among the other adrenal component (steroidogenic tissue). In lungfishes they are along the dorsal aorta.

In most tetrapods the two components are interspersed. However, in some lizards and snakes the aminogenic tissue tends to aggregate, forming an almost complete capsule around the steroidogenic tissue (Fig. 17-10, lizard). This is the reverse of the condition in mammals. But even in mammals there are species in which the steroidogenic tissue does not form a complete cortex. In sea lions, for example, cortical tissue is scattered in the medulla, and medullary tissue is scattered in the cortex.

The adrenal glands of anurans are flattened, elongated masses on the ventral surface of the kidneys (Fig. 14-15). In urodeles they form small bright flecks and nodules along the postcava and are difficult to locate without a lens. In amniotes the adrenals are at or near the cephalic pole of the kidney. For this reason, in erect mammals they are also called **suprarenal** glands.

Epinephrine has a number of endocrine roles (Fig. 17-11). The most prominent are to increase the amount of blood sugar in

times of sudden metabolic need and to stimulate increased production of adrenocortical hormones in times of prolonged stress. Norèpinephrine, another amine produced by medullary tissue, is concerned chiefly with maintaining the tonus of the circulatory system through its vasoconstrictor effect.

The aminogenic cells of the adrenal complex and the cell bodies of postganglionic neurons in sympathetic ganglia are the same kind of cells. Both arise from neural crests, both are stimulated by preganglionic neurons, and both produce catecholamines in response to stimulation. The cells of the medulla are potential neurons that fail to sprout processes. They secrete higher ratios of epinephrine than do most postganglionic neurons.

■ ENDOCRINE ORGANS DERIVED FROM MESODERM
□ Adrenal cortex and steroidogenic tissue

We have just seen that the adrenal complex of vertebrates is composed of two entirely different components, aminogenic and steroidogenic, which may be intimately associated or spatially separated. The aminogenic component becomes the adrenal medulla in mammals, and the steroidogenic component becomes the cortex.

The steroidogenic component is derived from mesodermal cells that arise from the coelomic mesoderm of the gonadal ridge (Fig. 14-16) and from the underlying nephrogenic mesoderm. Therefore steroidogenic cells are closely associated with kidneys from their first appearance. In elasmobranchs the steroidogenic cells aggregate to form clusters or elongate masses lying between the kidneys and called, appropriately, **interrenal bodies.** In teleosts the steroidogenic cells most frequently aggregate in the pronephric region near the aminogenic cells, and in tetrapods they are interspersed among the aminogenic cells to form a more or less discrete adrenal

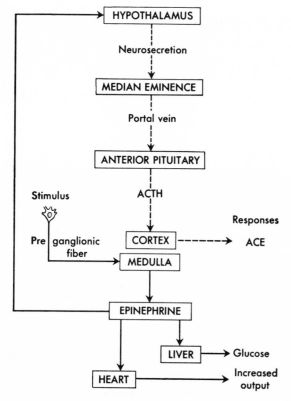

Fig. 17-11. Some regulatory functions of the adrenal medulla. When presented with a suitable neural stimulus (left center), a preganglionic neuron of the sympathetic nervous system stimulates the medulla to release epinephrine. The latter elicits many responses, three of which are indicated at the right. **ACE,** Adrenal corticoids.

gland on or near the kidney (Fig. 17-10). All such steroidogenic masses are homologous with the adrenal cortex of mammals. Although the term "cortical tissue" is appropriate for these masses in mammals only, it is frequently used to designate all steroidogenic masses that produce "corticoids" (steroids similar to those of the mammalian adrenal cortex).

Corticoids regulate sodium levels by acting on the gills and kidneys of fishes and on the kidneys of terrestrial vertebrates. The most potent sodium-regulating steroid is **aldosterone**. The predominant effect of other corticoids such as **cortisone, cortisol** and **corticosterone** is to stimulate the conversion of proteins into sugar.

The origin of steroidogenic tissue from the same mesothelium that gives rise to gonads is interesting because corticoid tissue and gonads both produce steroid hormones, and they are the only vertebrate tissues that do so. The mammalian adrenal cortex produces some fifty different steroids, including small amounts of male and female sex hormones. The bearded lady is an example of what may happen when the adrenal cortex of a female produces excessive quantities of male hormones.

□ **Corpuscles of Stannius**

Embedded in the posterior part of the mesonephric kidneys or attached to the mesonephric ducts of ray-finned fishes are spherical epithelioid bodies, the corpuscles of Stannius. They are easily mistaken for interrenal bodies, but their embryonic origin is different, since they arise as evaginations of the pronephric duct. In most teleosts there are two, but in *Amia* there are forty to fifty. In large salmon the corpuscles may reach 0.5 cm in diameter.

The corpuscles perform an endocrine role distinct from that of the steroidogenic component of the adrenal complex, but their precise role remains to be clarified. They are capable of converting one steroid into another in certain species, but no steroids have been demonstrated in the corpuscles. Ablation of the corpuscles has been followed by changes in calcium and sodium within body fluids. Ablation also caused a fall in arterial blood pressure in freshwater eels, and extracts of the corpuscles raised the blood pressure of rats as well as eels. There seems to be a functional relationship, direct or indirect, between corpuscles and steroidogenic tissue, since ablation of the latter stimulated the corpuscles and vice versa. The secretions of the corpuscles probably supplement other hormones in the maintenance of electrolyte homeostasis and osmoregulation.

□ **Gonads as endocrine organs**

Gonads arise from the coelomic mesothelium as a pair of gonadal ridges medial to the kidneys (Fig. 14-16). Ovaries and testes of most vertebrates produce three types of steroid hormones—**estrogens, androgens,** and **progestogens.** Collectively, these hormones are essential for reproduction.

Estrogens produced by ovarian follicles and androgens produced by interstitial cells of the testes affect most prominently the accessory sex organs, including the reproductive tracts. Differentiation of muellerian ducts to become uteri and oviducts is partly an expression of the effects of estrogens, and failure of muellerian ducts to develop in males may be ascribed, in some lower vertebrates at least, to the dominance of androgens. These steroids are also responsible for secondary sex characteristics, such as mammary glands in female mammals and large muscles and skeleton in males. They also regulate reproductive behavior.

Progesterone is an intermediate precursor in the synthesis of estrogens and androgens. In female mammals it has achieved an independent role—maintaining the uterus in a progestational state—a state supporting pregnancy. By negative feedback to the hypothalamus (Fig. 17-4), it also inhibits the formation of a new wave of ovarian follicles and hence delays the next ovulation. Much of the progesterone in mammals is a product of the corpus luteum. This is the name given to an ovarian follicle after the cells have undergone luteinization (chemical and morphological change) under the influence of luteinizing hormone from the pituitary.

In addition to steroid hormones, the mammalian ovary produces **relaxin,** a peptide. Relaxin produced during pregnancy softens the ligaments of the pubic symphysis and sacroiliac joints before birth, thus enlarging the birth canal for easier delivery of the fetus.

■ ENDOCRINE ORGANS DERIVED FROM ENDODERM

The walls of the embryonic pharyngeal pouches develop thickenings that become parathyroids, thymus, and ultimobranchial bodies, and the pharyngeal floor evaginates to form thyroid tissue. Except in a few lower vertebrates, these organs separate from the pharynx, sink into the surrounding mesenchyme, and migrate some distance from their origin. All have known endocrine functions except the thymus, which is still under study. The endocrine pancreas is also endodermal, and will be discussed first.

□ Pancreatic islets

The pancreas of most tetrapods and of some teleosts contains a large number of islands of endocrine tissue that are interspersed among the exocrine pancreas cells. The islands secrete the hormones **insulin** and **glucagon.** These, along with other hormones, help to regulate blood sugar levels—insulin by preventing an excess, glucagon by preventing a deficit. They both have other metabolic activities also.

In most teleosts the endocrine cells that synthesize insulin and glucagon are aggregated into a few large distinct nodules—frequently two or three "principal islets"—in the mesenteries not far from the intestine.

These same mesenteries support strands of exocrine pancreas. In elasmobranchs the endocrine cells are within the pancreas, but they are in the epithelium of the tiny pancreatic ductules instead of in islets. In adult lampreys they are in the lining of the bile duct close to the intestine, but in larval lampreys they are still embedded in the intestinal wall.

The location of the endocrine islets in different vertebrates becomes comprehensible when their embryogenesis is known. Potential endocrine cells lie at first in the endodermal lining of the embryonic foregut. Later, they are displaced, along with the exocrine cells, in the growing pancreatic buds. Still later, endocrine cells usually bud off from the endodermal lining and become isolated islands.

□ Thyroid gland

Early in vertebrate evolution some of the cells of the pharyngeal floor acquired the capacity to accumulate iodine and to bind it to a protein, which is what thyroid cells do. Cells of the hypobranchial groove (endostyle) of the amphioxus (Fig. 2-10, C) perform this synthesis, as does the subpharyngeal gland (endostyle) of larval lampreys (Fig. 17-12). In both organisms the iodinated protein is secreted into the pharynx and absorbed farther along the digestive tract. But when the duct of the subpharyngeal gland of a larval lamprey closes at metamorphosis, the iodinated protein is thereafter secreted into the circulation from thyroid follicles that become isolated beneath the pharyngeal floor (Fig. 17-13). The follicles in hagfishes are similarly located, although hagfish larvae have no subpharyngeal gland. A thyroid follicle consists of a cuboidal epithelium surrounding a colloid that stores, temporarily, the iodinated protein.

The thyroid gland of gnathostomes, like endostyles, arises as an unpaired evagination from the midventral pharyngeal floor,

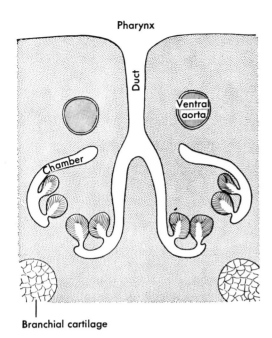

Fig. 17-12. Cross section of subpharyngeal gland (endostyle) of a larval lamprey at site of its duct. All chambers empty into the duct. At metamorphosis the duct closes and some of these glandular cells form thyroid follicles.

at approximately the level of the second pharyngeal pouches (Fig. 17-14). After the thyroid evagination has reached its adult location, it organizes thyroid follicles. The stalk of embryonic cells connecting the thyroid to the pharyngeal floor usually disappears, isolating the thyroid in its adult site. However, a duct remains in the elasmobranch *Chlamydoselache.* It perforates the basihyal cartilage and empties into the pharyngeal floor. In other fishes a solid stalk may persist. In mammals the foramen cecum, a small pit on the caudal surface of the tongue, marks the site of the embryonic evagination. Remnants of the stalk may persist in mammals as a cystlike **thyroglossal duct,** which occasionally requires surgical removal in man.

In teleosts the follicles are usually scattered singly or in small groups along the ventral aorta and along some of the afferent branchial arteries. They may even accompany the arteries into the base of the gills. In a few teleosts they follow the dorsal aorta caudad and even invade the mesonephroi.

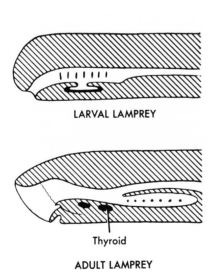

Fig. 17-13. Subpharyngeal gland of the larval lamprey (black) and thyroid follicles in the adult.

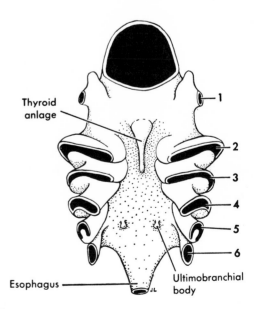

Fig. 17-14. Pharynx of shark embryo, viewed from below. **1,** Spiracle; **2** to **6,** gill slits. (From Camp.[4])

Sometimes they form one or two compact masses between the bases of the first gill pouches.

Except in cyclostomes and teleosts, the unpaired thyroid anlage develops into either a single compact gland or a pair of glands. Sharks have a median thyroid located just behind the mandibular symphysis between two muscles. Unpaired thyroids are also characteristic of snakes, turtles, a few lizards, and *Echidna*. Most other adult vertebrates have paired thyroid glands.

In amphibians the two glands lie in the

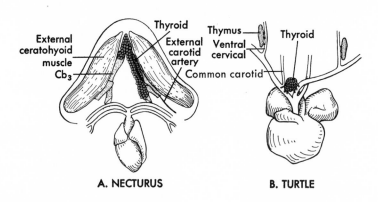

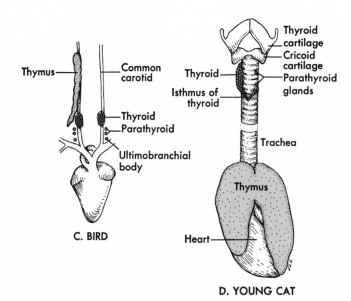

Fig. 17-15. Thymus (light red), thyroid (dark red), parathyroids, and ultimobranchial body in selected vertebrates. The thymus of necturus lies in the angle between the posterior ends of the masseter and external ceratohyoid muscles and is not illustrated. **Cb₃,** The ceratobranchial cartilage of the third pharyngeal arch. In turtles the parathyroids are embedded in the thymus.

floor of the pharynx under cover of the my-lohyoid muscle (Fig. 17-15, *A*). In amniotes the glands migrate caudad varying distances from the pharynx, taking a position close to the common carotid arteries (from which they receive a rich arterial supply) and tra-chea (Fig. 17-15, *B* to *D*). The gland was named "thyroid" because of its position close to the thyroid cartilage in mammals.

The capacity to combine iodine into an organic molecule is not restricted to ani-mals, but only vertebrate thyroid cells are known to synthesize the hormones **thy-roxin** and **triiodothyronine.** The synthesis is stimulated by thyroid-stimulating hormone from the adenohypophysis. The immediate role of thyroid hormone is not known, but the most evident ultimate effect is an in-crease in the rate of cellular respiration.

The thyroid gland produces a third hor-mone, **calcitonin,** which impedes bone re-sorption and thus reduces serum calcium levels. Calcitonin is a product of parafollic-ular, or "C" cells that have migrated into the thyroid gland from the ultimobranchial glands.

Parathyroid glands

Parathyroid glands arise as evaginations from one to three pairs of pharyngeal pouches and are so named because they usually lie beside or embedded in the thy-roid gland (Fig. 17-15, *C* and *D*). Parathyroid glands have not been identified in fishes or larval or neotenous amphibians. A few rep-tiles have three pairs that arise as endoder-mal outgrowths from pharyngeal pouches II, III, and IV, but most tetrapods have two pairs, since the pouch II anlagen usually fail to mature. When there is only one pair (as in a few urodeles, crocodilians, chickens and some mammals), they may have devel-oped from pouch III or IV, depending on the species, or the single gland on each side may have contributions from both pouches.

The parathyroid glands produce **parathy-roid hormone** and **calcitonin,** which regulate the levels of calcium and phosphate in the blood. Low levels of serum calcium evoke release of parathyroid hormone from the gland, and the hormone promotes release of calcium and phosphate from bone and other storage sites. Calcitonin lowers serum calcium by impeding resorption of bone.

Although fishes have no parathyroids, they do not lack a calcium-regulating factor. It is produced by ultimobranchial glands, to be discussed next.

Ultimobranchial glands

Ultimobranchial glands develop from the epithelium of the last pair of pharyngeal pouches and, like parathyroids, produce calcitonin (Figs. 17-14, 17-15, *C,* and 17-16). In fishes they lie not far from their em-bryonic origin near the esophagus. In tetra-pods they lie close to the thyroid gland. Adult mammals have no ultimobranchial glands because the calcitonin-producing cells of the embryonic ultimobranchial gland migrate into the developing thyroid where they become C cells.

In birds and mammals that have been studied thus far the C cells of the ultimo-branchial body arise from neural crests and migrate into the ultimobranchial gland as it is differentiating from pharyngeal pouch epithelium. If this proves to be true in lower vertebrates also, we might conclude that neural crests were the primitive source of calcium-regulating factors in vertebrates. No ultimobranchial glands have been found in cyclostomes. It may be that an undis-covered hypocalcemic factor is being se-creted in some neural crest derivative in these lowest vertebrates. On the other hand, scales and skeleton are the major storage sites of calcium in fishes, and cyclostomes have no scales and little skeleton. Since they lack these reservoirs of calcium it would not be unreasonable to think that a hypocalce-mic factor might have no survival value in cyclostomes and might even be a disadvan-tage.

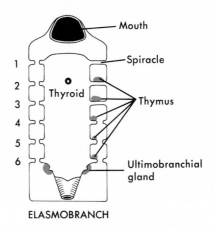

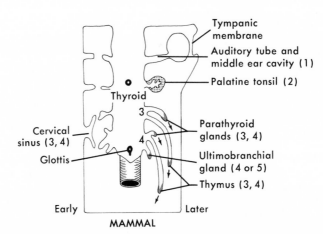

Fig. 17-16. Pharyngeal derivatives of sharks and mammals (diagrammatic frontal sections looking down onto pharyngeal floor). Left sides are earlier in ontogeny than the right. Numbers identify ectodermal grooves or pharyngeal pouches. Arrows indicate caudal growth of anlagen. Whether the ultimobranchial gland of mammals is from pouch 4 or from a vestige of pouch 5 is not certain. Endocrine anlagen, except thyroid, are red.

□ Thymus

Thymus glands arise as thickenings of the linings of several pharyngeal pouches (Fig. 17-16). The epithelia become invaded by lymphocytes that migrate from the embryonic yolk sac and, later, from the fetal liver. The thymic masses subsequently become separated from the pouches and take a position nearby or remote from their origin.

Primitively, thymus probably developed from all pouches. In lampreys it is said to differentiate from all seven. In jawed fishes and tailed amphibians it usually develops from all pouches except the first, and transient thymus tissue has been described from the first. In caecilians the first six pouches participate, and in salamanders, thymus arises from pouches III, IV, and V. Thymus origin is more restricted in some vertebrates. In amniotes pouches III and IV are the sole contributors, and in a few mammals

III is the sole source. In frogs pouch II is frequently the sole source.

There is a tendency for successive thymus anlagen to unite during development so that there may be fewer adult thymus masses than embryonic anlagen. In fishes, excepting elasmobranchs, all anlagen more or less fuse to form a single, elongated gland lying above the branchial chambers. In amphibians and reptiles they may fuse or remain separate. The single gland on each side in frogs usually lies just behind the tympanic membrane. In reptiles and birds the thymus may consist of a series of large nodes in the neck extending caudad as far as the thyroid (Fig. 17-15, bird). In newborn and young mammals the thymus is usually a large bilobed mass in the thoracic cavity dorsal to the sternum and mostly anterior to the heart. After puberty it becomes infiltrated by fat. In the larvae of at least one elasmobranch *(Heptanchus cinereus)*, ducts lead from the first six lobes of the thymus into the pharynx. These ducts may persist in young adults.

Interest in the thymus as an endocrine organ has alternately waxed and waned over a span of many years as bits of evidence for an endocrine role have been reported in the literature, only to fail the test of reproducibility. Among the facts that had to be taken into account is that the thymus is largest in fetal, neonatal, and young mammals and undergoes fatty infiltration as the animal attains sexual maturity. Recent studies by immunologists have provided one explanation. The thymus is large just before birth and early in life because at that time it is housing and processing stem cells that are released into the general circulation and which, along with their descendants, participate in cell- and antibody-mediated immune reactions. The stem cells are immigrants into the pharyngeal epithelium from hemopoietic tissues of the embryonic yolk sac. A hormone, thymosin, has been attributed by some workers to the thymus, but the status of the thymus as an endocrine organ is still not widely accepted.

In birds the role of the thymus is supplemented by the **bursa of Fabricius,** an organ that arises as an evagination of the cloaca. It resembles the thymus in structure and disappears completely at sexual maturity.

☐ Chapter summary

1. Hormones are metabolic products of specific groups of cells that regulate the metabolism of specific nearby or remote cells of a different nature.
2. Neurosecretions are small polypeptide hormones produced by neurosecretory neurons and released into the circulatory channels of neurohemal organs.
3. Known vertebrate neurohemal organs are the median eminence, the posterior lobe of the pituitary, and the urophysis of fishes.
4. The hypothalamus is the major source of vertebrate neurosecretions. It is subject to regulation in part by the external environment. The caudal end of the spinal cord contains neurosecretory neurons in fishes.
5. Endocrine organs derived from ectoderm are the pituitary, pineal, and aminogenic chromaffin tissue including the mammalian adrenal medulla.
6. Endocrine organs derived from mesoderm are the steroidogenic component of the adrenal complex, corpuscles of Stannius, and gonads.
7. Endocrine organs derived from endoderm are the pancreatic islets, thyroid, parathyroids, and ultimobranchial glands. The thymus is also considered an endocrine organ by some investigators.
8. The pituitary consists of a neurohypophysis derived from the diencephalic floor and an adenohypophysis derived from the roof of the stomodeum.
9. The parts of the neurohypophysis are the median eminence, infundibulum, and pars nervosa (posterior lobe), the latter beginning with lungfishes. Hypothalamic neurosecretions released from the posterior lobe have antidiuretic and smooth muscle–stimulating effects.
10. The adenohypophysis typically consists of a pars distalis, pars intermedia, and pars tuberalis. The pars distalis secretes growth-promoting, thyroid-stimulating, adrenal cortical–regulating, and gonad-regulating hormones and prolactin. The pars intermedia produces intermedin, which causes darkening of the skin of ectotherms. Elasmobranchs have an inferior lobe also.
11. The pineal produces melatonin, which blanches the skin of ectotherms and has a regulatory effect on reproduction in some species.
12. The role of the urophysis is not yet clear. It may have osmoregulatory effects.
13. Aminogenic chromaffin tissue synthesizes hormones that are amines.
14. Corticoid-producing tissues arise from the coelomic mesothelium. They are the interrenal bodies of fishes, are usually interspersed among aminogenic cells in amphibians, reptiles, and birds, and form an adrenal cortex in mammals. Corticoids regulate sodium excretion and elevate blood sugar levels.
15. Corpuscles of Stannius are derivatives of the pronephric duct. They are found in ray-finned fishes and affect electrolyte homeostasis.
16. Endocrine pancreas is less intimately asso-

ciated with exocrine pancreas in fishes than in tetrapods. It produces insulin, which lowers blood sugar, and glucagon, which elevates it.

17. The gonads of both sexes produce numerous steroids including androgens and estrogens, the former predominating in males, the latter in females. The ovary also produces progesterone and relaxin, which affect the female reproductive tract.

18. The midventral pharyngeal floor evaginates to produce thyroid follicles. Except in cyclostomes and teleosts the follicles aggregate into discrete paired or unpaired thyroid glands. They produce thyroxin and triiodothyronine that increase metabolic rates and the utilization of oxygen. The thyroid also produces calcitonin.

19. Parathyroid glands are absent in fishes and larval and neotenic amphibians. They are derivatives of pharyngeal pouches II to IV in some reptiles, III and IV in birds and mammals. They produce parathyroid hormone, which elevates serum calcium levels, and calcitonin, which depresses them.

20. Ultimobranchial bodies develop from the last pharyngeal pouches. They are absent in adult mammals. They contribute calcitonin-producing cells to the thyroid and parathyroid.

21. The thymus is lymphoidal tissue derived from most or all pharyngeal pouches of fishes, pouches III and IV in amniotes. It is the source of stem cells that migrate to other lymphoid organs and participate in immune reactions.

LITERATURE CITED AND SELECTED READINGS

1. Barrington, E. J. W.: An introduction to general and comparative endocrinology, London, 1975, Oxford University Press.
2. Birch, M. C., editor: Pheromones, New York, 1974, Elsevier/Excerpta Medica/North-Holland.
3. Bentley, P. J.: Comparative vertebrate endocrinology, New York, 1975, Cambridge University Press.
4. Camp, W. E.: The development of the suprapericardial (postbranchial, ultimobranchial) body in *Squalus acanthias*, Journal of Morphology **28**:369, 1917.
5. Harris, G. W., and Donovan, B. T., editors: The pituitary gland (3 vols.), Berkeley, 1966, University of California Press.
6. Holmes, R. L., and Ball, J. N.: The pituitary gland, a comparative account, New York, 1974, Cambridge University Press.

Symposia in American Zoologist

Comparative aspects of parathyroid function, **7**:882, 1967.
Comparative endocrinology of the pineal, **10**:189, 1970.
Comparative aspects of the endocrine pancreas, **13**:565, 1973.
The current status of fish endocrine systems, **13**:710, 1973.
Prolactin in the lower vertebrates, **15**:865, 1975.
Endocrine role of the pineal gland, **16**:1, 1976.

18

Organic evolution and some other words

In this chapter we will reduce the theory of organic evolution to its basic premise, look briefly at what Darwin meant by "Natural Selection," learn how Lamarck explained changes in the structure of successive generations, and be reminded that just because an animal needs a structural change does not mean that this change will come about. We will also examine some words that are used to represent certain evolutionary concepts.

The theory of organic evolution: what it is and
   what it isn't
Natural selection and the synthetic theory
The Lamarckian doctrine
Need as a basis for mutation
Teleology
Ontogeny, phylogeny, and the biogenetic law
Homology: an idea that triggered a word
Words that trigger ideas

■ THE THEORY OF ORGANIC EVOLUTION: WHAT IT IS AND WHAT IT ISN'T

The theory of organic evolution is a single, simple, easy to understand, unequivocal, yet widely misstated theorem, which is: *The plants and animals on earth have been changing,* and *the ones around us today are descendants of those that were here earlier.*

The conclusion that the animals and plants have been changing is based on geological evidence. It indicates that, if a human being from today could be transported backward in time several hundred million years, the plants and animals he would see would be exotic, alien, and unfamiliar. Five hundred million years ago he would see mostly water, and on the land there would be no land plants. If he happened to bring a fishing pole (!) he would probably catch no fish because the fish would be ostracoderms, and they were filter feeders. Four hundred million years ago fishing would have improved but there would be no trout, perch, or salmon. The land would be higher and drier, there would be mosses and other simple land plants, and labyrinthodonts would be lumbering in and out of swampy waters, but there would be no frogs or toads. Three hundred million years ago the land would be still higher, and the swamps would be forested with seed ferns and conifers, but no flowering plants. Cotylosaurs would be basking in the sun, but there would be no lizards or snakes. One hundred fifty million years ago the birds would have teeth, dinosaurs would be large and specialized, and hairy little animals with pouches to carry their babies in would be scurrying around in the forest, staying out of the way, but there would be no cats, rats, or monkeys. The traveller might be happy to return to the age of the mammals, if only because he is home, and home is familiar. It is in these geological findings that the theory of organic evolution has its roots, but by no means all of them.

The second part of the theorem, that the animals and plants around us today are offspring of the animals of yesterday, is axiomatic. Life comes from preexisting life.

This is the theory of organic evolution, stripped of all satellite theories. It doesn't say that multicellular organisms came from protozoa; it doesn't say that man came from a monkey. It doesn't say *where* man came from—that is another theory, or several other theories. Any one of them could be called *the theory of where man came from,* but it is not the theory of organic evolution. Neither does the theory say what caused, or what causes, species to change. That, too, is another theory or several theories. The theory does not state how life began, or how the universe began. There are theories about beginnings, but they are not the theory of organic evolution. These satellite or ancillary theories belong to scientific disciplines beyond the intended scope of this book. Finally, it is not a theory about a Supreme Intelligence. Science can neither affirm nor deny the existence of a Supreme Intelligence because it lacks the tools needed to collect the data. And without data science cannot conclude.

The theory of organic evolution might also be called the **theory of the mutability of species.** There is only one alternative, the **theory of the immutability of species.** Proponents of the latter theory must insist that every species on earth today is precisely like it was when it first appeared, and that the first member of the species appeared *de novo,* coming from no previous living organism. To admit even a single change ("fishes appeared and from these came all kinds of fishes") is to abandon the theory of the immutability of species in favor of organic evolution. If it is accepted that *one* change can occur, it must be accepted that *two* changes can occur. And once this step in

logic has been taken, there is no limit to the number of changes that could occur.

■ NATURAL SELECTION AND THE SYNTHETIC THEORY

The theory of natural selection verbalized by Charles Darwin (1809-1882) is a satellite to the theory of organic evolution. The latter is independent of Darwin's hypothesis—it is neither supported nor negated by it. Natural selection, an example of which is cited on p. 77, was Darwin's attempt to *account for* what he considered the fact of evolution. Variations in organisms, said Darwin, result in varying degrees of success in competition between individuals and with the physical conditions of life. The resulting "preservation of favorable variations [that is, survival of the fittest] and the rejection of injurious variations, I call Natural Selection."[4] Darwin did not know, in 1859, about genetic mutations and recombination of genes whereby hereditary variations arise. Certainly he would have been pleased to know about them.

The discussion entitled "The theory of organic evolution: what it is and what it isn't," reduced the theory to a basic premise in order to separate the *premise* from the *mechanics.* The so-called synthetic theory of evolution is a synthesis of the basic premise, current genetic insights (which are subject to change), and the theory of natural selection. Mutation and genetic recombination are the presently known sources of hereditary variability, and natural selection seems to dictate evolutionary trends, although it may not be the only directive force. This synthesis combines the premise with some of the satellite theories.

■ THE LAMARCKIAN DOCTRINE

Jean Baptiste de Lamarck (1744-1829) had an explanation for the changes that occur in species. He stated that when a part is employed in successive generations it becomes stronger and better adapted for its role. The animal's use of various structures brings about internal changes that help to perfect these structures. Conversely, when a part is neglected it tends to become vestigial. That is how Lamarck would account for the fact that the olfactory nerves of whales are vestigial. A whale cannot use its sense of smell under water any more than a human being can, because inhaling would lead to drowning! For that reason, according to Lamarck, the whale's olfactory apparatus has become vestigial. The doctrine is not acceptable to biologists because the present state of our knowledge provides no explanation as to how use or disuse of a part in any individual can be translated into alteration in the hereditary code stored within the sperm and eggs, which are set aside as "germ plasm" early in embryonic life.

NEED AS A BASIS FOR MUTATION

There is no scientific basis for the widespread misconception that, because an animal "needs" a specific structure, the structure will appear as a mutation. Populations that are locked into an environment and that need something in order to remain successful in that environment and do not acquire it become extinct. Some explanation other than need accounts for adaptive modifications. The best current explanation is to attribute the fulfillment of such needs to chance genetic mutations.

TELEOLOGY

That a population acquires a structure because it needs it is an example of teleological reasoning. Another example is, "Birds have wings *in order that* they may fly." The alternative is, "Birds have wings and therefore they *can* fly." Teleology is the philosophy that natural phenomena take place not by chance but in accordance with a preconceived purpose, intention, or conscious design. It necessitates an intelligence embodied in, or guiding, natural phenomena. Because there are no supporting *scientific*

data for this philosophy, science (a body of knowledge and a method based on observable data) cannot adopt it.

ONTOGENY, PHYLOGENY, AND THE BIOGENETIC LAW

All vertebrates exhibit a basic architectural pattern that is dramatically expressed during ontogeny, that is, during development of the individual organism. This concept is incorporated in a generalization formulated in 1828 by the embryologist von Baer that became known as **Baer's law.** It states that *general features* common to all members of a group of animals—in this case, vertebrates—*develop earlier in ontogeny* than do the special features that distinguish the various subdivisions of the group (orders, genera, species). For instance, all vertebrates exhibit very early in ontogeny a notochord, dorsal nervous system, pharyngeal pouches, and aortic arches, and these become modified in the direction of the species as development progresses. These modifications of the basic pattern constitute an important part of the subject matter of this book.

The concept of organic evolution resulted in an addition to Baer's law, namely, that *features that develop earliest are the oldest phylogenetically,* having been inherited from early common ancestors and that features that develop later in ontogeny are of more recent phylogenetic origin. This is known as the **biogenetic law.** It implies that ontogeny should provide *some indication* of the phylogeny (evolutionary history) of any group. Although still a valid generalization, the time has probably passed when it was a valuable tool in theoretical research.

HOMOLOGY: AN IDEA THAT TRIGGERED A WORD

The idea of homology evolved slowly but persistently in anatomical thought during several centuries before evolutionary theory began to crystallize. As formulated during the seventeenth century, a homologue is "the same organ in different animals under every variety of form and function."[3] Thus the incus in the middle ear of mammals and the quadrate process of the upper jaw cartilage of sharks are homologous structures (homologues). So are the precava of cats and the right common cardinal vein of lower vertebrates. So, too, are the intermaxillary bone of the human embryo (not an independent bone in adults) and the premaxilla of apes.

How can one be sure two organs, or parts, in two different species are the same organs phylogenetically, especially if there has been a change in function? Among vertebrates, at least, evidence with a high probability of validity comes from embryology and, for muscles, from innervation, also. Two parts in two different vertebrates are homologous if they come from the same embryonic precursor. This is equivalent to saying that they are homologous if they have a common ancestry. Some biologists take exception to this definition, justifiably, because it oversimplifies the problem. de Beer presents an easily read analysis of the concept from the viewpoint of an embryologist.[6]

■ WORDS THAT TRIGGER IDEAS

Words are a necessity for conceptual thought. The more words one can command, the greater the variety of ideas one may entertain. The following words stimulate thinking in relation to evolutionary concepts. You will read them, hear them, and use some of them. There will be differences of opinion with reference to the connotations of most of them. However, if calling attention to these abstract terms stimulates discussion, inclusion of them will have been justified.

Primitive is a relative term. It refers to a beginning or origin. A primitive trait is one that appears in a stem ancestor from which arose an array of subsequent species, some of which may retain the trait. The notochord

is primitive, since it occurred in the first chordates. The placoderms were primitive fish in that they gave rise to an array of later fish. Ancient insectivorous mammals were primitive *placentals* because they gave rise to an array of later placentals. However, they were not primitive *vertebrates*. Somewhere in phylogeny there was a primitive primate and a primitive species of man. However, one cannot always be certain that a given structure is primitive. For example, the lateral neural cartilages of lampreys are primitive only if they reveal an original condition from which typical vertebrae later evolved.

Generalized refers to structural complexes that, at least in some of the descendants, have undergone subsequent adaptation to a variety of conditions. The hand of an insectivore was, and remains, a generalized mammalian hand. It was competent to evolve into the wing of a bat, the hoof of a horse, the flipper of a seal, and the hand of a primate. A generalized group of animals has demonstrated that it was genetically suitable for divergent evolution, that is, evolution in many directions. Labyrinthodonts were generalized tetrapods. The terms "generalized" and "primitive" come into contrast in that generalized connotes a state of potential adaptability and primitive connotes a state of being ancestral.

A **specialized** condition is one that represents an adaptive modification. Vertebrate wings are specializations of anterior limbs, and beaks are specialized upper and lower jaws. Beaks (Fig. 18-1) may be needlelike for extracting nectar from flowers (hummingbirds), chisellike for drilling holes (woodpeckers), hooked for piercing and tearing captured prey (raptorial birds, such as hawks and eagles), long and pointed for capturing moving fish, lizards, and other prey (herons), or recurved for extracting grubs from burrows (the female huia of New Zealand). Increased specialization connotes increased adaptation. The greater the spe-

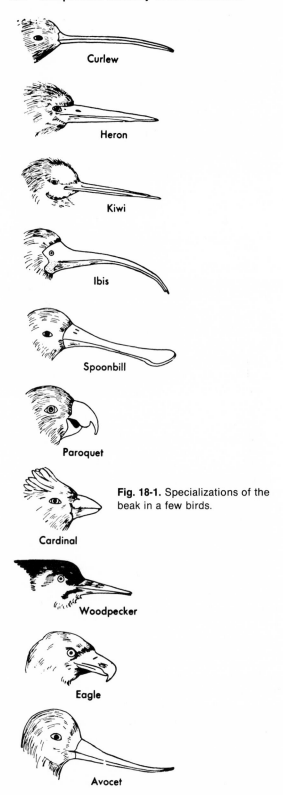

Fig. 18-1. Specializations of the beak in a few birds.

cialization, the less may be the potential for further adaptive changes.

Modified connotes any state of change from a previous condition, a mutated state. If the presence of bone is a primitive trait, wholly cartilaginous skeletons are modifications of the condition. The modification (loss of the potential to form bone) was a specialization if it adaptively modified the animal. Modifications are not necessarily adaptive (students of speciation disagree strongly about this); if they are not, they may portend the demise of the species, since any change is statistically more likely to make the animal less competitive. An exception would be a modification that proved to be preadaptative. (Preadaptations are fortuitous modifications that enable descendants to enter new niches or to cope with conditions not existing at the time of the modification.)

The terms **higher** and **lower** express the relative position of major taxa on a phylogenetic "scale." Birds and mammals evolved from cotylosaurs, and hence are said to be higher than cotylosaurs. In this context the words have some meaning. Sometimes the terms are used to express relative mutational distance of a given taxon from a common ancestor when compared with some other taxon—but are mammals to be considered higher than birds? In this context the term may be misleading. The terms may also be meaningless when used to compare a genus within one taxon with a genus within another taxon, as when comparing a modern frog with a perch, or man with a hummingbird.

Simple is a relative term connoting a lack of complexity of component parts. A simple state is not necessarily a primitive one. The skull of a human being is simple compared to that of a teleost, but it is not primitive. The primitive may also be far from simple.

The term **advanced** should connote a modification in the direction of further adaptation. Unfortunately the word has overtones connoting progress and hence is subjective and misleading. It is a matter of opinion of one species—man—whether or not a modification in another species represented or represents progress. The phrase "more recent" or "more specialized" may be more informative than the term "advanced."

Degenerate is another value-judgment word. For example, it is sometimes applied to cyclostomes by those who think cyclostomes have lost jaw skeletons, paired appendages, bone in the skin, and other characteristics of typical vertebrates. However, the condition of the cyclostomes represents an adaptation to a semiparasitic state, and as such may better be characterized as "specialized." These agnathans may have specialized themselves right into a state of neosimplicity! To call them "degenerate" would seem to discount the value of adaptive modification. Degenerate seems to be a term that should be avoided.

The words **vestigial** and **rudimentary** require explanation. A phylogenetic remnant that was better developed in an ancestor is vestigial. The pelvic girdle of whales is said to be vestigial, since ancestors of whales were tetrapods with functional tetrapod appendages. The yolk sac of the mammalian embryo is vestigial. The term rudimentary is used in two different senses, phylogenetic and ontogenetic. In the phylogenetic sense structures that became more fully exploited in descendants are said to be rudimentary in the phylogenetic precursor. For example, the lagena of the inner ear of fish is sometimes referred to as a rudimentary cochlea, since it evolved into a cochlea in later forms. In the ontogenetic sense, a structure that is undeveloped or not fully developed is said to be rudimentary. The muellerian duct may be considered rudimentary in most male vertebrates. It is not always possible to be certain whether a structure should be called rudimentary or vestigial. The pseudobranch of the shark *Squalus acanthias* is vestigial if it represents a gill that, in ancestral sharks,

was a full-fledged functional gill surface. However, if the pseudobranch is a potential future gill surface, then it is rudimentary. The majority opinion is that it is vestigial.

If the foregoing thoughts trigger discussion, no matter how dissonant, the space employed in presenting them will have been well utilized. It must never be forgotten that words are noises made by man to connote a concept. Or, do you wish to object, on a semantical basis, to the word noise?

LITERATURE CITED AND SELECTED READINGS

1. Ayala, F. J.: Teleological explanations in evolutionary biology, Philosophical Society (London) **37**:1, 1970.
2. Blum, H. F.: Time's arrow and evolution, New York, 1962, Harper & Row, Publishers.
3. Boyden, A.: Homology and analogy: a century after the definitions of "homologue" and "analogue" of Richard Owen, The Quarterly Review of Biology **18**:228, 1943.
4. Darwin, C. R.: On the origin of species by means of natural selection, or the preservation of favoured species in the struggle for life, London, 1859, John Murray Publishers, Ltd. (Republished by numerous publishers.)
5. de Beer, G. R.: Embryos and ancestors, rev. ed., London, 1951, Oxford University Press.
6. de Beer, G. R.: Homology, an unsolved problem, London, 1971, Oxford University Press. (Reprinted in 1974 as Oxford Biology Readers series, #11, Head, J. J., and Lowenstein, O. E., editors.)
7. Eaton, T. H., Jr.: Evolution, New York, 1970, W. W. Norton & Co., Inc.
8. Gregory, W. K.: Evolution emerging, 2 vols., New York, 1951, The Macmillan Co.
9. Mayr, E.: Animal species and evolution, Cambridge, Mass., 1963, Harvard University Press.
10. Stanley, S. M.: A theory of evolution above the species level, Proceedings of the National Academy of Sciences **72**:646, 1975.
11. Van Valen, L.: A new evolutionary law, Evolutionary Theory **1**:1, 1973.
12. Volpe, E. P.: Understanding evolution, ed. 3, Dubuque, Iowa, 1977, William C. Brown Co., Publishers.
13. Williams, G. C.: Adaptations and natural selection, Princeton, N.J., 1966, Princeton University Press.

Symposium in American Zoologist

Models and mechanisms of morphological change in evolution, **15**:294, 1975.

ABRIDGED CLASSIFICATION
OF THE VERTEBRATES

An abridged classification has been provided so that students may readily determine the relationships of animals they meet in the text. If they consult it at once when an unfamilar group of vertebrates is first mentioned, they will find they are gaining a working knowledge of vertebrate relationships without resorting to rote memorization.

Authorities are not in complete accord in matters of classification. This is not unhealthy, since it fosters continual inquiry into the validity of existing schemes. For example, Agnatha are sometimes considered a superclass, or even a subphylum in recognition of the chasm that separates them from jawed fishes, which are then placed in either a superclass or subphylum Gnathostomata. Since no single classification is universally accepted, the primary consideration in adopting this one was to achieve maximal convenience and usefulness.

All classes are included, but entire subclasses have been omitted if all members are extinct and no representative has been cited in the text. All widely recognized orders containing living members are included except those of Osteichthyes and Aves. Inclusion of the many taxa in those groups would not be in accord with the purpose of the abridgement. Totally extinct groups are indicated by an asterisk.

PHYLUM CHORDATA

Subphylum Urochordata (Tunicata)

Class Ascidiacea
Class Larvacea (Appendicularia)
Class Thaliacea

Subphylum Cephalochordata. *Branchiostoma, Asymmetron,* sole genera
Subphylum Vertebrata (Craniata)

Class Agnatha. Jawless fishes.
  *Order Heterostraci
  *Order Osteostraci
  *Order Anaspida   } Ostracoderms
  *Order Thelodonti (Coelolepida)

✓**Order Petromyzontiformes.** Lampreys. ⎱
     Order Myxiniformes. Hagfishes. ⎰ . Cyclostomes

***Class Placodermi.** Armored Paleozoic gnathostome fishes.
    ***Order Acanthodii.** Acanthodians.
    ***Order Arthrodira.** Arthrodires.
    ***Order Antiarchi.** Antiarchs.
    Additional extinct orders.

Class Chondrichthyes. Cartilaginous fishes.
    Subclass Elasmobranchii. Naked gill slits.
       ***Order Cladoselachii.** Primitive Paleozoic sharks.
       Order Selachii. Sharks.
       Order Batoidea. Sawfish, skates, rays.
    Subclass Holocephali. Gill slits covered by an operculum.
       Order Chimaeriformes. Chimaeras.

Class Osteichthyes. Higher bony fishes.
    Subclass Sarcopterygii (Choanichthyes). Lobe-finned fishes, many with internal nares.
       Order Crossopterygii. Chiefly Paleozoic.
         ***Suborder Rhipidistia.** Probable ancestors to amphibians.
         Suborder Coelacanthini. Specialized crossopterygians, internal nares absent; *Latimeria* sole living crossopterygian.
       Order Dipnoi. Lungfish; *Lepidosiren, Neoceratodus, Protopterus* sole living genera.
    Subclass Actinopterygii. Ray-finned fishes.
       Superorder Chondrostei. Chiefly Paleozoic; sturgeons, spoonbills (paddlefish), *Polypterus*, and *Calamoichthys* extant. Ganoids.
       Superorder Holostei. Dominant Mesozoic fishes; *Amia* (bowfin), *Lepidosteus* (gar) sole living genera. Ganoids.
       Superorder Teleostei. Recent bony fishes; 95% of all living fishes.
         Order Clupeiformes. Herringlike fishes.
         Order Cypriniformes. Goldfish, carp, North American catfish, buffalo fish; minnows; exhibit weberian apparatus.
         Order Anguilliformes. Eels.
         Order Gadiformes. Codfish, etc.
         Order Perciformes. Perchlike fish.
         And up to thirty-five additional living orders.

Class Amphibia. Highest anamniotes.
    ***Subclass Labyrinthodontia.** Stem amphibians, precursors of reptiles. *Ichthyostega, Eusthenopteron.*
    ***Subclass Lepospondyli.** Paleozoic; relationships unclear.
    Subclass Lissamphibia. Modern amphibians.
       Order Anura. Frogs and toads.
          Family Pipidae. Tongueless frogs.
          Family Bufonidae. Toads.
          Family Hylidae. Tree frogs.
          Family Ranidae. True frogs.
          And fourteen additional living families.

Order Caudata (Urodela). Tailed amphibians.

Family Hynobiidae	Family Salamandridae
Family Proteidae	Family Amybystomatidae
Family Amphiumidae	Family Plethodontidae
Family Cryptobranchidae	Family Sirenidae

Order Apoda (Gymnophiona). Caecilians; 160 species.

Class Reptilia. Lowest amniotes, mostly extinct; 5,000 species.

  Subclass Anapsida

    *Order Cotylosauria. Stem reptiles.

    Order Chelonia. Turtles and tortoises; 335 species.

  *Subclass Euryapsida (Synaptosauria). Large marine reptiles, including plesiosaurs and ichthyosaurs.

  Subclass Lepidosauria

    Order Rhynchocephalia. *Sphenodon punctatum* sole living species.

    Order Squamata. Lizards, geckos, etc., 2,700 species; snakes, 3,000 species; amphisbaenians (wormlike, burrowing squamates), 130 species.

  Subclass Archosauria. Diapsids; includes bird stock.

    *Order Thecodontia. Stem archosaurs.

    *Order Pterosauria. Flying reptiles.

    *Order Saurischia. Dinosaurs with reptilelike pelvis.

    *Order Ornithischia. Dinosaurs with birdlike pelvis.

    Order Crocodilia. Crocodiles, alligators, caimans, gavials; 21 species.

  *Subclass Synapsida. Mammallike reptiles.

    *Order Pelycosauria. Early synapsids.

    *Order Therapsida. Late synapsids.

Class Aves. Feathered vertebrates.

  *Subclass Archaeornithes. Earliest birds, derived from bipedal archosaur; *Archaeopteryx, Archaeornis* sole genera, each with one species.

  Subclass Neornithes. All other extinct and living birds.

    *Superorder Odontognathae. Extinct marine birds, some with teeth.

    Superorder Paleognathae. Ratites; mostly nonflying, ostrichlike birds.

    Superorder Neognathae. Carinates; twenty-two living orders, 8,600 species.

      Order Columbiformes. Doves.

      Order Pelecaniformes. Pelicans.

      Order Anseriformes. Ducks, geese, other waterfowl.

      Order Falconiformes. Hawks, eagles, vultures.

      Order Galliformes. Chickens, grouse, quail.

      Order Psittaciformes. Parrots, paroquets.

      Order Passeriformes. Perching birds—up to sixty-four families, including songbirds.

      And fifteen other living orders.

Class Mammalia. Vertebrates with hair.

  Subclass Prototheria. Egg-laying mammals.

    Order Monotremata. Duckbilled platypuses and echidnas (Australian spiny anteaters).

Subclass Theria. Give birth to, and suckle, their young; all except marsupials have a chorioallantoic placenta.

> **Order Marsupialia.** Yolk sac serves as placenta; opossums, etc.
> **Order Insectivora.** Moles, shrews, tenrecs, hedgehogs, etc.
> **Order Dermoptera.** Gliding lemurs; one genus, two species.
> **Order Chiroptera.** Bats.
> **Order Primates**
>> **Suborder Prosimii.** Lemurs, lorises, tarsiers.
>> **Suborder Anthropoidea.** Monkeys, apes, man.
>>> Superfamily Ceboidea. New world monkeys.
>>> Superfamily Cercopithecoidea. Old world monkeys.
>>> Superfamily Hominoidea. Apes and man.
>>>> Family Pongidae. Apes.
>>>> Family Hominidae. Man.
>>>>> **Australopithecus africanus*
>>>>> **Homo erectus.* Java man.
>>>>> **Homo neanderthalensis.* Sometimes cited as *Homo sapiens neanderthalensis.*
>>>>> *Homo sapiens.* Modern and **Cro-Magnon man.

> **Order Carnivora**
>> **Suborder Fissipedia.** Land carnivores. Cats, dogs, bears, mink, hyenas, etc.
>> **Suborder Pinnipedia.** Marine carnivores. Seals, sea lions, walruses.
> **Order Cetacea.** Whales, dolphins, porpoises.
> **Order Edentata.** Sloths, armadillos, South American anteaters.
> **Order Tubulidentata.** Aardvarks. Insectivorous. One species.
> **Order Pholidota.** Pangolins. Toothless, insectivorous: *Manis*, sole genus.
> **Order Rodentia.** Gnawing mammals other than lagomorphs.
> **Order Lagomorpha.** Rabbits, hares.
> **Order Perissodactyla.** Ungulates with mesaxonic foot; usually odd-toed. Horses, tapirs, rhinoceros.
> **Order Artiodactyla.** Ungulates with paraxonic foot; usually even-toed.
>> **Suborder Suina.** Pigs, hippopotamuses, peccaries—relatively primitive artiodactyls.
>> **Suborder Ruminantia.** Cud chewers with complex stomachs.
>>> Family Camelidae. Camels, llamas.
>>> Family Cervidae. Deer, caribou (reindeer).
>>> Family Giraffidae. Giraffes.
>>> Family Antilocapridae. American pronghorn antelopes sole species.
>>> Family Bovidae. Cattle, sheep, goats, antelopes (except pronghorns).
>>> Family Tragulidae. Chevrotains.
> **Order Proboscidea.** Elephants, *mastodons. Subungulates.
> **Order Hyracoidea.** Conies. *Hyrax*, sole genus. Subungulate.
> **Order Sirenia.** Dugongs, manatees.

ANATOMICAL TERMS*

Following are some of the components of anatomical terms used in the text, with examples. Familiarity with the entries should enable a reader to deduce, within useful limits, the meanings of many additional words, such as hemangioepithelioblastoma, which otherwise may be a meaningless jumble of letters. The list is no substitute for a general unabridged or standard medical dictionary, but it is sufficiently long (nearly 800 entries) and varied to motivate a reader toward habitual use of those standard works should he be so inclined. Such a practice will extend the reader's intellectual horizons far beyond the boundaries of comparative anatomy. Pragmatically, it will result in better recognition and recall and even in more accurate spelling of technical terms.

Meanings are those relevant to the subject matter of the text. Headings that are stems (**acanth-**) are usually the smallest combinations of letters that are common to the derivatives. The symbol > means *hence* and separates a classical from a derived meaning.

a- lacking, without; see *acelous, Agnatha, alecithal, azygos.*

ab- from, away from.
    abduct to move a part away from the longitudinal axis, as in raising the arm laterally.
    abducens a nerve that abducts (q.v.) the eyeball.

*Meanings of terms not included here can be located in the text by referring to the index.

acanth- spine, spiny.

acelous lacking a cavity.

acetabulum a cup for holding vinegar; the socket in the innominate bone.

acr- extremity, highest.
    acrodont tooth on the summit of the jaw.
    acromion process at proximal extremity of the shoulder *(-omo).*

actin- ray.
    Actinopterygii ray-finned fish (see *pter-*).

ad- to, toward, upon
    adduct to draw toward.
    adrenal a gland on the kidney.

aden- gland.
    adenohypophysis glandular part of the pituitary.
    adenoid resembling a gland; a nasal tonsil.

-ae nominative plural ending, as in chordae tendineae (tendinous chords); genitive singular ending as in radix aortae (root of the aorta).

aestivate to spend summer in a state of lowered metabolism.

af- same as *ad-* (to, toward); the *ad-* is changed to *af-* to make the word easier to pronounce.
    afferent carrying something to or toward something else (see *-ferent*).

Agnatha lacking jaws (see *gnath-*).

ala- wing; see also *ali-*.
    alar winglike.

alba white.

alecithal lacking yolk.

ali- wing.
    alisphenoid wing of the sphenoid.

alveolus a small chamber or pit.

amel- enamel.
    ameloblast a cell that produces enamel.

amphi- both.
    amphicelous having concavities at both ends.

amphioxus an animal with both ends pointed (*-oxy*).

amphiarthrosis a joint with severely limited movement.

ampulla a flask > a small dilation.

an- without.

anamniote an animal lacking an amnion.

anapsid lacking an arch.

ana- up, upward.

anadromous *ana-* + *-dromos* (course) > a migrant from the sea up into freshwater streams.

anastomose to unite end to end.

anatomy *ana-* + *-tome* (q.v.).

anch- gill.

andr- male.

androgen a hormone that induces maleness.

angi- vessel.

ankyl- a growing together of parts.

ankylose to fuse in an immovable articulation.

anlage an embryonic rudiment or precursor of a developing structure.

annulus a ring.

annulus tympanicus bony ring to which the eardrum is attached.

ante- before.

antebrachium the forearm; the part before the brachium.

anthrop- refers to human beings.

anti- against, opposite.

antidiuretic inhibiting loss of water via kidneys (diuresis).

antrum a cavernous space.

Anura *an-* + *uro* (q.v.); tailless amphibians.

apical at the apex.

Apoda *a* + *pod-*; without legs.

apophysis an outgrowth or process.

aponeurosis *apo-* (away from) + *neuron* (a tendon); a broad, flat, tendinous sheet; the meaning of the word cannot be deduced from its parts.

apsid refers to an arch.

arch- first, primary, ancient.

archenteron primitive gut.

archetype an early model.

archinephros hypothetical primitive kidney.

archipallium first roof of the telencephalon.

arcuate arched.

arrector pili (pl., **arrectores pilorum**) muscle that erects a hair.

arthro- joint.

arthrodire placoderm with joints, because of dermal plates, in the neck (*-dire*).

artio- an even number.

artiodactyl having an even number of digits.

arytenoid resembling a ladle.

ataxia *a-* (lacking) + *taxia* (order); a disorder of the neuromuscular system.

-ate having the property of, as septate: with septa.

atlas the vertebra that supports the head like the mythical Atlas holds up the earth.

atrium the courtyard of a Roman home > a cavity that has entrances and exits.

auricle an ear or earlike flap.

auto- self.

autostyly a condition in which the upper jaw braces (*-styly*) itself against the skull.

autotomy cutting one's self, as when a lizard breaks off the end of its tail.

axial in the longitudinal axis.

azygos *a-* (lacking) + *zyg-* (a yoke) > on one side only.

baro- pressure.

baroreceptor a sense organ that monitors pressure.

basi- most ventral; pertaining to a basal location.

basihyal basal element of hyoid skeleton.

basisphenoid basal element of sphenoid complex.

bi- two.

bicornuate having two horns (*cornua*).

bicuspid having two cusps.

bipartite having two parts.

bio- life.

blast- an embryonic precursor; a germ of something.

blastema an embryonic concentration of mesenchyme.

blastocoel cavity of the blastula.

blastocyst the mammalian blastula.

blastula a little (*-ula*) embryo.

brachi- arm.

brachiocephalic associated with the arm and head.

branchi- gill.

branchiomeric referring to parts of the branchial arches.

bucco- cheek.

bulbus bulb

bulbus arteriosus muscular swelling on ventral aorta.

bulbus cordis term sometimes applied to conus arteriosus in lungfishes, amphibians, and mammalian embryos; part of the heart.

bulla a bubble > a bubblelike part such as the tympanic bulla.

bursa a sac or pouch.

caecum a blind pouch.

calamus a stem or reed; the stem of a feather.

canaliculus a little canal.

capitulum a little head.

caput head.

cardi- heart.

cardinal of basic importance; chief.

carina a keel.

carn- flesh.

   carnivore a flesh-eating animal.

carotid from a word meaning heavy sleep; compression of the carotid artery cuts off blood to the brain.

carpo- wrist.

cat- down.

   catarrhine having a nose (*-rhin*) with nares directed downward.

cauda tail.

   caudad toward (*-ad*) the tail.

cecum see *caecum.*

cel- see *-coel-.*

cephal- head.

cerat- horn.

   ceratohyal horn of the hyoid.

   ceratotrich a horny, hairlike fin support.

-cercal tail.

cerumen wax.

cervical pertaining to the *cervix* (q.v.).

cervix neck.

cheir-, chir- hand.

   Chiroptera mammals in which the hand is modified as a wing; bats.

chiasma shaped like Greek letter chi (X) > a crossing.

choana a funnel-shaped opening.

chole- bile.

chondr- cartilage.

chorion a membrane.

choroid, chorioid resembling a membrane (see *chorion*).

chrom-, chromato- color.

circum around.

clava club.

clavo-, cleido- clavicle.

clavotrapezius a shoulder muscle attached to clavicle.

cleidomastoid a neck muscle attached to clavicle.

cleidoic closed, locked up > a reptilian, bird, or monotreme egg with much yolk and a shell.

cloaca a sewer > common terminus for digestive and urinary tracts.

cochlea a snail with a spiral shell > spiral labyrinth of inner ear.

-coel- hollow, a cavity.

   coelom body cavity

collagen gelatinous, gluelike material.

columella a little (*-ella*) pillar or column.

com-, con-, cor- with, together.

conch- shell.

   concha the pinna of the ear, shaped like a clamshell.

contra- opposite.

   contralateral on the opposite side.

conus a cone.

   conus arteriosus chamber of heart after the ventricle.

copr- feces.

   coprodeum fecal passage derived from cloaca.

cor- see *com-.*

coracoid shaped like a crow's beak.

corn- horn (keratinized tissue); also, a horn-shaped structure.

   corniculate like a little horn.

   cornified changed to horn by keratinization.

   cornu (pl., **cornua**) horn of the hyoid.

corona a wreath or crown.

   coronary sinus forms a "wreath" around the heart.

corpus (pl., **corpora**) body.

   corpora quadrigemina the two pairs of twins (*-gemini*) or four bodies of the roof of the mesencephalon of mammals.

   corpus luteum a yellow body of the ovary.

   corpus spongiosum spongy body.

costa rib.

   costal cartilage a cartilage at the ventral end of a rib.

cotyledonary cup shaped; resembling half a bean seed.

coxa hip.

cribriform sievelike.

cricoid resembling a ring.

crista a ridge or crest.

crus leg.

cten- comb.

> **ctenoid** resembling a comb.

cucullaris from a word meaning a hood; the two cucullaris (trapezius) muscles resemble collectively a hood or shawl.

cuneiform wedge-shaped.

cusp a peak or point.

cutaneous referring to skin.

cyclo- circular.

> **cyclostome** agnathan with round mouthlike funnel.

cyst fluid-filled sac.

cyt- cell.

dactyl- finger, toe.

decidu- fall off or be shed.

> **deciduous placenta** one in which the uterine wall of the mother is partly shed at parturition.

deltoid resembling the Greek letter delta (Δ).

demi- half.

> **demibranch** gill on one face of a gill arch.

dent- tooth.

> **dentin** bone like that in teeth.

derm- skin.

> **dermatome** layer of a somite giving rise to skin.
>
> **Dermoptera** mammals in which the skin forms a wing membrane (see *-pter*).

dermato- referring to skin.

> **dermatocranium** skull bones phylogenetically derived from skin.

-deum, -daeum a passageway.

deuter- two.

> **deuterostome** an animal that uses the blastopore as an anus and forms a second mouth.

di- two.

> **diapophysis** one of two lateral processes.
>
> **diapsid** having two arches.
>
> **diarthrosis** freely movable joint between two bones or cartilages.
>
> **Dipnoi** fish with two breathing apertures (external and internal).

dia- through, apart.

> **diameter** a measurement through the width of a circle.
>
> **diaphragm** a separation (*phragma*) between two parts.

-didym- twins > the testes.

digiti- fingers or toes.

> **digitigrade** walking (*-grade*) on the digits.

dino- fearful, terrible.

dinosaur a fear-inspiring reptile.

diphy- double.

> **diphyodont** having two successive sets of teeth.

diplo- double, two.

> **diplospondyly** two vertebrae in each body segment.

dis- separation, taking apart.

> **dissect** to disassemble.

diverticulum an outpocketing.

dorsum the back.

> **dorsad** toward the back.

duodenum twelve; the length of the human duodenum is about the breadth of twelve fingers.

dura tough, hard.

dys- bad, faulty, painful.

e- without.

> **Edentata** an order of mammals lacking teeth.

ect- outer.

> **ectopterygoid** the outer pterygoid bone.
>
> **ectotherm** an animal whose temperature varies with the environment.

-ectomy *ex-* (out of) + *tome* (cut), as in appendectomy.

ectopic *ex-* (out of) + *topo-* (place).

> **ectopic pregnancy** a pregnancy in which the fetus is implanted elsewhere than in the uterus (in the coelom, for example).

Edentata see *e-*.

ef- variant of *ex-*.

> **efferent** that which carries away from; efferent branchial arteries carry blood out of the gills.

elasmo- plate.

> **elasmobranch** cartilaginous fish with gills composed of flat plates.

-ella a diminutive, as in columella.

en- in, into.

> **encephalon** the brain, a structure in the head.

endo- within, inner; see also *ento-*.

> **endochondral** within cartilage.
>
> **endolymph** the inner lymph within the membranous labyrinth.
>
> **endotherm** an animal that maintains a relatively constant body temperature regardless of environmental fluctuations.

enteron the gut.

ento- within, inner; see also *endo-*.

> **entoglossal** within the tongue.

ep-, epi- upon, above, over.

> **epaxial** above an axis.

ependyma an outer garment > the membrane covering the part of the central nervous system exposed to the neurocoel.

epididymis (pl., **epididymides**) a structure lying on the testis (named from the position in man).

epiglottis a skeletal flap over the glottis.

epimere the upper part of the embryonic mesoderm; dorsal mesoderm.

epiphysis *epi* + *-physis* (q.v.), pineal complex.

epiploic relating to greater omentum.

epithelioid resembling an epithelium.

erythro- red.

eso- carrier.

esophagus carrier of substances that have been eaten.

estr- female.

ethmoid seivelike.

eu- true.

eury- wide.

euryhaline able to live in waters with a wide range of salinity (*halo-*).

ex-, exo- out, out of, away from, outer.

excurrent pore a pore for the exit of a current of water.

exoskeleton a skeleton in the skin.

extra- beyond, outside of.

extraembryonic outside of the embryo.

falciform shaped like a sickle.

fauces throat.

fenestra a window > an aperture.

ferent, -ferous carrying, as in *afferent* (q.v.).

fil thread.

fimbria fringe.

foramen a small opening, usually transmits something.

foramen magnum the large foramen in the occipital region.

fore- before, in front.

forearm the part of the arm before the upper arm.

fossa a pit, cavity, depression, vacuity.

frenulum a little bridle > the membrane that bridles (ties) the tongue to the floor of the oral cavity.

frug- fruit.

frugivore a fruit eater.

fundus the bottom or base of the cavity of a hollow organ.

gan- bright.

ganglion a swelling > a group of cell bodies outside of the central nervous system.

gastr- a belly, a stomach; the digastric muscle has two bellies.

gastralia ventral abdominal ribs.

gastric refers to stomach.

gastrula a little stomach > a stomachlike embryo.

gen- origin.

genio- chin.

genu knee.

geo- earth.

glans acorn > the tip of the penis.

glenoid resembling a socket.

glia glue.

glomerulus a little glomus (ball or skein) > a tiny plexus of blood vessels.

gloss- tongue.

gnath- jaw.

-gnosis- knowledge.

gon- seed > generative, as in glucagon (giving rise to glucose).

gonad the source of gametes.

gubernaculum a rudder > a governor, as the ligament that (partly) governs the position of the testis.

gula throat.

gular fold a fold at the throat of some tetrapods.

gustatory related to gustation (taste).

gymn- naked.

gyrus a ridge between grooves.

haem-, hem- blood.

hemopoiesis formation of blood.

hamate having a hook.

hamulus a little hook.

hemi- half; equal to *demi-*.

hemo see *haem-*.

hepat- liver.

hept- seven.

herbivore an animal that devours grasses (herbs).

hetero- other, different; opposite to *homo-*.

heterodont having different kinds of teeth.

hex six.

hilum a notch.

hipp- horse.

hist- tissue.

histogenesis the formation of a tissue.

holo- entire, whole.

holonephros a kidney extending the length of the coelom.

hom-, homeo-, homo-, homoio- like, similar.

homeostasis maintenance of a constant internal environment.

homeotherm an animal that maintains a steady body temperature despite ambient (external) temperature; an endotherm.

homodont teeth all alike.

hyaline clear, glassy.

hyoid shaped like the capital Greek letter upsilon (Y).

hyostyly see -*styly*.

hyp-, hypo- under, below, less than ordinary.

hypaxial below a given axis.

hypophysis a growth under the brain; the pituitary body.

hyper- above, beyond the ordinary.

ichthy- fish.

-iform having the shape of.

ileo- pertaining to the ileum of the intestine.

ilio- pertaining to the ilium of the pelvis.

impar unpaired.

in- not.

innominate not named.

incus an anvil.

infra- beneath, under.

infraorbital beneath the orbit.

infundibulum a little funnel.

inguen the groin.

inguinal in the region of the groin.

inter- between.

intercalary plate part of neural arch between neural plates in fishes.

interrenal steroidogenic tissue between the kidneys.

intersegmental between segments.

intra- within.

intramembranous within membrane.

intrasegmental within a segment.

ipsi- the same.

ipsilateral on the same side.

irid- iris of the eye.

iridophore a pigment cell containing refractory bodies that result in iridescence.

ischi- hip, pelvis.

iso- equal, alike.

isolecithal egg an egg having even distribution of yolk.

-issimus a superlative ending; the longissimus dorsi muscle is the longest muscle of the back.

-itis inflammation of.

jejunum empty; part of intestine that is often empty at death.

juga- yoke > something that joins; the jugal bone in mammals is a yoke uniting maxilla and temporal bones.

jugular pertaining to the neck.

juxta- next to, near.

juxtaglomerular near glomeruli.

kat- down; same as *cat-*.

keratin from a word meaning horn; a relatively insoluble substance in the stratum corneum.

kinetic capable of moving.

labium lip.

labyrinth a maze.

labyrinthodont an early tetrapod with greatly folded dentin in the teeth.

lac-, lact- milk.

lacrimal pertaining to tears; from lachryma (a teardrop).

lacuna a lake.

lag- hare.

lagena a flask > flask-shaped part of inner ear.

lambdoidal having the shape of the Greek letter lambda (λ).

lamina a thin sheet, plate, or layer.

laryng- larynx.

latissimus the broadest; see -*issimus*.

lecith- yolk.

lemmo- sheath or envelope.

lemur from a word meaning a nocturnal being or ghost.

lepid-, lepis- scale.

Lepidosteus a ganoid fish with bony scales.

lepto weak, thin, delicate.

leuco-, leuko- white, colorless.

levator that which elevates or raises.

lien- spleen.

linea alba a whitish line.

lingua the tongue.

lip- fat.

liss- smooth.

longissimus the longest; see -*issimus*.

lumen light > an opening that light can pass through; the cavity in a tube.

lunar, lunate moon-shaped.
luteo- yellow.

macula a spot.
magnus, -a, -um large.
malleus a hammer.
mandibula a jaw.
manu- hand.
manubrium a handle.
marsupium a pouch.
mater mother.
maximus, -a, -um largest.
meatus a canal or passageway.
medulla bone marrow.
    medulla spinalis the marrow of the backbone
    > spinal cord.
meg- great, very large.
melan- dark, black.
    melanocyte a cell containing melanin granules
    that cannot aggregate or disperse.
    melanophore bearing (*-phore*) dark pigment.
meninx (pl., **meninges**) a membrane.
mento- chin.
    mental foramen foramen on mandible near chin.
-mer- a segment, a part, one of a series.
mes- middle, midway, intermediate.
    mesaxonic foot one in which the weight-bear-
    ing axis passes through the middle toe.
    mesenchyme a tissue (*enchyme*) that is not yet
    differentiated.
    mesentery associated with the midline of the
    enteron.
    mesonephros an intermediate embryonic kid-
    ney.
met- after, last in succession.
    metacarpal a bone after (distal to) a carpal.
    metamorphosis the final change in shape.
    metanephros hindmost kidney.
mimetic capable of mimicking; having the characteristics of a mime.
mitral refers to a bishop's mitre or headdress;
the mitral valve is the bicuspid valve of the
mammalian heart.
mono- one.
    monotreme a mammal with one caudal open-
    ing.
morph- shape, structure, form.
    morphogenesis development of shape or struc-
    ture.
    morphology study of form; anatomy.
morula a mulberry.

myel- marrow.
    myelencephalon the marrow inside the skull
    > the medulla.
    myelin fatty material.
mylo- from a word meaning a millstone > applicable to molar (grinding) teeth.
    mylohyoid muscle a muscle attached near the
    molar teeth.
myo- muscle.
    myocardium muscle of heart.
    myomere one of a series of muscle segments.
    myotome part of somite giving rise to muscle.

neo- new, recent.
nephr- kidney.
    nephrogenic giving rise to kidney.
    nephron a functional kidney unit.
neur- nervous tissue.
    neuroblast an undifferentiated neuron.
nomen- (pl., **nomina**) name.
noto- back.
    notochord cordlike skeleton of the back.
nuchal refers to the nape of the neck.

occiput the part of the head surrounding the
foramen magnum; in mammals especially, the
back of the head.
ocul- eye.
odon-, odont- tooth.
    odontognath a bird with teeth on the jaws.
    odontoid resembling a tooth.
-oid like, having a resemblance to, as in hominoid
(manlike).
-ole small, as in arteriole.
-oma swelling.
omentum a free fold of peritoneum.
omni- all.
    omnivore an animal that eats both plants and
    animals.
omo- shoulder.
ontogenesis *onto* (individual) + *genesis* (origin);
the development of an individual.
oö- egg (pronounced oh-oh).
    oöcyte egg cell.
    oöphoron *oö* + *-phore-* (q.v.); ovary.
operculum a cover or lid.
ophthalm- eye.
-opia sight.
opisth- at the rear, at the end.
    opisthocelous with a cavity at the caudal
    end.

opisthonephros kidney behind the pronephros of anamniotes.

orb- a circle.

    orbit cavity for the eyeball.

ornith- bird.

oro- mouth.

-orum of the, as in branchiorum (of the gills).

os bone.

    ossicle a small bone.

    ossify to become bony.

os mouth.

    os uteri entrance to uterus from vagina or urinogenital sinus.

osmoregulation electrolyte homeostasis.

oste- bone.

    Osteichthyes bony fish.

    osteon unit of bone in concentric lamellae (layers).

ostium an entranceway > a mouth.

ostraco- shell.

oto- ear.

    otic refers to the ear.

    otocyst a vesicle that becomes the inner ear.

-ous having the characteristic of.

ovale shaped like an ovum or egg; oval.

ovi-, ovo- egg.

    oviparous egg laying.

    ovipositor a structure for laying eggs.

-oxy- sharp, acute, acid.

pachy- thick.

    pachyderm a thick-skinned animal.

paed-, ped- child.

    paedogenesis reproducing without attaining full maturity.

palae- see *pale-*.

palatine referring to the palate.

pale- old, ancient.

pallium a cloak > a roof.

panniculus a small piece of cloth > a layer of tissue.

papilla a nipple > a nipple-shaped structure.

par-, para- beside, near.

    parasphenoid parallel to the sphenoid.

    parotid near the ear.

parie- wall; a parietal artery supplies the body wall.

-parous bearing, giving birth to.

pars (pl., **partes**) part.

pectoral refers to the chest.

pedicel a slender stalk.

pelvis a basin.

penta- five.

perennibranchiate having permanent gills.

peri- around.

    perilymph fluid surrounding membranous labyrinth.

periss- odd.

peritoneum something that stretches over or around > the coelomic lining.

petro- stone, rock.

    petrosal bone the bone surrounding the inner ear, which resembles a steep, rugged rock.

phag- eat.

phalanx (pl., **phalanges**) a line of soldiers > a bone of a digit.

pharyng- pharynx.

-phil loving > having an affinity for.

-phore- bearing, one that bears, as in photophore (an organ that emits light).

phrenic refers to diaphragm.

phylo- tribe.

    phylogeny evolutionary history of a group.

-physis that which grows.

physo- bellows > lung or air bladder.

    physoclistous lacking a duct from air bladder.

    physostome a fish that can get air to the swim bladder via the mouth.

pia tender, kind.

    pia mater the delicate meninx of the brain.

pilo- hair.

pisci- fish.

placo- thick, flat, platelike.

    placode in embryology, an ectodermal thickening that gives rise to something.

    placoderm fish with (bony) plates in the skin.

planta- the sole of the foot.

    plantigrade a flatfooted stance.

platy- flat, wide, broad.

pleur- rib, side.

    pleural refers to ribs.

    pleurapophysis apophysis (process) of vertebral column that is fused with a short rib.

plexus a network.

pneumo- lung.

pneumato- air.

pod- foot.

poikilotherm an ectotherm (q.v.).

poly- many, much.

pons a bridge.

post- after, behind.

    posttrematic behind a trema or slit.

pre- before, in front of; see also *pro-*.
    pretrematic in front of a trema or slit.
prim- first, earliest.
    primate first in rank.
pro- in front of, before, preceding.
    procelous with a cavity at the cephalic end.
    prostate a gland standing *(stat-)* at the beginning of the urethra.
pro- favoring, on behalf of.
    prolactin hormone necessary for milk production.
proboscis a trunk such as that of an elephant.
procto- anus.
proprio- one's own.
    proprioception reception of stimuli from muscles, joints, tendons.
pros- toward, near.
    prosencephalon the anterior end of the embryonic brain.
proto- early, first.
    protothrix (pl., **prototriches**) prototype of hair.
pseudo- false.
pter-, pteryg- wing, feather.
    pterosaur winged reptile.
    pterotic wing of the otic complex.
    pterygoid resembling a wing.
pulmo- lung.
pyg- rump.

quadrate square.
quint- five.

rachi- vertebral column.
    rachis supporting "spine" for feather.
    rachitomous vertebra a vertebra consisting of several pieces; see *-tome.*
radix (pl., **radices**) a root.
ramus a branch.
rectus, rectum straight.
renal pertaining to the kidney.
rete (pl., **retia**) a network.
    rete mirabile a remarkable *(mirabilis)* network of vessels.
    reticulum a little network.
retro- behind.
rheo- current, flow.
rhin- nose.
    rhinencephalon an olfactory part of the brain.
    rhinoceros an animal with a horn *(cerato-)* on the nose.
rhomb- rhomboid.

rhynch- snout.
risorius a muscle named from the Latin word meaning to laugh.
ruga (pl., **rugae**) a wrinkle.

sacculus a little sac.
sagittal from a word meaning an arrow.
sangui- blood.
sarco- flesh.
    Sarcopterygii fish with fleshly lobe at base of fin.
saur- lizard > reptile.
scalene a triangle with sides and angles unequal.
scler- hard, skeletal.
    sclerotome part of somite giving rise to skeletal components.
-sect- cut, divide.
sebum grease, wax.
    sebaceous having an oily secretion.
sella a seat.
    sella turcica a structure shaped like a turkish saddle.
semi- half, partial.
    semilunar shaped like a half-moon.
seminal pertaining to seed > to semen.
    seminiferous carrying sperm.
serrate notched or toothed along the edge; the serratus muscle is serrate.
sex- six.
sigmoid S-shaped.
sinus a cavity.
    sinusoid a thin-walled, sinuslike vascular channel.
soma-, somato- body.
    somite a body segment.
sphenoid wedge-shaped.
spiracle a breathing hole.
splanchn- viscera.
splen- spleen.
spondyl- vertebra.
squam- scale.
    squamous scalelike, flattened; squamate.
stapes a stirrup.
    stapedial associated with the stapes.
stato- standing, fixed.
stellate star-shaped.
stom- mouth.
stratum layer.
strept- twisted, curved.
stria a stripe.

-**style** pillar > an elongated process such as the urostyle.

styloid having an elongated shape.

-**styly** braced; in hyostyly the jaws are braced against the hyoid.

sub- under, below, to an inferior degree.

subclavian under the clavicle.

subunguis under the nail.

subungulate not quite an ungulate.

sudor sweat.

sulcus a groove.

super- over, above, in addition.

supra- equivalent to super.

suprarenal a gland just above the mammalian kidney.

sur over, above; equivalent to super.

surangular bone a bone above the angular.

sym-, syn- together.

symphysis a growing together; see *-physis*.

synarthrosis an immovable suturelike joint.

synapse a junction.

synsacrum sacrum united with other vertebrae.

tarsus ankle; also, a connective tissue plate in the eyelid.

tax- order.

taxon a formal taxonomic unit such as a phylum or species.

taxonomy the orderly arrangement of named groups; classification.

tectum a roof.

tel-, teleo-, telo- end, complete.

telencephalon anterior end of the brain.

temporal refers to the temple or the side of the skull behind the eye.

teres round.

tetra- four.

theco- a case.

thecodont having socketed teeth.

therio- an animal with hair, a beast.

thyroid shield-shaped.

-**tome** cut; also, the result of cutting, as a section or thin sheet.

trabecula a little beam > a strand, ridge, rod, or bundle.

trans- across.

transect to cut across.

trapezoid a four-sided plane with two parallel sides.

trapezius muscle named for its shape in man.

-**trema** a slit.

tri- having three parts.

trigeminal from word meaning triplets; a nerve with three primary branches.

trich- hair.

trochlea a pulley; the trochlear nerve of man passes through a pulley at its attachment.

troph- nourishment.

truncus trunk.

truncus arteriosus ventral aorta.

tuber a swelling or knob > a tuberosity, tubercle, or protuberance.

tuberculum a little tubercle.

tunic a coat or wrap.

tympanum a drum > eardrum.

ulna elbow > the bone at the elbow.

ultimobranchial a gland derived from the last (*ultimo-*) branchial pouch.

-**ulus, -ula, -ulum** diminutive endings denoting tiny.

uncus a hook.

uncinate hooked.

unguis nail, claw, hoof.

ungula hoof.

ungulate hooved.

uro- tail.

urophysis a growth at the base of the tail.

uropygium the rumplike tail of a bird.

utricle a little sac or vesicle.

vagina a sheath.

vagus wandering.

vas (pl., **vasa**) vessel.

vasa vasorum vessels of the blood vessels.

velum a veil > a thin membrane.

venter abdomen; the part opposite the dorsum or back.

ventricle a cavity in an organ.

vesica a bladder or vesicle.

vestibule an antechamber or entranceway.

vitelli- yolk.

vitreous having a glassy appearance.

vivi- alive.

vomer a ploughshare; the mammalian vomer bone resembles a plowshare.

-**vorous** eating, devouring, as in insectivore, an animal that eats insects.

Xanth- yellow.

xiph- sword.

xiphoid process a swordlike process of the sternum opposite the manubrium or handle.

ypsiloid shaped like the Greek letter upsilon (Y).

zyg- yoke > something that links two things.
   zygomatic arched.
   zygote result of union of gametes.
   zygapophysis a vertebral process that articulates with a more anterior or posterior one.

General references

The following are supplemental to the selected readings at the end of each chapter.

COMPREHENSIVE

Cole, J. F.: A history of comparative anatomy, London, 1944, The Macmillan Co., Ltd.

Getty, R., editor: Sisson and Grossman's The anatomy of the domestic animals, ed. 5, 2 vols., Philadelphia, 1975, W. B. Saunders Co.

Goodrich, E. S.: Studies on the structure and development of vertebrates, London, 1930, The Macmillan Co., Ltd. (Reprinted by Dover Publications, Inc., New York, 1958.)

Grassé, P.-P., editor: Traité de zoologie, Anatomie, systématique, biologie, vols. 11-17, Paris, 1948-1970, Masson et Cie.

Harmer, S. F., and Shipley, A. E., editors: The Cambridge natural history, vols. 7-10, London, 1898-1902, The Macmillan Co., Ltd. (Reprinted by Hafner Press, New York, 1958-1960.)

Nickel, R., Schummer, A., Seiferle, E., and Sack, W. O.: The viscera of the domestic mammals, New York, 1973, Springer-Verlag.

Parker, T. J., and Haswell, W. A.: Textbook of zoology, vol. II (revised by Marshall, A. J.), ed. 7, London, 1962, The Macmillan Co., Ltd.

Vertebrate structures and functions: readings from Scientific American, San Francisco, 1955-1974, W. H. Freeman & Co. Publishers.

Young, J. Z.: The life of vertebrates, New York, 1962, Oxford University Press.

SELECTED BIOLOGIES OR PHYSIOLOGIES

Bellairs, A.: The life of reptiles, 2 vols., New York, 1970, Universe Books.

Farner, D. S., and King, J. R.: Avian biology, 4 vols., New York, 1971-1974, Academic Press, Inc.

Gans, C., editor: Biology of the reptilia, 8 vols., New York, 1969-1976, Academic Press, Inc.

Goin, C. J., Goin, O. B., and Zug, G.: Introduction to herpetology, San Francisco, 1978, W. H. Freeman and Co., Publishers.

Hoar, W. S., and Randall, D. J., editors: Fish physiology, 6 vols., New York, 1969-1970, Academic Press, Inc.

Prosser, C. L., editor: Comparative animal physiology, ed. 3, vol. 2, Philadephia, 1973, W. B. Saunders Co.

Schmidt-Nielsen, K.: How animals work, New York, 1972, Cambridge University Press.

Sturkie, P. D., editor: Avian physiology, ed. 3, New York, 1976, Springer-Verlag.

BIOMECHANICS

Alexander, R. M.: Animal mechanics, Seattle, 1968, University of Washington Press.

Frost, H. M.: An introduction to biomechanics, Springfield, Ill., 1967, Charles C Thomas, Publisher.

Gans, C.: Biomechanics: an approach to vertebrate biology, Philadelphia, 1974, J. B. Lippincott Co.

Hildebrand, M.: Analysis of vertebrate structure, New York, 1974, John Wiley & Sons, Inc.

ANATOMICAL PREPARATIONS

Hildebrand, M.: Anatomical preparations, Berkeley, 1968, University of California Press.

MEDICAL DICTIONARIES AND ADOPTED ANATOMICAL NAMES

Dorland's illustrated medical dictionary, ed. 25, Philadelphia, 1974, W. B. Saunders Co.

Nomina anatomica, ed. 3, Amsterdam, 1966, Excerpta Medica Foundation.

Nomina anatomica veterinaria, Department of Anatomy, Ithaca, N.Y., 1973, State Veterinary College.

Stedman's medical dictionary, ed. 22, Baltimore, 1972, The Williams & Wilkins Co.

Index